INTRODUCTION TO THE MECHANICS OF VISCOUS FLUIDS

PAU-CHANG LU
The University of Nebraska-Lincoln

HOLT, RINEHART AND WINSTON, INC.

New York · Chicago · San Francisco · Atlanta · Dallas
Montreal · Toronto · London · Sydney

To the Memory
of
My Parents

Library of Congress Catalog Card Number: 72–9355
ISBN: 0-03-084686-2

Printed in the United States of America
3456 038 123456789

PREFACE

Fifteen years ago it was by no means a common practice to teach a course of viscous flow theory to upperclassmen or first-year graduate students in an engineering college. It was usually unnecessary and impossible. But a lot of things have changed since then. An engineering student of today is raised on basic principles and fundamental mathematical tools instead of on detailed applications and handbooks. He is aware of the distinction between his profession and that of a technician. He looks forward to a career that is ever-changing and ever-challenging. He is, therefore, motivated and qualified enough for a rational study of the flow of viscous fluids in the later part of his academic training. On the other hand, modern engineering demands a lot more from its practitioners than its pre-World War II counterpart. Design problems become more critical; control and optimization use more stringent criteria; new fields open up in rapid succession. All these presuppose wide as well as deep understanding of a variety of disciplines, of which viscous flow theory is an important member. In all the traditional and modern engineering curricula—the mechanical, the aeronautical, the astronautical, the electrical, the chemical, the biomedical, the oceanographical, the environmental, and the hydraulic—a rational study of viscous flow has always a relevant, if not a key, place.

This book, then, is designed mainly to furnish a modern engineering student with a guide to this desired or required study. For those who will major in the applications and theory of modern fluid dynamics, this book will also serve as a bridge between the elementary and the advanced courses in the field. This dual role is played conscientiously; it is a designed, not an incidental, feature of the book. The exposition is careful, self-contained within the set scope, quite precise; and yet, simple. This is achieved by narrowing down the scope drastically: In Part 1, only plane flows of incompressible fluids are treated; and only rectangular Cartesian coordinates are used. Generalizations to three-dimensional flows of compressible fluids in a general coordinate system are not indicated until Part 3. The intermediate Part 2 covers more complicated flows of incompressible fluids. To some, this three-part gradation may seem a step backward in a sophisticated and rapidly advancing world. But, it is not. After all, this is a pedagogical text, not a scholarly treatise.

With this arrangement, an upperclassman should find no difficulty at all in reading Part 1. A course can then be designed for him, covering Part 1 and selected topics from Part 2. A course for first-year graduate students should emphasize Part 3, but it should start somewhere in the middle of Part 2, with the rest of the book as required reading.

In a book of this nature, the author's indebtedness to others is everywhere. He has no claim, of course, to any originality, except on certain pedagogical points. He can only point out, with self-satisfaction, the logical thread that goes through the entire work; the anticipation of difficulties and misunderstandings on the part of the student; the cautions against possible mistakes, misapplications, jumps to conclusion, and wrong generalizations. He especially takes pride in his introduction, to a book at this level, of the distinction between "ansatz" (that is, trial form of solution) and assumption, the Friedrichs problem (with a physical interpretation) as the prototype of the boundary layer, the rudiments of perturbation methods, the search for similarity solutions, the elements of matched inner-outer limits, the modern interpretation of low-Reynolds-number flows (in Part 2), and the coordinate-invariant formulation (in Part 3). All these topics, and many more, he learned from other authors, especially J. Serrin, C. Truesdell, K. O. Friedrichs, P. A. Lagerstrom, S. Goldstein, and L. M. Milne-Thomson.

As an ancient Chinese sage once asserted, the sickness of man lies in his fondness for playing teacher to others. Thus, in a world full of gurus, I claim merely the title of a guide, to be dismissed at the end of the journey.

March 1973
Lincoln, Nebraska

PAU-CHANG LU

ACKNOWLEDGMENTS

I am deeply grateful to Dr. Gustav Kuerti of Case Western Reserve University, Dr. Chang-Lin Tien of University of California at Berkeley, Dr. Ching-Jen Chen of the University of Iowa, and an anonymous reviewer for valuable suggestions and general encouragement, as well as helpful criticism. I am, of course, solely responsible for mistakes and defects that remain.

I am also indebted to Dr. Donald R. Haworth of the University of Nebraska for official support.

Mrs. Janet C. Lu compiled the index.

CONTENTS

LIST OF SYMBOLS

General Notations

$\mathbf{q}, \mathbf{F}, \mathbf{H}, \ldots$	vectors
$\hat{\mathbf{n}}, \hat{\mathbf{x}}, \hat{\boldsymbol{\theta}}, \hat{\mathbf{X}}_1, \ldots$	unit vectors in the n-, x-, θ-, X_1-, . . . directions
$q, F, \ldots$	magnitudes of $\mathbf{q}$, $\mathbf{F}$, . . .
$q_r, q_l, g_x, F_\theta, \ldots$	components of $\mathbf{q}$, $\mathbf{g}$, $\mathbf{F}$ in the r-, l-, x-, θ-, . . . directions
$\nabla\mathbf{q}, \mathbf{BC}, \mathbf{aA}, \ldots$	dyadics (in the form of dyadic products of vectors)
$\tilde{\boldsymbol{\sigma}}, \tilde{\mathbf{I}}, \ldots$	dyadics
$\sigma_{xx}, \sigma_{R\theta}, \ldots$	xx- and $R\theta$-components of $\tilde{\boldsymbol{\sigma}}$, . . .
$\mathbf{q}\overleftarrow{\nabla}$	conjugate of $\nabla\mathbf{q}$
$f\langle x, y\rangle, \mathscr{G}\langle t\rangle, \ldots$	functions of x and y, of t, . . .
$\cdot$	multiplication; dot-multiplication, between two vectors (Part 3)
$\times$	multiplication; cross-multiplication, between two vectors (Part 3)

AB	dyadic product of **A** and **B**
⁎	ordinary, dot-, cross-, or dyadic multiplication as the case may be
:	double dot product of two dyadics
D/Dt	differentiation following the motion, material differentiation
$\overline{(\ )}$	a "ceiling" of a quantity
$O(\ )$	of same or smaller order of
$o(\ )$	of smaller order of
$\square(\ )$	of same order of
$\overline{(\ )}$	an average (mean) quantity; conjugate of a complex number (Section 13.6)
$\sphericalangle(\ ,\ )$	angle between two directions
$[\![\mathcal{F}]\!]$	$\mathcal{F}_2-\mathcal{F}_1$, a jump of $\mathcal{F}$ across a surface of discontinuity
$\oint$	integration over a closed surface or curve
$\oint\!\!\!\!/$	integration along a path through B

Abbreviations

curl	curl of
$\stackrel{\text{def}}{=}$	equality by definition
div	divergence of
F–S	Falkner-Skan
$fct\langle\ \rangle$	a function of
grad	gradient of
K–P	Kármán-Pohlhausen
$\lim_{(P)}$	limiting process as a volume or area shrinks to zero, while always surrounding or passing through the point P
N-S	Navier-Stokes
$\parallel$	parallel to
$\perp$	perpendicular to

Superscripts

o	outer flow

$'$	dimensionless
$''$	dimensionless (usually stretched)
$+$	fluctuations
$*$	characteristic quantities

Subscripts

c	conjugate of a dyadic
P	evaluated at point P
s	scalar of a dyadic
v	vector of a dyadic
w	on the wall
∞	at infinity; at $y = Y$ (Section 13.8)

Latin letters

A	a constant; a parameter; an area; a function; a coefficient
$\mathbf{a}$	acceleration vector
a	a constant; a parameter; an index; a function
B	a constant; a function
B'	dimensionless Bernoullian; see Eq. (6.32)
b	a constant; a parameter; an index; a coefficient
C	a constant; a parameter; an index; a shear rate; a closed spatial circuit (Part 3)
$\mathscr{C}$	a material curve
δC, dC	an infinitesimal closed spatial circuit
C_D	drag coefficient
c	a constant; a function; a coefficient
c_{in}	specific heat capacity of an incompressible fluid
c_p	specific heat capacity at constant pressure

c_v	specific heat capacity at constant volume
D	a constant
d	a constant; a coefficient
E	a constant
Eu	Euler number
e	a constant; a parameter; specific internal energy (Part 3)
$\mathbf{F}$	force vector; a vector
F	a constant; a parameter; a function
$\mathscr{F}$	a parameter; a function; a scalar, vector, or dyadic, as the case may be (Part 3); a fluid property (Part 3)
Fr	Froude number
$\mathbf{f}$	force intensity vector
$\mathbf{f}\langle\hat{\mathbf{n}}\rangle$	$\mathbf{f}$ on an area element with unit normal vector $\hat{\mathbf{n}}$
f	a function
$\mathbf{G}$	velocity of advancement of a surface
G	a function
$\mathscr{G}$	a function; a scalar, vector, or dyadic, as the case may be (Part 3)
$\mathbf{g}$	gravitational (body) force per unit mass
g	a function
$\mathbf{H}$	heat flux vector (per unit area per unit time)
h	a constant; a height; a function; specific enthalpy (Part 3)
h_1, h_2, h_3	scale factors for ζ_1-, ζ_2-, ζ_3-coordinates (Part 3)
h_r	recovery specific enthalpy
$\tilde{\mathbf{I}}$	idemfactor
I_1, I_2, I_3	see Eqs. (13.61a,b,c)
i	$\sqrt{-1}$ (Section 5.4); an index
J	a parameter
K	a parameter; thermal conductivity (Part 3)

k	a constant; an index; a function
L	a length
$\hat{\mathbf{l}}$	unit vector along a curve with a chosen sense
l	distance along a curve in the direction of $\hat{\mathbf{l}}$; an index
dl	line element along a curve
$d\mathbf{l}$	vectorial line element along a curve
$d\mathbf{l}_1$	$d\mathbf{l}$ rotated 90° clockwise
ℓ	an index; a length
M	a constant; a length; a parameter; an index; Mach number (Part 3)
m	an index
$\hat{\mathbf{N}}$	a unit normal vector
N	an index; coordinate in N direction (Part 3)
$\hat{\mathbf{n}}$	unit normal vector; a unit vector
$\hat{\mathbf{n}}_{12}$	unit normal vector of a surface, pointing from side 1 to side 2
n	an index; a constant; coordinate in n-direction (Part 3)
P	a point; a parameter
Pr	Prandtl number
p	average pressure
$\mathring{p}$	hydrostatic pressure
$\tilde{p}$	a pressure difference
Q	volume flow rate or mass flow rate; heat generation per unit volume per unit time (Part 3)
$\mathbf{q}$	velocity vector; a vector
$\mathbb{R}$	gas constant
R	radial coordinate in the cylindrical or the plane polar coordinate system
$\mathscr{R}(\ )$	real part of
Re	Reynolds number
Re_L	Re based on length L
Re^*	$u^* Y/\nu$ (Section 10.3)
$\mathbf{r}$	position vector in space (Part 3)

r	radial coordinate in the spherical coordinate system
S	a spatial surface; a spatial surface area
$\mathscr{S}$	a material surface
δS, dS	an area element (magnitude only)
$\hat{\mathbf{n}}\,\delta S$, $\hat{\mathbf{n}}\,dS$	vectorial area element
St	(generalized) Strouhal number
s	specific entropy
T	temperature (in degrees Kelvin or Rankine)
T_r	recovery temperature
t	time
$\mathbf{U}$	$\mathbf{q}-\mathbf{G}$
U	a speed
u	x-component of velocity
V	a speed; a spatial volume (Part 3)
$\mathscr{V}$	a material volume
dV, δV	a volume element
v	y-component of velocity
W	a speed
w	z-component of velocity
X	a constant; a length
$\hat{\mathbf{X}}_1$, $\hat{\mathbf{X}}_2$, $\hat{\mathbf{X}}_3$	three orthogonal unit vectors
x	a rectangular Cartesian coordinate
x_1, x_2, x_3	the three principal axes
Y	a constant; a length
y	a rectangular Cartesian coordinate
$\underset{\sim}{y}$	$Y-y$
y_1, y_2	constants; lengths
Z	a constant; a length
z	a rectangular Cartesian coordinate
z_1	a constant; a length

Greek letters

α	a parameter; an auxiliary quantity; an angle

$\mathring{\alpha}$	an "incompressible" pressure
β	a constant; a coefficient; an auxiliary quantity
Γ	circulation
γ	a parameter; a constant; c_p/c_v (Part 3)
Δ	a thickness, a parameter; a determinant (Part 3)
$\Delta(\ )$	difference of
δ	boundary-layer thickness; a parameter; a width
δ_1	displacement thickness
δ_2	momentum thickness
δ_3	energy thickness
$\tilde{\boldsymbol{\epsilon}}$	rate-of-strain dyadic
$\epsilon_{11}, \epsilon_{22}, \epsilon_{33}$	principal rates of strain
$\boldsymbol{\zeta}$	a vectorial distance
ζ	a parameter; an independent variable
$\zeta_1, \zeta_2, \zeta_3$	three orthogonal curvilinear coordinates
$\boldsymbol{\eta}$	a vector
η	a parameter
$\eta\langle \mathrm{II} \rangle$	cross-viscosity function
Θ	an angle
ϑ	an angle
$\underline{\theta}$	$\Theta - \theta$
θ	angular coordinate in the cylindrical or the plane polar coordinate system; θ-coordinate in the spherical coordinate system, see Figure 9.7
κ	a coefficient
Λ	a parameter; a general volume, not necessarily material or spatial (Part 3)
λ	a parameter; a constant; second viscosity (Part 3)
μ	viscosity
$\mu\langle \mathrm{II} \rangle$	viscosity function (of a non-Newtonian fluid)

μ_t	see Eq. (10.40)
ν	kinematic viscosity
ν_t	μ_t/ρ
$\hat{\boldsymbol{\xi}}$	a unit vector
ξ	a parameter; an independent variable; a function
Π	thermodynamic pressure
ρ	density
Σ	a general surface (not necessarily material or spatial); summation sign
$\tilde{\boldsymbol{\sigma}}$	stress dyadic; a dyadic
$\sigma_{11}, \sigma_{22}, \sigma_{33}$	principal stresses
$\sigma^+_{xx'}, \ldots$	Reynolds stresses
$\tilde{\boldsymbol{\tau}}$	a dyadic
τ	a parameter, a constant (Section 3.3); shear stress
Υ	a speed
Φ	energy dissipation (Part 3)
ϕ	velocity potential; a scalar
φ	φ-coordinate in the spherical coordinate system, see Figure 9.7
ψ	(d'Alembert's) stream function for a plane flow
$\widehat{\psi}$	(Stokes') stream function for an axisymmetric flow
$\boldsymbol{\Omega}$	vorticity vector
Ω	vorticity
ω	a constant; circular frequency; rotational speed

Other symbols

∇	del operator
$\overleftarrow{\nabla}$	del operator operating *to its left*
∇_A	∇ differentiating on $\mathbf{A}$ or A only
$\overleftarrow{\nabla}_A$	$\overleftarrow{\nabla}$ differentiating on $\mathbf{A}$ or A (to its left) only
∇^2	$\nabla \cdot \nabla$ (the Laplacian)
II	see Eqs. (2.17) and (8.26)

TO THE STUDENT

Objects

This introductory text is designed with the following objects in mind:

1. *To serve as a bridge between elementary and advanced courses in fluid dynamics* so that the jump in the level of sophistication will not be too big.
2. *To consolidate the preliminary knowledge* you have gained from a course in elementary fluid mechanics. You will find some old topics re-presented, probably from a different angle. Some of the old concepts are probably sharpened up so as to be more precise. Some methods of solution may have been generalized.
3. *To furnish*, as the title suggests, *some fundamental information on the flow of viscous fluids*. More specifically, you will find out how the problem is formulated in mathematical language, how it is solved exactly for some situations, and how approximations are introduced in a rational manner for other more complicated cases; you will also learn about the crowning achievement of the twentieth century fluid dynamics – the boundary-layer theory.

Scope

To achieve these objects within a reasonable period of time, we have to select judiciously the items to be covered and limit our scope. Thus, it is decided to keep to the following three limitations in Part 1 of the book:

1. Only plane flows are treated.
2. Only the rectangular Cartesian coordinate system is used.
3. Only incompressible fluids are treated.

In Part 2 the first two limitations are relaxed to some extent. The general formulation of three-dimensional flow problems for compressible fluids in an arbitrary co-ordinate system is introduced in Part 3; but a comprehensive effort in treating the overall picture is definitely out of our scope. Furthermore, with the exception of a single chapter in Part 2, only laminar flows are discussed; and with the exception of two sections and some occasional paragraphs, only Newtonian fluids are treated.

Nature

The nature of the book can be explained under three headings: organization, content, and language.

I will first say something about the organization of the material. Within the chosen scope, the book is reasonably self-contained, and *follows a logical sequence*. Most problems either illustrate a point, or fill in a gap, or point out a possible generalization, and thus form an integral part of the text. You will find comparatively little drill material incorporated. Of the three parts, each part is an extension of its predecessor so that there is, by design, a three-leveled gradation clearly marked out.

As to the content, I shall only say that you will *not* find experimental data, detailed information in the form of tables and charts, irrational and empirical makeshifts, extensive numerical calculations, complicated situations that do not throw light on the basic principles, and so on.

In keeping with the three-leveled gradation, you will also notice a change in language and pace as you proceed from Part 1 to Part 3. As a matter of fact, Part 1 tends to be *lecturelike*. It is informal, leisurely, minutely detailed, intense (but never flowery), reiterative (probably overemphasizing at times), and somewhat verbose. On the other hand, Part 3 is written in the dry, impersonal, and formal language usually associated with scientific reports. The pace there is brisk; the statements are concise and precise; and, it is no longer the practice to explain every step however minute or obvious. The style must change not only because time is short, but also because too much attention to hair-splitting details breeds boredom.

Mathematical Level

This book is designed to use the minimum possible amount of mathematics. For example, except in Part 3, vector calculus is not used, nor are tensors or dyadics even mentioned.[1] Yet, the spirit behind the exposition is definitely that

[1] Vector operations are reviewed, and dyadics introduced, at the beginning of Part 3.

of mathematical model building. Thus, although we might be able to avoid certain mathematical complications, a minimum level of mathematical maturity is still required of you. It is not a question of what specific topics you have learned, or what manipulative ability you possess, or what tools you have hidden in your bag of "Advanced Engineering Mathematics." It is your general outlook on mathematics as a whole and its relation to engineering science that counts. For those of you who are uncertain or even suspicious of the role of mathematics, the next section may be of help.

Mathematical Models

As an engineer, you are undoubtedly appreciative of the importance of laboratory models. But conceptual models using mathematical language are far more important. Such mathematical models are very ancient and natural to the human race, contrary to the common belief. To recognize that all round things have something conceptual in common (that is, roundness) and describe them as circular (that is, with the periphery a fixed distance away from a certain point) is building a mathematical model for all these things. To understand this mathematical model is to understand an aspect of all the round things. Herein lies the basic difference between pure mathematicians and engineering scientists: Pure mathematicians investigate the properties of a circle per se, whereas we do that with round things in mind. In other words, although our approach may be "abstract," our aim is concrete. (At this juncture, I wish to recommend highly to you "A dialogue on the applications of mathematics" by Alfréd Rényi in his *Dialogues on Mathematics*, Holden-Day, San Francisco, 1967. And, since you are at it, read also the third and the first dialogues, in that order.)

A mathematical model overreaches by far the usefulness of laboratory models because of the following inherent characters:

1. A mathematical model can be made *very general.*
2. It can be made *more realistic*. (For example, a model airplane in a wind tunnel will always feel the presence of the confining wall, which is not there in the case of a real airplane. Yet, a mathematical model can easily be built without this confining wall.)
3. It can be built even when a laboratory model is simply out of question. (For example, in space exploration, astrophysics, weather forecasting.)
4. It has all the available mathematical tools at its disposal so that it can be observed or explored thoroughly in an orderly (logical) fashion.
5. Its initial cost is almost nil, and it is cheap (even considering the use of electronic computers) to draw information from it.
6. It can be easily modified, improved, abandoned, or resurrected without financial repercussions.

But I hasten to add that, just as the laboratory models, mathematical models must be built and used with constant close contact with physical reality. In this respect, it is instructional to regard our subject as a branch of (applied) mathematical physics and display the following self-explanatory chart:

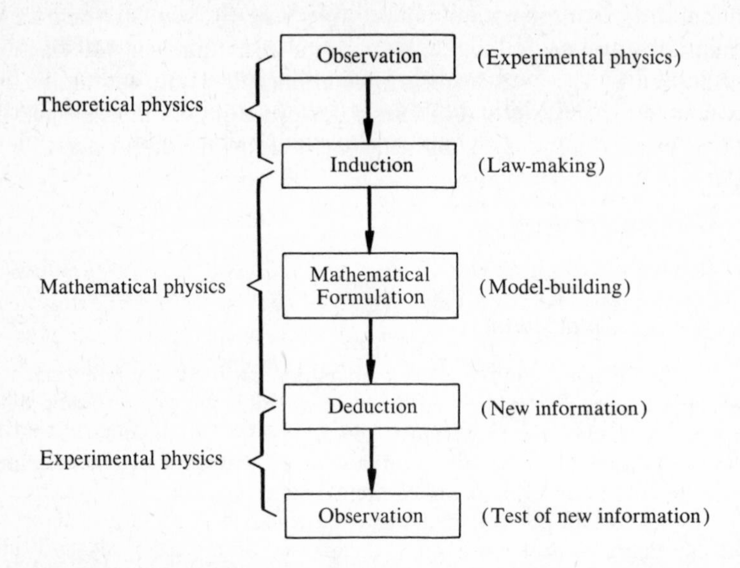

The important thing is that, before the model building, key physical observations must be performed judiciously to serve as guidelines in making the model; and that, afterwards, information obtained from the model must be tested by critically chosen physical observations. Only in this way can theory and experiment be interwoven to the benefit of both. (Read here "Mathematical models – uses and limitation," by S. W. Golomb, *Astronautics and Aeronautics*, pp. 57–59, January 1968.) In this book, however, we will start only with the formulation, based on established physical laws, and end up with drawing conclusions. As to the preliminary and verifying observations, we will just consider them done (by others, in other books).

Role of Computer

The modern electronic computer plays no role whatsoever in this book. This is not to belittle its importance in viscous flow theory. We are simply at a stage here where direct numerical computations on a computer do not add to our understanding of the underlying principles.

Footnotes

A large portion of the footnotes in Part 1 is used to inject some humor and human interest (or gossip, if you wish) into our dry and serious subject. You are urged to read them, if only to get a few moments of relaxation. They are *not* historical notes, although all came from reliable (second-hand) sources.

Matters of Notation

The most unusual notation adopted in this book is that for the *general* function; thus, a function f of x is denoted by $f\langle x\rangle$, not $f(x)$. At the beginning you might feel uncomfortable about the symbol ⟨ ⟩, but later on its use will reduce the number of possible misunderstandings on your part. (However, when we deal with *definite* functions, we still follow the common usage; thus, sin x or sin (x), but not sin $\langle x \rangle$.) Another symbol that needs explanation is the unit vector. In this book, the unit vector in the positive-x direction, for example, is denoted just by $\hat{\mathbf{x}}$, not the usual $\mathbf{i}$, $\mathbf{i}_x$, or $\mathbf{e}_x$; similarly, $\hat{\mathbf{n}}$ is the unit vector in the $+n$ direction.

Decimal System

Sections and equations (including conditions and inequalities) within a chapter are numbered consecutively in the decimal system; for example, Section 3.1 is the first section of Chapter 3, and Eq. (8.83c) is the third member equation in a *set* known as the 83rd equation of Chapter 8. If a *capital* letter follows an equation number, it denotes an alternative or modified form of that equation; for example, Eqs. (8.16A) and (8.16B) denote two alternative or modified forms of Eq. (8.16). Problems *within a section* are also arranged in this manner; thus, Problem 4.1.2 is the second problem of Section 4.1.

PART 1

FLOWS OF INCOMPRESSIBLE FLUIDS – 1

CHAPTER 1

KINEMATICS OF THE FLOW FIELD

1.1 INTRODUCTION

At the beginning let us state that Part 1 deals only with the plane flows of incompressible, homogeneous, and Newtonian fluids with constant viscosities, unless otherwise indicated. We will explain all these terms as we proceed. But first, a word about the concept of flow field.

Suppose we are looking at a portion of our physical world that is occupied by a fluid. To describe the motion of the fluid, we will erect a rectangular Cartesian coordinate system[1] (x, y, z). If we look at the fluid at any instant t_a, we will see that a certain fluid particle is passing by the space point (x_1, y_1, z_1) with an instantaneous velocity whose three components are (u_1, v_1, w_1), another certain fluid particle is passing by the space point (x_2, y_2, z_2) with velocity (u_2, v_2, w_2), and so forth (Fig. 1.1). If we look at it at another instant t_b, the fluid particles passing by the same space points will in general move with different velocity components. Thus, we will have a complete knowledge of the motion of the fluid for all times only when we know the three velocity components

$$\begin{cases} u = u\langle x, y, z; t\rangle & \textbf{(1.1a)} \\ v = v\langle x, y, z; t\rangle & \textbf{(1.1b)} \\ w = w\langle x, y, z; t\rangle & \textbf{(1.1c)} \end{cases}$$

[1] For simplicity, only rectangular Cartesian coordinates are used in Part 1.

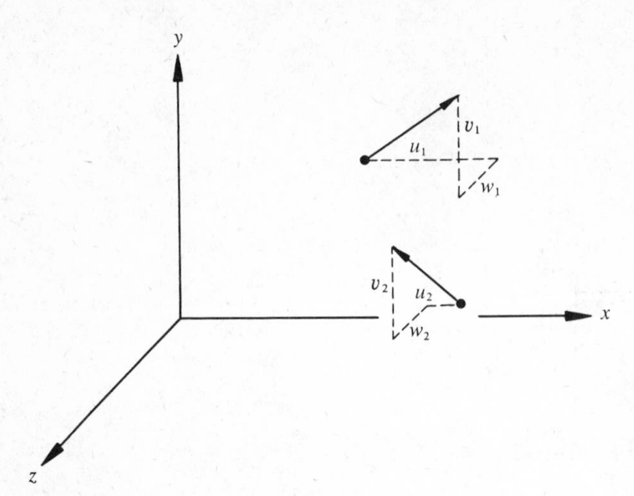

Figure 1.1 Instantaneous velocity components of fluid particles passing through two space points.

as functions of the position (x, y, z) and time t. This distribution of velocity components over the space points is known as a time-dependent velocity field. Since the components u, v, w form the velocity vector $\mathbf{q}$, we have actually a so-called *vector field (time-dependent)*:

$$\mathbf{q}\langle x, y, z; t\rangle = \begin{cases} u\langle x, y, z; t\rangle \\ v\langle x, y, z; t\rangle \\ w\langle x, y, z; t\rangle \end{cases}$$

This field description can be visualized as in Figure 1.2. We can imagine that, at every instant, all space points carry vectors to denote the velocities of the fluid particles that pass by these points at that instant.

Of course, to describe the behavior of the fluid completely, the velocity field alone is not enough. The space points must also carry density values, temperature values, stress and other force values, and the like (Fig. 1.3). The totality

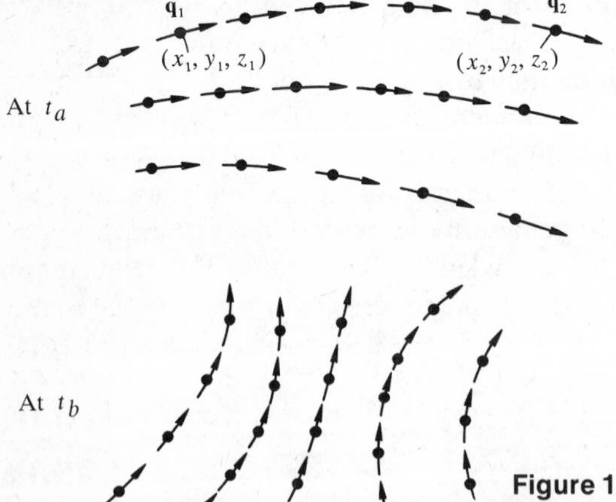

Figure 1.2 A time-dependent velocity field at two different instants.

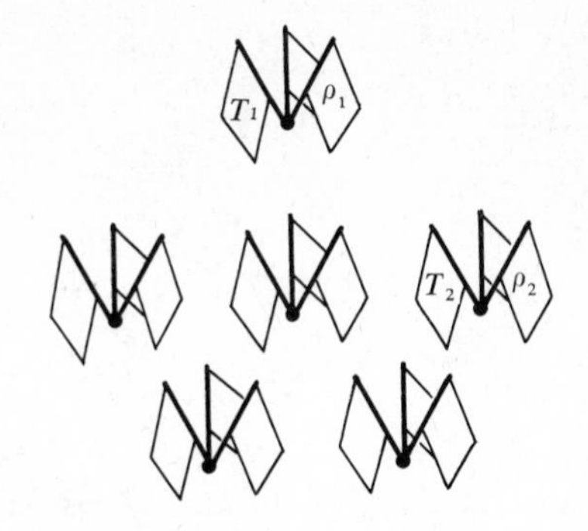

At t_a

Figure 1.3 A flow field at a particular instant.

of all these fields, vectorial or scalar, is called the *flow field* of the fluid:

$$\text{flow field} \begin{cases} \mathbf{q}\langle x, y, z; t\rangle \\ \rho\langle x, y, z; t\rangle \\ T\langle x, y, z; t\rangle \\ \cdot \\ \cdot \\ \cdot \end{cases}$$

where ρ denotes the density; T, the temperature; and so on.

In conclusion, a flow field is the description of the flow of a fluid by prescribing all fundamental flow properties as functions of position in space and time. It is essential to note that such a description is not concerned with the fate of an individual particle—that is, which particular particle happens to be at (x_1, y_1, z_1) at t_a, or where this particle was before t_a, or where it will be after t_a. It tells you how the entire field of fluid evolves, not how each fluid particle moves. It gives you *a series of* "snapshots" of all the particles without identifying one particular particle (by dyeing it red, for example).

In this chapter, we will study only the kinematics of the flow field—that is, a description of the motion and the mass distribution—without asking how the motion came about. The only physical principle investigated will be that of mass conservation.

As an illustration, the following information is sufficient to describe the kinematics of a fluid flow:

$$\begin{cases} \rho = \text{constant} \\ u = Cy \\ v = 0 \\ w = 0 \end{cases}$$

From it we learn that the fluid density does not change; the flow is parallel to the x-axis; and the speed is independent of time, x, and z, but proportional to y (Fig. 1.4). It gives us, therefore, a kinematic flow field.

We notice, with interest, that this illustrative flow has

$$\begin{cases} w = 0 \\ \text{all flow properties independent of } z \end{cases}$$

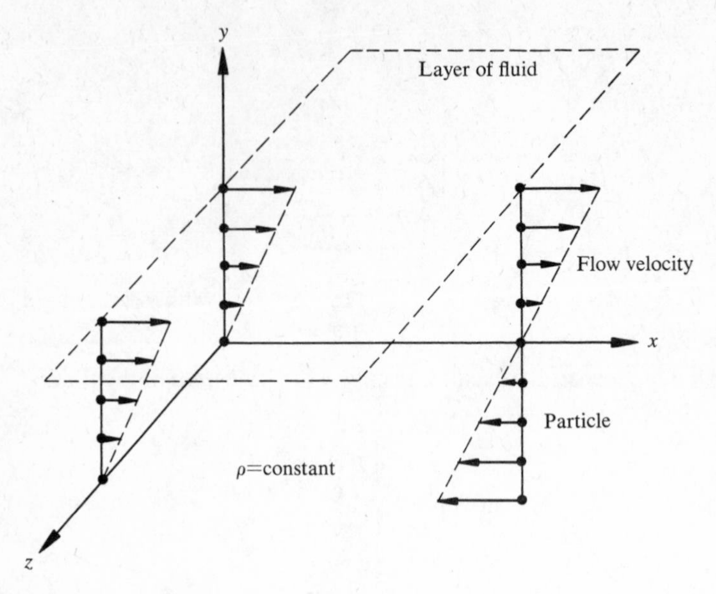

Figure 1.4 An example of plane flow.

The first of these characteristics makes the velocity vector **q** lie in planes perpendicular to the z-axis; the second makes the flow field exactly the same from one plane (perpendicular to the z-axis) to another (Fig. 1.5). Thus, the whole spatial flow field can be represented in its entirety by one of these planes perpendicular to the z-axis, say $z = 0$. Flows like this, where one can represent the entire flow situation by that on a *representative plane* and where there is no flow out of the representative plane, are called *plane flows*. (If we know the flow picture on the representative plane, the entire flow can be produced by simply shifting the representative plane perpendicularly to itself.)

In practical situations, there is no such a thing as a plane flow, to be exact. But, approximately, there are many flow situations where plane flows occur at least in a portion of the apparatus. For example, in the flow past an airplane whose wings are long and not tapered, the flow in the middle portion of the wing is approximately plane (Fig. 1.6). It is important to notice here that, even if the wing is not long, or is tapered, an investigation of the flow regarded as being plane anyway can still be useful in the sense that a tip correction, a fuselage-interference correction, and/or a tapering correction may be applied to the result to compensate for slight errors. As another example, consider a flow through a duct with a rectangular cross section. If the height of the duct is very small compared to the width, again an approximately plane flow can be seen in the middle region far away from the side walls (Fig. 1.7). The region near the side walls is then to be taken care of by side-wall corrections.

In Part 1 we shall restrict overselves to these plane flows for simplicity. We shall always denote the representative plane by $z = 0$. The flow velocity vector **q** then becomes a plane vector in the xy-plane with components $u\langle x, y; t\rangle$ and

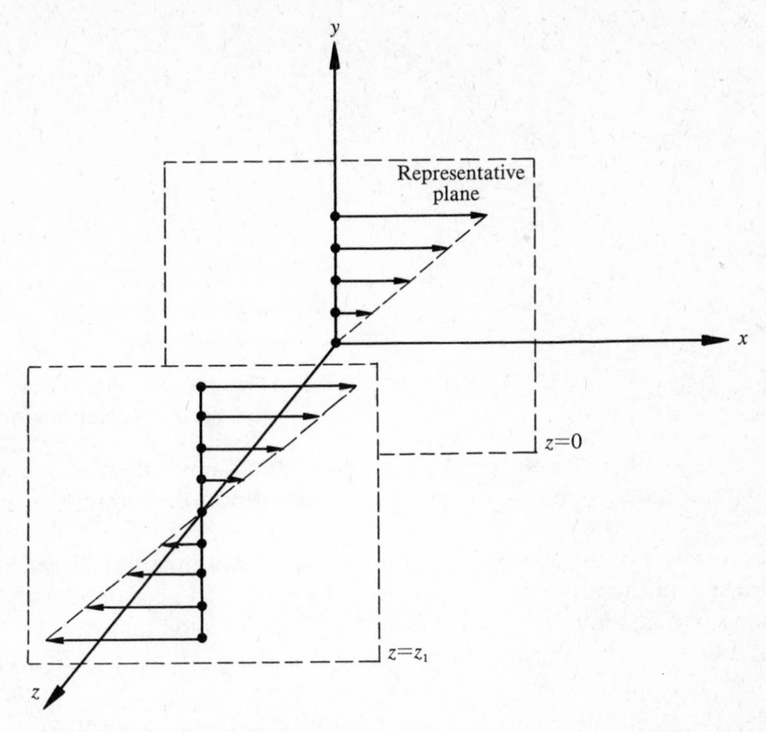

Figure 1.5 A plane flow is generated from its representative plane.

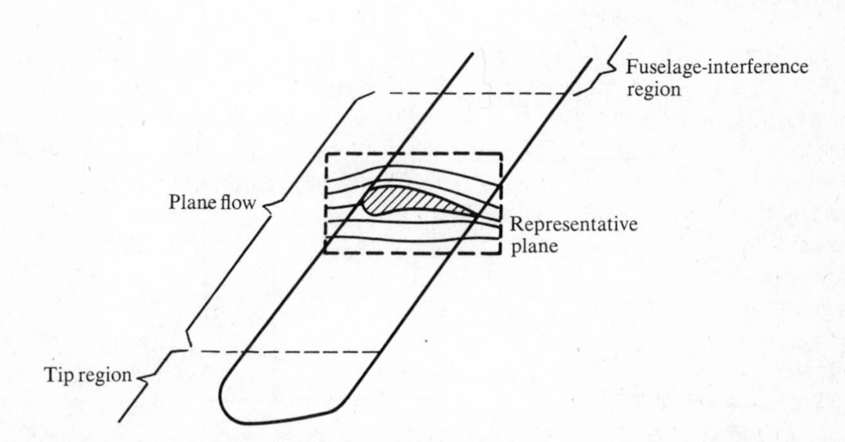

Figure 1.6 Plane-flow and nonplane-flow regions of a nonswept-back airfoil.

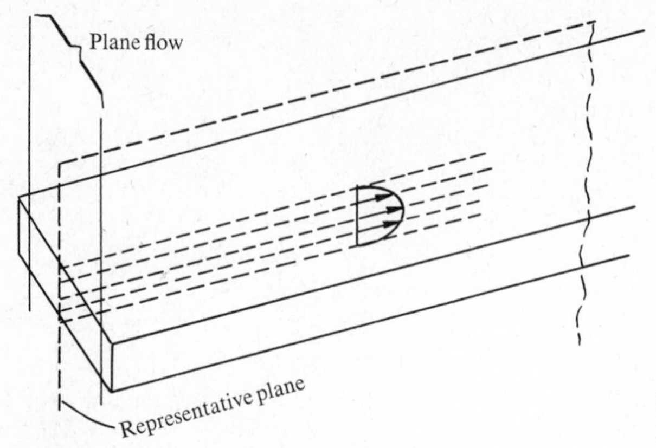

Figure 1.7 Plane-flow region in a narrow rectangular duct.

$v\langle x, y; t\rangle$. All the other flow properties will also be functions only of x, y, and t, but not of z. The simplifications introduced in the plane flow are, therefore, $w \equiv 0$ and $\partial(\ \ )/\partial z \equiv 0$.

In interpreting the plane flow physically, we must not forget that the picture we see on the representative plane is to be shifted perpendicularly to the plane in order to generate the true space flow. Thus, it should always be kept in mind that a point on the representative plane is actually a line; a fluid particle, a rodlike fluid element; a curve, a cylindrical surface (see Fig. 1.8a). On the other hand, since the situation is simply repeated from one plane to another, it is convenient to consider only the portion of the true space flow between $z = 0$ and $z = -1$ (1 meter, say). We then imagine that we are dealing with the flows per unit depth (of a depth of 1 meter, for example) and can refer to all flow quantities on per-unit-depth basis, wherever applicable. For example, in Figure 1.8b the force F on the profile in the representative plane is actually that per unit of depth of the cylindrical body whose trace is the profile in the representative plane. (If we wish to apply the result to a cylindrical body of a length of M units, the total force would be MF.)

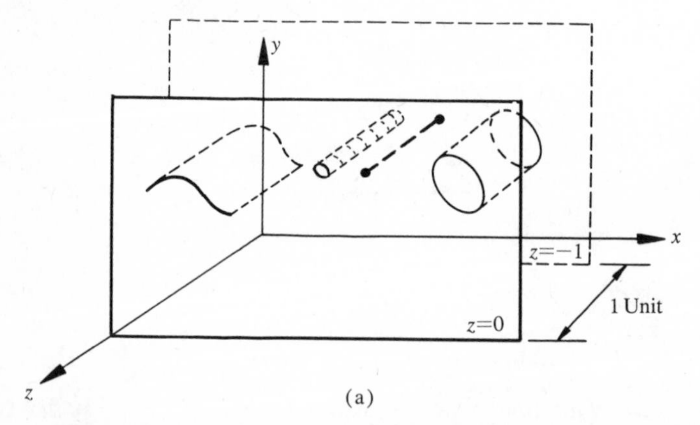

Figure 1.8 (a) Traces on the representative plane.

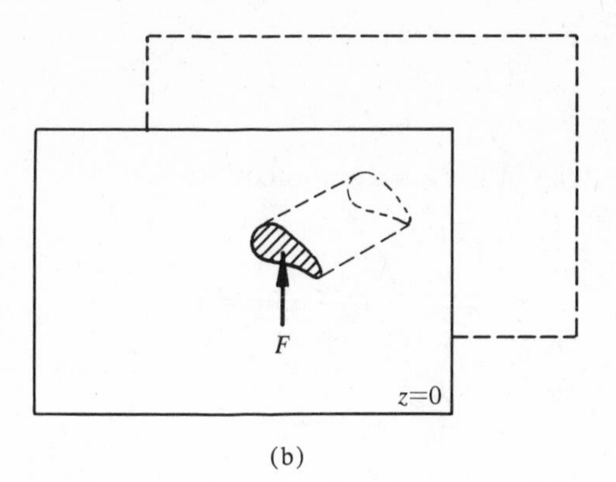

Figure 1.8 (b) Force F per unit depth.

1.2 MASS CONSERVATION, STREAM FUNCTION, AND STREAMLINE

If a flow is such that *all* the properties in the field description are time-independent—that is, $\mathbf{q} = \mathbf{q}\langle x, y\rangle$, $\rho = \rho\langle x, y\rangle$, and so on—we say that the flow is steady. (If *only* $\mathbf{q}$ is time-independent, we say that it is a steady-$\mathbf{q}$ flow, and so on.) For a steady flow, the situation never changes as time goes on. It would be wrong, however, to conclude that a fluid particle is not accelerating in a steady flow. As a matter of fact, the fluid particle that was at (x, y) at the instant t will have moved to a new position $(x + \delta x, y + \delta y)$ at a later instant $t + \delta t$, where $\delta x = u\,\delta t$, $\delta y = v\,\delta t$ (Fig. 1.9). The original flow velocity was (u, v), at $(x, y; t)$. The new velocity at

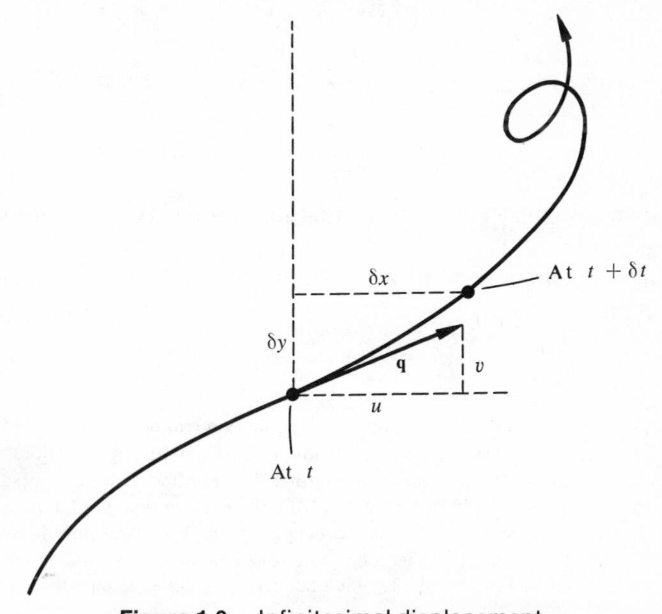

Figure 1.9 Infinitesimal displacement.

$t+\delta t$ is

$$\left(u+\frac{\partial u}{\partial x}\delta x+\frac{\partial u}{\partial y}\delta y+\frac{\partial u}{\partial t}\delta t,\quad v+\frac{\partial v}{\partial x}\delta x+\frac{\partial v}{\partial y}\delta y+\frac{\partial v}{\partial t}\delta t\right)$$

The rate of change of velocity of the particle, or the acceleration of the particle, is

$$\mathbf{a}=\begin{cases}\lim\limits_{\delta t\to 0}\dfrac{1}{\delta t}\left(\dfrac{\partial u}{\partial x}\delta x+\dfrac{\partial u}{\partial y}\delta y+\dfrac{\partial u}{\partial t}\delta t\right)\\[2ex]\lim\limits_{\delta t\to 0}\dfrac{1}{\delta t}\left(\dfrac{\partial v}{\partial x}\delta x+\dfrac{\partial v}{\partial y}\delta y+\dfrac{\partial v}{\partial t}\delta t\right)\end{cases}$$

or, since $\delta x = u\,\delta t$ and $\delta y = v\,\delta t$,

$$\mathbf{a}=\left(u\frac{\partial u}{\partial x}+v\frac{\partial u}{\partial y}+\frac{\partial u}{\partial t},\ u\frac{\partial v}{\partial x}+v\frac{\partial v}{\partial y}+\frac{\partial v}{\partial t}\right)\tag{1.2}$$

which is equal to

$$\left(u\frac{\partial u}{\partial x}+v\frac{\partial u}{\partial y},\ u\frac{\partial v}{\partial x}+v\frac{\partial v}{\partial y}\right)\neq 0$$

even if the flow is steady.

It is customary to introduce here the shorthand

$$\frac{D}{Dt}=\frac{\partial}{\partial t}+u\frac{\partial}{\partial x}+v\frac{\partial}{\partial y}\tag{1.3}$$

(which is just $u(\partial/\partial x)+v(\partial/\partial y)$ for a steady flow). Then,

$$\mathbf{a}=\lim_{\delta t\to 0}\frac{\delta\mathbf{q}}{\delta t}=\frac{D\mathbf{q}}{Dt}\tag{1.4}$$

$$=\begin{cases}\dfrac{Du}{Dt} & \text{(1.5a)}\\[2ex]\dfrac{Dv}{Dt} & \text{(1.5b)}\end{cases}$$

D/Dt is known as the (time) differentiation *following the motion*,[2] or just "capital dee-dee-tee."

The same argument can also be applied to other flow properties as well as $\mathbf{q}$. For example,

$$\frac{D\rho}{Dt}=\frac{\partial\rho}{\partial t}+u\frac{\partial\rho}{\partial x}+v\frac{\partial\rho}{\partial y}\tag{1.6}$$

---

[2] The first implicit use of differentiation following the motion can be found in a 1755 paper (published in 1757) of Euler, "General principles of fluid motion" (in French). See A List of Pre-1900 Works Cited at the end of this book. It is altogether fitting that the very first historical personage we meet should be Leonhard Euler (1707–1783), for he was truly the first *rational* fluid dynamicist. You will see, as we go on, that we owe most of the basic concepts to him. Euler was born in Switzerland, but spent some 30 years in Russia. After he lost the sight of one eye, he claimed that it only served to reduce distraction; and, after he became totally blind, he merely switched to dictating and continued to pour out his ideas on paper.

is the time rate of change of the fluid density ρ following the motion. It is equal to $(u(\partial\rho/\partial x)+v(\partial\rho/\partial y))$ even for a steady-ρ flow. In general,

$$\frac{D(\ )}{Dt}: \text{ Time differentiation of ( ) following the motion}$$

$$\frac{\partial(\ )}{\partial t}: \text{ Time differentiation of ( ) at a fixed space point}$$

$$\frac{D(\ )}{Dt}=0: \text{ ( ) of a particle does not change as it moves}$$

$$\frac{\partial(\ )}{\partial t}=0: \text{ ( ) of the flow field is steady}$$

At this juncture, we can introduce the precise definition of an incompressible fluid: If the density of *every* particle of a fluid is constant following the motion, that is, if

$$\frac{D\rho}{Dt}=0 \tag{1.7}$$

it is said to be incompressible. This, however, does not rule out the possibility that different particles possess different *constant* densities. For example, a fluid may be made up of a number of layers of different liquids (Fig. 1.10). To rule out this complexity in the present book, we will restrict ourselves from now on to homogeneous fluids only. For these fluids, we have not only $D\rho/Dt=0$, but also

$$\rho = \text{constant, } \textit{throughout the flow field and at all times} \tag{1.8}$$

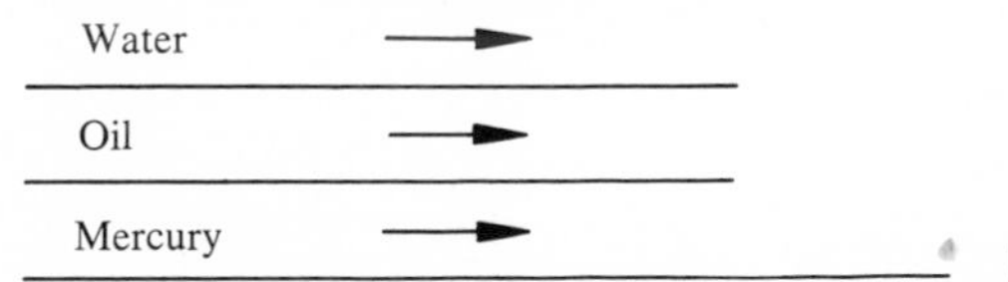

Figure 1.10 Flow of three liquid layers.

Of course, there is no real fluid that is truly incompressible; but experience tells us that ordinary liquids can be approximately looked upon as incompressible, except in special problems such as sound propagation, underwater explosion, water hammer, and so on. (Gas flows at very low speeds *in regions where viscous stresses and heat fluxes are negligible* can also be treated approximately as incompressible.)

One important simplification because of the incompressibility shows up in the following investigation of *mass conservation.*

Consider an infinitesimal rectangle (really a rectangular column perpendicular to the representative plane; or rather, a portion of it of unit depth) *fixed* in the xy-plane (Fig. 1.11). At any instant t, fluid with constant density ρ flows into the rectangle (that is, the rectangular column of unit depth) at a rate (per

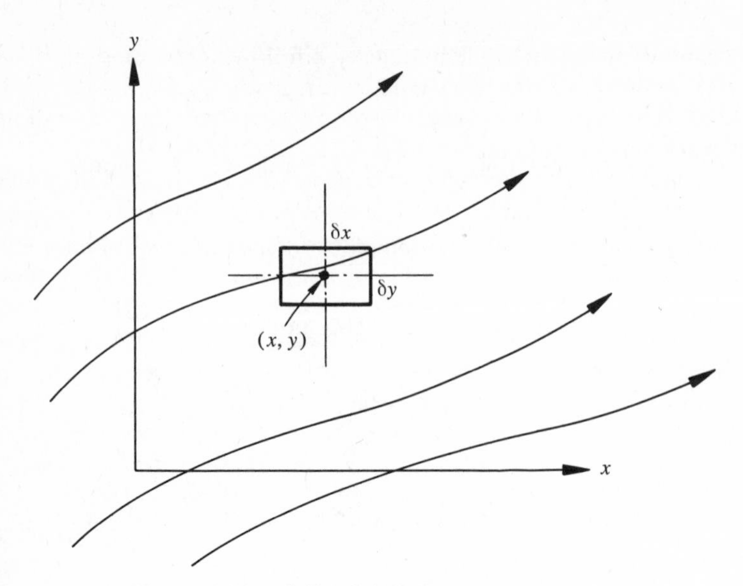

Figure 1.11 A fixed, infinitesimal rectangle.

second, say):

$$\rho\left[u\langle x, y; t\rangle - \frac{1}{2}\frac{\partial u\langle x, y; t\rangle}{\partial x}\delta x\right]\delta y \text{ (across the left face)}$$

$$+\rho\left[v\langle x, y; t\rangle - \frac{1}{2}\frac{\partial v\langle x, y; t\rangle}{\partial y}\delta y\right]\delta x \text{ (across the bottom face)}$$

and also out of the rectangle at a rate:

$$\rho\left[u\langle x, y; t\rangle + \frac{1}{2}\frac{\partial u\langle x, y; t\rangle}{\partial x}\delta x\right]\delta y \text{ (across the right face)}$$

$$+\rho\left[v\langle x, y; t\rangle + \frac{1}{2}\frac{\partial v\langle x, y; t\rangle}{\partial y}\delta y\right]\delta x \text{ (across the top face)}$$

for infinitesimal δx and δy. Now, the law of mass conservation demands that, in the absence of mass creation and annihilation, the *net* mass flow into the infinitesimal rectangle must be zero at all locations (x, y) and at all times t. It is easy to see that we must then have

$$\frac{\partial u}{\partial x} + \frac{\partial v}{\partial y} = 0 \tag{1.9}$$

at all locations and at all times.

Equation (1.9) is usually called the equation of continuity.[3] One must

[3] It would have been much better if this were simply known as the equation of mass conservation. The word "continuity" is inappropriate, but unfortunately it seems here to stay. Equation (1.9) first made its appearance in d'Alembert's long *Essay on a New Theory of Fluid Resistance* (in French, submitted to the Berlin Academy in 1749 and published privately in 1752). Jean le Rond d'Alembert (1717–1783) was a brilliant, if not always reliable, mathematician. His name came from the chapel of St. Jean le Rond in Paris, on the steps of which his mother abandoned him. As befitting the coeditor of the famous Encyclopedia, he contributed to many fields, including music, literature, and philosophy; but he was still unquenchably thirsty for priority, especially over Euler.

remember that it was derived for the plane flow of an *incompressible fluid*, steady or not. The absence of any $\partial/\partial t$ term in Eq. (1.9) should not mislead you into thinking that it is true only for steady flows. It is valid in exactly the same form for time-dependent flows. (See also Problems 1.2.7 and 1.2.8.)

Equation (1.9) has many immediate consequences, one of which is the existence of a *stream function*. We will approach the subject of stream function from two angles: its mathematical introduction and its physical interpretation.

Before we embark on a formal, mathematical discussion, however, let us sum up the highlights of the physical interpretation here: Choose a fixed reference point $A(x_0, y_0)$ on the representative plane, and consider an arbitrary point $P(x, y)$. The volume flow rate across all curves connecting A and P must be the same because of mass conservation, and is, therefore, a function of (x, y) independently of the curves from A to P. This is called the stream function. The most important use of the stream function is that it yields, on differentiation, the flow velocity (u, v) that automatically satisfies the equation of continuity.

With the foregoing physical anticipation in mind as our motivation, we will now indulge in a little mathematical argument before we come back to physics. This mathematical introduction not only will help in making our physical interpretation more precise, but also will show the existence of other kinds of stream or "streamlike" functions later on (for example, in Sections 6.3, 9.2, and 12.7).

(1) The Mathematical Introduction of Stream Function

It is obvious that if we can find a function $\psi\langle x, y; t\rangle$ such that

$$u\langle x, y; t\rangle = \frac{\partial\psi\langle x, y; t\rangle}{\partial y} \tag{1.10a}$$

$$v\langle x, y; t\rangle = -\frac{\partial\psi\langle x, y; t\rangle}{\partial x} \tag{1.10b}$$

then Eq. (1.9) is automatically satisfied. This function $\psi\langle x, y; t\rangle$, if we can find it, is called the stream function of the velocity field (u, v). Its introduction helps us in two ways: (a) the equation of continuity is already taken care of; (b) instead of dealing with two functions u and v, there is now only one function, ψ.

The question that remains is: Can we determine ψ for any velocity field *that satisfies* Eq. (1.9)? The answer is yes, since we can readily present a procedure for determining ψ. Starting with the requirement of Eq. (1.10a), we can integrate both sides partially with respect to y and obtain

$$\psi\langle x, y; t\rangle = \int u\langle x, y; t\rangle\, dy + f\langle x; t\rangle \tag{1.11}$$

where $f\langle x; t\rangle$ came out of the partial integration and must be determined with the help of the second requirement, Eq. (1.10b). Differentiating Eq. (1.11) with respect to x, we have

$$\frac{\partial\psi}{\partial x} = \frac{\partial}{\partial x}\left(\int u\langle x, y; t\rangle\, dy\right) + \frac{\partial f\langle x; t\rangle}{\partial x}$$

But the left-hand side is just the minus of $v\langle x, y; t\rangle$ by Eq. (1.10b); therefore,

$$\frac{\partial f\langle x; t\rangle}{\partial x} = F\langle x, y; t\rangle$$

where

$$F\langle x, y; t\rangle = -\frac{\partial}{\partial x}\left(\int u\langle x, y; t\rangle\, dy\right) - v\langle x, y; t\rangle \tag{1.12}$$

is a function calculable from the given (u, v). Integrating partially with respect to x, then, gives

$$f\langle x; t\rangle = \int F\langle x, y; t\rangle\, dx + G\langle t\rangle \tag{1.13}$$

where $G\langle t\rangle$ is an arbitrary function of t. It must be realized here that Eq. (1.13) demands that $F\langle x, y; t\rangle$ be actually independent of y. But this is guaranteed by the equation of continuity since, from Eq. (1.12),

$$\frac{\partial F}{\partial y} = -\frac{\partial}{\partial x}\left(\frac{\partial}{\partial y}\int u\, dy\right) - \frac{\partial v}{\partial y} = -\frac{\partial u}{\partial x} - \frac{\partial v}{\partial y} = 0$$

Thus, it may be better to write $F\langle x; t\rangle$ in Eqs. (1.12) and (1.13).

Finally, substituing Eq. (1.13) into Eq. (1.11), we have

$$\psi\langle x, y; t\rangle = \int u\langle x, y; t\rangle\, dy + \int F\langle x; t\rangle\, dx + G\langle t\rangle \tag{1.14}$$

Thus, we have shown that ψ can be determined up to an arbitrary function of t (an arbitrary constant if the flow is steady). This arbitrary function $G\langle t\rangle$ is not of any real interest to us, since, after all, it is the velocity field, $(\partial\psi/\partial y, -\partial\psi/\partial x)$, that is important. It is clear that the contribution of $G\langle t\rangle$ here is nil, no matter what form it assumes.

On the other hand, of course, we would also like to fix $G\langle t\rangle$ so that there are no extraneous "floating" terms. To do this, we can simply refer the value of ψ to an arbitrarily chosen datum value ψ_0 assigned to an arbitrarily selected reference point (x_0, y_0); that is, we set

$$\psi\langle x_0, y_0; t\rangle = \psi_0, \text{ a constant}$$

Then in accordance with Figure 1.12,

$$\psi_0 = \left(\int u\langle x, y; t\rangle\, dy + \int F\langle x; t\rangle\, dx\right)\Big|_{\substack{x=x_0\\ y=y_0}} + G\langle t\rangle$$

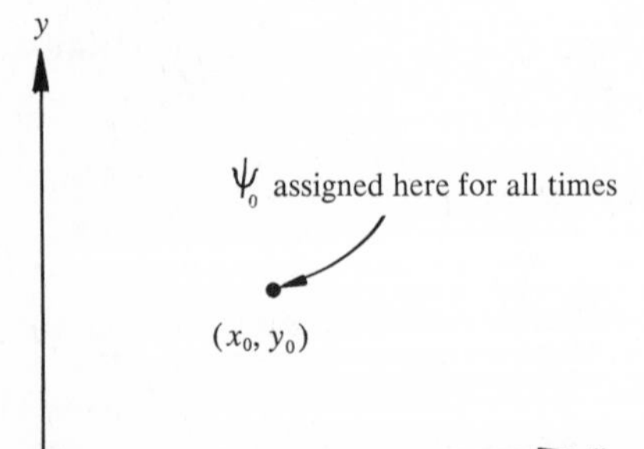

Figure 1.12 Datum point of ψ.

that is,

$$G\langle t\rangle = \psi_0 - \left(\int u\langle x, y; t\rangle\, dy + \int F\langle x; t\rangle\, dx \right)\Bigg|_{\substack{x=x_0 \\ y=y_0}}$$

If the flow is steady, we can go one step further and choose

$$\psi_0 = \left(\int u\langle x, y\rangle\, dy + \int F\langle x\rangle\, dx \right)\Bigg|_{\substack{x=x_0 \\ y=y_0}}$$

thus forcing the arbitrary constant involved to vanish for our convenience in writing.

At any rate, Eq. (1.14) shows how we can calculate $\psi\langle x, y; t\rangle$ in all its essentials. We may therefore conclude here that the equation of continuity yields a stream function ψ such that $(u, v) = (\partial\psi/\partial y, -\partial\psi/\partial x)$.

EXAMPLE 1

Given the velocity field

$$\begin{cases} u = Ut \\ v = x \end{cases}$$

find $\psi\langle x, y; t\rangle$.

SOLUTION It is always simpler to follow the procedure just described, not to substitute into the general result, Eq. (1.14). We start with Eq. (1.10a) which is here

$$\frac{\partial\psi}{\partial y} = u = Ut$$

Integrating with respect to y, we have

$$\psi = Uty + f\langle x; t\rangle$$

Then

$$\frac{\partial\psi}{\partial x} = \frac{\partial f}{\partial x}$$

On the other hand,

$$\frac{\partial\psi}{\partial x} = -v = -x$$

Therefore

$$f\langle x; t\rangle = -\frac{x^2}{2} + G\langle t\rangle$$

or

$$\psi = Uty - \frac{x^2}{2} + G\langle t\rangle$$

If we assign value 0 to ψ at $x = 0$, $y = 1$, we have

$$G\langle t\rangle = -Ut$$

or

$$\psi = Ut(y-1) - \frac{x^2}{2}$$

But it would be more convenient to assign a zero value to ψ at $(0, 0)$, which yields $G\langle t\rangle = 0$, or

$$\psi = Uty - \frac{x^2}{2}$$

Both choices yield the same velocity field, so it does not really matter which form you use.

EXAMPLE 2

Given the velocity field (a uniform flow in the x-direction)

$$\begin{cases} u = U \\ v = 0 \end{cases}$$

find $\psi\langle x, y; t\rangle$.

SOLUTION Since the field is steady, we are really searching for $\psi\langle x, y\rangle$. We start with

$$\frac{\partial\psi}{\partial y} = u = U$$

$$\psi\langle x, y\rangle = Uy + f\langle x\rangle$$

Then

$$\frac{\partial\psi}{\partial x} = \frac{df}{dx}$$

but also

$$\frac{\partial\psi}{\partial x} = -v = 0$$

and therefore

$$f\langle x\rangle = C, \text{a constant}$$

and

$$\psi\langle x, y\rangle = Uy + C$$

Choosing $\psi = 0$ on $y = 0$, we have $C = 0$; and $\psi = Uy$.

COUNTEREXAMPLE

Given the velocity field

$$\begin{cases} u = \dfrac{x}{1+t} \\ v = 1 \end{cases}$$

find $\psi\langle x, y; t\rangle$.

SOLUTION Starting with

$$\frac{\partial\psi}{\partial y} = \frac{x}{1+t}$$

we have

$$\psi\langle x, y; t\rangle = \frac{xy}{1+t} + f\langle x; t\rangle$$

Then

$$\frac{\partial \psi}{\partial x} = \frac{y}{1+t} + \frac{\partial f\langle x; t\rangle}{\partial x}$$

also

$$\frac{\partial \psi}{\partial x} = -v = -1$$

and therefore

$$\frac{\partial f\langle x; t\rangle}{\partial x} = -1 - \frac{y}{1+t}$$

But we cannot go on from here, since the left-hand side is free of y and yet the right-hand side has y in it. This inconsistency stems from the fact that the given velocity field does *not* satisfy the equation of continuity. The flow is, therefore, not mass-conserving, and thus no stream function exists at all.

(2) The Physical Interpretation of Stream Function

The physical meaning of the equation of continuity is mass conservation. Now consider the mass flow across a curve ABP (actually a cylindrical surface of unit depth) as shown in Figure 1.13. The flow across an element dl of ABP from left to right is obviously $\rho(u\,dy - v\,dx)$ per unit time at any instant. (The flow across a portion of ABP may actually be from right to left, depending on the signs of u and v.) The total *net* flow across ABP from left to right is then the *line integral*

$$\rho \int_{A}^{P} (u\,dy - v\,dx)$$

per unit time at the same instant. Similarly, that across ACP is

$$\rho \int_{A}^{P} (u\,dy - v\,dx)$$

In the above the paths B and C are arbitrary except that they are both from A to P. Now mass conservation of an incompressible fluid demands that, at any instant, a

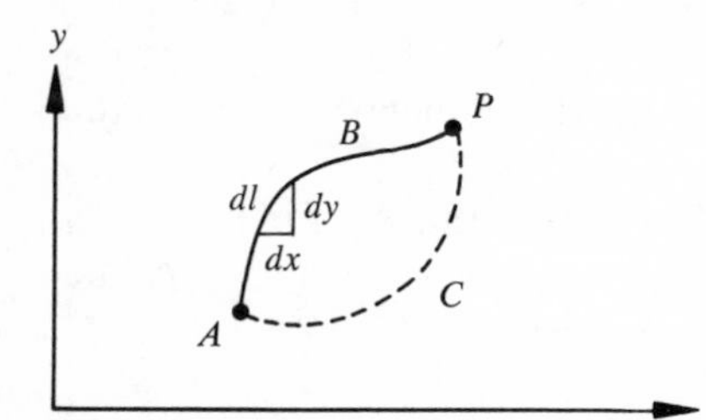

Figure 1.13 Two paths of integration.

mass flow across ABP must push the same amount of mass flow across ACP; that is,

$$\int_A^P (u\,dy - v\,dx) = \int_A^P (u\,dy - v\,dx)$$

for arbitrary paths between A and P. So, the line integral

$$\int_A^P (u\,dy - v\,dx)$$

depends only on the end points, not on the path. If we choose $A(x_0, y_0)$ as our fixed reference point, we have

$$\begin{aligned}\int_{A(x_0,y_0)}^{P(x,y)} (u\,dy - v\,dx) &= \text{a function of } \langle x, y; t\rangle \\ &= \psi\langle x, y; t\rangle \qquad \textbf{(1.15)}\end{aligned}$$

This immediately assigns a zero value of ψ to the point (x_0, y_0). The function is exactly the same as our stream function introduced before, since, from Eq. (1.15),

$$d\psi = u\,dy - v\,dx \qquad \textbf{(1.16)}$$

at any t; that is,

$$\begin{cases} u = \dfrac{\partial\psi}{\partial y} \\ v = -\dfrac{\partial\psi}{\partial x} \end{cases}$$

at any t. The arbitrary function $G\langle t\rangle$ is now embedded in the arbitrary choice of the zero-ψ point $A(x_0, y_0)$.

Equation (1.15) thus supplies a physical meaning to the stream function: $\psi\langle x, y; t\rangle$ is the rate of volume flow from left to right past *any* curve (cylindrical surface of unit depth) connecting the point $P(x, y)$ and a chosen reference point $A(x_0, y_0)$ at the instant t. Furthermore, the difference of ψ between two points (x_1, y_1) and (x_2, y_2) at any instant t is

$$\begin{aligned}\psi\langle x_2, y_2; t\rangle - \psi\langle x_1, y_1; t\rangle &= \int_A^{(x_2,y_2)} (u\,dy - v\,dx) - \int_A^{(x_1,y_1)} (u\,dy - v\,dx) \\ &= \int_{(x_1,y_1)}^{(x_2,y_2)} (u\,dy - v\,dx) \qquad \textbf{(1.17)}\end{aligned}$$

which is the volume flow from left to right across any curve connecting (x_1, y_1) and (x_2, y_2). (See Fig. 1.14.) This shows, in another juncture, the triviality of the arbitrary function $G\langle t\rangle$ or the arbitrary reference point A, since it is always the difference in ψ that counts.

If, at an instant, we plot the locus of $\psi\langle x, y; t\rangle = \text{constant}$ for a series of constants, we will have a family of curves (actually cylindrical surfaces) called equi-ψ lines. Along an equi-ψ line, we have, from Eq. (1.16),

$$d\psi = u\,dy - v\,dx = 0 \qquad \textbf{(1.18)}$$

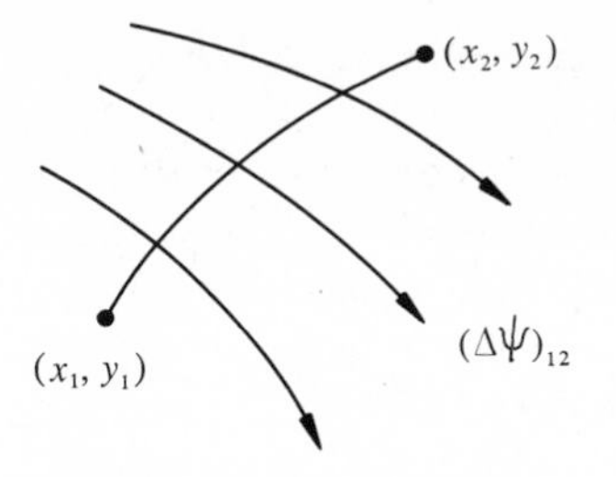

Figure 1.14 $\Delta\psi$ as volume flow.

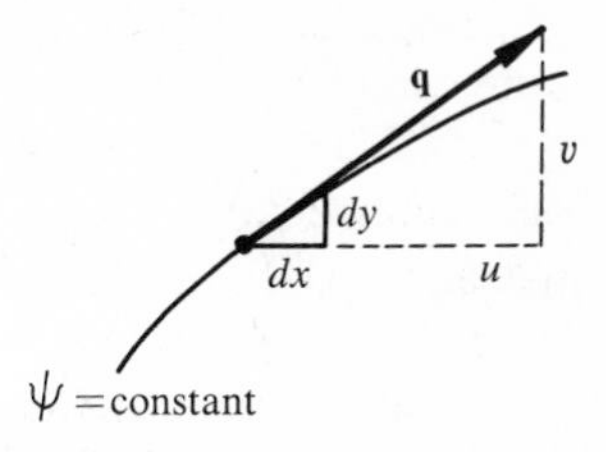

Figure 1.15 Equi-ψ line and local velocity.

that is,

$$\psi = \text{constant:} \quad \frac{dy}{dx} = \frac{v}{u} \tag{1.19}$$

Equation (1.19) means that the local tangents of equi-ψ lines are in the directions of local velocities at all times (Fig. 1.15). Thus, a flow is always *along* equi-ψ lines, never across them (Fig. 1.16). One must notice, however, that the equi-ψ lines for one instant are different from those for the next instant, unless the flow is steady.

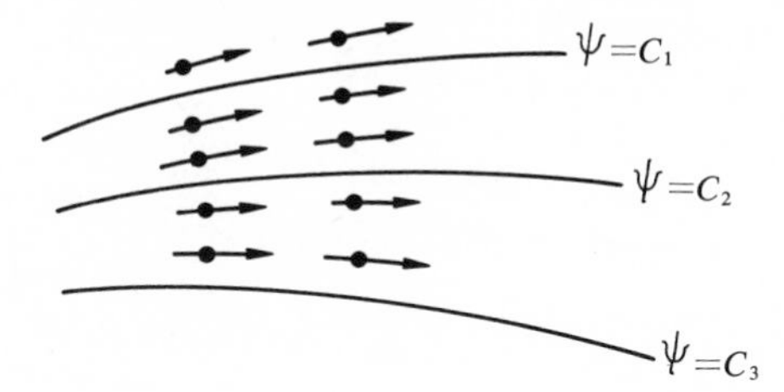

Figure 1.16 Flow along equi-ψ lines.

EXAMPLE 3

Given the velocity field

$$\begin{cases} u = U \\ v = Vt \end{cases}$$

find the equi-ψ lines.

SOLUTION Here,

$$\psi = Uy - Vtx + G\langle t \rangle$$

Assigning 0 to $\psi\langle 0, 0; t\rangle$, we have

$$\psi = Uy - Vtx$$

Then equi-ψ lines are

$$Uy - Vtx = \text{constants}$$

These are easily seen to be straight lines, which are parallel to the x-axis at $t = 0$, become inclined as t increases, and are finally parallel to the y-axis as $t \to \infty$.

As the last topic of this section, we will define and discuss the streamlines and give the connection between streamlines and equi-ψ lines.

We have stated previously that a given velocity field can be viewed at any instant as many space points carrying velocity vectors of the fluid particles passing by (Fig. 1.17). In such a representation we can draw a series of (dashed) curves tangent to all these vectors. The result is known as the family of streamlines of the given field at that particular instant. In an unsteady flow the resulting curves will change from instant to instant. But for steady flows the streamlines will not change with time. (One way of visualizing streamlines is to label *a large number* of fluid particles with dye or aluminum powder and take a series of *snapshots*. The results are a series of fields of directed segments from each of which a bunch of streamlines can be easily recognized.) Streamlines indicate the general flow directions (there cannot be any flow across a streamline), but not the speed of the flow.

Figure 1.17 Streamlines.

Mathematically, the above idea requires that, at time t,

$$\frac{dy}{dx} = \frac{v\langle x, y; t\rangle}{u\langle x, y; t\rangle} \tag{1.20}$$

For any given u and v, one can solve the ordinary differential equation (1.20) to get the family of streamlines. The constant of integration serves to identify individual lines in the family. Since Eq. (1.20) includes the time t as a parameter, the result will be time-dependent as well, unless the field is steady.

EXAMPLE 4

Given the velocity field in the first quadrant

$$\begin{cases} u = tx \\ v = t^2 y \end{cases}$$

where $t \neq 0$, find the streamlines.

SOLUTION From the definition of streamlines,

$$\frac{dy}{dx} = \frac{t^2 y}{tx} = \frac{ty}{x}$$

Therefore

$$\frac{1}{t}\ln y = \ln x + \ln C$$

that is,

$$y = (Cx)^t \qquad (x \geqslant 0, y \geqslant 0; t \neq 0)$$

is the family of streamlines. A few plots are shown in Figure 1.18.

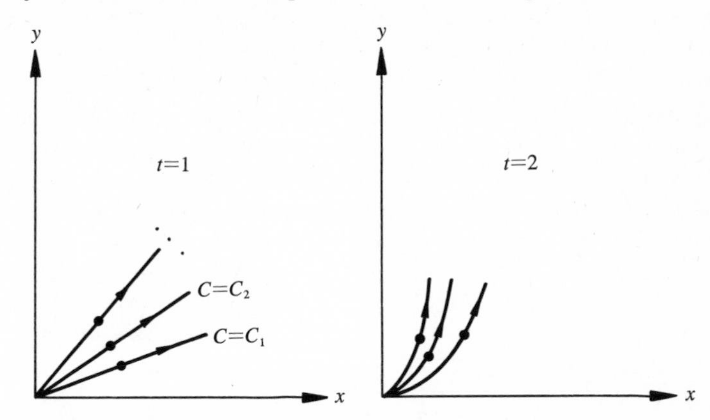

Figure 1.18 Streamlines in Example 4 for two representative instants. (To avoid imaginary streamlines for fractional t-values, C must not be negative here.)

It should be emphasized here that the foregoing definition of streamline holds whether the velocity field satisfies the equation of continuity or not. As a matter of fact, the field in Example 4 is *not* mass-conserving, and yet it has streamlines. However, for flow fields that do satisfy the equation of continuity, the stream function exists and Eq. (1.19) for the equi-ψ lines is in effect. *In such a situation* we of course realize that Eq. (1.20) is actually identical to Eq. (1.19). Therefore, *equi-ψ lines are just streamlines for flows that are mass-conserving.*

Since *non*mass-conserving flows are of no physical interest, we are not going to include them in our study from now on. Streamlines, then, furnish one more physical significance to the stream function ψ; in other words, equi-ψ lines show the general directions of the flow.

To summarize, no (mass-conserving) flow can at any instant go across the instantaneous streamlines (equi-ψ lines). If $\psi\langle x, y; t\rangle$ is given, the flow velocity as well as the streamlines can be determined. On the other hand, if the geometric pattern of the streamlines is given, neither the flow velocity nor the stream function can be determined; all we know is that ψ represents constants on these given curves, but we do not know the values of the constants.

1.3 ROTATION AND DEFORMATION

A velocity field tells us many things, not just the shape of the streamlines. Some of these are hidden and have to be dug out. In this section we will investigate the rotation and deformation, or rather the rate of deformation, of fluid particles. We will see that these are hidden in the space variation of the velocity $\mathbf{q}$.

Consider an arbitrary fluid particle P that is located at (x, y) *at the instant* t. Its velocity components at that instant are

$$\begin{cases} u\langle x, y; t\rangle \\ v\langle x, y; t\rangle \end{cases}$$

Now, at the same instant, a fluid particle Q, (dx, dy) away from P, is moving with a velocity (Fig. 1.19)

$$\begin{cases} u\langle x+dx, y+dy; t\rangle = u\langle x, y; t\rangle + \left.\dfrac{\partial u}{\partial x}\right|_P dx + \left.\dfrac{\partial u}{\partial y}\right|_P dy \\ \\ v\langle x+dx, y+dy; t\rangle = v\langle x, y; t\rangle + \left.\dfrac{\partial v}{\partial x}\right|_P dx + \left.\dfrac{\partial v}{\partial y}\right|_P dy \end{cases}$$

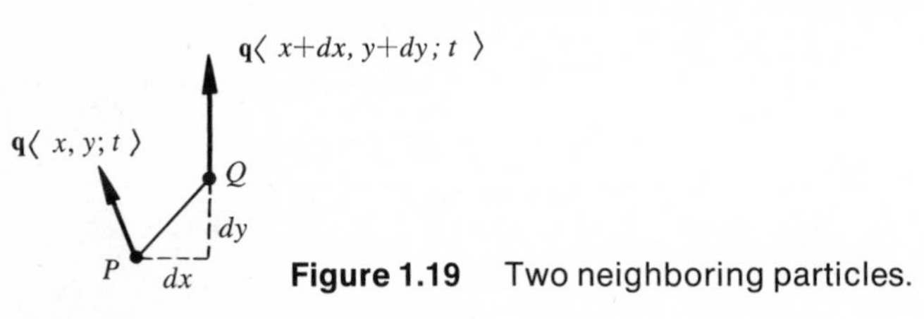

Figure 1.19 Two neighboring particles.

So a fluid particle in the neighbourhood of the particle P will move relatively to P with a relative velocity

$$\begin{cases} \left.\dfrac{\partial u}{\partial x}\right|_P dx + \left.\dfrac{\partial u}{\partial y}\right|_P dy & \textbf{(1.21a)} \\ \\ \left.\dfrac{\partial v}{\partial x}\right|_P dx + \left.\dfrac{\partial v}{\partial y}\right|_P dy & \textbf{(1.21b)} \end{cases}$$

which is completely determined by the four space derivatives $\partial u/\partial x$, $\partial u/\partial y$, $\partial v/\partial x$, $\partial v/\partial y$ at $P(x, y)$, and at t. (Note that we are interested here in the relative velocity, or the rate of relative displacement, in the neighborhood of P. In fluid dynamics we are always interested in the rate of relative displacement, whereas in elasticity we would be interested in the relative displacement itself.) If we could place ourselves on particle P, we would see a small fluid lump surrounding P stretching and rotating in every direction. Let us now examine the exact details of what we would see.

As they stand, the four derivatives $\partial u/\partial x$, $\partial u/\partial y$, $\partial v/\partial x$, $\partial v/\partial y$ can hardly tell us anything physical in terms of the (rate of) relative motion. First, let us split up the relative velocity components (1.21)[4] and rewrite them in the following

[4] Decomposition of this kind is found in Cauchy's 1841 "Memoir on the expansion, compression and rotation produced by a change in form of a system of particles" (in French). Augustin-Louis (Baron) de Cauchy (1789–1857), French mathematician, was the true successor to Euler whose work on continuum mechanics had no effect at all on the pre-Cauchy fluid dynamicists.

equivalent form:

$$du = -\frac{1}{2}\left(\frac{\partial v}{\partial x} - \frac{\partial u}{\partial y}\right) dy + \left[\frac{\partial u}{\partial x}\, dx + \frac{1}{2}\left(\frac{\partial v}{\partial x} + \frac{\partial u}{\partial y}\right) dy\right] \tag{1.22a}$$

$$dv = \frac{1}{2}\left(\frac{\partial v}{\partial x} - \frac{\partial u}{\partial y}\right) dx + \left[\frac{1}{2}\left(\frac{\partial v}{\partial x} + \frac{\partial u}{\partial y}\right) dx + \frac{\partial v}{\partial y}\, dy\right] \tag{1.22b}$$

To see the physical significance of Eqs. (1.22a,b), we start with an investigation of the rotational speed of a fluid particle:

(1) Rotation[5]

Consider the motion of fluid particles situated on an infinitesimal circle of radius $dR = [(dx)^2 + (dy)^2]^{1/2}$ around the particle $P(x, y)$ at a fixed instant (Fig. 1.20). We then ask: What is the rotational or angular speed of the line $\overline{PQ}$ about P?

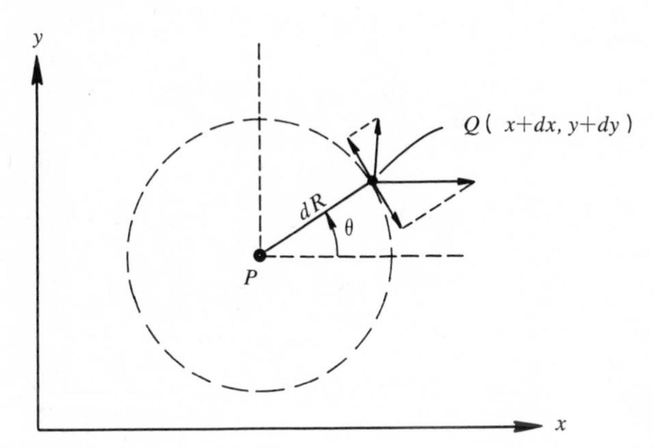

Figure 1.20 A circle of neighboring particles.

To answer this, we first notice that the velocity of Q relative to P is

$$\mathbf{q}_{QP} = \begin{cases} \dfrac{\partial u}{\partial x}\, dx + \dfrac{\partial u}{\partial y}\, dy \\[2ex] \dfrac{\partial v}{\partial x}\, dx + \dfrac{\partial v}{\partial y}\, dy \end{cases}$$

The contribution of the x-component of $\mathbf{q}_{QP}$ to the rotational speed of $\overline{PQ}$ about P (+, counterclockwise) is its projection perpendicular to $\overline{PQ}$ divided by $\overline{PQ}$; that is,

$$-\left(\frac{\partial u}{\partial x}\, dx + \frac{\partial u}{\partial y}\, dy\right)\frac{\sin\theta}{dR}$$

[5] The following discussion is attributable originally to Cauchy (footnote 4). Here, however, we have adopted R. von Mises' argument (*Mathematical Theory of Compressible Fluid Flow*, Academic Press, New York, 1958).

or

$$-\left(\frac{\partial u}{\partial x}\sin\theta\cos\theta+\frac{\partial u}{\partial y}\sin^2\theta\right)$$

Similarly, the contribution of the y-component of $\mathbf{q}_{QP}$ is

$$\frac{\partial v}{\partial x}\cos^2\theta+\frac{\partial v}{\partial y}\sin\theta\cos\theta$$

Therefore,

rotational speed of $\overline{PQ}$ about P

$$=\frac{\partial v}{\partial x}\cos^2\theta-\frac{\partial u}{\partial y}\sin^2\theta+\left(\frac{\partial v}{\partial y}-\frac{\partial u}{\partial x}\right)\sin\theta\cos\theta \tag{1.23}$$

The result, Eq. (1.23), is dependent on the angular position of $\overline{PQ}$. However, from this we can easily calculate an average value by integrating it with respect to θ from 0 to 2π and dividing the result by 2π. This average value will characterize the rotation in the neighborhood of P. In this averaging, we must not forget that the partial derivatives in Eq. (1.23) are all evaluated at $P(x, y)$, and hence act as constants in the integration. Thus,

average of rotational speeds of all $\overline{PQ}$'s about P

$$=\frac{1}{2\pi}\left[\left(\frac{\partial v}{\partial x}\right)\int_0^{2\pi}\cos^2\theta\,d\theta-\left(\frac{\partial u}{\partial y}\right)\int_0^{2\pi}\sin^2\theta\,d\theta+\left(\frac{\partial v}{\partial y}-\frac{\partial u}{\partial x}\right)\int_0^{2\pi}\sin\theta\cos\theta\,d\theta\right]$$

$$=\frac{1}{2}\left(\frac{\partial v}{\partial x}-\frac{\partial u}{\partial y}\right) \tag{1.24}$$

Conclusion For a given velocity field, the quantity $\frac{1}{2}((\partial v/\partial x)-(\partial u/\partial y))$ evaluated at $P(x, y)$ represents the instantaneous, average, angular speed of all line segments $\overline{PQ}$ within an infinitesimal circle of center P, or, briefly, the instantaneous, mean, angular speed of the fluid element around P.

This conclusion can be visualized in the following manner:[6] If the particle P in the flow field can be rigidified instantaneously at t, and the surrounding fluid can be instantaneously annihilated, then the particle will go on rotating with a speed equal to the local value of $\frac{1}{2}((\partial v/\partial x)-(\partial u/\partial y))$ at t.

On the other hand, since the fluid particle around P is rotating on the

[6] Suggested by Stokes in his 1845 paper, "On the theories of the internal friction of fluids in motion, and the equilibrium and motion of elastic solids." The particle is meant to be circular in shape (that is, a circular cylinder perpendicular to the representative plane). Sir George Gabriel Stokes (1819–1903) was born in Ireland but became Lucasian Professor of Mathematics at Cambridge University. His name will recur many times throughout this book, since without him there would be no mechanics of viscous fluids.

average (in addition to performing other movements, which will be explained later) with the above speed, it will induce a relative velocity, at a point $Q(x+dx, y+dy)$ in the neighborhood, that is equal to (Fig. 1.21):

$$\begin{cases} -\dfrac{1}{2}\left(\dfrac{\partial v}{\partial x}-\dfrac{\partial u}{\partial y}\right) dy \\ \dfrac{1}{2}\left(\dfrac{\partial v}{\partial x}-\dfrac{\partial u}{\partial y}\right) dx \end{cases}$$

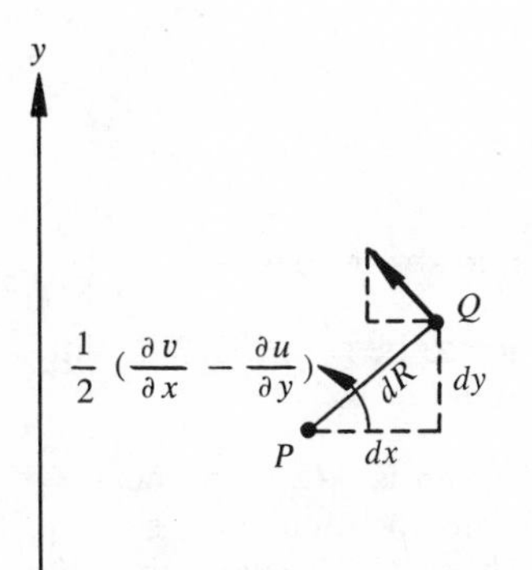

Figure 1.21 Velocity due to rotation.

In other words, the first part of the split-up form, Eqs. (1.22a,b), is exactly the (rate of) relative motion due to the (average) fluid rotation.

We dwell on these because the true nature of the rotation of a fluid particle is often not realized by beginners. It is important to keep in mind that *fluid rotation is a local affair* and *it is average in nature.*

Incidentally, for plane flows, the expression $((\partial v/\partial x)-(\partial u/\partial y))$ is called the *vorticity* and is often denoted by symbol Ω:

$$\Omega\langle x, y; t\rangle = \frac{\partial v}{\partial x}-\frac{\partial u}{\partial y} \quad \textbf{(1.25)}$$

$$= \text{twice the mean angular speed} \quad \textbf{(1.26)}$$

If at a point, we have $\Omega = 0$, we say that the flow is *irrotational*[7] *at that point*. If $\Omega = 0$ at *every* point inside a region, we say that the flow is irrotational *in that region.*

Please remember that irrotationality is basically a local affair. Whether the flow field as a whole appears to be rotating or not has nothing to do with irrotationality. The following examples should be examined carefully.

[7] This word was first used by Kelvin in his 1869 paper, "On vortex motion." It must be emphasized that the lines $\overline{PQ}$ may still rotate at a point of irrotationality, as long as their rotational speeds average out, yielding zero Ω. William Thomson (Lord Kelvin) (1824–1907), like his teacher Stokes, was a native of Ireland, but for 53 years was professor of natural philosophy (that is, physics) at the University of Glasgow.

1. The *simple shear flow*[8]—that is, the viscous flow between one stationary and one sliding plate (Fig. 1.22), having the velocity $\mathbf{q} = (Cy, 0)$—is rotational, since

$$\Omega = -C \neq 0$$

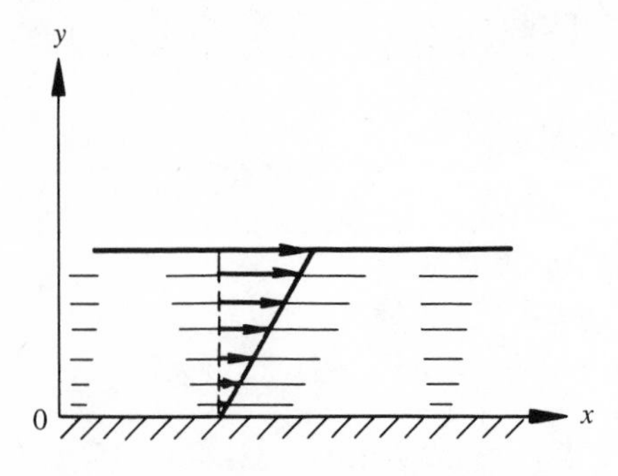

Figure 1.22 Simple shear flow.

Note that the flow field as a whole appears not to be rotating. The fluid particles, however, are rotating[9] with a (mean) angular speed of $-C/2$.

It is also instructive to realize that the angular speed is $-C/2$, and not $-C$. Note that a small fluid element moves in such a fashion that the horizontal line (see Fig. 1.23) does not rotate, whereas the vertical line rotates with speed $-C$. This would indicate that the (mean) angular speed of the element cannot be $-C$. If you really take the speed of all lines, from 0 to 2π in inclination, and average them out, you will find that the result is $-C/2$. (To count and average only the speeds of two lines is dangerous.[10] To be exact, we must count all lines. As an example, we might draw your attention to the fact that the originally vertical line is actually turning more and more slowly as the element moves downstream. If you couple this with the fact that the horizontal line does not turn at all, and conclude that the fluid element must be rotating more and more slowly, you would be com-

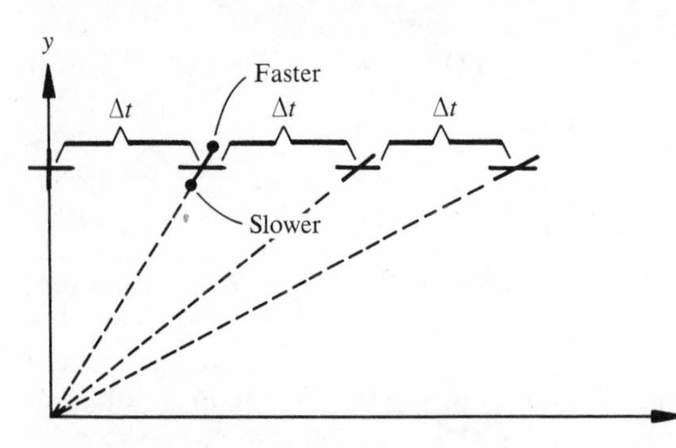

Figure 1.23 Simple shear flow is rotational.

---

[8] This flow was introduced and analyzed by Euler in his 1755 paper, cited in footnote 2.

[9] This observation is exactly what Helmholtz used in shutting up a French hydrodynamicist (Joseph Bertrand) in an acrimonious argument carried on between 1867 and 1868. The battle was duly recorded in the *Comptes Rendus* of the Academy of Sciences, Paris. Hermann Ludwig Ferdinand von Helmholtz (1821–1894) was a German physician-physicist.

[10] Unless the two lines are perpendicular at the instant of reckoning (as you will show in one of the Problems).

pletely wrong. What actually happens is that some of the originally inclined lines are rotating faster and faster; on the average, the fluid element is turning at a constant speed.)

2. The "*free*"-*vortex flow*[11]—that is, the circular flow with speed inversely proportional to the radius) (Fig. 1.24)—has the velocity

$$\mathbf{q} = \left(-\frac{Cy}{x^2+y^2}, \frac{Cx}{x^2+y^2}\right)$$

This flow is irrotational everywhere except at the origin, since

$$\Omega = \frac{C(x^2-y^2)}{(x^2+y^2)^2} - \frac{C(x^2-y^2)}{(x^2+y^2)^2}$$
$$= 0; \text{ unless } x, y = 0$$

(At $x, y = 0$, Ω is undefined.)

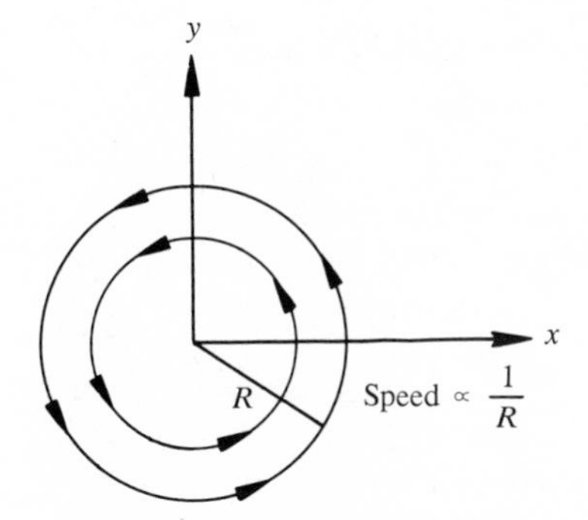

Figure 1.24 Simple irrotational vortex.

Note that the flow as a whole appears to be rotating, although the fluid elements are not. (Do not let the word "vortex" give you mistaken ideas.) What really happens to a fluid element is explained in Figure 1.25. The fluid element on top deforms as it moves; but the two dashed lines maintain their orientation, and do not rotate at the moment, because the upper particles are slower than the lower ones. This then indicates that the element is not rotating as it moves across the top (Problem 1.3.4). For any other element, merely reorient the coordinate lines so that the element is on top.

The name "*free*" *vortex* is an unfortunate one. A better name is *simple irrotational vortex*.

3. The "*forced*"-*vortex flow*—that is, the circular flow with speed proportional to the radius (Fig. 1.26)—has the velocity $\mathbf{q} = (-Cy, Cx)$. It is rotational since

$$\Omega = (C + C) = 2C \neq 0$$

---

[11] Analysis of this kind of flow was initiated by Newton in his 1687 *Mathematical Principles of Natural Philosophy* (in Latin). Sir Isaac Newton (1642–1727) needs no introduction. See, however, "Reproduction of prints, drawings and paintings of interest in the history of physics, 61. Caricatures of Sir Isaac Newton by two famous artists," by E. C. Watson, *Am. J. Phys.*, vol. 22, May 1954, pp. 247–249.

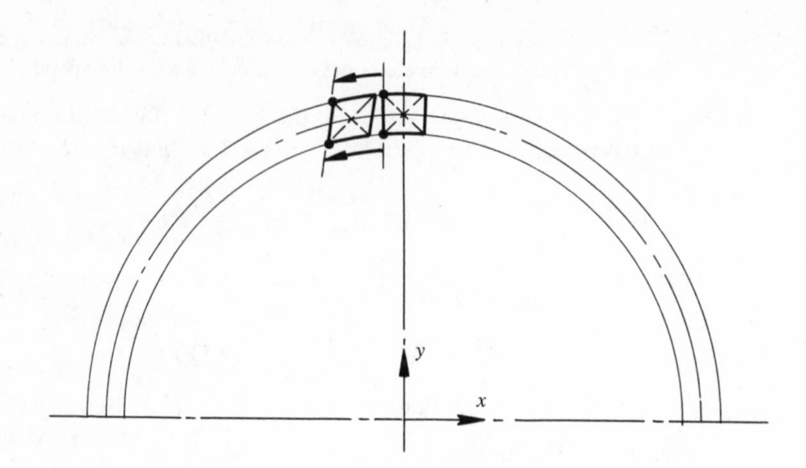

Figure 1.25 A fluid element in the simple irrotational vortex.

> The fluid element is now moving like the moon around the earth—that is, rotating (Fig. 1.27).[12]

The name *"forced" vortex* is not a good one either. It is better referred to as the *wheel flow* or, simply, *rigid-body rotation.*

To summarize, *a curved flow is not necessarily rotational*; *a rectilinear flow, not necessarily irrotational.*

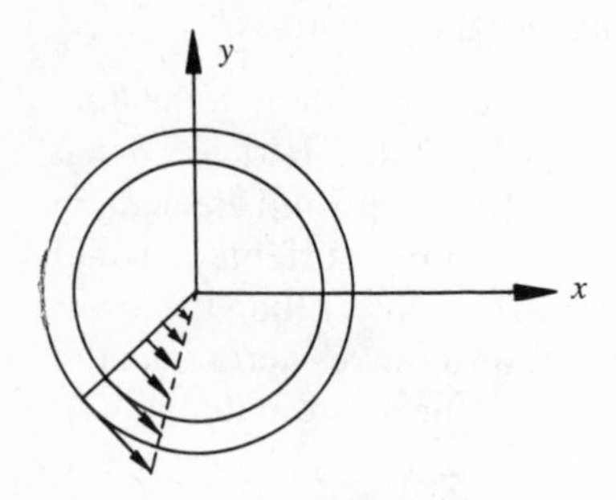

Figure 1.26 Wheel flow.

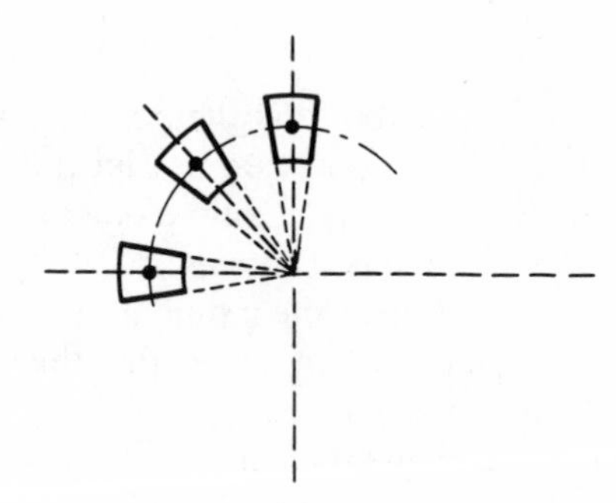

Figure 1.27 Wheel flow is rotational.

(2) Deformation

Now let us investigate the quantities $\partial u/\partial x$, $\partial v/\partial y$, and $\frac{1}{2}((\partial v/\partial x) + (\partial u/\partial y))$ involved in the second part of the splitting-up; see Eqs. (1.22a,b). We will show that these represent physically the deformation of the fluid particle or,

[12] In common language, the fact that we always see one side of the moon is referred to as: "The moon is not rotating by itself." In our language, *the moon is really rotating* by itself.

more precisely, the rates of strain[13] of the particle. It is customary to introduce here the symbols

$$\begin{cases} \epsilon_{xx} = \dfrac{\partial u}{\partial x} & \text{(1.27a)} \\ \epsilon_{yy} = \dfrac{\partial v}{\partial y} & \text{(1.27b)} \\ \epsilon_{xy} = \dfrac{1}{2}\left(\dfrac{\partial v}{\partial x} + \dfrac{\partial u}{\partial y}\right) & \text{(1.27c)} \end{cases}$$

To see the physical meaning of these three quantities, let us take a small rectangle of fluid with the point P as a corner (Fig. 1.28). Referring to the enlarged view, Figure 1.29, we see that the four corners of the rectangle are moving

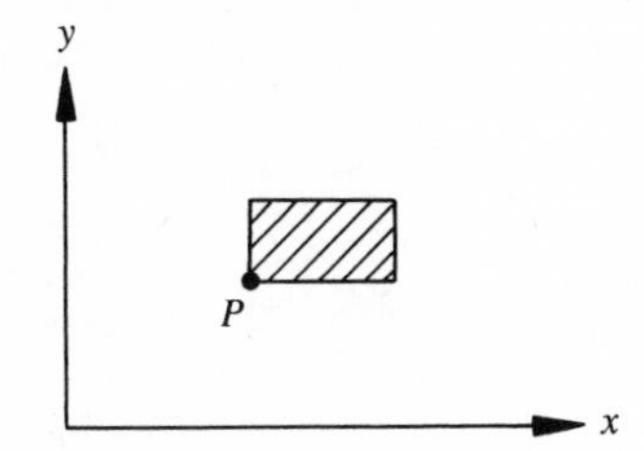

Figure 1.28 A rectangular fluid element.

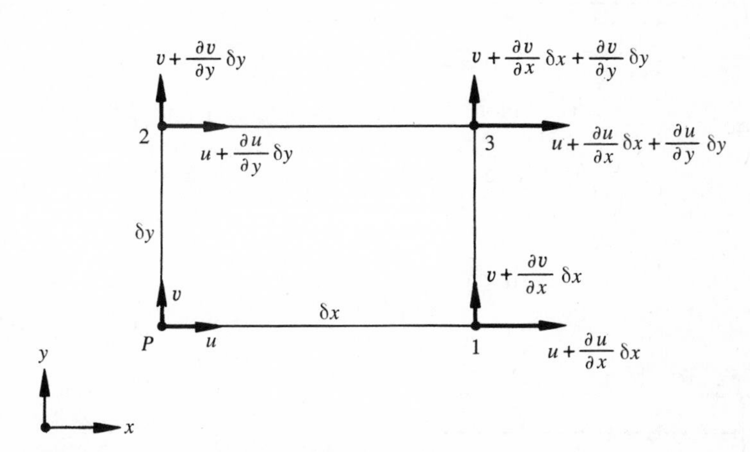

Figure 1.29 Relative velocities of the corners of a rectangular fluid element.

---

[13] Euler obtained rates of strain in his 1769 paper "Second section (of a projected Hydrodynamic Treatise), principles of fluid motion" (in Latin, it was presented in 1766, wrongly dated 1759 by the printer, and actually appeared in 1770); but the work remained unnoticed until Stokes rediscovered everything in his 1845 paper, cited in footnote 6. In the meantime, Cauchy also worked out the counterpart in solid bodies—that is, the *strains*; see his 1827 paper "On the compression and expansion of solid bodies" (in French).

with the velocities shown. Now consider the motion of the other three corners relative to the point P during a time interval δt:

1. The relative velocity $(\partial v/\partial x)\ \delta x$ would have pushed points 1 and 3 up a little, namely $(\partial v/\partial x)\ \delta x\ \delta t$, and $(\partial u/\partial y)\ \delta y$ would have pushed 2 and 3 to the right by $(\partial u/\partial y)\ \delta y\ \delta t$. The combined result is then an angular or shear deformation (Fig. 1.30). The total shearing strain is measured by the change of half of the angle $(\beta_1+\beta_2)$, or

$$\frac{1}{2}\left(\frac{\partial v}{\partial x}+\frac{\partial u}{\partial y}\right)\delta t = \epsilon_{xy}\ \delta t$$

   Dividing by δt, we have ϵ_{xy} as the rate of shearing strain in the plane of flow.

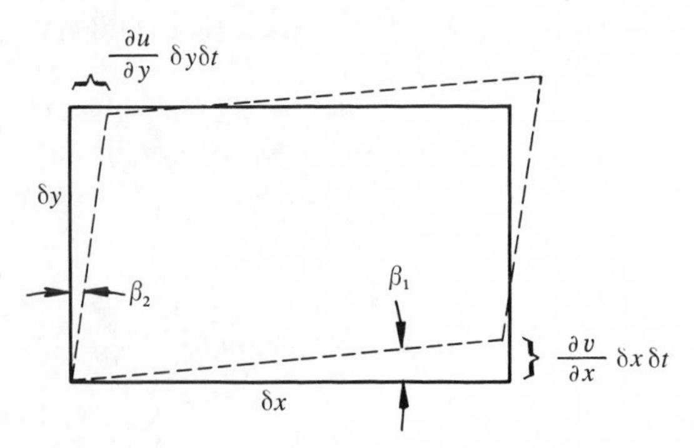

Figure 1.30 Shearing of a fluid element.

2. The relative velocity $(\partial u/\partial x)\ \delta x$ pushes 1 and 3 to the right by an amount $(\partial u/\partial x)\ \delta x\ \delta t$ during δt and $(\partial v/\partial y)\ \delta y$ pushes 2 and 3 up by $(\partial v/\partial y)\ \delta y\ \delta t$. The result is as shown in Figure 1.31; that is,

$$\text{normal strain in}\begin{cases} x\text{-direction} = \dfrac{\partial u}{\partial x}\ \delta t \\[2ex] y\text{-direction} = \dfrac{\partial v}{\partial y}\ \delta t \end{cases}$$

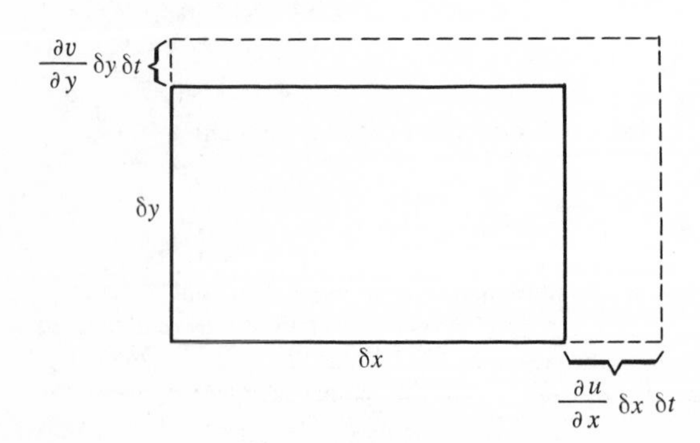

Figure 1.31 Stretching of a fluid element.

Therefore

$$\epsilon_{xx} = \text{rate of normal strain in } x\text{-direction}$$

$$\epsilon_{yy} = \text{rate of normal strain in } y\text{-direction}$$

For example, take the simple shear flow again:

$$\mathbf{q} = (Cy, 0)$$

Here

$$\begin{array}{ll|ll} du = 0 \cdot dx + \tfrac{1}{2}C\,dy & & +0 \cdot dx + \tfrac{1}{2}C\,dy \\ dv = -\tfrac{1}{2}C\,dx + 0 \cdot dy & & +\tfrac{1}{2}C\,dx + 0 \cdot dy \\ \text{rotation} & & \text{pure shear} \end{array}$$

The situation can be seen very clearly if we consider a small square fluid element (Fig. 1.32) with sides 2 dx and 2 dy $(dx = dy)$. First, du and dv due to pure shear are seen to combine into a pull in the 45° direction, and a push in the −45° direction. The result is a squashed square—that is, a rhombus (Fig. 1.33). Now the du and dv due to rotation gives a turning with speed $-C/2$ about an axis perpendicular to the xy-plane (Fig. 1.34). This turning then brings the squashed square back to the original orientation to yield the final result (Fig. 1.35).

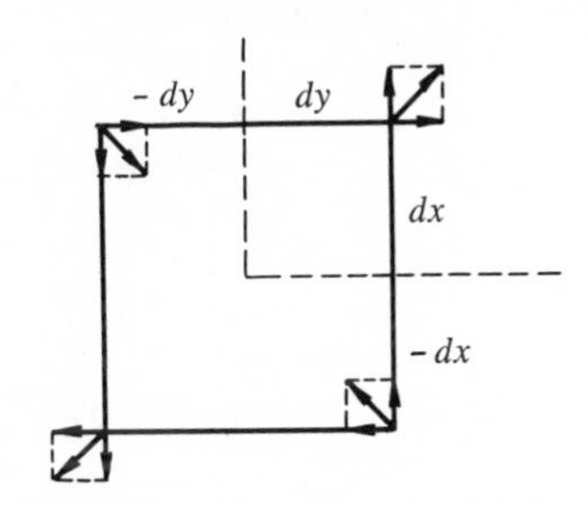

Figure 1.32 Pure shear of a square fluid element in simple shear flow.

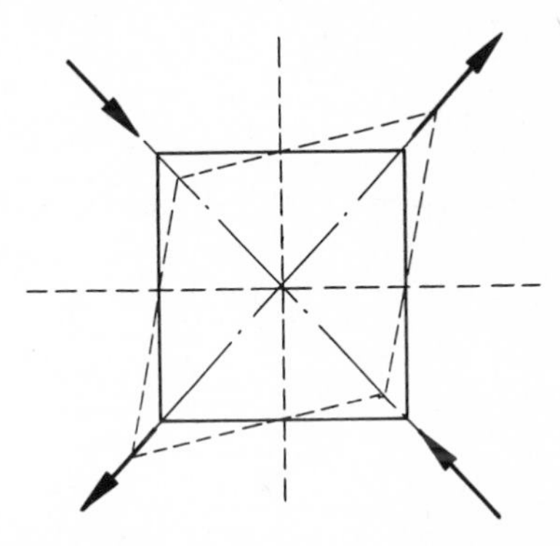

Figure 1.33 A squashed square or rhombus.

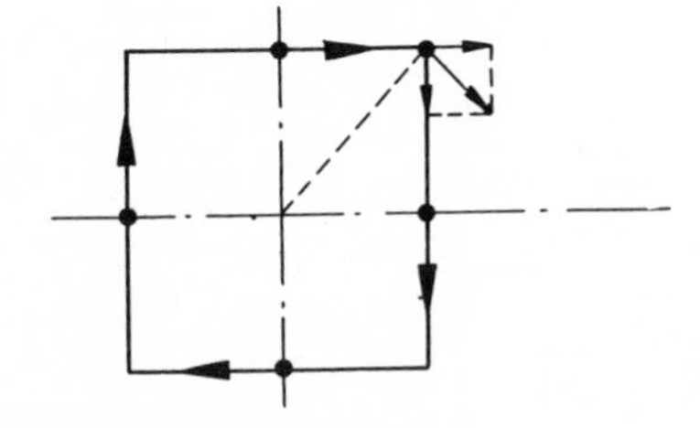

Figure 1.34 Rotation of a square element in simple shear flow.

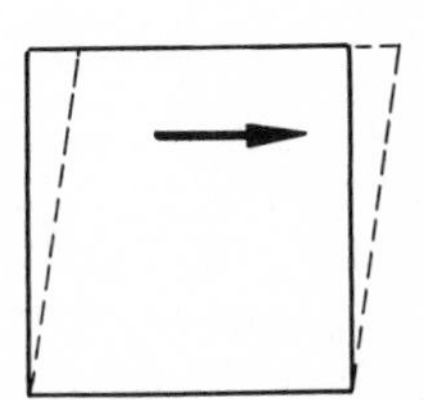

Figure 1.35 Squashed square rotated back.

Physical Summary

$$\text{Equations (1.22a,b)} \underset{\text{(imply)}}{\Longrightarrow}$$

$$\text{relative velocity} = \text{that due to rotation} + \text{that due to rate of strain}$$

In short, we have just demonstrated that the velocity field in the immediate neighborhood of a fluid particle, relative to the particle, is made up of two parts, caused respectively by the rotation and the (rate of) straining of the particle. In the next chapter, it will be shown (Section 2.3) that a relation between the rates of strain on the one hand and the stresses on the other is needed in order to complete the dynamic formulation of a flow field. It is in anticipation of this relation that the rates of strain are separated from the rotation in the present section. The rotation plays no *direct* role in the general flow; its appearance (or, rather, its disappearance) will be heralded again when the so-called potential flow is considered (for example, in Section 6.3).

CHAPTER 2

DYNAMICS OF THE FLOW FIELD

2.1 FORCES AND STRESSES

In the previous chapter we only investigated a flow field per se, without asking how it got that way. To see the reason behind a certain flow, or to ask whether a certain flow is dynamically possible, we have to study its dynamics by inquiring into the nature of the forces and applying the Newtonian laws of motion.

There are two kinds of forces acting on a fluid particle: (a) *Body forces* such as gravity.[1] In this Part we will denote the resultant body force per unit mass by

$$\mathbf{g}\langle x, y; t\rangle = \begin{cases} g_x\langle x, y; t\rangle \\ g_y\langle x, y; t\rangle \end{cases}$$

so a rectangular fluid element with the point (x, y) as its centroid at t is acted upon by the body force $\rho\langle x, y; t\rangle \mathbf{g}\langle x, y; t\rangle\ \delta x\ \delta y$ (Fig. 2.1). (The reader should keep in mind that we are actually talking about a rodlike element with unit depth.)

[1] It is very possibly true that a falling apple *did* induce Newton to think about gravity, according to William Stukeley's 1752 *Memoirs of Sir Isaac Newton's Life*. (See E. C. Watson's Reproduction of prints, etc., cited in footnote 11 of Chapter 1.) But Stukeley never implied that the apple struck Newton. For an amusing verse about Newton and his farmer-neighbor, see G. Gamow's *Biography of Physics*, Harper & Row, New York, 1961, 1964.

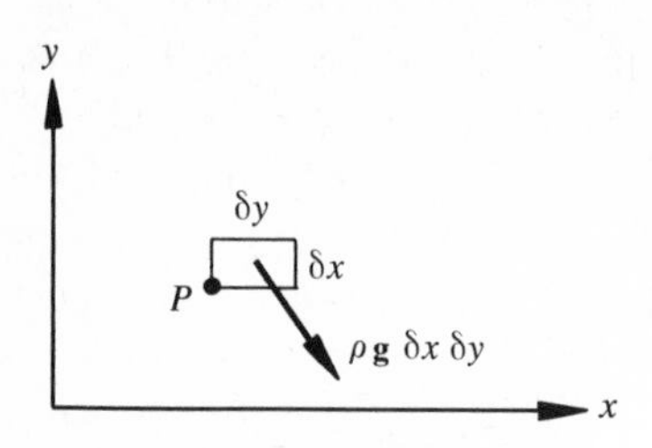

Figure 2.1 Body force.

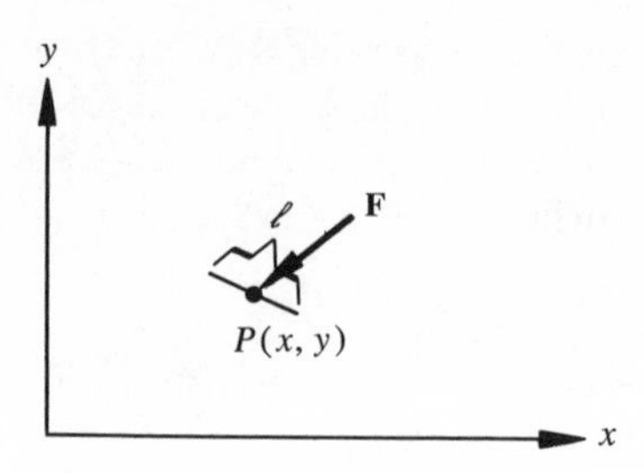

Figure 2.2 Inclined surface force.

Although the only body force we will ever meet is the gravity whose **g** is approximated by a constant, we will nevertheless consider here a general $\mathbf{g}\langle x, y; t\rangle$ just for the fun of it. (b) *Surface forces* exerted by the surrounding fluid. These are contact forces and have to be examined in more detail.

Consider an inclined plane represented by an inclined line of length ℓ in the representative plane passing through $P(x, y)$ (Fig. 2.2). The resultant force (per unit depth) acting on the fluid to the left of this line exerted by the fluid to the right is **F**, *which is not necessarily perpendicular to the plane.*[2] The average resultant force per unit area is $\mathbf{F}/\ell$. Now, let $\ell \to 0$ towards the point while *keeping the inclination unchanged.* The resulting limit

$$\lim_{\substack{\ell\to 0\\ \text{(same inclination at } P)}} \left(\frac{\mathbf{F}}{\ell}\right) = \mathbf{f}\langle x, y; t; \text{inclination}\rangle$$

is the local, instantaneous, intensity of the contact force on the fluid to the left of the infinitesimal line ℓ. Note that **f** is a function not only of x, y, and t, but also of the inclination of the infinitesimal line ℓ. Even at the same location and instant, **f** on infinitesimal lines of different inclinations would, in general, vary (Fig. 2.3). The inclination of the line (or plane) can be represented by a unit vector (that is, of unit length) $\hat{\mathbf{n}}$ perpendicular to it, and pointing away from the fluid portion we are interested in (Fig. 2.4). The corresponding **f** is then written as $\mathbf{f}\langle x, y; t; \hat{\mathbf{n}}\rangle$. It is in-

Figure 2.3 Surface force depends on the surface orientation.

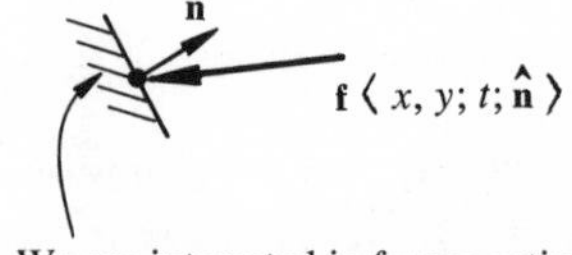

Figure 2.4 The unit normal vector.

[2] This decisive discovery was made by Cauchy in his 1823 paper, "Researches on the internal equilibrium and motion of solid bodies or fluids, elastic or nonelastic" (in French).

structive to note that the line (plane) has two sides. The positive-n side (with $\hat{\mathbf{n}}$ pointed as shown) serves as the bounding surface of the fluid portion to the left on which $\mathbf{f}\langle x, y; t; \hat{\mathbf{n}}\rangle$ acts. The negative-n side (the opposite side) is the bounding surface of the fluid portion to the right on which $\mathbf{f}\langle x, y; t; -\hat{\mathbf{n}}\rangle$ acts (Fig. 2.5).

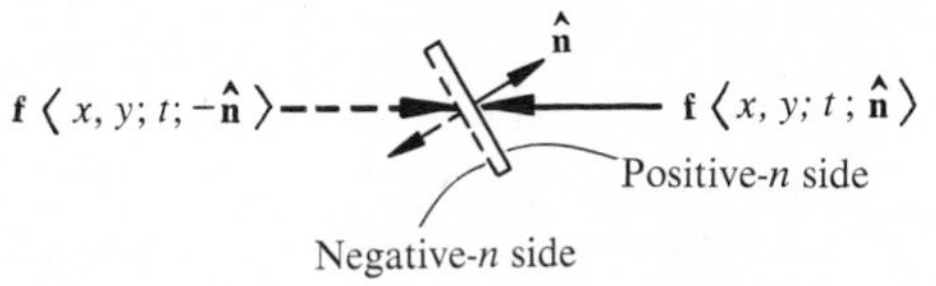

$\mathbf{f}\langle x, y; t; -\hat{\mathbf{n}}\rangle$ $\hat{\mathbf{n}}$ $\mathbf{f}\langle x, y; t; \hat{\mathbf{n}}\rangle$
Positive-n side
Negative-n side

Figure 2.5 Surface forces on the two sides of a surface.

$\mathbf{f}\langle x, y; t; -\hat{\mathbf{n}}\rangle$ is, of course, never the same as $\mathbf{f}\langle x, y; t; \hat{\mathbf{n}}\rangle$. However,

$$\mathbf{f}\langle x, y; t; -\hat{\mathbf{n}}\rangle = -\mathbf{f}\langle x, y; t; \hat{\mathbf{n}}\rangle \tag{2.1}$$

This is very easy to see from the figure, where the plane is exaggerated to possess a certain thickness. In reality the plane is of no thickness and therefore possesses no mass. It is thus not acted upon by any body force (such as gravity); the only forces are the two **f**'s acting on its two faces. Then the two **f**'s must balance one another; otherwise, Newton's second law of motion would yield (with zero mass, and yet, nonzero resultant force) an infinite acceleration for the plane of fluid, which is physically impossible.

Conclusion An infinitesimal plane of fluid is always acted on its two faces, by the rest of the fluid, with equal but opposite force intensities.[3]

Next, consider a fluid element in the shape of a right-angled triangle (that is, a prism of unit depth) shown in Figure 2.6 (in which dependence on x, y, and t is suppressed for ease of writing). The mass of the element is (per unit depth)

$$\rho\langle x, y; t\rangle \cdot \tfrac{1}{2} h \ell \cos \measuredangle (n, x)$$

where $\measuredangle(n, x)$ is the angle in between the positive-n and positive-x directions. Assuming that the element is acted upon by a body force per unit mass $\mathbf{g}\langle x, y; t\rangle$, and has an acceleration $\mathbf{a}\langle x, y; t\rangle$, we can apply Newton's second law of motion

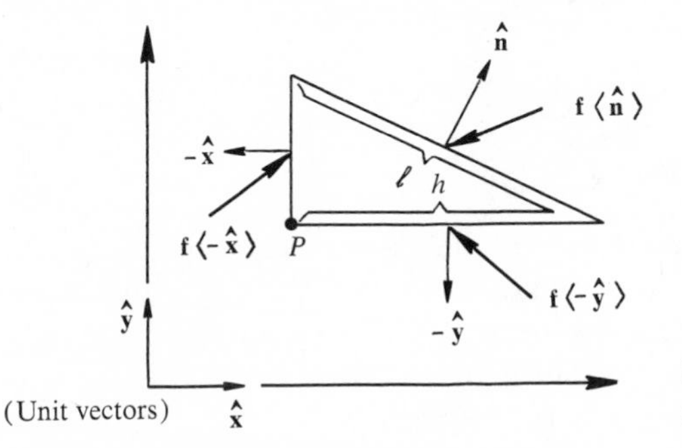

Figure 2.6 Surface forces on a fluid element.

[3] This is known as Cauchy's lemma. It will be used presently in establishing Eq. (2.3).

and write (again suppressing x, y, and t):

$$\mathbf{f}\langle-\hat{\mathbf{x}}\rangle \cdot \ell \cos \measuredangle(n, x) + \mathbf{f}\langle-\hat{\mathbf{y}}\rangle \cdot \ell \cos \measuredangle(n, y) + \mathbf{f}\langle\hat{\mathbf{n}}\rangle \cdot \ell + \mathbf{g} \cdot \frac{\rho}{2} h \ell \cos \measuredangle(n, x)$$

$$= \frac{\rho}{2} h \ell \cos \measuredangle(n, x) \cdot \mathbf{a}$$

As $h \to 0$ (the element approaches a particle), we have, since $\mathbf{g}$ and $\mathbf{a}$ are both to remain finite,

$$\mathbf{f}\langle\hat{\mathbf{n}}\rangle = -\mathbf{f}\langle-\hat{\mathbf{x}}\rangle \cos \measuredangle(n, x) - \mathbf{f}\langle-\hat{\mathbf{y}}\rangle \cos \measuredangle(n, y) \tag{2.2}$$

Using Eq. (2.1), we have

$$\mathbf{f}\langle\hat{\mathbf{n}}\rangle = \mathbf{f}\langle\hat{\mathbf{x}}\rangle \cos \measuredangle(n, x) + \mathbf{f}\langle\hat{\mathbf{y}}\rangle \cos \measuredangle(n, y) \tag{2.3}$$

In the above argument, we have used the expressions $\ell \cos \measuredangle(n, x)$, and so on, for the extent of side areas of the element. This is in order only when $\cos \measuredangle(n, x)$ and $\cos \measuredangle(n, y)$ are both positive as shown in the figure. For elements where $\cos \measuredangle(n, x)$ or/and $\cos \measuredangle(n, y)$ is negative, one cannot use $\ell \cos \measuredangle(n, x)$ or/and $\ell \cos \measuredangle(n, y)$ for the extent of side areas because it is negative. (Areas must be positive in reckoning total forces from the **f**'s.) Instead, $-\ell \cos \measuredangle(n, x)$ or/and $-\ell \cos \measuredangle(n, y)$ should be used.

You should now satisfy yourself that, for other orientations of $\hat{\mathbf{n}}$, similar arguments as before (with modifications just indicated) always lead to the same formula, Eq. (2.3). In other words, Eq. (2.3) is always true, the specific sketch notwithstanding!

Now Eq. (2.3) embodies a great simplification to the study of fluid dynamics. Originally, the resultant force intensity due to fluid contact can only be specified at a point by giving $\mathbf{f}\langle\hat{\mathbf{n}}\rangle$ for all directions of $\hat{\mathbf{n}}$ (that is, for all inclinations of the plane of action). This requires an infinite number of **f**'s. But Eq. (2.3) says that this is unnecessary. All we need to know is **f** on two planes of chosen inclination $\hat{\mathbf{n}} \| \hat{\mathbf{x}}$ and $\hat{\mathbf{n}} \| \hat{\mathbf{y}}$; that is, $\mathbf{f}\langle\hat{\mathbf{x}}\rangle$ and $\mathbf{f}\langle\hat{\mathbf{y}}\rangle$. Once we have these two **f**'s, all the other **f**'s at the point on different planes can be calculated through Eq. (2.3). In other words, two vectorial quantities $\mathbf{f}\langle\hat{\mathbf{x}}\rangle$ and $\mathbf{f}\langle\hat{\mathbf{y}}\rangle$ determine completely the local contact force intensities.

Furthermore, the two vectors $\mathbf{f}\langle\hat{\mathbf{x}}\rangle$ and $\mathbf{f}\langle\hat{\mathbf{y}}\rangle$ can also be written in their component forms:

$$\mathbf{f}\langle\hat{\mathbf{x}}\rangle = \sigma_{xx}\hat{\mathbf{x}} + \sigma_{xy}\hat{\mathbf{y}} \tag{2.4a}$$

$$\mathbf{f}\langle\hat{\mathbf{y}}\rangle = \sigma_{yx}\hat{\mathbf{x}} + \sigma_{yy}\hat{\mathbf{y}} \tag{2.4b}$$

where the first subscript denotes the inclination of the plane element under force—that is, the direction of the unit vector perpendicular to it and pointing away from the fluid portion under force; the second subscript denotes the particular component. Thus, σ_{xx} is the x-component of the contact force intensity acting on a plane perpendicular to the x-axis from the right (Fig. 2.7a). In this figure, σ_{xx} happens to be a negative number. Similarly, σ_{xy} is the y-component of the force

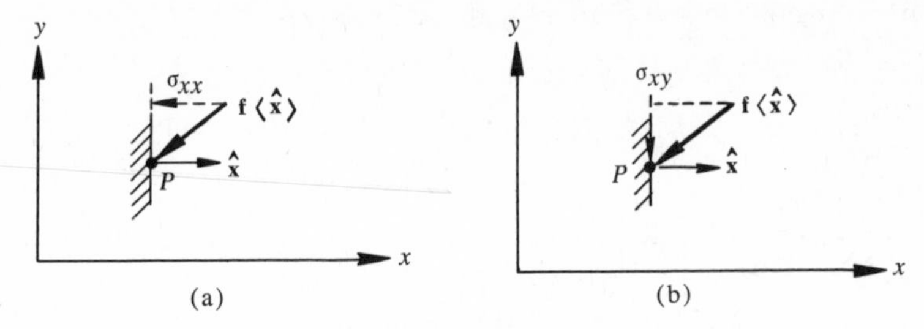

Figure 2.7 (a) Normal stress component. (σ_{xx} happens to be negative here.) (b) Shear stress component.

intensity acting on a plane perpendicular to the x-axis from the right (Fig. 2.7b). Here, σ_{xy} happens also to be a negative number and is seen to be a shear on the fluid to the left of the plane exerted by the fluid on the right.

The four quantities σ_{xx}, σ_{xy}, σ_{yx}, and σ_{yy} are called the (local and instantaneous) stresses. The four of them serve to determine completely the (local and instantaneous) contact force situation. The stresses with two identical subscripts (σ_{xx} and σ_{yy}) represent components perpendicular (or normal) to the plane elements, and are called the normal stresses. The other two with different subscripts are components parallel to (or in shear with) the plane elements, and are called the shear stresses. (Incidentally, these stresses also occur in elasticity or strength of material, as you probably know; they actually occur in mechanics of all continua.)

Conclusion[4] At every point in the fluid, and at every instant, there are four stresses σ_{xx}, σ_{xy}, σ_{yx}, and σ_{yy} such that the contact force intensity

---

[4] Cauchy's fundamental theorem. At this juncture, it is probably necessary to review again the steps taken: Equation (2.2) is derived by considering a fluid element of the shape of a prism; then, with the aid of Eq. (2.1), Eq. (2.3) is obtained. But, what about the following interesting observation? Applying Eq. (2.2) to $\hat{\mathbf{n}} = \hat{\mathbf{x}}$, we have

$$\begin{aligned}\mathbf{f}\langle\hat{\mathbf{n}}\rangle &= -\mathbf{f}\langle-\hat{\mathbf{x}}\rangle \cos \measuredangle (x, x) - \mathbf{f}\langle-\hat{\mathbf{y}}\rangle \cos \measuredangle (x, y)\\ &= -\mathbf{f}\langle-\hat{\mathbf{x}}\rangle\\ &= -\mathbf{f}\langle-\hat{\mathbf{n}}\rangle\end{aligned}$$

which is just Eq. (2.1). Therefore, can we not say that Eq. (2.1) is a consequence of Eq. (2.2)? Well, if $\hat{\mathbf{n}} = \hat{\mathbf{x}}$, the inclined line in Fig. 2.6 will turn vertical and the fluid elementwill collapse into a piece of plane of two faces. The arguments leading to Eq. (2.2) simply *do not apply*; in other words, to be strictly accurate, Eq. (2.2) should have carried the caution that $\hat{\mathbf{n}} \neq \pm\hat{\mathbf{x}}$ or $\pm\hat{\mathbf{y}}$. So, some argument similar to the one used in deriving Eq. (2.1) in the text is needed after all. The only possible alternative would be to *state* explicitly that, although Eq. (2.2) is established only with $\hat{\mathbf{n}} \neq \pm\hat{\mathbf{x}}$ or $\pm\hat{\mathbf{y}}$, it must *also* hold for $\hat{\mathbf{n}} = \pm\hat{\mathbf{x}}$ or $\pm\hat{\mathbf{y}}$ since otherwise the phenomenon will exhibit a discontinuity at these $\hat{\mathbf{n}}$-orientations (see p. 134 of J. Serrin's "Mathematical principle of classical fluid mechanics," *Encyclopedia of Physics*, **VIII/1**, Springer-Verlag, Berlin, 1959).

$\mathbf{f}\langle x, y; t; \hat{\mathbf{n}}\rangle$ acting on a plane element with inclination $\hat{\mathbf{n}}$ is

$$\mathbf{f}\langle\hat{\mathbf{n}}\rangle = [\sigma_{xx}\cos\measuredangle(n,x) + \sigma_{yx}\cos\measuredangle(n,y)]\hat{\mathbf{x}} + [\sigma_{xy}\cos\measuredangle(n,x) + \sigma_{yy}\cos\measuredangle(n,y)]\hat{\mathbf{y}}$$

or

$$\mathbf{f}\langle\hat{\mathbf{n}}\rangle = \begin{cases}\sigma_{xx}\cos\measuredangle(n,x) + \sigma_{yx}\cos\measuredangle(n,y)\\ \sigma_{xy}\cos\measuredangle(n,x) + \sigma_{yy}\cos\measuredangle(n,y)\end{cases} \tag{2.5}$$

Finally, Eq. (2.1) with $\hat{\mathbf{n}} = \hat{\mathbf{x}}$ yields

$$\mathbf{f}\langle-\hat{\mathbf{x}}\rangle = -\mathbf{f}\langle\hat{\mathbf{x}}\rangle$$

or

$$\sigma_{(-x)x}\hat{\mathbf{x}} + \sigma_{(-x)y}\hat{\mathbf{y}} = -\sigma_{xx}\hat{\mathbf{x}} - \sigma_{xy}\hat{\mathbf{y}}$$

that is,

$$\sigma_{(-x)x} = -\sigma_{xx} \tag{2.6a}$$

$$\sigma_{(-x)y} = -\sigma_{xy} \tag{2.6b}$$

Similarly,

$$\sigma_{(-y)x} = -\sigma_{yx} \tag{2.6c}$$

$$\sigma_{(-y)y} = -\sigma_{yy} \tag{2.6d}$$

On the other hand, it is obvious that

$$\sigma_{(\ )x} = -\sigma_{(\ )(-x)} \tag{2.7a}$$

$$\sigma_{(\ )y} = -\sigma_{(\ )(-y)} \tag{2.7b}$$

Equations (2.6a,b,c,d) and (2.7a,b) combined give

$$\sigma_{(-x)(-x)} = \sigma_{xx} \tag{2.8a}$$

$$\sigma_{(-x)(-y)} = \sigma_{xy} \tag{2.8b}$$

$$\sigma_{(-y)(-x)} = \sigma_{yx} \tag{2.8c}$$

$$\sigma_{(-y)(-y)} = \sigma_{yy} \tag{2.8d}$$

The meaning of Eq. (2.8a), for example, is shown in Figure 2.8.

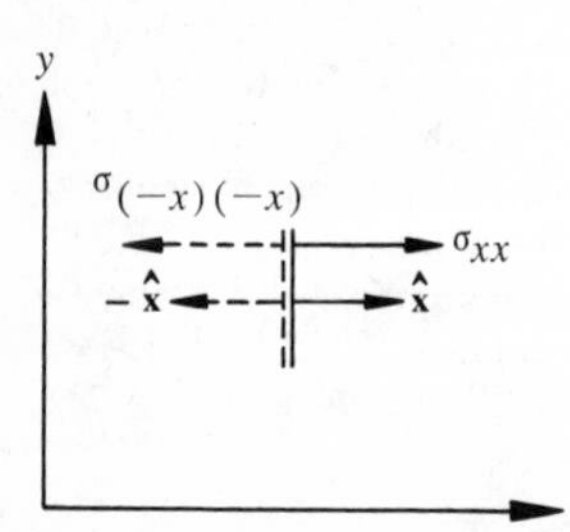

Figure 2.8 Normal stresses on the two sides of a surface. (The two *numbers* σ_{xx} and $\sigma_{(-x)(-x)}$ are the same, so $\sigma_{xx}\hat{\mathbf{x}} = \sigma_{(-x)(-x)}\hat{\mathbf{x}} = -\sigma_{(-x)(-x)}(-\hat{\mathbf{x}})$.)

2.2 EQUATIONS OF MOTION[5]

Consider a fluid particle in the shape of a rectangle with acceleration **a** and acted upon by the body force per unit mass **g** (Fig. 2.9). The inclinations of the four faces of the particle are indicated by the unit normal vectors as shown.

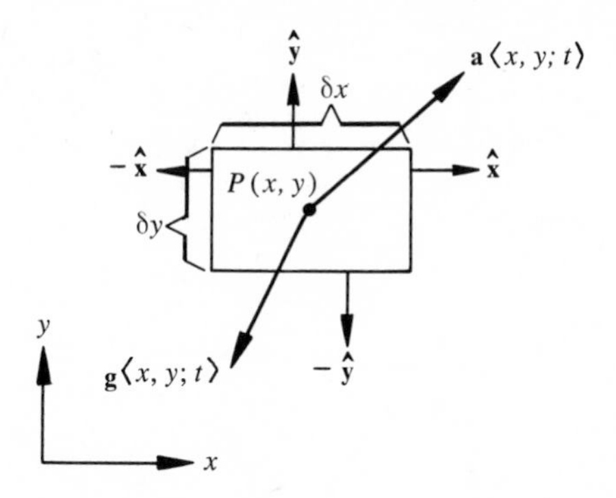

Figure 2.9 Body force on an accelerating fluid element.

The stress situation at the point P is completely determined by $\sigma_{xx}\langle x, y; t\rangle$, $\sigma_{xy}\langle x, y; t\rangle$, $\sigma_{yx}\langle x, y; t\rangle$, and $\sigma_{yy}\langle x, y; t\rangle$. The situations on the four faces are then easily seen to be as shown in Figure 2.10. An explanation of the

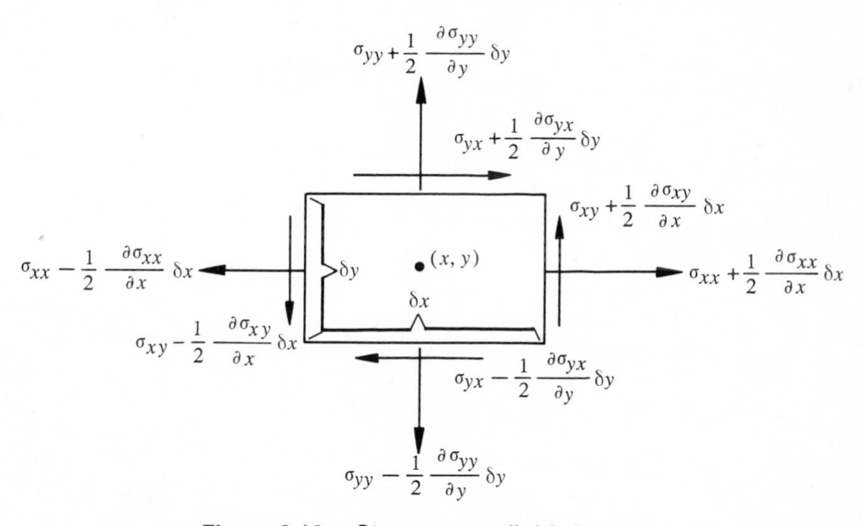

Figure 2.10 Stresses on a fluid element.

[5] At this point let us state, once and for all, that we always choose nonaccelerating *frames of reference* in describing the motion. (Being nonaccelerating means either resting, or moving in a straight line with a constant speed.) But, nonaccelerating with respect to what? To avoid the difficult discussion of the so-called inertial frame, we will answer this question *empirically* thus: with respect to the local earth surface for the usual, localized, engineering flow problems; with respect to (roughly speaking) the plane that passes through the center of the sun and the axis of the earth for geological (for example, oceanographical and atmospherical) flow problems.

stresses on the lower face only should be sufficient here. Originally, the stresses shown on the lower face with the directions indicated should be

$$\sigma_{(-y)(-y)} - \frac{1}{2}\frac{\partial \sigma_{(-y)(-y)}}{\partial y}\,\delta y$$

and

$$\sigma_{(-y)(-x)} - \frac{1}{2}\frac{\partial \sigma_{(-y)(-x)}}{\partial y}\,\delta y$$

They become as shown in the figure because of Eqs. (2.8d) and (2.8c).

The total contact force[6] (per unit depth) in the x-direction is clearly

$$\left(\sigma_{xx} + \frac{1}{2}\frac{\partial \sigma_{xx}}{\partial x}\,\delta x\right)\delta y - \left(\sigma_{xx} - \frac{1}{2}\frac{\partial \sigma_{xx}}{\partial x}\,\delta x\right)\delta y$$

$$+ \left(\sigma_{yx} + \frac{1}{2}\frac{\partial \sigma_{yx}}{\partial y}\,\delta y\right)\delta x - \left(\sigma_{yx} - \frac{1}{2}\frac{\partial \sigma_{yx}}{\partial y}\,\delta y\right)\delta x$$

$$= \left(\frac{\partial \sigma_{xx}}{\partial x} + \frac{\partial \sigma_{yx}}{\partial y}\right)\delta x\,\delta y$$

Similarly, that in the y-direction is

$$\left(\frac{\partial \sigma_{xy}}{\partial x} + \frac{\partial \sigma_{yy}}{\partial y}\right)\delta x\,\delta y$$

The mass of the element is

$$\rho\langle x, y; t\rangle\,\delta x\,\delta y$$

and the acceleration is

$$\mathbf{a} = \begin{cases} \dfrac{Du}{Dt} \\[2ex] \dfrac{Dv}{Dt} \end{cases}$$

by Eqs. (1.5a,b). Therefore, Newton's second law of motion[7] gives

$$\rho\cancel{(\delta x\,\delta y)}\frac{Du}{Dt} = \rho g_x\cancel{(\delta x\,\delta y)} + \left(\frac{\partial \sigma_{xx}}{\partial x} + \frac{\partial \sigma_{yx}}{\partial y}\right)\cancel{(\delta x\,\delta y)} \tag{2.9a}$$

$$\rho\cancel{(\delta x\,\delta y)}\frac{Dv}{Dt} = \rho g_y\cancel{(\delta x\,\delta y)} + \left(\frac{\partial \sigma_{xy}}{\partial x} + \frac{\partial \sigma_{yy}}{\partial y}\right)\cancel{(\delta x\,\delta y)} \tag{2.9b}$$

---

[6] Euler was the first to assert, without the slightest vagueness, that the mechanical action on a fluid element (set apart *in our imagination*) due to the rest of the fluid is equipollent to a distribution of contact forces, although he did not realize that these forces are not necessarily normal to the bounding surface of the element.

[7] This interpretation of Newton's law is attributable to John Bernoulli; it appeared in his 1743 *Hydraulics* (in Latin). Euler immediately hailed it as the true and genuine method. John Bernoulli (1667–1748), his brother James (1655–1705), and son Daniel (1700–1782), are the three prime members of a Swiss family of geniuses. John is considered today the founder of mathematical physics. It was the son, Daniel, strangely enough, who lit the way. Equally strange is the fact that Daniel later turned completely against mathematical methods in favor of physical arguments and the so-called "physical insight." For information on James, see next footnote.

In these equations, four stresses are involved. A simplification can be achieved by applying also the balance of angular momentum[8] to the same rectangular element about $P(x, y)$. The body force and the normal stresses contribute nothing to the torque since they go through the point P. The torque due to the shear stresses on the horizontal faces (+, counterclockwise) is

$$-\left(\sigma_{yx}+\frac{1}{2}\frac{\partial\sigma_{yx}}{\partial y}\,\delta y\right)(\delta x)\left(\frac{1}{2}\,\delta y\right)-\left(\sigma_{yx}-\frac{1}{2}\frac{\partial\sigma_{yx}}{\partial y}\,\delta y\right)(\delta x)\left(\frac{1}{2}\,\delta y\right)=-\sigma_{yx}(\delta x\,\delta y)$$

Similarly, that due to the shear stresses on the vertical faces is

$$\sigma_{xy}(\delta x\,\delta y)$$

The moment of inertia of the element is known from elementary mechanics:

$$\tfrac{1}{12}[(\delta x)^2+(\delta y)^2]$$

Therefore,

$$(\sigma_{xy}-\sigma_{yx})\cancel{(\delta x\,\delta y)}=\rho\cancel{(\delta x\,\delta y)}\cdot\frac{(\delta x)^2+(\delta y)^2}{12}\cdot(\text{angular acceleration})$$

Letting δx and $\delta y \to 0$ and demanding that the angular acceleration remain finite on physical grounds, we have

$$\sigma_{xy}-\sigma_{yx}=0$$

or

$$\sigma_{yx}=\sigma_{xy} \tag{2.10}$$

So, out of the four stresses, only three are independent.

Using Eq. (2.10) in Eqs. (2.9a,b), we finally obtain

$$\rho\frac{Du}{Dt}=\rho g_x+\left(\frac{\partial\sigma_{xx}}{\partial x}+\frac{\partial\sigma_{xy}}{\partial y}\right) \tag{2.11a}$$

$$\rho\frac{Dv}{Dt}=\rho g_y+\left(\frac{\partial\sigma_{xy}}{\partial x}+\frac{\partial\sigma_{yy}}{\partial y}\right) \tag{2.11b}$$

where

$$\frac{D}{Dt}=\frac{\partial}{\partial t}+u\frac{\partial}{\partial x}+v\frac{\partial}{\partial y} \tag{1.3}$$

Equations (2.11a,b) are known as the equations of motion.

Together with the equation of continuity

$$\frac{\partial u}{\partial x}+\frac{\partial v}{\partial y}=0 \tag{1.9}$$

---

[8] James Bernoulli was probably the first to discover this law of balance as a generalization of the ancient law of the lever (in 1686, one year earlier than Newton's *Principles*, cited in footnote 11 of Chapter 1). But it was Euler who in 1744 first used it as an independent law to a system of *n* linked bars. It is dependent on Newton's laws *only* for systems of discrete particles with central interparticle forces. (If interested, see C. Truesdell's "Whence the law of moment of momentum?" *Mélanges Alexandre Koyré – L'Aventure de la Science*, 588–612, Hermann, Paris, 1964.)

Equations (2.11a,b) govern the mechanics of plane flows of incompressible fluids.[9] In these governing equations, ρ, g_x, and g_y are given with the problem. We are then supposed to solve them for the unknowns u, v, σ_{xx}, σ_{xy}, and σ_{yy}, in terms of x, y, and t. But this is plainly impossible, since we have more unknowns than equations. Some additional information that relates σ_{xx}, σ_{xy}, and σ_{yy} with u and v is needed before we can go on.

Unfortunately, fluid mechanics itself does not supply us with this additional information. We must turn to the inner structure of the fluids in order to establish the needed additional relation.

2.3 NEWTONIAN FLUIDS AND THE NAVIER-STOKES EQUATIONS

We will start our discussion with the special case where the fluid is in a rigid-body(-like) motion in a certain region. (It is not necessarily at rest, or non-accelerating; it may translate or rotate with time-dependent velocities as long as it moves like a rigid body.) Then, by the very definition of a fluid (that is, the substance that does not support the existence of any shear stress when in rigid-body motion[10]), the shear stress vanishes everywhere in this region. Equation (2.5) then becomes (Figure. 2.11)

$$\mathbf{f}\langle\hat{\mathbf{n}}\rangle = \sigma_{nn}\hat{\mathbf{n}}$$

$$= \sigma_{xx}\cos\measuredangle(n,x)\,\hat{\mathbf{x}} + \sigma_{yy}\cos\measuredangle(n,y)\,\hat{\mathbf{y}}$$

But

$$\hat{\mathbf{n}} = \cos\measuredangle(n,x)\,\hat{\mathbf{x}} + \cos\measuredangle(n,y)\,\hat{\mathbf{y}}$$

Therefore,

$$\sigma_{nn} = \sigma_{xx} = \sigma_{yy}$$

---

[9] Except for the shear stresses, which had to wait for Cauchy, these three equations already appeared in Euler's 1755 paper, cited in footnote 2 of Chapter 1. As a matter of fact, the later writings of Euler are perfectly modern. No wonder his contemporaries could not appreciate him, charging that his work on fluids consisted only of mathematical exercises and was practically useless. But it was the same Euler who in 1754 presented a detailed design of a hydraulic turbine, which, when finally built in 1944 (!), reached an amazing efficiency of 71 percent. (J. Ackeret, "Untersuchung einer nach den Euler'schen Vorschlagen (1754) gebauten Wasserturbine," *Schweiz. Bauzeitung*, **123**, 9–15, 1944.) Euler's secret weapon was, of course, mathematical physics (or, mathematical model-building, in today's engineering jargon) which he learned from John and Daniel Bernoulli. It is interesting to note here that d'Alembert, being a mathematician, distrusted mathematics to such an extent that he either belittled, or ignored, or disbelieved, or tried to present in a pseudo-physico-philosophical fashion many of his *correct* findings. On the other hand, D. Bernoulli started as a friend and benefactor of both Euler and mathematical physics but later changed into an enemy, with the result that he would only communicate with Euler indirectly through his nephew John III. This left Euler alone to the field, which was to bloom only after Cauchy.

[10] This definition is more *liberal* than usual; but it is all there in the definition of a Newtonian fluid, Eqs. (2.12a,b,c), which will be introduced in a moment. Furthermore, it was also implied by Stokes in his 1845 paper, cited in footnote 6 of Chapter 1, which in our times served as the starting point of the theory of non-Newtonian fluids.

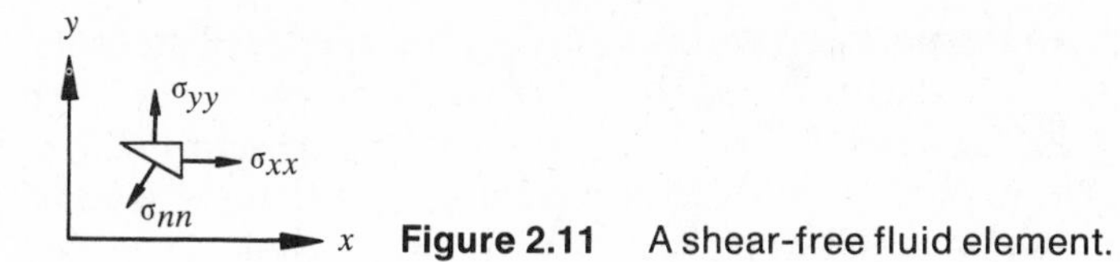

Figure 2.11 A shear-free fluid element.

that is, the normal stresses on plane elements of any inclination are the same at any one point and one instant. It is convenient to denote this common (local and instantaneous) normal stress by one symbol, $-\mathring{p}\langle x, y; t\rangle$; a negative sign is attached because of another basic characteristic of fluids in rigid-body motion, that $\sigma_{nn} < 0$ on plane elements of any inclination (that is, the stress is always opposite to $\hat{\mathbf{n}}$, pressing on the element instead of pulling on it). The negative of the common normal stress, $\mathring{p}$, is called the local *hydrostatic* pressure. It can be easily measured by a manometer with a small enough opening.

Conclusion In a region of rigid-body motion,[11] $\sigma_{xy} = 0$, and

$$\begin{cases} \sigma_{xx} = -\mathring{p} \\ \sigma_{yy} = -\mathring{p} \end{cases}$$

This is fine. But once the flow is not rigid-body-like, the foregoing discussion callapses completely. The hydrostatic pressure $\mathring{p}$ simply *ceases* to exist at points in such a flow.

To tighten up the discussion a bit, let us notice here that, in a region of rigid-body motion, the deformation as measured by the rates of strain, Eqs. (1.27a, b,c), vanishes at every point in the region. We will from now on call a point where *all* rates of strain vanish *a point of no deformation.* With this new terminology, we can localize the above conclusion, thus: At points of no deformation, $\sigma_{xy} = 0$ and $\mathring{p}$ exists.

Then, at *points of deformation,* $\mathring{p}$ is *not* defined. It is meaningless at these points. You cannot measure it by inserting a manometer, no matter how small the opening is! You would be hunting for some nonexisting animal.

But, on the other hand, we all know that many people do go out and *hunt* in exactly this manner. For example, it is a common practice to tap a pipe wall and connect a manometer to it. Here the flow at the wall is definitely with deformation (Fig. 2.12). What is measured here is certainly not the hydrostatic pressure. What this value actually is will be revealed a little later.

The nonexistence of a hydrostatic pressure is not the only worry we have for a flow with deformation. Remember that we are still without a relation between the stresses and the velocity field. (As a contrast, we do know that, in a deformationless velocity field, the shear stresses vanish and the normal stresses

[11] To be more specific, we may mention the following easily realizable cases: a region of constant velocity; a region of uniform velocity, speed varying linearly with time; and a region where the fluid is in rigid-body rotation with constant angular velocity. The name hydro*static* pressure is improper; but tradition prevails.

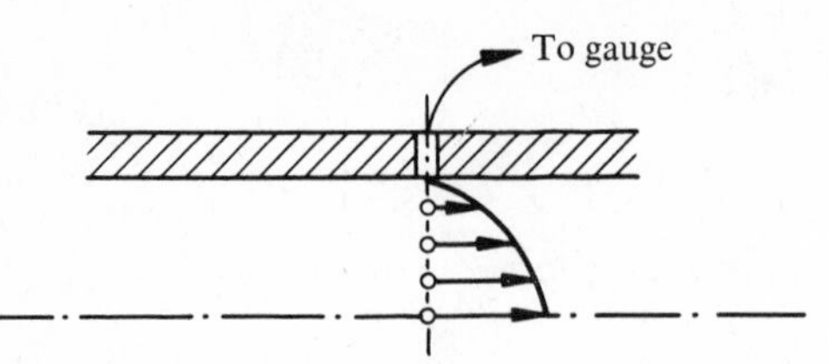

Figure 2.12 Tapping a pipe wall.

are all the same and negative.) Such a relation can only come from two sources: statistical mechanics and experimental observations. The first not only is beyond the reach of this book but also is not in a definitive form, especially not for the liquid state. We therefore can rely only on the experimentalists here. The situation is quite similar to the determination of the equations of state for various substances in thermodynamics; experimentalists must furnish the exact forms of those equations. A closer analogy is provided by elasticity, where there is a linear relationship between the stresses and the strains, obtained from material testing; any body that behaves in accordance with this relation is said to be elastic. Similarly, in fluid mechanics, experiments show that there is a linear relation between the stresses σ_{xx}, σ_{yy}, σ_{xy} and the rates of strain ϵ_{xx}, ϵ_{yy}, ϵ_{xy}:

$$\begin{cases} \sigma_{xx} = -\alpha + 2\mu\epsilon_{xx} = -\alpha + 2\mu \dfrac{\partial u}{\partial x} & \textbf{(2.12a)} \\ \sigma_{yy} = -\alpha + 2\mu\epsilon_{yy} = -\alpha + 2\mu \dfrac{\partial v}{\partial y} & \textbf{(2.12b)} \\ \sigma_{xy} = 2\mu\epsilon_{xy} = \mu\left(\dfrac{\partial v}{\partial x} + \dfrac{\partial u}{\partial y}\right) & \textbf{(2.12c)} \end{cases}$$

(The two quantities α and μ will be discussed in a moment.)

Fluids that satisfy Eqs. (2.12a,b,c) are called Newtonian. In this book, as already stated at the beginning of Section 1.1, we deal mainly with Newtonian fluids. One of the implications of Eqs. (2.12a,b,c) is that there exist two additional quantities α and μ. This is news to us. But if the experimentalists say so, we will accept it as a fact. It is not difficult to see the physical significance of α: Adding Eqs. (2.12a) and (2.12b), we have

$$\alpha = -\frac{1}{2}(\sigma_{xx} + \sigma_{yy}) + \mu\left(\frac{\partial u}{\partial x} + \frac{\partial v}{\partial y}\right)$$

But, the fluid is incompressible; so Eq. (1.9) demands that

$$\begin{aligned} \alpha &= -\tfrac{1}{2}(\sigma_{xx} + \sigma_{yy}) \\ &= -(\textit{mean}\text{ normal stress, based on normal stresses on two plane elements } \perp x \text{ and } y) \end{aligned} \qquad \textbf{(2.13)}$$

Another of the implications of Eqs. (2.12a,b,c) is that they are valid no matter what directions you choose to be the $+x$ and $+y$ directions (Fig. 2.13). So Eq. (2.13) is more generally stated as

$$\alpha = -(\text{mean normal stress based on any two perpendicular plane elements})$$

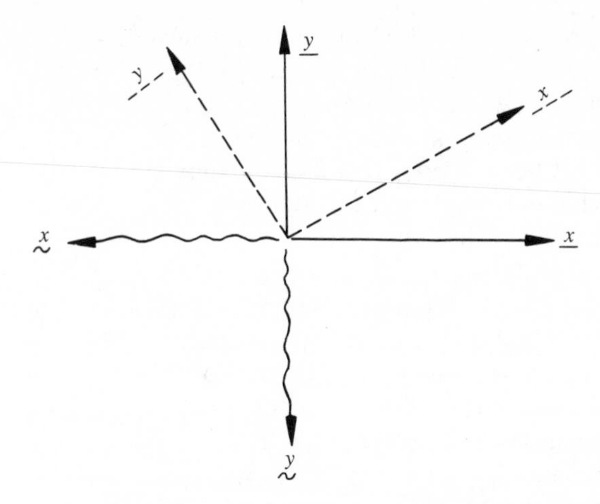

Figure 2.13 Different orientations of coordinates.

It is customary therefore to call α the average pressure; and denote it by the symbol p. Although it is true that in a region of uniform flow or rigid-body rotation, or at rest, the average pressure p is identical with the hydrostatic pressure $\mathring{p}$, p is definitely *not* $\mathring{p}$ in a flow region with deformation. Although p always exists, $\mathring{p}$ exists only in regions of uniform or zero flow, or of rigid-body rotation. There is *no* such animal as a hydrostatic pressure in a flow region with deformation:

$$p = -\tfrac{1}{2}(\sigma_{xx} + \sigma_{yy}), \text{ always}$$

$$p = \mathring{p} \text{ only where } \epsilon_{xx}, \epsilon_{yy}, \text{ and } \epsilon_{xy} \text{ vanish (including zero flow)}$$

From now on, in this book, we will mean the average pressure p whenever we say "pressure" without any modifier. Equations (2.12a,b,c) now read

$$\begin{cases} \sigma_{xx} = -p + 2\mu \dfrac{\partial u}{\partial x} & \textbf{(2.14a)} \\[2ex] \sigma_{yy} = -p + 2\mu \dfrac{\partial v}{\partial y} & \textbf{(2.14b)} \\[2ex] \sigma_{xy} = \mu \left(\dfrac{\partial v}{\partial x} + \dfrac{\partial u}{\partial y} \right) & \textbf{(2.14c)} \end{cases}$$

The remaining quantity μ is purely experimentally given to us. It is known as the viscosity[12] of the fluid under consideration. For a given homogeneous fluid, it is independent of $\mathbf{q}$, x, y, t, ϵ_{xx}, ϵ_{yy}, and ϵ_{xy}. It may, however, depend on the temperature. But, as stated at the beginning of Section 1.1, μ is regarded as a known constant in Parts 1 and 2 of this book.

[12] The germ of the concept of this kind of viscosity appeared in Newton's 1687 *Principles* (cited in footnote 11 of Chapter 1), Book II; hence the term Newtonian fluid. Equations (2.14a,b,c) were obtained by the French engineer Navier in his 1822 "Memoir on the laws of motion of fluids" (in French). Claude-Louis-Marie-Henri Navier (1785–1836) wrote a two-volume treatise on bridges and designed a suspension bridge across the Seine at Paris. The fact that the bridge promptly failed after erection because of the settlement of a pier should not be blamed on Navier, since soil mechanics is such a modern science.

At this point, the author must confess to some cheating in stating that Eqs. (2.12a,b,c) are to be looked upon as an experimental fact. In the first place, no experimental work can supply such a general relation; all experimentalists can do is to verify it in specially simplified apparatus, such as viscometers. In the second place, Eqs. (2.12a,b,c) *can* actually be derived within fluid dynamics[13] (by way of a few general principles such as homogeneity and isotropy; then, the only thing that is supplied by experiments would be the numerical values of μ, and their dependence on temperature, of various fluids).

This "cheating" was resorted to for purposes of simplification at this point. The complete story is given later in Part 3. For our present purpose, we will continue to look upon Eqs. (2.14a,b,c) as an experimental fact. For those of you who are familiar with the postulational view, it is highly recommended that you look upon Eqs. (2.14a,b,c) as a postulation whose validity lies in the fact that all its deductions are borne out by experiments.

We will now continue to discuss more of the implications of Eqs. (2.14a,b,c).

First, let us bring it down to more familiar ground by considering two elementary situations:

1. Where there is no deformation, Eqs. (2.14a,b,c) yield simply

$$\begin{cases} \sigma_{xx} = \sigma_{yy} = -\mathring{p} \\ \sigma_{xy} = \sigma_{yx} = 0 \end{cases}$$

   in keeping with hydrostatics for fluids at rest, in uniform motion, or in rigid-body rotation.

2. In a flow with straight and parallel streamlines (Fig. 2.14), that is,

$$\mathbf{q} = \begin{cases} u\langle y\rangle \\ 0 \end{cases}$$

   Equations (2.14a,b,c) yield

$$\begin{cases} \sigma_{xx} = \sigma_{yy} = -p \\ \sigma_{xy} = \sigma_{yx} = \mu \dfrac{du}{dy} \end{cases}$$

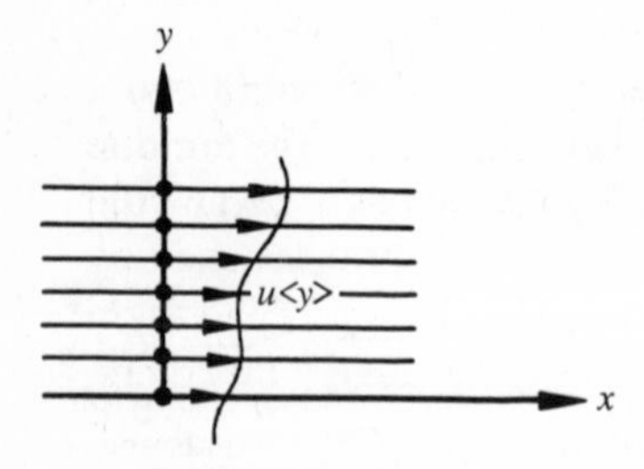

Figure 2.14 Parallel flow.

[13] In a linear *constitutive* framework.

This second case is in keeping with the elementary idea of viscosity presented in elementary texts (usually with $u \propto y$) except for the following:

1. It is the average pressure that plays a role, not the nonexistent hydrostatic pressure. (To state that $\sigma_{xx} = \sigma_{yy} = -\mathring{p}$ in such a flow is a common mistake committed by many authors. You should not make the same mistake. Note also that, here, $\sigma_{nn} \neq -p$ for general $\hat{\mathbf{n}}$.)
2. There is also a σ_{xy} (that is, shear stress on a plane element perpendicular to the flow), which is equal to σ_{yx}; see Figure 2.15. (The existence of a shear stress on planes perpendicular to a flow is often not emphasized enough, if not altogether ignored, in elementary texts, leaving the reader the deadly wrong impression that shear stresses exist only between gliding fluid layers. The crude kinetic theory often presented in these texts, supposedly to explain the phenomenon of viscosity on a molecular basis, does not help either. It only serves, if anything, to convince the reader that there cannot be any shear stress on planes perpendicular to the flow, since there is seemingly no molecular mechanism to account for it![14] Once this harm is done to the reader, it is rather difficult to undo it without going very deep into the *real* kinetic theory. As a last effort, let it be stated once more that there is always shear stress on a plane element perpendicular to the flow if there is one on a plane element parallel to the flow, merely because of the conservation of angular momentum, Eq. (2.10).)

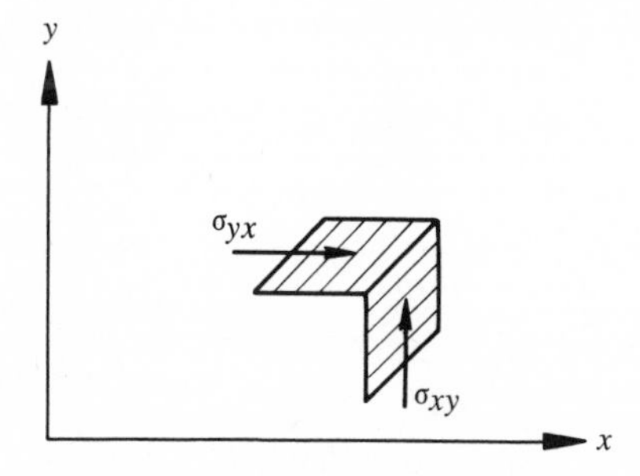

Figure 2.15 Shear stress on a plane perpendicular to the flow.

Next, just by looking at Eqs. (2.14a) and (2.14b), we realize (but many engineers and textbook writers do not!) that parts of the normal stresses are due to viscosity (unless $\partial u/\partial x$ and $\partial v/\partial y$ happen to be zero). Now, many students would find this hard to believe, again because they have been implicitly brainwashed by the little picture (see Fig. 2.16) and the crude kinetic theory, which mislead them into always associating viscosity with shearing forces between gliding fluid layers.

[14] For an attempt in straightening this out on a relatively elementary level, see G. M. Volkoff and D. S. Carter, "On the shearing stress in a viscous fluid across a surface normal to the lines of flow," *Am. J. Phys.*, **17**, 37–40, 1949.

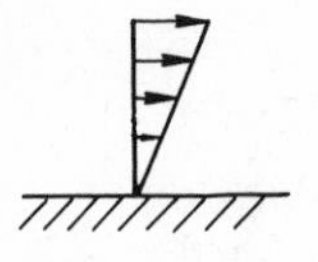

Figure 2.16 A "little picture."

They then find it impossible to allow a viscous contribution to a normal stress. To get out of this straitjacket, you must either heed the lessons taught by Eqs. (2.14a, b,c), or study *real* kinetic theory. In any event, you have to warn yourself constantly that what you have learned in an elementary course may need serious reexamination. The elementary texts are seldom out-and-out wrong; only you should not have read too much into them.

This viscous contribution to the normal stresses has many practical significances. The most important one is seen in the following reexamination of the usual laboratory operation known as "pressure measurement." You are certainly all familiar with this basic measuring process. It is very instructive to ask here what is it that you really measure in such a process. Or, more precisely, we ask here two questions: (a) What *kind* of pressure is it? (b) Is it *really* some kind of pressure (that is, something that applies on planes of all inclinations with the same intensity)?

As basic requirements, let us take for granted that the probe used is small enough that the flow field is not perturbed by it, and the reading is really local and instantaneous. It is also taken for granted that the opening of the probe is parallel to the local instantaneous streamline (Fig. 2.17). The opening, whose unit

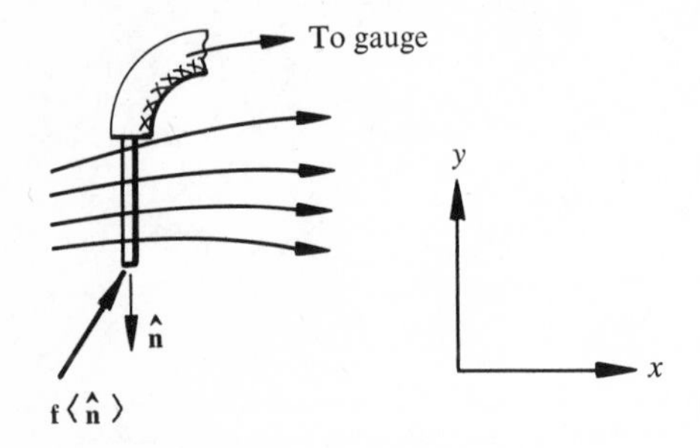

Figure 2.17 A probe.

normal vector is $\hat{\mathbf{n}}$, is acted on by the fluid with a force per unit area $\mathbf{f}\langle\hat{\mathbf{n}}\rangle$. Let us also grant that, because $\mathbf{f}\langle\hat{\mathbf{n}}\rangle$ is to be transmitted by some fluid in the connection, only the normal component of $\mathbf{f}\langle\hat{\mathbf{n}}\rangle$ is read on the gauge. We then conclude that what we measure is σ_{nn} (+, if pulling on the opening) which in the figure is just

$$\sigma_{(-y)(-y)} = \sigma_{yy} = -p + 2\mu\frac{\partial v}{\partial y}$$

$$= -(\text{average pressure}) + (\text{some contribution from viscosity})$$

Conclusion The reading on a so-called pressure guage is some kind of normal stress with a viscous contribution to it.

That, then, answers both of our questions: It is *not* a pressure at all!

However, in ordinary situations, the viscosity of the fluid and/or the derivatives $\partial u/\partial x$, $\partial v/\partial y$ are so small that the viscous contribution $[\mu(\partial u/\partial x), \mu(\partial v/\partial y)]$ is negligible; then, in the previous figure, for example,

$$\sigma_{(-y)(-y)} \cong -p$$

The measurement is then really some kind of pressure. (These ordinary situations are the things that saved pressure measurement from an untimely demise.) But even here, we see that it is the average pressure that we measure, *never* the hydrostatic pressure, which does not even exist *unless* the flow happens to be without deformation there.

Thus, we have both attacked and saved the practice of pressure measurement. But it is conceivable that in special situations the viscous contribution is so large that a "pressure measurement" becomes meaningless. In such special cases one must use two gauges with openings perpendicular to each other (Fig. 2.18); the two *normal stresses* (with *non*negligible viscous contributions) read can be averaged to yield the negative of the average pressure. This is then the basic procedure to be used in the measurement of (average) pressure. But it is seldom, if ever, used by the experimentalists, because of two inconveniences: (a) The double probe may be too large to use. (b) The opening that is inclined to the flow will convert a part of the kinetic energy of the fluid into the reading; this has to be compensated. However, whenever one is in doubt about the size of the viscous contribution, this basic procedure is to be used.

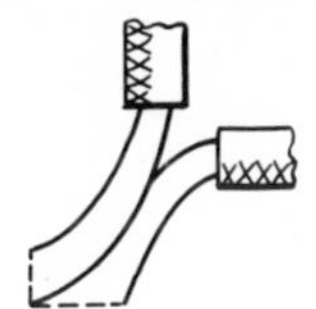

Figure 2.18 A double probe.

Now we are ready to go back to our equations of motion, Eqs. (2.11a,b). For a Newtonian fluid with constant μ, a substitution of Eqs. (2.14a,b,c) into Eqs. (2.11a,b) yields:

$$\begin{cases} \dfrac{\partial u}{\partial t}+u\dfrac{\partial u}{\partial x}+v\dfrac{\partial u}{\partial y}=g_x-\dfrac{1}{\rho}\dfrac{\partial p}{\partial x}+\nu\left(\dfrac{\partial^2 u}{\partial x^2}+\dfrac{\partial^2 u}{\partial y^2}\right) & \textbf{(2.15a)} \\ \dfrac{\partial v}{\partial t}+u\dfrac{\partial v}{\partial x}+v\dfrac{\partial v}{\partial y}=g_y-\dfrac{1}{\rho}\dfrac{\partial p}{\partial y}+\nu\left(\dfrac{\partial^2 v}{\partial x^2}+\dfrac{\partial^2 v}{\partial y^2}\right) & \textbf{(2.15b)} \end{cases}$$

where the equation of continuity, Eq. (1.9), has been used; and where $\nu = \mu/\rho$ (a given constant for the fluid considered) is called the kinematic viscosity.

Equations (2.15a,b) are known as the Navier-Stokes (abbreviated N-S) equations.[15] Together with the equation of continuity, Eq. (1.9), the N-S equations

[15] Navier first obtained these equations in his 1822 memoir (see footnote 12) by constructing a molecular model for the inner structure of the fluid. His model was soon abandoned as wrong; but his result turned out to be correct. Our derivation here is essentially based on Stokes' work, as given in his 1845 paper, cited in footnote 6 of Chapter 1. Stokes was reportedly an excellent lecturer. He once lectured for three hours; but he also was known to have sat by the fire with a visitor for two whole hours without uttering a word!

now describe completely any plane flow of incompressible, Newtonian fluid with constant viscosity. The system of three nonlinear partial differential equations, Eqs. (1.9) and (2.15a,b) (nonlinear because of products of u, v, and their derivatives on the left-hand side of Eqs. (2.15a,b)), involves exactly three unknowns, u, v, and p. So, we are now in business! All we have to do from now on is to solve it.

2.4 NON-NEWTONIAN FLUIDS

Although the main concern of this book is with Newtonian fluids, we should, however, be aware of the modern development in non-Newtonian fluids.

It is true that ordinary water, oils, glycerine, liquid metals, and so on, are all incompressible Newtonian fluids. But experience with such materials as synthetic latices, high polymer solutions, molten polymers, and protein solutions (more specifically, blood, paints, printing ink, starch suspensions, solution of polyisobutylene in decalin, solution of polyethylene oxide in water, and so on) shows that, as a rule, they do not obey Eqs. (2.12a,b,c) and are thus non-Newtonian fluids. The scope of such fluids is clearly very wide; and a comprehensive theory that encompasses all of them is of necessity rather complicated. In this book, we will only touch on a special class of such fluids known as Reiner-Rivlin[16] fluids (or visco-inelastic fluids).

Without further ado, we will now state that a fluid is said to be Reiner-Rivlin if it satisfies the following stress-rate-of-strain relations:

$$\left\{\begin{aligned} \sigma_{xx} &= -\beta\langle \mathrm{II}\rangle + 2\mu\langle \mathrm{II}\rangle \frac{\partial u}{\partial x} + \frac{1}{2}\eta\langle \mathrm{II}\rangle \mathrm{II} && \textbf{(2.16a)} \\ \sigma_{yy} &= -\beta\langle \mathrm{II}\rangle + 2\mu\langle \mathrm{II}\rangle \frac{\partial v}{\partial y} + \frac{1}{2}\eta\langle \mathrm{II}\rangle \mathrm{II} && \textbf{(2.16b)} \\ \sigma_{xy} &= \sigma_{yx} = \mu\langle \mathrm{II}\rangle \left(\frac{\partial v}{\partial x} + \frac{\partial u}{\partial y}\right) && \textbf{(2.16c)} \end{aligned}\right.$$

where $\beta\langle \mathrm{II}\rangle$, $\mu\langle \mathrm{II}\rangle$, and $\eta\langle \mathrm{II}\rangle$ are all functions of

$$\mathrm{II} = 2\left[\left(\frac{\partial u}{\partial x}\right)^2 + \frac{1}{4}\left(\frac{\partial u}{\partial y} + \frac{\partial v}{\partial x}\right)^2\right] \qquad \textbf{(2.17)}$$

and, possibly, of temperature. In addition, we have to supplement the above with

$$\sigma_{zz} = -\beta\langle \mathrm{II}\rangle \qquad \textbf{(2.16d)}$$

where σ_{zz} is obviously the normal stress a manometer would register if its opening were parallel to the representative plane and facing us. Equation (2.16d) is needed since our average pressure p should really be defined physically as

$$p = -\tfrac{1}{3}(\sigma_{xx} + \sigma_{yy} + \sigma_{zz})$$

---

[16] M. Reiner, contemporary Israeli rheologist, and R. S. Rivlin, professor of applied mathematics at Lehigh University.

that is, an average over three manometer readings with mutually perpendicular openings. For the Newtonian fluids, we did not bother to drag in σ_{zz} but took $-(\sigma_{xx}+\sigma_{yy})/2$ as p. This was done because it did not make any difference there. For Reiner-Rivlin fluids, however, $-(\sigma_{xx}+\sigma_{yy})/2$ does not have the status of the (physical) average pressure any more. We have to add up Eqs. (2.16a), (2.16b), and (2.16d) to get, with the help of the Eq. (1.9),

$$-3p=-3\beta\langle \text{II}\rangle+\eta\langle \text{II}\rangle \text{II}$$

that is,

$$\beta\langle \text{II}\rangle = p+\tfrac{1}{3}\eta\langle \text{II}\rangle \text{II}$$

In terms of p, Eqs. (2.16a,b,c,d) can be rewritten as

$$\begin{cases} \sigma_{xx}=-p+2\mu\langle \text{II}\rangle \dfrac{\partial u}{\partial x}+\dfrac{1}{6}\eta\langle \text{II}\rangle \text{II} & \text{(2.18a)} \\ \sigma_{yy}=-p+2\mu\langle \text{II}\rangle \dfrac{\partial v}{\partial y}+\dfrac{1}{6}\eta\langle \text{II}\rangle \text{II} & \text{(2.18b)} \\ \sigma_{xy}=\sigma_{yx}=\mu\langle \text{II}\rangle\left(\dfrac{\partial v}{\partial x}+\dfrac{\partial u}{\partial y}\right) & \text{(2.18c)} \\ \sigma_{zz}=-p-\tfrac{1}{3}\eta\langle \text{II}\rangle \text{II} & \text{(2.18d)} \end{cases}$$

where $\mu\langle \text{II}\rangle$ and $\eta\langle \text{II}\rangle$ are called the viscosity function and cross-viscosity function, respectively. To understand the complicated Eqs. (2.18a,b,c,d), we will point out the following special cases:

1. If $\mu\langle \text{II}\rangle$ = constant (but may still depend on temperature), and $\eta\langle \text{II}\rangle = 0$, we get back to the Newtonian fluids.
2. If the flow is without deformation, we have again

$$\begin{cases} \sigma_{xx}=\sigma_{yy}(=\sigma_{zz})=-\mathring{p} \\ \sigma_{xy}=\sigma_{yx}=0 \end{cases}$$

3. In a simple shear flow, that is, where $\mathbf{q}=(Cy,0)$, Eqs. (2.18a,b,c,d) yield

$$\begin{cases} \sigma_{xx}=\sigma_{yy}=-p+\tfrac{1}{12}\eta\langle |C|\rangle C^2 & \text{(2.19a)} \\ \sigma_{xy}=\sigma_{yx}=\mu\langle |C|\rangle C & \text{(2.19b)} \\ \sigma_{zz}=-p-\tfrac{1}{6}\eta\langle |C|\rangle C^2 & \text{(2.19c)} \end{cases}$$

(Note that μ and η originally depend on C^2, and *hence* on $|C|$.)

Equation (2.19b) states that the plot of shear stress vs. shear rate (C) is a certain curve, instead of a straight line as in the Newtonian case. (μ is then said to be nonlinear.) Figure 2.19 shows a few typical examples. The shapes of these plots suggest various rheological formulas. For example, there is the power law

$$\sigma_{xy}=m|C|^{n-1}C$$

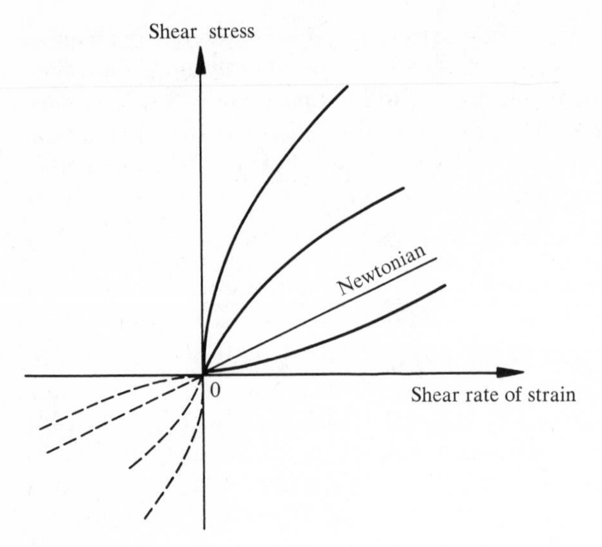

Figure 2.19 Shear stress versus shear rate for various fluids.

that is,

$$\mu\langle|C|\rangle = m|C|^{n-1} \tag{2.20}$$

We may mention that, for a 0.67 percent (by weight) carboxymethyl-cellulose preparation in water, this law holds with $n = 0.716$, $m = 0.00634$ [$lb_f(sec)^{0.716}/ft^2$]; for a 4 percent (by weight) paper pulp in water, $n = 0.575$, $m = 0.418$ [$lb_f(sec)^{0.575}/ft^2$]. Many other laws have also been proposed to fit various kinds of fluids. This practice is all right if one keeps three things in mind: (a) Each law fits probably only one narrow class of fluid. (b) It says nothing about the normal stresses, which happen to be very important in the flows of non-Newtonian fluids. (c) We are talking here only about the simple shear flow.

But when the rheologists start to use the same kind of laws in the general flows—for example, writing Eq. (2.20) in the form of

$$\mu\langle \mathrm{II}\rangle = m|\sqrt{\mathrm{II}}|^{n-1} \tag{2.21}$$

and inserting it into Eqs. (2.18a,b,c,d)—we have to be wary. Although Eq. (2.21) is very popular with rheologists, chemical and biomedical engineers, and multiphase-flow researchers (they have even introduced the tag "power law fluids"). we have to drop it as not in keeping with the spirit of this book.

As a matter of fact, even Reiner-Rivlin relations themselves are not faring very well in all respects. Take the simple shear again as our example; we see that

$$\begin{cases} \sigma_{xx} = \sigma_{yy} \\ \sigma_{xx} - \sigma_{zz} = \sigma_{yy} - \sigma_{zz} = \frac{1}{4}\eta\langle|C|\rangle C^2 \end{cases}$$

Originally, the second relation could be used to find $\eta\langle|C|\rangle$ from measured difference in normal stresses. (Incidentally, this also shows that, in order to realize the simple shear flow, additional forces must be applied across the flow—a strange fact characteristic of non-Newtonian fluids.) However, this happy note got muffled up by the first requirement $\sigma_{xx} = \sigma_{yy}$. As of now, no non-Newtonian fluid has been discovered that satisfies this. Although there is still the possibility that some new fluids of the future are Reiner-Rivlin, one must say flatly that no existing non-Newtonian fluid satisfies all of Eqs. (2.18a,b,c,d). To fit the existing fluids, other more complicated relations must be proposed. As examples, we may mention the very successful Rivlin-Ericksen[17] fluids (for liquid crystals such as solutions of polypeptides), and, above all, the fantastically successful simple fluids of Noll.[18] (It is probably worth noticing here that the theory of simple fluids is anything but simple.)

Going back to the Reiner-Rivlin fluids, we must state that the faults involved there should not be put in the same category as the so-called "power-law fluid." A Reiner-Rivlin fluid is, after all, a rational model; being such, it is possible to locate the real source of the trouble and to remedy it once and for all. This is exactly how Rivlin-Ericksen fluids and simple fluids got started—as improvements of Reiner-Rivlin fluids.

In the rest of the book, we will continue to comment on Reiner-Rivlin fluids from time to time as a contrast to the Newtonian fluids. The comparison should be qualitatively enlightening; and the conclusions, at least qualitatively representative of the non-Newtonian behavior. As to the other fluids, you will not meet them again.

The equations of motion, Eqs. (2.11a,b), are still valid. It is senseless here to substitute Eqs. (2.16a,b,c) and (2.17) into them. Let us just say that Eqs. (2.11a,b), with Eqs. (2.16a,b,c) and (2.17), together with the equation of continuity, Eq. (1.9), govern the plane flows of (incompressible) Reiner-Rivlin fluids.

Our discussion on the viscous contribution to the normal stresses in the last section applies here to non-Newtonian fluids *a fortiori* because of the cross-viscosity. It is safe to say that no experimentalist in the field of non-Newtonian fluids strives to measure the (average) pressure. He measures, as a rule, only normal stresses. This is just fine if one remembers that the normal stress varies with the orientation of the gauge opening, unless the flow is without deformation at the point.

---

[17] J. L. Ericksen, professor of mechanics, Johns Hopkins University.

[18] W. Noll, professor of mathematics, Carnegie-Mellon University.

CHAPTER 3

SIMPLE EXACT SOLUTIONS

3.1 FLOWS BETWEEN PARALLEL PLATES

To solve Eqs. (2.15a,b) together with Eq. (1.9) is easier said than done. Known exact solutions are few in number, and they all involve rather simple geometrical situations. In this chapter we will present the elementary ones. More complicated ones will be taken up in the next chapter.[1] After that, we will search for approximate solutions to continue our study.

The flows studied in this chapter are all confined in between two parallel lines, $y = 0$ and Y (Fig. 3.1), which represent two rigid plates (permeable or impermeable), or one such plate and one liquid surface. We will consider only the cases in which the plates are either stationary, or are moving in their own planes in the x-direction. At every point between $y = 0$ and Y, there are two velocity components and one (average) pressure to be determined at all times. The N-S equations and the continuity equation are to be solved between $y = 0$ and Y in such a way that all fluid particles that happen to be on the plates $y = 0$ and Y move with the plates in the direction tangential to the plates. This is known as the no-slip condition. It is a directly observable fact for liquids, no matter how small the viscosity

[1] These solutions, besides their obvious practical uses, also furnish you implicitly with a great deal of general information.

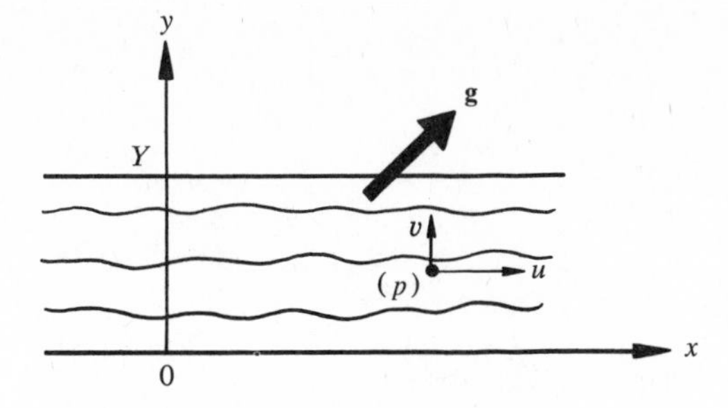

Figure 3.1 Flow region between two parallel lines.

may be; even when the liquid does not wet the plate (for example, mercury on glass).[2] The fluid at the plates must also stay with them in the direction normal to the plates, if they are impermeable. Otherwise the fluid will either enter the plate or will leave the plate, creating a vacuum; both are physically impossible. The case of permeable plates will be discussed later in this section. At present, let us stipulate that the plates are impermeable.

Mathematically, we then say that the solutions to the equations are subject to the *boundary conditions*: $\mathbf{q}\langle x, 0; t\rangle$, $\mathbf{q}\langle x, Y; t\rangle$ = given velocities of the plates $y = 0$ and Y, respectively.

From what we said just a while ago, the exact forms of the boundary conditions for our present situation are

$$\begin{cases} u\langle x, 0; t\rangle = F_1\langle t\rangle & \textbf{(3.1a)} \\ u\langle x, Y; t\rangle = F_2\langle t\rangle & \textbf{(3.1b)} \\ v\langle x, 0; t\rangle = 0 & \textbf{(3.1c)} \\ v\langle x, Y; t\rangle = 0 & \textbf{(3.1d)} \end{cases}$$

where F_1 and F_2 are given functions of time.

The final correct solution of the problem must of necessity satisfy boundary conditions (3.1c) and (3.1d). A flash of thought suddently enters our mind: $v = 0$ *everywhere between* $y = 0$, Y *and at all times* certainly satisfies these two conditions. Is it possible that the solution *is* indeed without the y-component of the velocity everywhere and at all instants? Let us find out by searching for a solution with $v\langle x, y; t\rangle \equiv 0$. If we succeed in this task, our brainstorm proves to be right. (If we fail, we will admit that we made a wrong speculation; there is no solution to the problem with $v \equiv 0$! We then start again, this time with special allowance for the possibility of a nonzero v.)

Formally, we would say that the boundary conditions (3.1c,d) suggest

[2] For gases, this no-slip condition is also *postulated.* Its justification then lies in the fact that, in all cases except where the gas is rarefied, no attributable discrepancies have ever turned up when comparing deductions based on this postulated condition with experiments. (Slip does occur in flows of rarefied gases and certain special non-Newtonian liquids.)

that we try the "ansatz" (or, trial form of solution)[3]

$$v\langle x, y; t\rangle \equiv 0$$

With this ansatz, Eqs. (1.9) and (2.15a,b) become

$$\frac{\partial u}{\partial x} = 0 \tag{3.2}$$

$$\frac{\partial u}{\partial t} + u\frac{\partial u}{\partial x} = g_x - \frac{1}{\rho}\frac{\partial p}{\partial x} + \nu\left(\frac{\partial^2 u}{\partial x^2} + \frac{\partial^2 u}{\partial y^2}\right) \tag{3.3a}$$

$$0 = g_y - \frac{1}{\rho}\frac{\partial p}{\partial y} \tag{3.3b}$$

If our ansatz is correct, we should be able to solve Eqs. (3.2) and (3.3a,b) subject to boundary conditions (3.1a,b) (and other conditions to be posed later for specific problems). First of all, Eq. (3.2) demands that

$$u = u\langle y; t\rangle \text{ only}$$

and Eq. (3.3a) should read

$$\frac{\partial u}{\partial t} = g_x - \frac{1}{\rho}\frac{\partial p}{\partial x} + \nu\frac{\partial^2 u}{\partial y^2} \tag{3.4}$$

The fact (if our ansatz later proves to be right) that $u = u\langle y; t\rangle$ means that the velocity $\mathbf{q} = \mathbf{q}\langle y; t\rangle$, since v is zero anyway in the ansatz. That is, the velocity distribution at any instant is the same at all x-sections (Fig. 3.2). We then say that the flow is *fully developed* (that is, $\partial\mathbf{q}/\partial x = 0$) in the x-direction.

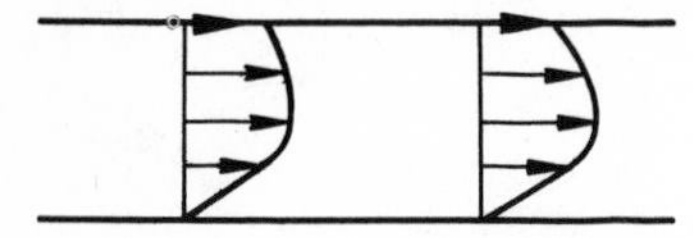

Figure 3.2 A fully developed flow.

In the beginning, we could have thought that boundary conditions (3.1a, b,c,d) suggest the ansatz $\partial\mathbf{q}/\partial x = 0$, since they do not involve x in any sense. (They do involve y since u at $y = 0$, Y are different.) If we had introduced this ansatz instead of $v \equiv 0$, we would have also ended up with $v \equiv 0$, together with Eqs. (3.3b) and (3.4).

From Eq. (3.3b), we can still go one step further if we agree to restrict ourselves from now on to constant g_y and g_x. (We naturally have gravity in mind all the time. From now on, let us explicitly regard $\mathbf{g}$ as a constant and refer to it as "the gravity.") Thus,

$$\frac{\partial p}{\partial y} = \rho g_y = \text{constant}$$

$$p = \rho g_y y + G\langle x; t\rangle$$

[3] In the following the German word *ansatz* will be treated as if it were English, with the same form for both singular and plural.

Then,

$$\frac{\partial p}{\partial x} = \frac{\partial G\langle x; t\rangle}{\partial x}$$

Substituting into Eq. (3.4), we have

$$\frac{\partial G\langle x; t\rangle}{\partial x} = \rho g_x + \rho\left(\nu\frac{\partial^2 u}{\partial y^2} - \frac{\partial u}{\partial t}\right)$$

where the second term on the right-hand side is just a fct $\langle y; t\rangle$. Then,

$$\frac{\partial^2 G\langle x; t\rangle}{\partial x^2} = 0$$

that is,

$$\frac{\partial G\langle x; t\rangle}{\partial x} = F\langle t\rangle$$

or

$$\frac{\partial p}{\partial x} = F\langle t\rangle \tag{3.5}$$

Furthermore,

$$G\langle x; t\rangle = F\langle t\rangle x + f\langle t\rangle \tag{3.6}$$

If we measure the difference of the local (average) pressure, Δp, between two points a distance L apart on the same y-line, we have (Fig. 3.3)

$$\Delta p = \Delta G = LF\langle t\rangle$$

Then, the function $F\langle t\rangle = \Delta p/L$ can be physically determined (measured) as the time variation of the pressure difference.

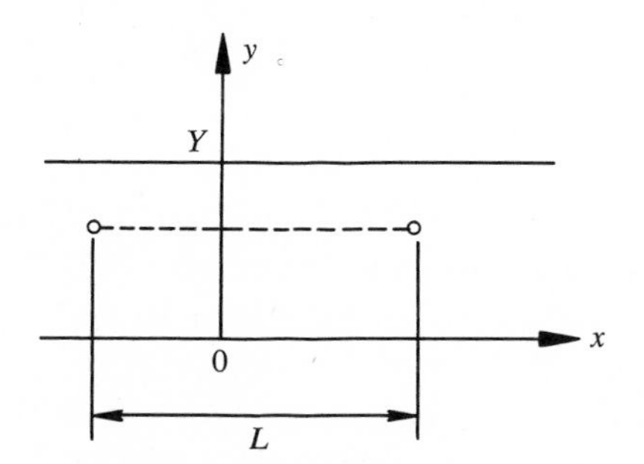

Figure 3.3 Measurement of $\Delta p/L$.

In the foregoing, both $f\langle t\rangle$ and $F\langle t\rangle$ are supposedly known from the description of the specific problem.

Summary

$$F\langle t\rangle = \frac{\Delta p}{L} = \frac{\partial p}{\partial x} \tag{3.7}$$

is a function of t only!

Conclusion

1. In order for our ansatz to work (or be able to pull through), the pressure difference between the sections of the channel (or the pressure gradient in the flow direction) must be a constant at any instant, and with known variation with time. (This is practically true for long channels whose ends are connected to a pumping device, if we deal only with the portion far away from the device.)
2. If we find that the flow in a portion of channel is parallel to the channel wall (which fact is equivalent to the flow being fully developed as indicated before), we can then be sure that the pressure gradient in the flow direction is constant at any instant, in that portion.

Substituting Eq. (3.7) into Eq. (3.4), we now have

$$\frac{\partial u}{\partial t}=\left[g_x-\frac{1}{\rho}F\langle t\rangle\right]+\nu\frac{\partial^2 u}{\partial y^2} \tag{3.8}$$

where $F\langle t\rangle$ is a given function associated with the performance of the pumping device.

The equation (3.3b) simply yields

$$p=\rho g_y y+F\langle t\rangle x+f\langle t\rangle \tag{3.9}$$

where

$$f\langle t\rangle = p\langle x_0, y_0; t\rangle-\rho g_y y_0-F\langle t\rangle x_0$$

is known if $p\langle x_0, y_0; t\rangle$ is given at any position (x_0, y_0). The key equation (3.8) is now to be solved for $u\langle y; t\rangle$ subject to the boundary conditions

$$u\langle 0; t\rangle = F_1\langle t\rangle \tag{3.10a}$$

$$u\langle Y; t\rangle = F_2\langle t\rangle \tag{3.10b}$$

and the initial condition

$$u\langle y; 0\rangle = F_3\langle y\rangle \tag{3.11}$$

(that is, the solution must link with the initial flow field known to exist at the start ($t=0$) of our investigation). If we succeed in this final task, our ansatz is correct for the specific problem at hand; if not, we have to go back to the very beginning and start all over again, probably by trying out another ansatz.

(We have taken painful care in arriving at this stage. In the existing texts, it usually takes only a few sentences to gloss over all the above points. The important thing to realize is that we have a physical phenomenon here we wish to investigate; we are not playing a mathematical game. Of course, we cannot keep this pace throughout; not only because time is short, but also because repetition kills interest. In the later part of this book, we will proceed more briskly. However, it is strongly recommended that you refer back to the present section and look upon it as an example of meticulous argument.)

We will presently list a number of cases where the ansatz really pulls us through. But before that, we wish to make two remarks. One, an ansatz is not an assumption. Two, numerous counterexamples exist where our ansatz would fail.

First, ansatz is no assumption. In sharp contrast with the language used in many texts, we have *never* assumed in this section that $v \equiv 0$, *nor* have we assumed that $\partial \mathbf{q}/\partial x \equiv 0$! We have assumed that the plates are parallel, rigid, impermeable; we have assumed that they move in a direction parallel to themselves; we have assumed that $\mathbf{g}$ is a constant; we have assumed that the fluid is incompressible, Newtonian, and with constant viscosity. But we have *not* assumed either $v \equiv 0$ or $\partial \mathbf{q}/\partial x \equiv 0$. What we really did was to look at the governing equations (1.9), (2.5a,b), and boundary conditions (3.1a,b,c,d); and to *conjecture* that the (still unknown) solution probably will have $v \equiv 0$ (or $\partial \mathbf{q}/\partial x \equiv 0$) because the geometry and the conditions seem to suggest it. It is a guessed property or aspect of the (still unknown) solution.

Assumptions are really a part of the description or specification of the given problem. You cannot change them without changing the problem: The flows of a compressible fluid between parallel plates constitute a totally different problem from the one we are now studying with the *assumption* of incompressibility. On the other hand, if an ansatz fails for a given problem, a new ansatz would not change the problem. An assumption always narrows down the scope of the problem, but an ansatz does no such thing, again in contrast with the implications of many texts.

Summary The generality of a given problem does not suffer from any ansatz.

Second, let us examine some counterexamples wherein our ansatz of $v \equiv 0$ (or, equivalently $\partial \mathbf{q}/\partial x \equiv 0$) fails.

1. *Channel with finite length*. If the channel has an inlet at $x = -a$ and an exit at $x = b$ (Fig. 3.4), the ansatz fails because v is not zero at the ends. This fact is known as the end effect. There is one possible practical exception here. If the height of the channel Y is very small compared with the length $a + b$, the middle portion of the flow, far away from the ends, will be approximately parallel (or fully developed). Our ansatz will then work to yield an approximate solution; but $\Delta p/L$ must be measured in this middle portion.

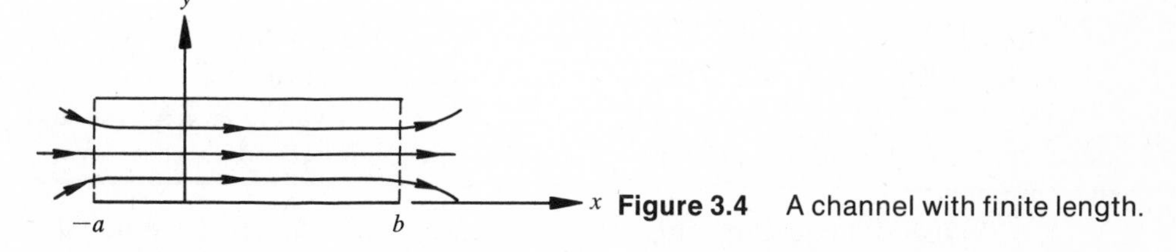

Figure 3.4 A channel with finite length.

2. *Nonuniform pumping*. The channel is long enough here, but the pumping device is effective only on a part of the fluid (Fig. 3.5).

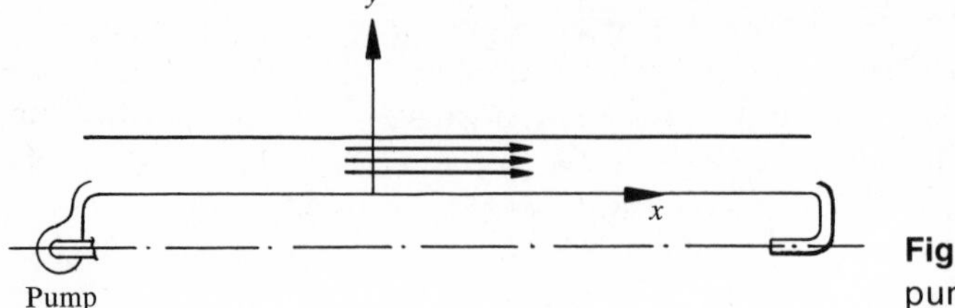

Figure 3.5 Nonuniform pumping.

Then, $\Delta p/L$ depends on y also, thus violating a consequence of our ansatz. If the nonuniform pumping is known to be smeared out in the mid-portion, an approximate use of our ansatz is again allowable, if $\Delta p/L$ is measured in the mid-portion.

3. *Channel with suction and blowing*. Suppose that the plates are permeable and there is (the same) fluid blowing into the channel at $y = 0$ with upward velocity $v_0\langle t\rangle$, and similarly sucked out at $y = Y$ with the same upward velocity $v_0\langle t\rangle$. Then, boundary conditions (3.1c,d) should be changed to

$$v\langle x, 0; t\rangle = v_0\langle t\rangle \tag{3.12a}$$

$$v\langle x, Y; t\rangle = v_0\langle t\rangle \tag{3.12b}$$

Now obviously the ansatz $v \equiv 0$ fails since it cannot take care of boundary conditions (3.12a,b). This, however, does not prevent us from trying another ansatz, $v\langle x, y; t\rangle \equiv v_0\langle t\rangle$. Again, this new ansatz is not an additional assumption. No generality is lost under it. We have only guessed that maybe the upward flow goes right through the channel, unchanged.

Let us see whether this new ansatz can possibly succeed. Equation (1.9) here again demands that the flow be fully developed; that is, $\partial u/\partial x \equiv 0$. Equations (2.15a,b) then become

$$\begin{cases} \dfrac{\partial u}{\partial t} + v_0\langle t\rangle \dfrac{\partial u}{\partial y} = g_x - \dfrac{1}{\rho}\dfrac{\partial p}{\partial x} + \nu \dfrac{\partial^2 u}{\partial y^2} & \text{(3.13a)} \\[2ex] \dfrac{dv_0\langle t\rangle}{dt} = g_y - \dfrac{1}{\rho}\dfrac{\partial p}{\partial y} & \text{(3.13b)} \end{cases}$$

Incidentally, if we adopt $\partial \mathbf{q}/\partial x \equiv 0$ as our new ansatz, we will end up with the same equations (3.13a,b). We will come back to these equations later. Here, we will only say that the two equations govern two dependent variables u and p; and it seems reasonable to expect that we can succeed in solving u and p under certain generous conditions.

In this last counterexample, we have also shown that, if an ansatz fails, we should not despair and give up; we should, instead, study the problem further and try new ansatz which may be successful.

3.2 STEADY CHANNEL FLOWS

To continue our study of Section 3.1, we will investigate here a class of steady flows—that is with $\partial(\quad)/\partial t \equiv 0$, where $(\quad)$ stands for all flow properties. To be exact, we wish to see whether we can solve the following equation:

$$\mu \frac{d^2u}{dy^2} = F - \rho g_x \tag{3.14}$$

where F is a known constant, subject to the boundary conditions

$$u\langle 0\rangle = 0 \tag{3.15a}$$

$$u\langle Y\rangle = U \tag{3.15b}$$

where U is another known constant (see Fig. 3.6).

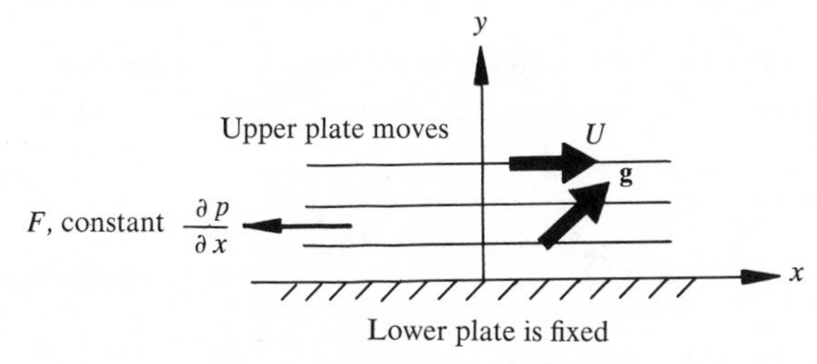

Figure 3.6 Steady channel flow. (Positive F means increasing p in the positive-x direction.)

The solution of Eq. (3.14) subject to boundary conditions (3.15a,b) is easily seen to be

$$u = \left(\frac{y}{Y}\right) U - \frac{Y^2}{2\mu}(F - \rho g_x)\left(\frac{y}{Y}\right)\left(1 - \frac{y}{Y}\right)$$

Or, if $U \neq 0$,

$$\frac{u}{U} = \frac{y}{Y} - \frac{Y^2}{2\mu U}(F - \rho g_x)\left(\frac{y}{Y}\right)\left(1 - \frac{y}{Y}\right) \tag{3.17}$$

$$= \frac{y}{Y} + P\left(\frac{y}{Y}\right)\left(1 - \frac{y}{Y}\right) \tag{3.18}$$

where

$$P = \frac{Y^2}{2\mu U}(\rho g_x - F) \tag{3.19}$$

is a dimensionless parameter. Also,

$$p = Fx + \rho g_y y + p_0 \tag{3.20}$$

where $p_0 = p\langle 0, 0\rangle$ is a measured value.

Our success here in finding u shows that our ansatz indeed works.

A plot of u/U versus y/Y for a number of P-values according to Eq. (3.18) is shown in Figure 3.7. It is seen that the velocity distribution is as follows:

$P = 0$: straight line
$P > 0$: curve showing a faster flow than the straight line

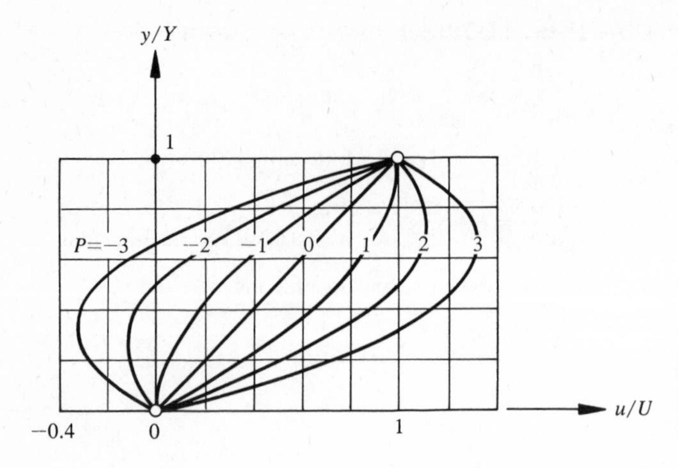

Figure 3.7 Velocity distributions for channel flows.

$P < 0$: curve showing an algebraically slower flow than the straight line, with possible region of backflow (that is, flow in the negative-x direction)

$-1 < P < 0$: curve showing no backflow

$P < -1$: curve showing always a region of backflow near the lower (stationary) plate

When $P < 0$, we have $F - \rho g_x > 0$. If we refer back to Eq. (3.8), we see that $\rho g_x - F$ is a force acting on the fluid element in the positive-x or the flow direction. Then a positive $F - \rho g_x$ is a force against the flow. There is, therefore, possible backflow for $P < 0$. Specifically, if $P < -1$, the magnitude of $F - \rho g_x$ against the flow is so large that the dragging of the upper plate in the flow direction is insufficient in counterbalancing it; and, a region of backflow really appears.

SPECIAL CASES

1. *Simple shear flow.* The case $P = 0$ is known as the simple shear flow, whose velocity distribution is seen to be a straight line.
2. *Plane Poiseuille flow.*[4] The case where $U = 0$ is called the plane Poiseuille flow. Here, from Eq. (3.16), we have

$$u = -\frac{Y^2}{2\mu}(F - \rho g_x)\left(\frac{y}{Y}\right)\left(1 - \frac{y}{Y}\right)$$

$$= A\left(\frac{y}{Y}\right)\left(1 - \frac{y}{Y}\right) \quad \textbf{(3.21)}$$

[4] The name *Poiseuille flow* without the modifier *plane* is reserved for the corresponding flow in a circular pipe, to be treated in Part 2. Although the following remarks really belong to the section on the *true* Poiseuille flow, we have moved them forward to here because they are of great importance, even to a beginner. Jean Louis Marie Poiseuille (1799–1869), a French physician, spent 11 years (1835–1846) observing blood flow in vessels and became thus the first biomedical fluid dynamicist. In his 1841 paper, "Experimental researches on the motion of liquids in tubes of very small diameters" (in French), Poiseuille proposed an empirical formula for the volume flow rate, which is now known as the Hagen-Poiseuille efflux formula. (Gotthilf Heinrich Ludwig Hagen, 1797–1884, was a German hydraulic

This is obviously a parabola. It represents a flow in the $\pm x$-direction if

$$A = -\frac{Y^2}{2\mu}(F - \rho g_x) \gtrless 0$$

In other words, the flow is entirely in one direction or the other. There is no longer a region of backflow (Fig. 3.8). If $A = 0$, $u \equiv 0$, and there is no flow at all; the pressure distribution is then just hydrostatic.

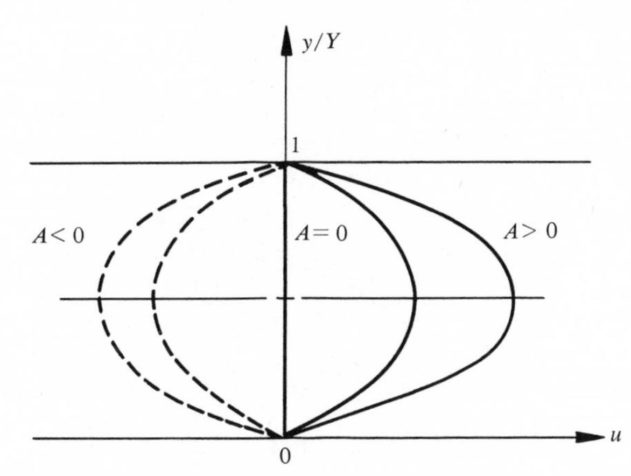

Figure 3.8 Channel with fixed walls.

3.3 STEADY FILM FLOWS OVER A FLAT PLATE

Closely associated with the flows between plates, there is one class of problems of practical interest—that is, flows of liquid layers (films) over a flat plate under gravity. For such problems, preliminary observations indicate that,

---

engineer who investigated water flow through circular pipes.) On the theoretical side, Stokes established the same formula based on the N-S equations (as will be shown in Part 2). It is interesting to learn that, not aware of Poiseuille or Hagen's measurements, Stokes compared his theoretical prediction with available experimental data and found no agreement whatsoever! Apparently very confident in his analysis (luckily for us), Stokes did not attach any importance to this disagreement, although he must have been disappointed. Today we know that the experimental data available to Stokes were all for turbulent flows; no wonder they did not fit. On the other hand, the agreement between Hagen-Poiseuille's empirical formula and Stokes' theoretical prediction should not be naively taken as experimental confirmation of the stress-rate-of-strain relation, Eqs. (2.12a,b,c), either. After all, other relations may give you the same formula for the volume flow rate through a circular pipe.

To summarize, Stokes did *not* know of any relevant or reliable experimental work before he formulated his theory; and today, it is difficult to single out any specific situation that holds the honor as the *crucial* test of Stokes' theory of viscous flow. Our faith in the N-S equations is actually based on the close correlation between theory and experience in millions of cases, of which the Poiseuille flow is but one. If you now refer back to the section on "Mathematical Models" in *To the Student*, preceding Chapter 1, you will see that the steps taken in model-building charted there, though very commendable, are not historically founded. As a matter of fact, should Stokes follow the steps, there would be no viscous-flow equations named after him.

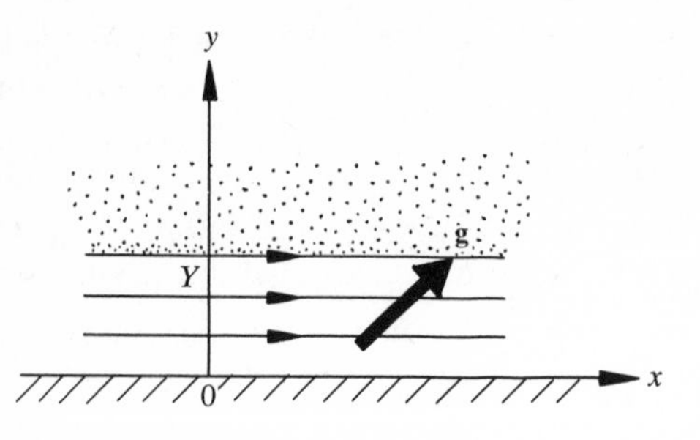

Figure 3.9 Steady film flow.

under certain conditions (essentially when the flow is slow), the liquid surface is (or approximately is) flat and parallel to the bottom plate (Fig. 3.9). We will assume that this is the case in this section. (Under other conditions, there may be large waviness on the surface; the result of this section is then invalid.)

To be exact, there is always some fluid above the liquid surface (for example, its own vapor or air). One should, originally, solve the flows of both fluids at the same time. But, for the usual cases, it turns out to be a good approximation to assume that the liquid layer is incapable of influencing the flow of the fluid above it. We then consider the flow of the fluid above the liquid, particularly the shear stress it exerts on the liquid layer, as known. The presence of this fluid is thus felt by the liquid layer through this known shear stress (and the fluid pressure). For example, if the air above the liquid layer is stationary or in negligible motion, this shear stress is zero; the liquid layer is then said to be in a free-surface flow with its top face free. If the air is in brisk motion (a strong wind), it is then assumed that we can measure the shear stress it exerts on the liquid. Both cases are now included in the following prescribed boundary conditions:

$$y = Y: \quad \mu \frac{\partial u}{\partial y} = \text{a given constant; or,} \quad \frac{\partial u}{\partial y} = \tau \tag{3.22}$$

where the given τ vanishes for a free-surface flow.

The assumption that the top face is flat and parallel to the bottom plate leads to the condition of no velocity component in the y-direction:

$$y = Y: \quad v = 0 \tag{3.23}$$

With these preliminary points cleared up, we can then proceed in exactly the same manner as in the previous sections; that is, an ansatz $v \equiv 0$ leads to the following key equation:

$$\nu \frac{d^2u}{dy^2} + g_x = \frac{F}{\rho} \tag{3.24}$$

subject to the boundary conditions

$$u\langle 0 \rangle = 0 \tag{3.25}$$

$$\left.\frac{du}{dy}\right|_{y=Y} = \tau \tag{3.26}$$

The major difference between the problem here and that of Section 3.2 lies in the boundary condition (3.26). (If the influence of the fluid over the liquid film is represented by the known velocity with which the liquid surface is being dragged along or if one measures or is given the air velocity instead of the air shear at the liquid surface, the present problem will be mathematically the same as that of Section 3.2. Physically, they would still be different, of course, since the top surface is not rigid here.)

The pressure distribution is, similarly to Eq. (3.20),

$$p = Fx + \rho g_y y + \text{constant}$$

or

$$p = \rho g_y y - \rho g_y Y + p\langle x, Y\rangle$$

But p in the liquid right at $y = Y$ is equal to p in the adjacent fluid at $y = Y$; that is,

$$p\langle x, Y\rangle = p_{\text{sur}}$$

where p_{sur} is the p value in the surroundings. This latter fact[5] holds here because (a) the liquid surface is not curved and thus exhibits no difference in normal stresses on the two sides of the surface due to surface tension and (b) $\partial v/\partial y$ vanishes, so the normal stress is just the negative of p, Eq. (2.14b). Usually, p in the surroundings (for example, the atmospheric pressure) is very nearly a constant; then $F = \partial p/\partial x \cong 0$.

The solution of Eq. (3.24), with $F = 0$, under conditions (3.25) and (3.26) is easily seen to be, with nonzero g_x,

$$\frac{u}{g_x Y^2/\nu} = -\frac{1}{2}\left(\frac{y}{Y}\right)^2 + \left(\frac{\nu\tau}{g_x Y} + 1\right)\left(\frac{y}{Y}\right) \tag{3.27}$$

which is sketched for several values of $(\nu\tau/g_x Y)$ in Figure 3.10.

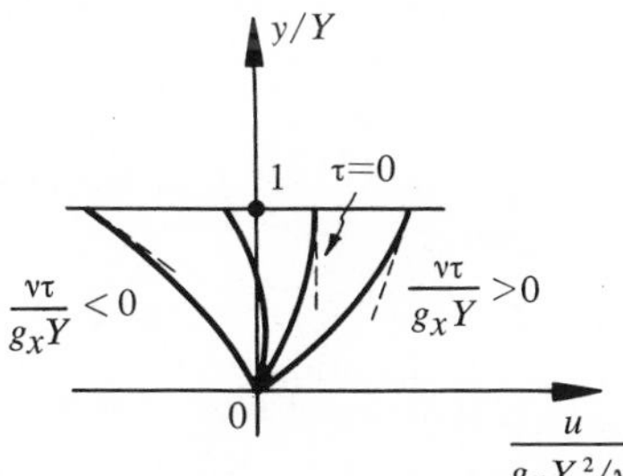

Figure 3.10 Velocity distributions for film flows.

If $g_x = 0$, Eq. (3.27) becomes just

$$u = \tau y \tag{3.28}$$

which is equivalent to a simple shear flow with $U = \tau Y$. If, in addition, $\tau = 0$, the film is at rest.

---

[5] We have incurred here implicitly the condition that both the shear and the normal (if no surface tension) stresses must vary continuously across an interface between two fluids.

The pressure distribution in the general case is given by

$$p = p_{sur} - \rho g_y (Y - y) \tag{3.29}$$

3.4 STEADY CHANNEL FLOWS WITH BLOWING AND SUCTION

In this section we wish to take up the steady case of the problem discussed as counterexample (3) in Section 3.1. To repeat, we consider the flow between two *permeable* plates, $y = 0$ and Y, with the same fluid blowing with constant velocity v_0 perpendicular to the plate at $y = 0$ (if v_0 is negative, the fluid is sucked out), and sucked out with the same v_0 at $y = Y$ (flowing downward if $v_0 < 0$). The upper plate moves with U in the x-direction; the lower one is fixed (Fig. 3.11).

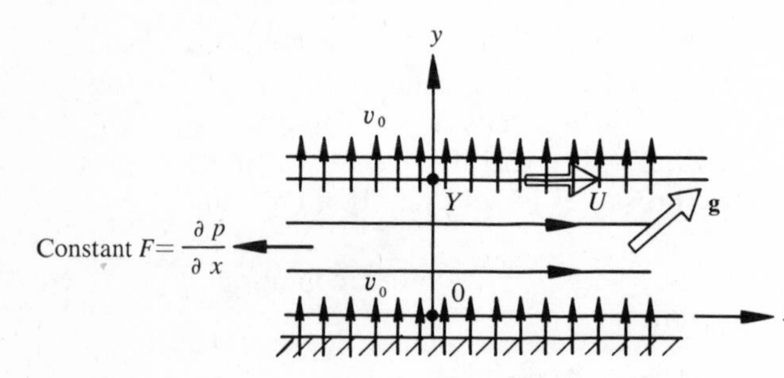

Figure 3.11 Channel flow with blowing and suction at the walls.

As indicated in Section 3.1, the ansatz $v = v_0$, a constant, seems to be promising. It actually leads to the following equations, which should be compared with Eqs. (3.13a,b):

$$v_0 \frac{du}{dy} = g_x - \frac{1}{\rho}\frac{\partial p}{\partial x} + \nu \frac{d^2 u}{dy^2} \tag{3.30a}$$

$$\frac{\partial p}{\partial y} = \rho g_y \tag{3.30b}$$

An argument similar to that used in relation to Eq. (3.7) yields again

$$\frac{\partial p}{\partial x} = \text{constant}, F$$

Then Eq. (3.30a) becomes

$$\mu \frac{d^2 u}{dy^2} - \rho v_0 \frac{du}{dy} = F - \rho g_x \tag{3.31}$$

Equation (3.30b) then gives

$$p = Fx + \rho g_y y + p_0 \tag{3.20}$$

where $p_0 = p\langle 0, 0\rangle$.

Now Eq. (3.31) is to be solved subject to the conditions

$$\begin{cases} u\langle 0\rangle = 0 & \text{(3.15a)} \\ u\langle Y\rangle = U & \text{(3.15b)} \end{cases}$$

If $v_0 = 0$ (that is, impermeable plates), we get right back to Eq. (3.14) and Section 3.2. If $v_0 \neq 0$, Eq. (3.30a) can be easily solved under boundary conditions (3.15a,b) to yield

$$\frac{u}{U} = \left[1 + \frac{C}{v_0/U}\right] \frac{1 - e^{Re(v_0/U)(y/Y)}}{1 - e^{Re(v_0/U)}} - \frac{C}{v_0/U}\left(\frac{y}{Y}\right) \tag{3.32}$$

where

$$C = \frac{F - \rho g_x}{\rho U^2/Y} \tag{3.33}$$

and

$$Re = \frac{\rho U Y}{\mu} \tag{3.34}$$

are two dimensionless parameters.

One must remember that Eq. (3.32) holds only where there is blowing at one plate and suction at another, and where the strengths of blowing and suction are the same. If there is blowing or suction at both plates, or where the strengths are not the same, the problem becomes totally different (and very complicated).

We will postpone the disçussion of this flow until we are ready to go into boundary-layer theory, since this simple flow happens to be a key to the understanding of the complicated theory of boundary layers. (See Section 6.1.)

CHAPTER 4

MORE EXACT SOLUTIONS

4.1 RAYLEIGH PROBLEM

In this chapter we will discuss further exact solutions, which are more involved than those of Chapter 3. As our first topic we will treat the famous Rayleigh problem (also known as Stokes' first problem[1]):

Consider the flow above a flat plate $y = 0$ (Fig. 4.1). At the start of our investigation, $t = 0$, the fluid and the plate are both at rest; that is, the initial condition to be satisfied is

$$t = 0: \quad \mathbf{q} = 0 \tag{4.1}$$

At $t = 0$, the plate is suddenly jerked parallel to itself in the x-direction so that it moves with a constant speed U for $t > 0$; that is, there is a boundary condition

$$y = 0: \quad u = U \qquad t > 0 \tag{4.2}$$

In this problem we will assume that the gravitational force is negligible.

[1] This was treated in Stokes' 1851 paper, "On the effect of the internal friction of fluids on the motion of pendulums," but was usually named after Rayleigh (John William Strutt, Lord Rayleigh, 1842–1919, English physicist and Nobel prize winner). Not being particularly priority-minded, we will use in this book the conventional association with Rayleigh.

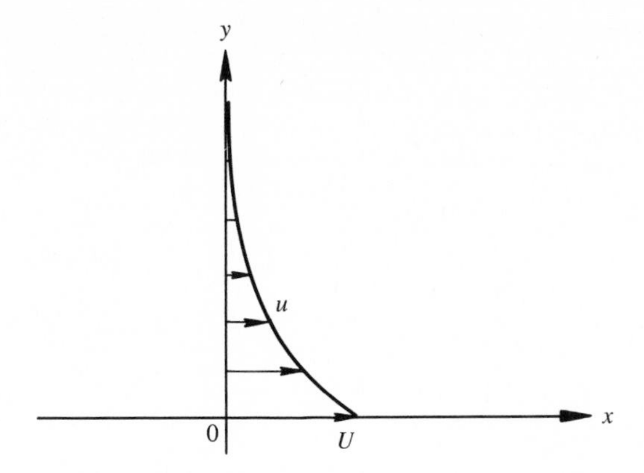

Figure 4.1 Rayleigh problem.

Similarly to Section 3.1, the ansatz of $v \equiv 0$ yields (see Eq. (3.4))

$$\begin{cases} \dfrac{\partial u}{\partial t} = -\dfrac{1}{\rho}\dfrac{\partial p}{\partial x} + \nu \dfrac{\partial^2 u}{\partial y^2} & \textbf{(4.3a)} \\ \dfrac{\partial p}{\partial y} = 0 & \textbf{(4.3b)} \end{cases}$$

Equation (4.3b) means that

$$p = p\langle x, t\rangle \text{ only}$$

being independent of y. So we can evaluate p at $y = \infty$ and obtain

$$p\langle x, t\rangle = \lim_{y \to \infty} p$$

Let us assume that p at $y = \infty$ is kept constant with respect to x and t (for example, exposed to atmospheric pressure). Then,

$$p\langle x, t\rangle = p_\infty, \text{ a constant} \tag{4.4}$$

(This means that the pressure throughout the fluid at all instants is a constant.)

Equation (4.3a) then becomes

$$\frac{\partial u}{\partial t} = \nu \frac{\partial^2 u}{\partial y^2} \tag{4.5}$$

to be solved subject to the conditions

$$\begin{cases} u\langle y, 0\rangle = 0 & \textbf{(4.6a)} \\ u\langle 0, t\rangle = U & \textbf{(4.6b)} \end{cases}$$

The success or failure of the ansatz $v \equiv 0$ now depends on whether we can solve for u here. To try our hands at this task, we will introduce still another ansatz:

$$u = g_1\langle t\rangle f\langle y g_2\langle t\rangle\rangle \tag{4.7}$$

With respect to this step, let us reiterate briefly the following points:

1. Ansatz (4.7) is not an assumption; it is only an *a priori* speculation.
2. If we can find all the three functions g_1, g_2, and f such that Eq. (4.5) and conditions (4.6a,b) are satisfied and such that $\lim_{y \to \infty} u$ behaves in a physically reasonable manner (for example, it does not increase without bounds), ansatz (4.7) is successful.
3. If ansatz (4.7) fails to go through, we should try to guess again; it does *not* necessarily mean that our previous ansatz $v \equiv 0$ is also wrong. After all, the solution of Eq. (4.5) under conditions (4.6a,b) and with reasonable behavior at $y = \infty$ may very well assume a form other than that shown in ansatz (4.7).
4. The form (4.7) is general enough to warrant your keeping it in your toolbox. You should constantly remember to give it a try when you are confronted with the task of solving a partial differential equation—any partial differential equation, not just Eq. (4.5). It is not guaranteed that it will always work; but it is always worth trying.
5. Again, ansatz (4.7) is sufficiently general that you have considerable leeway in its use. The procedure of trying it, as will be shown, can be applied by anybody on totally different problems. So, it is not pulling a rabbit out of a hat. (If we had said "try the ansatz $u = f\langle y/\sqrt{t}\rangle$," that would be pulling a rabbit out of a hat!) It tells you, rather, *how to* pull a rabbit out of a hat. It even has a name: *similarity solution.*[2] To try to find a solution under such an ansatz is known as the *method of similarity solutions.* To see whether there are solutions of this form is known as *searching for similarity solutions.*

With Rayleigh's problem as an example, let us now attempt to search for similarity solutions. But, to exhibit the power of this ansatz, we will generalize the problem slightly so that the plate moves with a time-dependent speed $U\langle t\rangle$; that is, we replace boundary condition (4.6b) by

$$u\langle 0, t\rangle = U\langle t\rangle \tag{4.6c}$$

We will leave the form $U\langle t\rangle$ unspecified; and will, in the end, *find out* exactly what kinds of $U\langle t\rangle$ admit similarity solutions.

---

[2] The Russians call it automodel solution. Other names are self-similar solution, and symmetric solution. It is better not to ask how these names came about. It only confuses instead of clarifies the issue at this stage. We will consider it a traditional name for this particular ansatz, and that is all. Just for completeness, we will also give here the most general definition of the similarity solution. A similarity solution is an ansatz that reduces the number of independent variables in the governing equations by one or more; *if* such an ansatz works for the problem, we say that a similarity solution exists. In our present problem, the governing equation, Eq. (4.5), has two independent variables, y and t; but after the introduction of ansatz (4.7), Eq. (4.5) will eventually be shown to reduce to an ordinary differential equation, with one independent variable $\eta = y/\sqrt{t}$. The ansatz is therefore a similarity solution that reduces the number of independent variables by one.

To begin, let us write

$$\eta = yg_2\langle t\rangle \tag{4.8}$$

for brevity. The ansatz (4.7) then becomes

$$u = g_1\langle t\rangle f\langle\eta\rangle \tag{4.9}$$

If solutions of such form exist, we will have

$$\frac{\partial u}{\partial t} = \frac{dg_1}{dt} f\langle\eta\rangle + g_1\langle t\rangle \frac{df}{d\eta}\left[y\frac{dg_2}{dt}\right] \tag{4.10a}$$

$$\frac{\partial u}{\partial y} = g_1\langle t\rangle \frac{df}{d\eta} \cdot g_2\langle t\rangle \tag{4.10b}$$

$$\frac{\partial^2 u}{\partial y^2} = g_1\langle t\rangle g_2^2\langle t\rangle \frac{d^2 f}{d\eta^2} \tag{4.10c}$$

Substituting Eqs. (4.10a) and (4.10c) into Eq. (4.5), we have

$$f\langle\eta\rangle\left[\frac{1}{g_2^2\langle t\rangle} \cdot \frac{dg_1/dt}{g_1\langle t\rangle}\right] + \underbrace{yg_2\langle t\rangle}_{\eta} \cdot \left[\frac{dg_2/dt}{g_2^3\langle t\rangle}\right] \cdot \frac{df}{d\eta} = \nu\frac{d^2 f}{d\eta^2} \tag{4.11}$$

Whether similarity solutions exist or whether the ansatz works depends on whether we can solve Eq. (4.11) *under the conditions*. First, let us see whether we can solve Eq. (4.11). If we can, we will go on and look at the conditions. (Be sure not to forget the conditions, though!)

We can certainly solve Eq. (4.11) if it turns out to be an ordinary differential equation for f with respect to η. We then ask: Under what circumstances will Eq. (4.11) be an ordinary differential equation? (This is the key question to ask in search for similarity solutions.) The answer is, obviously, when

$$\left[\frac{1}{g_2^2} \cdot \frac{dg_1/dt}{g_1}\right] \quad \text{and} \quad \left[\frac{dg_2/dt}{g_2^3}\right] \quad \text{are constants} \tag{4.12}$$

We then examine these two requirements:

1. $\dfrac{dg_2/dt}{g_2^3} = \text{constant}$

Therefore

$$\frac{1}{g_2^2} = C_1 t + C_2$$

or

$$g_2 = \frac{\pm 1}{\sqrt{C_1 t + C_2}} = \frac{C_4}{\sqrt{t + C_3}}$$

But the coefficient $C_4 = \pm 1/\sqrt{C_1}$ can always be absorbed into the still-to-be-determined function f without any loss of generality.[3] So let us conclude that

$$g_2 = \frac{1}{\sqrt{t + C_3}}$$

that is,

$$\eta = \frac{y}{\sqrt{t + C_3}} \tag{4.13}$$

2. $$\frac{dg_1/dt}{g_1 g_2^2} = \frac{(dg_1/dt)(t + C_3)}{g_1} = \text{constant, } n$$

Therefore

$$\ln |g_1| = n \ln |t + C_3| + \ln C_5$$

or

$$|g_1| = C_5 |t + C_3|^n$$

that is,

$$g_1 = C_6 |(t + C_3)^n|$$

Again, the coefficient $C_6 = \pm C_5$ can be absorbed into $f\langle\eta\rangle$ without any loss of generality. So

$$g_1 = |(t + C_3)^n|$$

that is,

$$u = |(t + C_3)^n| \cdot f\langle\eta\rangle \tag{4.14}$$

Next, let us investigate the various conditions as per the ansatz (4.14):

1. $u\langle y, 0\rangle = 0$. This requires that

$$\eta = \frac{y}{\sqrt{C_3}}: \quad |C_3^n| f = 0$$

that is,

$$\eta = \frac{y}{\sqrt{C_3}}: \quad f = 0$$

If this is used in solving Eq. (4.11), the result would be $f\langle\eta; y\rangle$ with y as a parameter. This is not consistent with ansatz (4.9), which surmises that f is a function of η with no strings attached! So, the ansatz apparently failed to work. Remember, *however*, that C_3 is so far an arbitrary constant. The above conclusion that the ansatz fails is valid *unless* we choose $C_3 = 0$. With $C_3 = 0$, it is only required that $f\langle\infty\rangle < \infty$, which is consistent.[4] If we turn the whole argument around we will

---

[3] To be more specific,

$$f\langle y(C_4/\sqrt{t + C_3})\rangle = F\langle y/\sqrt{t + C_3}\rangle$$

is, anyway, a still-to-be-determined function F of $y/\sqrt{t + C_3}$. If we agree, through laziness, to denote F by our old symbol f, we will have absorbed C_4 into f.

[4] $f\langle\infty\rangle < \infty$ means only that $f\langle\infty\rangle$ is a *finite* constant; this constant is fixed by the behavior at $y = \infty$. However, $f\langle\infty\rangle = 0$ here if $n \leqslant 0$.

have the following, more appealing finding: The initial condition forces us to try[5]

$$\begin{cases} \eta = \dfrac{y}{\sqrt{t}} & \text{(4.15)} \\ u = t^n f\langle \eta \rangle & \text{(4.16)} \end{cases}$$

If the ansatz (4.9) works at all, it must work only in this particular form.

2. $u\langle 0, t\rangle = U\langle t\rangle$. Against the above finding, this requires that

$$\eta = 0: \quad f = \frac{U\langle t\rangle}{t^n}$$

This is compatible with $f = f\langle \eta \rangle$, without t involved as a parameter, *only* when $U\langle t\rangle$ is proportional to a power of t:

$$U\langle t\rangle = At^n \tag{4.17}$$

That is, the ansatz may work only when the plate moves according to a power law of time. We will dutifully assume that this is so. (If the plate moves in other fashions, no similarity solutions exist, and Eq. (4.5) must be solved by other methods.) Then, the boundary condition becomes

$$\eta = 0: \quad f = A, \text{ a given constant} \tag{4.18}$$

As a special case n may be zero. This is, of course, the original Rayleigh problem.

Summary For a plate speed $\propto t^n$ (n may be zero),[6] a solution of the form

$$u = t^n f\langle \eta \rangle$$

with

$$\eta = \frac{y}{\sqrt{t}}$$

may exist where f is the solution of the ordinary differential equation

$$\nu \frac{d^2 f}{d\eta^2} + \frac{1}{2}\eta \frac{df}{d\eta} - nf = 0 \tag{4.19}$$

subject to the conditions

$$\eta = 0: \quad f = A \tag{4.20a}$$

$$\eta = \infty: \quad f < \infty \tag{4.20b}$$

We inserted the word "may" because there is still the physically reasonable behavior of $\lim_{y\to\infty} u$ to be checked. If this also passes, we can remove this word

[5] The symbol of absolute value around t^n is now omitted. We stipulate that, in the following, t^n always denote the positive branch whenever ambiguity occurs (for example, when $n = \frac{1}{2}$ or $\frac{1}{4}$).

[6] Note that, if $n < 0$, the plate starts with an infinite speed. Although the situation is of no practicability, it is not necessarily of no value since it may still yield important information of a qualitative nature.

from our summary. If this does not pass at this very last stage, and the ansatz fails after all, we can only say: "What a pity!"

We have allowed U to be a function of t just to show you how admissible conditions can be determined as a by-product of our search for similarity solutions. If we had insisted on the original Rayleigh problem, that is,

$$u\langle 0, t\rangle = U, \text{ a constant}$$

this boundary condition would have forced $n = 0$ upon us. That is to say, the solution of the Rayleigh problem must be

$$u = f\langle \eta \rangle \tag{4.21}$$

with

$$\eta = \frac{y}{\sqrt{t}} \tag{4.22}$$

where f is the solution of

$$\nu \frac{d^2f}{d\eta^2} + \frac{1}{2}\eta \frac{df}{d\eta} = 0 \tag{4.23}$$

subject to the conditions

$$\eta = 0: \quad f = U \tag{4.24a}$$

$$\eta = \infty: \quad f = 0 \tag{4.24b}$$

Now, Eq. (4.23) can be easily integrated once to yield

$$\ln \left| \frac{df}{d\eta} \right| = -\frac{\eta^2}{4\nu} + \ln C$$

that is,

$$\frac{df}{d\eta} = \pm C e^{-\eta^2/4\nu}$$

Integrating once more, we have, writing D for $\pm C$,

$$f = D \int_0^\eta e^{-\eta^2/4\nu}\, d\eta + E$$

Condition (4.24a) then gives $E = U$, and

$$f = D \int_0^\eta e^{-\eta^2/4\nu}\, d\eta + U$$

Writing $\xi = \eta/2\sqrt{\nu}$, we have

$$f = D\sqrt{\nu\pi} \left\{ \frac{2}{\sqrt{\pi}} \int_0^{\eta/2\sqrt{\nu}} e^{-\xi^2}\, d\xi \right\} + U$$

The integral

$$\frac{2}{\sqrt{\pi}} \int_0^\xi e^{-\xi^2}\, d\xi$$

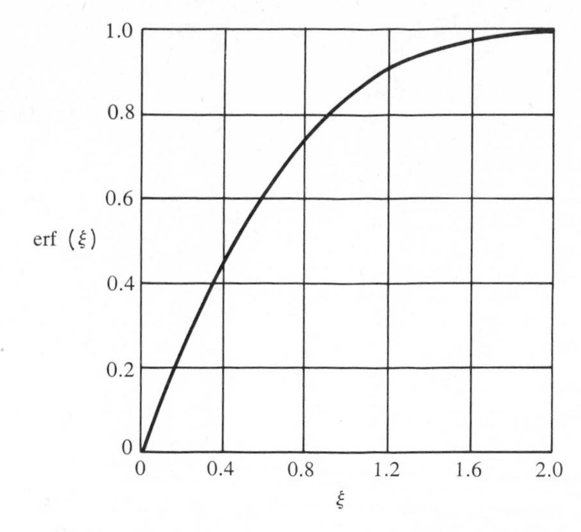

Figure 4.2 Error function.

as a function of ξ has already been calculated by others; and is called the error function, denoted by the symbol erf (ξ). (See Fig. 4.2.) Thus

$$f = D\sqrt{\nu\pi}\,\mathrm{erf}\left(\frac{\eta}{2\sqrt{\nu}}\right) + U$$

From the same source (see figure), it is also known that erf $(\infty) = 1$. So, condition (4.24b) demands that

$$0 = D\sqrt{\nu\pi} + U$$

that is,

$$D\sqrt{\nu\pi} = -U$$

and therefore

$$f = U\left[1 - \mathrm{erf}\left(\frac{\eta}{2\sqrt{\nu}}\right)\right]$$

This would then give

$$u = f\langle\eta\rangle = U\left[1 - \mathrm{erf}\left(\frac{y}{2\sqrt{\nu t}}\right)\right] \tag{4.25}$$

as the solution if $\lim_{y\to\infty} u$ behaves properly. But, indeed,

$$\lim_{\substack{y\to\infty \\ t<\infty}} u = U[1 - \mathrm{erf}(\infty)] = 0 \tag{4.26}$$

This means physically that the fluid at infinity is at rest at any finite time, or that the motion of the plate is damped down by viscosity as one moves away from the plate. So the last test on the behavior at infinity is passed. And we conclude that Eq. (4.25) is the velocity distribution for the Rayleigh problem. It is sketched for a few values of t in Figure 4.3. We see that the velocity damps down exponentially, away from the plate; but it takes a larger and larger distance, as t increases, to

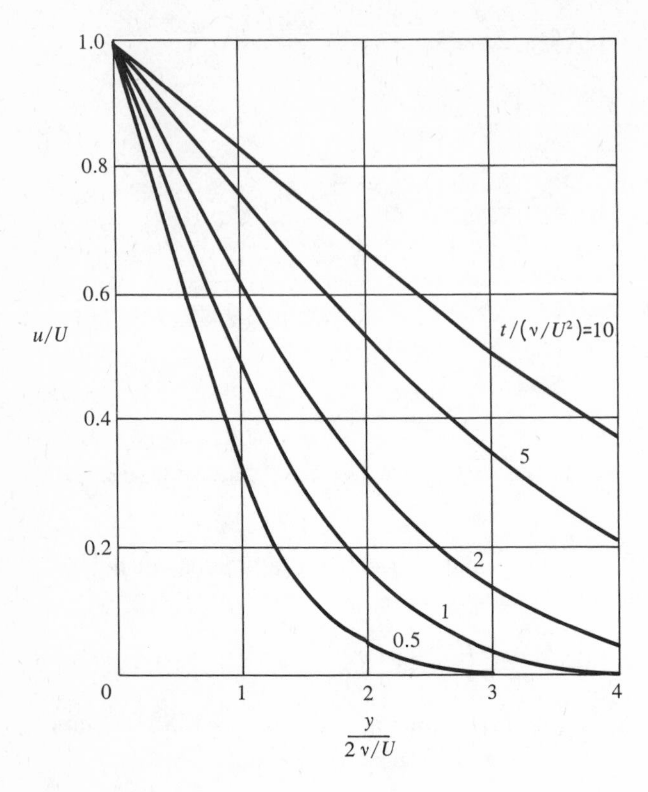

Figure 4.3 Velocity distributions for Rayleigh problem.

damp the flow to zero effectively. Right at $t = 0$, only the plate is in motion. But, as time goes on, the plate drags by viscosity more and more fluid with it. As $t \to \infty$, the entire fluid in $y > 0$ will eventually move with the plate; that is,

$$\lim_{\substack{t\to\infty \\ y<\infty}} u = U$$

4.2 STEADY STAGNATION-POINT FLOW

In this section we will investigate the steady flow impinging on a fixed, flat plate—that is, a flow where definitely $v \not\equiv 0$. Here we must start with Eq. (1.9) and the steady form of Eqs. (2.15a,b), but with gravity assumed negligible:

$$\begin{cases} \dfrac{\partial u}{\partial x} + \dfrac{\partial v}{\partial y} = 0 & \textbf{(1.9)} \\ u\dfrac{\partial u}{\partial x} + v\dfrac{\partial u}{\partial y} = -\dfrac{1}{\rho}\dfrac{\partial p}{\partial x} + \nu\left(\dfrac{\partial^2 u}{\partial x^2} + \dfrac{\partial^2 u}{\partial y^2}\right) & \textbf{(2.15c)} \\ u\dfrac{\partial v}{\partial x} + v\dfrac{\partial v}{\partial y} = -\dfrac{1}{\rho}\dfrac{\partial p}{\partial y} + \nu\left(\dfrac{\partial^2 v}{\partial x^2} + \dfrac{\partial^2 v}{\partial y^2}\right) & \textbf{(2.15d)} \end{cases}$$

The boundary conditions on the fixed plate are

$$y = 0: \quad u = 0 \tag{4.27a}$$

$$v = 0 \tag{4.27b}$$

The behavior of the flow at $y = \infty$ is left open for the time being. That is to say, we do not know as yet exactly what kind of flow impinges on the plate.

Since we cannot hope to succeed with the ansatz $v \equiv 0$ (there must be some vertical flow so that there is an impinging), we will try the next best thing suggested by the boundary condition (4.27b), which requires that $v = 0$ for $y = 0$ independently of x. That is, we will try the ansatz

$$v = -f\langle y \rangle \tag{4.28}$$

which means that v is independent of x for all y. (The use of the negative sign in the ansatz is of no particular importance, but it is a common convention.) Let us now try the ansatz out and see whether it works.

Substituting ansatz (4.28) into Eq. (1.9) yields

$$\frac{\partial u}{\partial x} = \frac{df}{dy}$$

or

$$u = x\left(\frac{df}{dy}\right) + G\langle y \rangle$$

But the situation now (see Fig. 4.4.) strongly suggests an ansatz of symmetry with

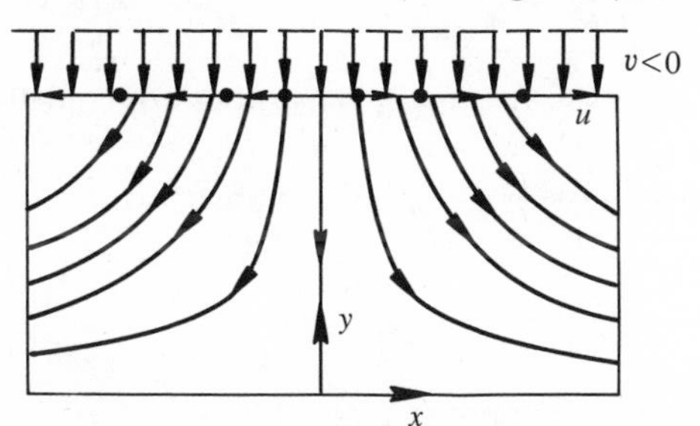

Figure 4.4 A stream approaching a flat plate.

respect to a vertical line, which may be chosen as our y-axis. So u must vanish on $x = 0$; that is,

$$G\langle y \rangle = 0$$

or

$$u = x\left(\frac{df}{dy}\right) \tag{4.29}$$

Substituting relations (4.28) and (4.29) into Eqs. (2.15c) and (2.15d), we obtain

$$-\left[\left(\frac{df}{dy}\right)^2 - f\frac{d^2f}{dy^2} - \nu\frac{d^3f}{dy^3}\right] = \frac{1}{\rho}\frac{\partial p/\partial x}{x} \tag{4.30a}$$

$$-\left(f\frac{df}{dy} + \nu\frac{d^2f}{dy^2}\right) = \frac{1}{\rho}\frac{\partial p}{\partial y} \tag{4.30b}$$

From Eq. (4.30b) it may be concluded that

$$\frac{p}{\rho} = -\int_0^y \left(f\frac{df}{dy} + \nu\frac{d^2f}{dy^2}\right) dy + \mathcal{G}\langle x\rangle \tag{4.31}$$

and

$$\frac{1}{\rho}\frac{\partial p}{\partial x} = \frac{d\mathcal{G}}{dx} \tag{4.32}$$

Substituting Eq. (4.32) into Eq. (4.30a), we conclude that

$$\begin{aligned}\left(\frac{df}{dy}\right)^2 - f\frac{d^2f}{dy^2} - \nu\frac{d^3f}{dy^3} &= a, \text{ a constant}\\ &= -\frac{d\mathcal{G}/dx}{x}\end{aligned} \tag{4.33}$$

since one side is a function of y only, whereas the other side is a function of x only, as can be seen from the derivation of Eq. (3.5). Equation (4.33) then yields

$$\frac{d\mathcal{G}}{dx} = -ax$$

or

$$\mathcal{G}\langle x\rangle = -\frac{ax^2}{2} + C$$

And, Eq. (4.31) gives

$$\frac{p}{\rho} = -\int_0^y \left(f\frac{df}{dy} + \nu\frac{d^2f}{dy^2}\right) dy - \frac{ax^2}{2} + C \tag{4.34}$$

where C is obviously $p\langle 0, 0\rangle/\rho$, which is supposedly known. The other constant a will be discussed later.

The key equation on which the success or failure of our ansatz depends is

$$\left(\frac{df}{dy}\right)^2 - f\frac{d^2f}{dy^2} - \nu\frac{d^3f}{dy^3} = a \tag{4.35}$$

This is to be solved, if possible, together with the boundary conditions

$$y = 0: \quad v = -f = 0, u = x\left(\frac{df}{dy}\right) = 0$$

that is,

$$\begin{cases} f\langle 0\rangle = 0 & \text{(4.36a)}\\ \left.\dfrac{df}{dy}\right|_{y=0} = 0 & \text{(4.36b)}\end{cases}$$

But what about the behavior at $y = \infty$?

In Figure 4.5, it is easy to see that if

$$\lim_{y\to\infty} \frac{d^nf}{dy^n} = A \neq 0$$

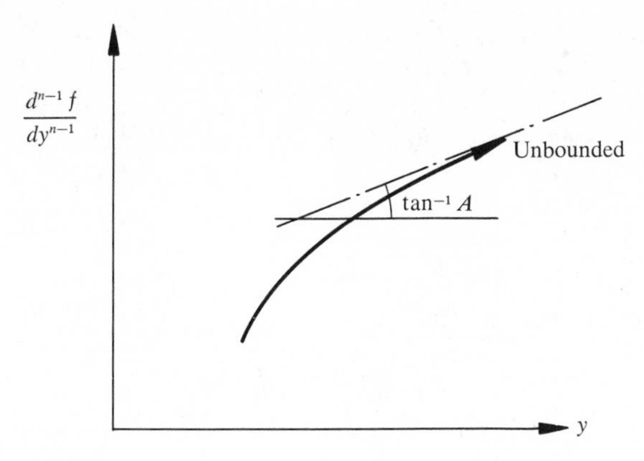

Figure 4.5 Proper behavior at infinity.

then

$$\lim_{y\to\infty}\frac{d^{n-1}f}{dy^{n-1}}$$

will "blow up." In Eq. (4.35), in order that no term blows up at $y=\infty$, we must have

$$\lim_{y\to\infty}\frac{d^3f}{dy^3},\frac{d^2f}{dy^2}=0 \tag{4.37}$$

which leaves

$$\left[\lim_{y\to\infty}\frac{df}{dy}\right]^2=a \tag{4.38}$$

Therefore, we see that $a>0$. For convenience let us write

$$a=b^2 \tag{4.39}$$

Then

$$\lim_{y\to\infty}\frac{df}{dy}=b \tag{4.40}$$

(a, or b, equal to zero is a trivial case since, then, Eq. (4.35) and boundary conditions (4.36) are satisfied by $f=0$; that is, there is no flow whatsoever.)

From condition (4.40), we have

$$\lim_{y\to\infty}u=bx \tag{4.41}$$

Is this behavior physically acceptable? Furthermore, condition (4.40) also asserts that $f\langle y\rangle$ at large y is almost a straight line:

$$\lim_{y\to\infty}f=by+b_0 \tag{4.42}$$

that is,

$$\lim_{y\to\infty}v=-by-b_0 \tag{4.43}$$

Now, is this behavior physically meaningful?

Certainly not, a purist would say, since condition (4.41) would give an unbounded u at infinite $|x|$, and (4.43) gives an unbounded v at $y=\infty$. But we cannot afford to be so pure all the time if only for the following reason: In the physical counterpart, the region does not extend to infinity anyway. Later on, we will investigate the associated physical picture more closely. But right now let us just say that the unbounded speed at infinity does not bother us because it happens outside our region of interest. Let us go ahead and try to solve Eq. (4.35) subject to the conditions (4.36) and (4.40), with b given as a description of what happens at infinity (or might happen, in view that we are not extending our interests to infinity).

Equation (4.35) is an ordinary differential equation, but unfortunately it is nonlinear. So to solve it is not a simple matter. Within the scope of this book, we can only say that it can be and has been solved numerically together with conditions (4.36) and (4.40) *if* $b > 0$. The result is plotted in Figure 4.6. As a by-product, the constant b_0 in relations (4.42) and (4.43) is also calculated and turns out to be

$$b_0 = -0.6479\sqrt{b\nu}$$

Thus

$$\lim_{y\to\infty} v = -by + 0.6479\sqrt{b\nu} \tag{4.44}$$

From Figure 4.6 it is seen that, at $y = 2.5\sqrt{\nu/b}$, u and v are already essentially bx and $-by - b_0$, respectively.

It is important, however, to point out that *no* solution to Eq. (4.35) with conditions (4.36) and (4.40) exists for $b < 0$. This can be seen as follows.

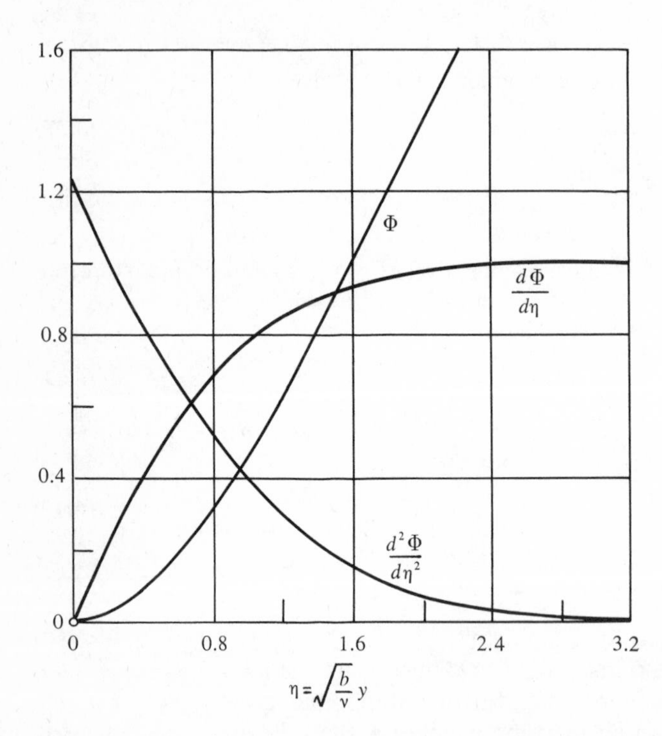

Figure 4.6 Solution of stagnation-point flow problem ($\Phi = f/\sqrt{b\nu}$).

We have already established the facts that for large y:

$$\frac{d^3f}{dy^3}, \frac{d^2f}{dy^2} \sim 0 \tag{4.37}$$

$$\frac{df}{dy} \sim b \tag{4.40}$$

$$f \sim by + b_0 \sim by \text{ (since } by \text{ is dominating)} \tag{4.45}$$

We would like to know, as one step further, how fast d^3f/dy^3 and d^2f/dy^2 approach zero as $y \to \infty$. To do this, let us substitute relations (4.40) and (4.45) into Eq. (4.35).

$$\text{For large } y\text{:} \quad b^2 - by\frac{d^2f}{dy^2} - \nu\frac{d^3f}{dy^3} (= a) = b^2$$

Solving for d^2f/dy^2, we have

$$\text{For large } y\text{:} \quad \frac{d^2f}{dy^2} \approx e^{-by^2/2\nu} \tag{4.46}$$

This is in keeping with requirement (4.37) only if $b > 0$. If $b < 0$, d^2f/dy^2 is not going to vanish as $y \to \infty$, but will grow unbounded. This means that it would be impossible for $\lim_{y\to\infty} df/dy$ to be bounded. Therefore, the solution being searched fails to exist if $b < 0$.

We have, at last, reached a stage where we can take a hard, physical look at what we have. What is the corresponding physical situation? Well, it is obviously some kind of flow against a flat plate, since

$$\lim_{y\to\infty} v = -by + 0.6479\sqrt{b\nu} \qquad b > 0 \tag{4.43}$$

indicates a dominant vertical speed toward the plate; that is, $-by$, far from the plate. (The corresponding flow away from the plate, $b < 0$, does not exist *under our ansatz* (4.28). But this does not mean that this flow fails to exist in some more general and complicated form.)

But exactly what kind of flow toward the plate is it? It may be better to start relating what kind of flow it is *not*, since wrong conclusions are easily made here. It is *not* a uniform flow at infinity (that is, at large y) against a flat plate since an x-velocity

$$\lim_{y\to\infty} u = bx \tag{4.41}$$

is always present.

It is often believed that this represents the viscous modification of an inviscid flow[7]

$$\begin{cases} u = bx & \text{(4.47a)} \\ v = -by & \text{(4.47b)} \end{cases}$$

---

[7] We will come back to more of this in Section 6.3, Case 3, which precedes Eq. (6.44), and in Section 7.3, Case 1.

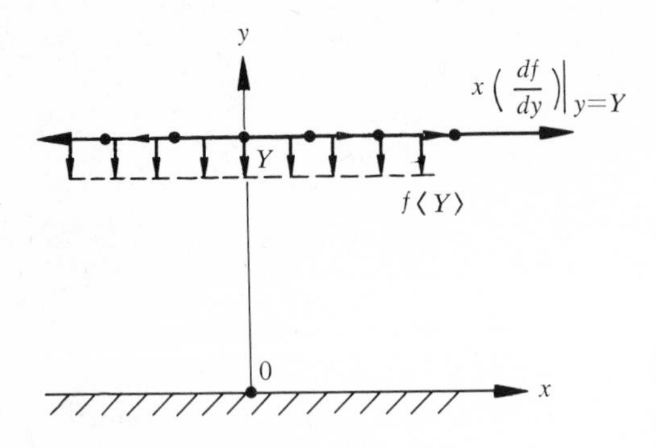

Figure 4.7 "Inlet plane" for the stagnation-point flow.

which satisfies Eq. (1.9), and Eqs. (2.15c,d) *with* $\nu = 0$, subject to boundary condition (4.27b) but not to (4.27a). It is thought that the flow field (4.47a,b) prevails at large y, while the solution presented in terms of $f\langle y\rangle$ comes in to save boundary condition (4.27a) near the plate. This interpretation is better, but is still *not* quite correct because of $b_0 = -0.6479\sqrt{b\nu}$. It is approximately good because b_0 is overwhelmed by by in relation (4.42) for large y, especially if ν is small. Since the flow represented by Eqs. (4.47a,b) is known as the (inviscid) stagnation-point[8] flow, the solution in terms of $f\langle y\rangle$ is traditionally also called the (viscous) stagnation-point flow.

The interpretation of the preceding paragraph would be exactly correct if the fluid at large y also recedes from the plate with a constant speed equal to $|b_0|$; that is, $0.6479\sqrt{b\nu}$, which is a very small number if ν is small.

Another correct interpretation would be to imagine another plane $y = Y$ and introduce the fluid at this "inlet plane" in such a fashion that

$$v = -f\langle Y\rangle \text{ (which depends on } b\text{)}$$

$$u = x\frac{df}{dy}\bigg|_{y=Y}$$

where $f\langle Y\rangle$ and $(df/dy)|_{y=Y}$ are read out of Figure 4.6. (See Fig. 4.7.) To realize such a situation is not so simple, however. One possible way would be to have a porous plate whose pores are judiciously inclined according to the ratio

$$\frac{f\langle Y\rangle}{[x(df/dy)_{y=Y}]}$$

The pores must actually be denser than shown

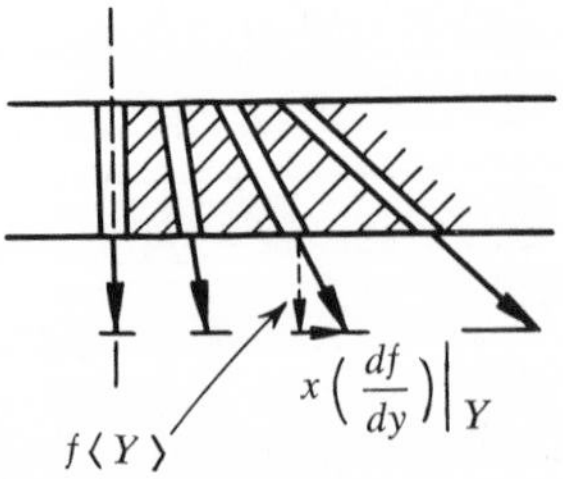

$f\langle Y\rangle$ $x\left(\frac{df}{dy}\right)\Big|_Y$

Figure 4.8 Realization of the "inlet plane."

[8] Referring to the point (0, 0), where $u, v = 0$.

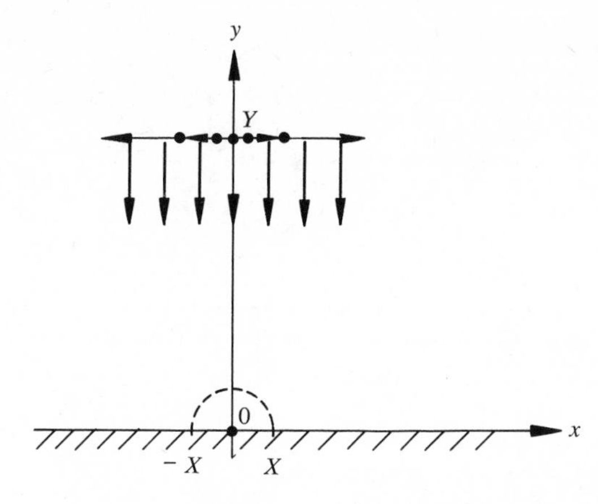

Figure 4.9 Approximate realization of the stagnation-point flow.

and blow with vertical strength $f\langle Y\rangle$. (See Fig. 4.8.) This is clearly very impractical, if not impossible. There is one approximation variation on this—that is, to restrict our interest only to the neighborhood of $(0, 0)$, say $-X < x < X$. Then if one blows hard *vertically* at $y = Y$ where $Y \gg X$, the flow within this neighborhood would be approximately the one we predicted by way of $f\langle y\rangle$, since the horizontal velocities required at $y = Y$ to make the correspondence exact are small. (See Fig. 4.9.) As a matter of fact, in practical applications it is always this approximate realization that is implicitly meant when one refers to the (viscous) stagnation-point flow.

CHAPTER 5

SOME APPROXIMATE SOLUTIONS

5.1 NONDIMENSIONALIZATION AND GOVERNING DIMENSIONLESS PARAMETERS

Exact solutions of the Navier-Stokes equations, together with the equation of continuity, are few in number. To learn more about the viscous flow, we must abandon the vain hope of solving all problems in an exact manner and instead take up the study of approximate solutions. To do this rationally and systematically we will adopt the following program: As a preliminary step, we will put the governing equations in dimensionless forms and recognize a number of dimensionless parameters. Then in the next section we will explain how this nondimensionalization should be handled in order that rational approximations can be introduced.

In the most primitive sense, nondimensionalization of our problem only means measuring our physical quantities with standards convenient to us. This is of course nothing fancy, being the first step in any physical investigation. The important thing here is that we do *not* find it convenient to measure length in meters, time in seconds, and so on. But instead, we find it convenient to measure our physical quantities in terms of some constant, *characteristic quantities* of the problem under investigation. For example, in a plane flow past a (long) circular cylinder, it is convenient to measure lengths with the diameter of the cylinder as a unit, the velocities by the speed far upstream, and so on. These characteristic quantities used as units need not even be concrete. For example, if we wish to investigate the flow around a certain instant (say, 15.6 seconds) after the start, we can use this *time of observation* as our characteristic time. As another example, we may use $t^* = \nu/L^{*2}$ as our characteristic time, where L^* is the characteristic length chosen and ν is the (constant) kinematic viscosity.

If we can solve a problem exactly, it really does not matter whether we nondimensionalize the problem or not. The nondimensionalization only systematizes the numerical work after the analysis is finished.[1] It also does not matter what characteristic quantities are used in the nondimensionalization. Nondimensionalization with *properly chosen* characteristic quantities becomes important only when we cannot solve a problem exactly, and wish to investigate instead its asymptotic behavior as some of its parameters become vanishingly small. We will discuss this in detail in the next section under the name *ordering process*.

For the present, we only have to introduce the format of nondimensionalization with respect to unspecified quantities $(\ )^*$. Let us introduce nondimensional quantities (all primed) as follows:

$t' = t/t^*$

$x' = x/x^*$ (that is, x^* is used in measuring length in the x-direction)

$y' = y/y^*$ (y^* may or may not be the same as x^*)

$u' = u/u^*$

$v' = v/v^*$

$p' = p/p^*$

Substituting these into Eqs. (1.9) and (2.15a,b), we have

$$\frac{\partial u'}{\partial x'} + \left(\frac{x^*}{y^*}\right)\left(\frac{v^*}{u^*}\right)\frac{\partial v'}{\partial y'} = 0 \tag{5.1}$$

$$\left(\frac{x^*}{u^*t^*}\right)\frac{\partial u'}{\partial t'} + u'\frac{\partial u'}{\partial x'} + \left(\frac{x^*}{y^*}\right)\left(\frac{v^*}{u^*}\right)v'\frac{\partial u'}{\partial y'}$$
$$= \frac{g_x x^*}{u^{*2}} - \left(\frac{p^*}{\rho u^{*2}}\right)\frac{\partial p'}{\partial x'} + \left(\frac{\nu}{u^*x^*}\right)\left[\frac{\partial^2 u'}{\partial x'^2} + \left(\frac{x^*}{y^*}\right)^2\frac{\partial^2 u'}{\partial y'^2}\right] \tag{5.2a}$$

[1] The practical influence of this systematization, however, is far-reaching; examples include *dimensional analysis* and *theory of modeling* (for wind tunnels, towing tanks, pilot plants, and the like).

$$\left(\frac{v^*}{u^*}\right)\left(\frac{x^*}{u^*t^*}\right)\frac{\partial v'}{\partial t'}+\left(\frac{v^*}{u^*}\right)u'\frac{\partial v'}{\partial x'}+\left(\frac{x^*}{y^*}\right)\left(\frac{v^*}{u^*}\right)^2 v'\frac{\partial v'}{\partial y'}$$

$$=\left(\frac{g_y}{g_x}\right)\frac{g_x x^*}{u^{*2}}-\left(\frac{x^*}{y^*}\right)\left(\frac{p^*}{\rho u^{*2}}\right)\frac{\partial p'}{\partial y'}+\left(\frac{v^*}{u^*}\right)\left(\frac{\nu}{u^*x^*}\right)\left[\frac{\partial^2 v'}{\partial x'^2}+\left(\frac{x^*}{y^*}\right)^2\frac{\partial^2 v'}{\partial y'^2}\right] \tag{5.2b}$$

We thus have, altogether, seven dimensionless governing parameters:[2]

$$\left(\frac{x^*}{y^*}\right)\quad\left(\frac{v^*}{u^*}\right)\quad\left(\frac{g_y}{g_x}\right)\quad\left(\frac{x^*}{u^*t^*}\right)\quad\left(\frac{g_x x^*}{u^{*2}}\right)\quad\left(\frac{p^*}{\rho u^{*2}}\right)\quad\left(\frac{\nu}{u^*x^*}\right)$$

The first three groups are just geometric, speed, and body-force ratios. Their physical significance is clear, being simply a description of the geometry, the flow inclination, and the body-force inclination. The last four have personal names associated with them: The first one is sometimes called the generalized Strouhal[3] number, and denoted by St; the second one, the *reciprocal of* the Froude[4] number, $1/Fr$ (that is, $Fr = u^{*2}/g_x x^*$); the third one is sometimes called the Euler number, Eu; the fourth one, the reciprocal of the Reynolds[5] number, $1/Re$ (that is, $Re = u^*x^*/\nu$). (Note that all these are defined here with respect to the x-direction, which is taken to be the main flow direction. The same names also apply to the numbers made up of corresponding quantities in the y-direction, if they appear.) Their physical interpretations will be discussed in the next section.

---

[2] Helmholtz was apparently the first to obtain in this manner the dimensionless parameters governing *dynamic similitude*. He presented the procedure in 1873 to the Berlin Academy of Sciences. You may have noticed that the famous Pi theorem is not needed here. It is not needed because we are already quite knowledgeable about the problem; in fact, we already know its governing partial differential equations. Resort to the Pi theorem is necessary only when our understanding is really primitive, and we know no more than what physical quantities are involved in the problem.

[3] V. Strouhal first used this parameter in an 1878 paper on singing wires (with $t^* = 1/\text{frequency}$).

[4] William Froude (1810–1879), an Englishman who "retired" at 36 to do research on ship rolling and resistance, built a 250-foot towing tank for the British Admiralty eight years before his death. His son, Robert Edmund (1846–1924), continued his work. It is ironic that Froude never used the dimensionless parameter now named after him.

[5] Osborne Reynolds (1842–1912), great English engineer, worked more like a physicist (or, nowadays, an engineering scientist). With a rare combination of experimental ability and conceptual insight, he contributed to many fields, including heat transfer and navigation. His three-year course of lectures was organized for all engineering students at the Victoria University in Manchester, regardless of their major fields; and was thus the originator of the *Engineering Core Course* of today. He first used the parameter now named after him in his 1883 paper, "An experimental investigation of the circumstances which determine whether the motion of water shall be direct or sinuous, and of the law of resistance in parallel channels." In two books published by the Cambridge University Press, *On the Inversion of Ideas as to the Structure of the Universe* (1902) and *The Sub-Mechanics of the Universe* (1903), Reynolds put forward his belief that the space is made of solidly packed spheres with a diameter of $1/(7\times10^{11})$ of the wavelength of light; somewhat like wet sand packed tight. He thought of material particles as bubbles of nothing, moving about in this beachlike universe, causing distortions in the pack; to him, gravitation is just the pressure force incurred by these distortions! This view endeared him to the soil mechanists, of course, and brought him an everlasting fame in general continuum mechanics; but it also earned him an entry into Martin Gardner's amusing book, *Fads and Fallacies in the Name of Science*, Dover, New York, 1957.

5.2 THE ORDERING PROCESS AND THE USE OF ASYMPTOTICS AS AN APPROXIMATION

As far as obtaining the relevant dimensionless parameters is concerned, it does not matter what characteristic values are chosen for the various quantities. However, if a little time is spent in finding out more about the problem on hand, one can have some feeling about the rough *orders of magnitude* of the various quantities involved. We can then choose the characteristic values according to the orders of magnitude. Such *nondimensionalization according to orders of magnitude* is called the *ordering process*. But what exactly is the order of magnitude of a variable quantity? We cannot answer the question "exactly," since the idea involved is inherently rather loose.

Roughly speaking, for the dependent variables, u, v, and p, if we see that the maxima of their absolute values in the flow field are *no greater* than $\overline{u}$, $\overline{v}$, and $\overline{p}$, respectively, we say that u, for example, is of the same or smaller order of $\overline{u}$, and write

$$u \sim \Box(\overline{u}) \text{ or smaller}$$

where "$\Box$" is read "square oh." ("Smaller" here means smaller in the order of "largeness." In the literature, the word "smaller" is often replaced by "higher." It then means higher in the order of smallness.) To shorten the writing, we will absorb the phrase "or smaller" and write $u \sim O(\overline{u})$ where O is read "big oh." (That is, $O(\ \ ) \equiv \Box(\ \ )$ *or* smaller. $u \sim O(\overline{u})$ then means "u is of the same or smaller order of $\overline{u}$.") In addition, we will also introduce $u \sim o\,(\overline{u})$, where o is the "small oh," to mean "u is of smaller order of $\overline{u}$." Then, $O(\ \ ) \equiv \Box(\ \ )$ *or* $o\,(\ \ )$. Similar statements also apply to v, p, and so on.

To summarize, "$u \sim \Box(\overline{u})$ or smaller" means that $|u(x, y; t)| \leqslant \overline{u}$ for all x, y, and t of interest; and similarly for the other quantities.

Then, in our nondimensionalization, if we choose $u^* = \overline{u}$, and so on, we will have

$$u' = \frac{u}{u^*} = \frac{u}{\overline{u}} \sim \Box(1) \text{ or smaller}$$

and so on.

Now, be sure that you really understand this. $u' \sim \Box(1)$, *without the "smaller" tagged along*, means that $|u'|$ may be as small as 0, but it *does* come to a value as big as 1, somewhere in the region of interest at some instant. Furthermore, it is not necessary to choose $u^* = \overline{u}$, and so on, exactly. u^* can be as high as $1.5\overline{u}$ or $2\overline{u}$, or as low as $0.5\overline{u}$ or $0.3\overline{u}$. In some instances, u^* can even be $10\overline{u}$ or $0.1\overline{u}$. As long as you do not use such outlandish figures as $100\overline{u}$ or $0.01\overline{u}$, it is *usually* all right; your u' is *still* $\sim \Box(1)$! Only, it means now that $|u'|$ may be as low as 0, but $|u'|_{max}$ *cannot be too far removed from 1* (usually from 0.5 to 3, but may be from 0.1 to 10). We admit that this is not a very precise notion. But such a rough idea is already useful.

Next, what about the independent variables x, y, and t? Well, we choose for them x^*, and so on, such that

$$\frac{\partial(\quad)'}{\partial x'}, \text{ etc.}, \sim \square(1) \text{ or smaller}$$

Since $\partial(\quad)/\partial x$, for example, may not have the same orders of magnitude for different ( ), one has to choose x^* in such a fashion that the largest of all $\partial(\quad)/\partial x$ reduces to $\square(1)$ after ordering. The other smaller $\partial(\quad)/\partial x$ may then, of course, assume orders smaller than $\square(1)$ after ordering.

Please note that the choice of x^*, and so on, here requires closer investigation (or guess) of the flow field in question, since one must know something about the first derivatives involved.

In Section 5.1 if we had chosen the characteristic quantities according to the plan just laid out, we would have obtained terms like $St(\partial u'/\partial t')$ *where*

$$\frac{\partial u'}{\partial t'} \sim \square(1) \text{ or smaller}$$

Therefore, proper ordering would assure us that the term

$$St\left(\frac{\partial u'}{\partial t'}\right) \sim \square(St) \text{ or smaller}$$

and similarly for the other terms.

In trying to say something along this line for every term in the dimensionless equations of the last section, we notice that many terms involve second derivatives. What about the orders of these terms then? Note that by now we are completely bound. We have chosen the characteristic values of dependent variables to make them of $\square(1)$ or smaller; we have also chosen the characteristic values of the independent variables to make the first derivatives of $\square(1)$ or smaller. There is nothing more we can do. But, fortunately, our experience assures us that, in fluid dynamics, the first-order derivatives do not vary faster than the flow variables themselves. That is to say,

$$\frac{\partial^2(\quad)}{\partial x^2}, \text{ etc.}, \sim \square\left[\frac{\partial(\quad)}{\partial x}\right], \text{ etc., or smaller}$$

Therefore, if the ordering makes $\partial(\quad)'/\partial x'$, and so on of $\square(1)$ or smaller, it automatically makes $\partial^2(\quad)'/\partial x'^2$, etc., of $\square(1)$ or smaller. As a consequence, we may now say that all the terms in the dimensionless equations of the last section, *using characteristic quantities properly chosen as explained*, are of order $\square(\star)$ or smaller, where $\star$ denotes the dimensionless parameters in front of the respective terms.

To explore this a little further, let us take Eq. (5.2a) as an example. For simplicity, let us treat only steady flow here. For the same reason, we will also assume that the same characteristic length and speed can be used in both the x- and the y-directions; that is, $y^* = x^*$ and $v^* = u^*$:

$$\underset{\substack{\text{(inertial}\\\text{"force")}}}{\left(u'\frac{\partial u'}{\partial x'}+v'\frac{\partial u'}{\partial y'}\right)} = \underset{\text{(gravity)}}{\frac{1}{Fr}} \underset{\substack{\text{(pressure}\\\text{force)}}}{-Eu\left(\frac{\partial p'}{\partial x'}\right)} + \underset{\text{(viscous force)}}{\frac{1}{Re}\left(\frac{\partial^2 u'}{\partial x'^2}+\frac{\partial^2 u'}{\partial y'^2}\right)} \tag{5.3}$$

The physical meaning of each term is that it is a measure of the correspondingly underwritten force (per unit mass) acting on a fluid particle. (The "force" on the left-hand side is really just the acceleration.) The terms in parentheses are all dimensionless and of $\square(1)$ or smaller, as guaranteed by the ordering process. *If* we are certain that these are really of $\square(1)$ *and not smaller* for a particular problem under investigation, we would have the following physical interpretations of the parameters:

$$\frac{\text{inertial "force"}}{\text{viscous force}} \sim \square(Re)$$

$$\frac{\text{gravity}}{\text{viscous force}} \sim \square\left(\frac{Re}{Fr}\right)$$

and so on. Thus, in *some* cases, Re may be looked upon as the relative importance of the inertial "force" as compared with the viscous force. Then, if Re is small, it merely means that the inertial "force" is small while the viscous force is large. However, it is very important to remember that this kind of interpretation is not always possible.

COUNTEREXAMPLE Take the steady channel flow in Section 3.2. Here, the governing system is

$$\begin{cases} \mu\dfrac{d^2u}{dy^2} = F - \rho g_x & \text{(3.14)} \\ u\langle 0\rangle = 0 & \text{(3.15a)} \\ u\langle Y\rangle = U & \text{(3.15b)} \end{cases}$$

Obviously there is no inertial "force" (that is, the fluid is not accelerating). Thus

$$\frac{\text{inertial "force"}}{\text{viscous force}} = 0$$

Yet if we choose $u^* = U$, $y^* = Y$, we have $Re = UY/\nu \neq 0$! Then, for such a problem, the physical interpretation that

$$\frac{\text{inertial "force"}}{\text{viscous force}} \sim \square(Re)$$

is definitely wrong. But why? If you refer back to the paragraph where this interpretation is stated, you will see that there is a premise standing before it—that is, "*if* the terms in parentheses are really of $\square(1)$ *and not smaller*." In Eq. (3.14) and boundary condition (3.15a,b), if we introduce $u^* = U$, $y^* = Y$, we have

$$\begin{cases} \underset{\text{(inertial)}}{(0)} = \underset{\text{(gravity)}}{\dfrac{1}{Fr}} - \underset{\text{(pressure)}}{Eu} + \underset{\text{(viscous)}}{\dfrac{1}{Re}\left(\dfrac{d^2u'}{dy'^2}\right)} & \textbf{(5.4)} \\ u'\langle 0\rangle = 0 & \textbf{(5.5a)} \\ u'\langle 1\rangle = 1 & \textbf{(5.5b)} \end{cases}$$

where $Fr = U^2/g_xY$, $Eu = FY/\rho U^2$, $Re = YU/\nu$. In Eq. (5.4) we have deliberately displayed the zero inertial term in parentheses. This term then is definitely not $\Box(1)$, *but much smaller*, namely zero! Since the premise is violated, naturally the interpretation is wrong. The correct interpretation here is

$$\frac{\text{pressure force}}{\text{viscous force}} \sim \Box(Re\,Eu)$$

$$\frac{\text{gravity}}{\text{viscous force}} \sim \Box\left(\frac{Re}{Fr}\right)$$

from Eq. (5.4).

The above discussion, though helpful, is not the main purpose of the ordering process. The main reason we carry out the process is to make approximations in a rational way. Let us look at Eq. (5.3) once more. Suppose that in a flow problem Fr turns out to be very large (that is, g_x is small compared to U^2/Y); then we may take $Fr \to \infty$ as a good approximation and investigate the limiting (or asymptotic) behavior of the problem as $Fr \to \infty$, which is simply governed by (among other things) Eq. (5.3), with the term $1/Fr$ dropped.[6]

Similarly, if Re is large, the viscous term can be dropped as a result of letting $Re \to \infty$. (See Chapter 6 for the associated trouble and remedy.) If Re is small, we first multiply Eq. (5.3) by Re and then let $Re \to 0$. We see then that the inertial term can be dropped. (Again, there is trouble associated with such a limiting process. But we will not have the opportunity of examining it in Part 1.) What happens to the other two terms in Eq. (5.3) depends on the circumstances. For example, although Re is small, Fr may also be small (strong gravity) such that $Re/Fr \sim \Box(1)$. Then the gravity term stays as $Re \to 0$. On the other hand, if Fr is not small enough, or is even large, the gravity term drops as $Re \to 0$ since Re/Fr is of $o\,(1)$ (much smaller than 1). As another example, when $Re \to \infty$, the pressure term may or may not be omitted, depending on whether $Eu \sim \Box(1/Re)$ or $o\,(1/Re)$—that is, whether $Eu\,Re \sim \Box(1)$ or $o\,(1)$. If in doubt, or if there is insufficient ground, in assigning a particular order to a term in an asymptotic investigation, *keep that term*! If, later on, that term turns out to be negligible after all, you only committed a stupidity in carrying it along, but no mistake!

There is, of course, no need to choose $x^* = y^*$, $u^* = v^*$ in the general case, Eqs. (5.1) and (5.2a,b). The above discussions apply then as well to the parameters (x^*/y^*), (v^*/u^*), and (g_y/g_x) (as well as St). In other words, one can

[6] We have deliberately used the symbol $\Box$ up to here. In the following, we will begin to use O and o.

also investigate the asymptotic behavior of a flow as $x^*/y^* \to 0$, or $v^*/u^* \to \infty$, and so on.

At this juncture[7] please note the following points:

1. Here it indeed does not make any difference whether the dimensionless terms before which the parameters appear are really $\square(1)$. They may actually be of smaller order. When in doubt, you can always play it safe by keeping the term. If later on you learn that it is really of smaller order, you can drop it at that stage. Even if you do not know better and have kept a small term, no harm would come of it (except that you may not be able to solve the problem with this term present). It would be like adding numbers in the thousands while keeping a term of 10^{-5}. It may be a cumbersome and even stupid thing to do; but it would never cause any mistake.
2. It is always a dimensionless parameter that is small or large, not one of its factors. For example, small *Re* means that u^*x^*/ν is small, not that u^* is small or ν is large! Although, one often refers to a low-*Re* flow sloppily as very slow, or highly viscous, you must keep in mind the correct basis for saying these things; that is, they all mean that *Re* is small.
3. The trouble here is that, so far, the process is still more an art than science. Its mastery, therefore, depends on personal experience to a certain extent. In other words, the more you know about fluid dynamics, the more you will know about ordering. Conversely, the more you know about ordering, the more you will know about fluid dynamics. This seeming paradox is exactly the exciting aspect of physical sciences—namely, how to approximate physical phenomena properly. If physics were free from things like these, it would cease to be physics. At this point, you should not feel helpless, because most of this book is devoted to teaching you just this fine art of making rational approximations through ordering and asymptotics.

---

[7] In the literature, the asymptotic investigation outlined above is often called the perturbation method (or *asymptotics*, which term was introduced by N. G. de Bruijn in his *Asymptotic Methods in Analysis*, Interscience, New York, 1961.) The method, complete with loose talks about quantities of the first, second, and higher orders, can already be found in Lagrange's 1781 "Memoir on the theory of motion of fluids" (in French) especially in relation to the establishment of shallow water theory. Joseph Louis (Count) de Lagrange (1736–1813), French mathematician, wrote his great 1788 treatise *Analytical Mechanics* (in French) and was very proud of the fact that there was not a single figure in it. Although encouraged by both Euler and d'Alembert in his career, Lagrange chose to side with d'Alembert in personal and political matters. Euler's influence on him was great but was rarely acknowledged, and even then, only reluctantly. His *Analitical Mechanics* contained only the arid skeleton of Euler's creation, although it was regarded as the storehouse of 18th century rational mechanics. As a result, Euler's work remained obscure, while many of his ideas were wrongly attributed to d'Alembert or Lagrange. Lagrange in fact usually sided with Euler in all mathematical and physical matters, but he did it in such a way as to be always flattering to d'Alembert; thus, the wrong impression was conveyed.

5.3 TRANSIENT CHANNEL FLOWS

We will start our long journey of approximation by going back to the transient counterpart of Section 3.2—that is, Eqs. (3.8) and (3.9), with conditions (3.10a,b) and (3.11):

$$\begin{cases} \dfrac{\partial u}{\partial t} = \left[g_x - \dfrac{1}{\rho} F\langle t\rangle\right] + \nu \dfrac{\partial^2 u}{\partial y^2} & \textbf{(3.8)} \\ p = xF\langle t\rangle + \rho g_y y + f\langle t\rangle & \textbf{(3.9)} \\ \text{where } f\langle t\rangle = p\langle 0, 0; t\rangle & \\ u\langle 0; t\rangle = 0 & \textbf{(3.10c)} \\ u\langle Y; t\rangle = U \qquad t > 0 & \textbf{(3.10d)} \\ u\langle y; 0\rangle = 0 & \textbf{(3.11A)} \end{cases}$$

where simplified forms of $F_1\langle t\rangle$, $F_2\langle t\rangle$ and $F_3\langle y\rangle$ are adopted for ease of argument.

To order the problem properly, we choose

$$u^* = U$$

$$x^* = y^* = Y$$

$$p^* = \rho U^2$$

For t^*, we will choose the time of observation—that is, the instant around which we focus our attention. (We will not bother to introduce another notation. Let us just say that here t^* = time of observation.) A nondimensionalization easily yields

$$\begin{cases} \left(\dfrac{Y}{Ut^*}\right) \dfrac{\partial u'}{\partial t'} = \left[\dfrac{g_x Y}{U^2} - F'\langle t'\rangle\right] + \dfrac{1}{Re} \dfrac{\partial^2 u'}{\partial y'^2} & \textbf{(5.6)} \\ p' = x'F'\langle t'\rangle + \left(\dfrac{g_y}{g_x}\right)\left(\dfrac{g_x Y}{U^2}\right) y' + f'\langle t'\rangle & \textbf{(5.7)} \\ u'\langle 0; t'\rangle = 0 & \textbf{(5.8a)} \\ u'\langle 1; t'\rangle = 1 \qquad t' > 0 & \textbf{(5.8b)} \\ u'\langle y'; 0\rangle = 0 & \textbf{(5.9)} \end{cases}$$

where

$$F'\langle t'\rangle = \frac{YF\langle t^* \cdot t'\rangle}{\rho U^2}$$

$$Re = \frac{UY}{\nu}$$

$$f'\langle t'\rangle = \frac{f\langle t^* \cdot t'\rangle}{\rho U^2}$$

Now we are particularly interested in (1) t^* very large (that is, more exactly, $Y/Ut^* \to 0$) and (2) t^* very small (that is, $Y/Ut^* \to \infty$); all the other terms and parameters are ordinary numbers or smaller—or, more precisely, of $O(1)$.

(1) Large Observation Time

As $Y/Ut^* \to 0$, our system becomes

$$\begin{cases} \dfrac{1}{Re}\dfrac{\partial^2 u'}{\partial y'^2} = F'\langle t'\rangle - \dfrac{g_x Y}{U^2} & (5.10) \\ p' = x'F'\langle t'\rangle + \left(\dfrac{g_y}{g_x}\right)\left(\dfrac{g_x Y}{U^2}\right) y' + f'\langle t'\rangle & (5.11) \\ u'\langle 0; t'\rangle = 0 & (5.8a) \\ u'\langle 1; t'\rangle = 1 \qquad t' > 0 & (5.8b) \\ u'\langle y'; 0\rangle = 0 & (5.9) \end{cases}$$

We immediately see that there is no derivative with respect to time in the differential equation (5.10). It is therefore an ordinary differential equation with t' as a *parameter* that can be treated as a constant in solving. As a matter of fact, the solution subject to boundary condition (5.8a,b) is, in the dimensional form,

$$u = \left(\frac{y}{Y}\right) U - \frac{Y^2}{2\mu}\left[F\langle t\rangle - \rho g_x\right] \cdot \left(\frac{y}{Y}\right)\left(1 - \frac{y}{Y}\right) \tag{5.12}$$

$$p = F\langle t\rangle x + \rho g_y y + p\langle 0, 0; t\rangle \tag{5.13}$$

These are almost the same results as Eqs. (3.16) and (3.20) for steady channel flows. We recognize the fact that, here, $t \sim \square(t^*)$; if the pumping parameter attained a constant (steady) value by this time, the results would be exactly the same. We thus see that the steady solution of Section 3.2 is actually the solution at a long time after the start. Our new solution, however, is a little more general than that of Section 3.2. It conveys the additional knowledge that, at large times, a varying pumping will produce at every instant a flow that is dictated by the instantaneous value of pumping (not the history of pumping). That is to say, at $t_0 \gg 0$, u and p are obtained simply by substituting $F\langle t\rangle$ in Eqs. (5.12) and (5.13), with $F\langle t_0\rangle$. This kind of situation is sometimes known as quasi-steady to distinguish it from the true steady case, in which F has attained a constant value.

The quasi-steady solution, Eq. (5.12), clearly cannot satisfy the initial condition (5.9) since it "lives only in the present, and has no memory of the past." Or we might say that what happens at $t = 0$ can influence the future $t > 0$ only for a while. It has no bearing on the long-time situation projected way into the future. Physically, this means that the initial situation is damped out fast (by viscosity) so that a steady or quasi-steady flow soon prevails. (Note that the above approximation is based on $\partial u'/\partial t' \sim O(1)$. If the pumping changes rather suddenly around t^*, the left-hand side of Eq. (5.6) may not be small.)

(2) Small Observation Time

Going back to the system (5.6) to (5.9), we have as a result of $Y/Ut^* \to \infty$ (that is, $Ut^*/Y \to 0$):

$$\begin{cases} \dfrac{\partial u'}{\partial t'} = 0 & \textbf{(5.14)} \\ p' = x'F'\langle t'\rangle + \left(\dfrac{g_y}{g_x}\right)\left(\dfrac{g_x Y}{U^2}\right) y' + f'\langle t'\rangle & \textbf{(5.7)} \\ u'\langle 0; t'\rangle = 0 & \textbf{(5.8a)} \\ u'\langle 1; t'\rangle = 1 \qquad t' > 0 & \textbf{(5.8b)} \\ u'\langle y'; 0\rangle = 0 & \textbf{(5.9)} \end{cases}$$

The solution of Eq. (5.14) subject to condition (5.9) is

$$u' = 0 \tag{5.15}$$

which does not say very much except the physically obvious fact[8] that there is hardly any flow at $t \cong 0$, if the flow is initially at rest.[9] (If the flow is initially not at rest, it will hardly change for $t \cong 0$.) This *short-time solution* $u \equiv 0$ cannot satisfy boundary condition (5.8b) because it takes some time for the motion of the upper plate to be felt by the fluid at rest. Also, this short-time solution obviously does not evolve into the long-time solution obtained before. (See Fig. 5.1.) The corresponding short-time pressure distribution is just (in dimensional form) Eq. (5.13) with $t \cong 0$ (that is, determined almost by the initial pumping).

This then brings in the interesting question of what happens at moderate time, where Ut^*/Y is neither small nor large. To answer this question will tend

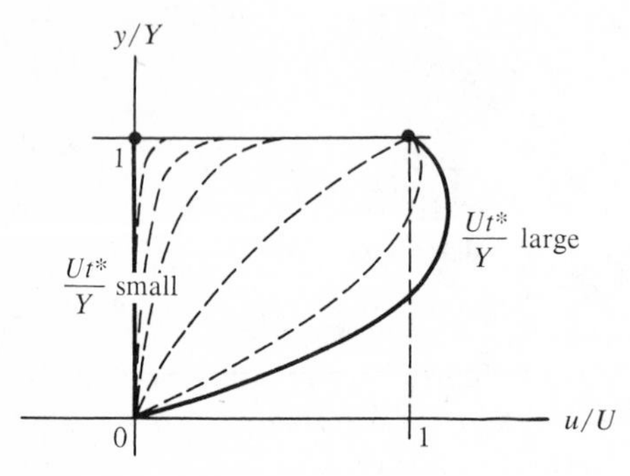

Figure 5.1 Short-time and long-time solutions.

[8] But the way that this fact is deduced is important.

[9] Here, we can ask a particularly nasty question: Now that we know $u' = 0$ and $\partial u'/\partial t' = 0$, had we really the right to throw the rest of the terms of Eq. (5.6) out? Isn't the term $\partial u'/\partial t'$ even smaller (being zero)? *Answer*: To be exact, Eq. (5.14) really means $\partial u'/\partial t' \sim O(Ut^*/Y)$ (that is, as small as Ut^*/Y), and Eq. (5.15) really means $u' = t'[O(Ut^*/Y)]$.

to sidetrack us. We will only say that the moderate-time solution must connect (1) and (2) as indicated tentatively by the dashed curves.

Actually, still more information is available for the flow at small observation time if we concentrate on the narrow neighborhood of the moving plate. To achieve this, it is more convenient to call the moving plate $y = 0$ and the stationary plate $y = Y$ (Fig. 5.2). The system (3.8) to (3.11A) applies, but with the boundary conditions corrected to

$$u(0; t) = U \qquad t > 0 \tag{3.10e}$$

$$u(Y; t) = 0 \tag{3.10f}$$

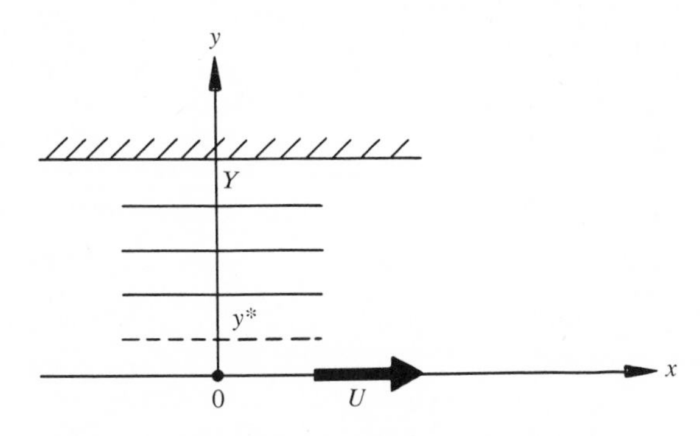

Figure 5.2 Neighborhood of the moving plate.

Now instead of Y, we will use the position of our *observational* station y^* as the characteristic length. This then makes the dimensionless $y'' \sim \square(1)$ [or $y \sim \square(y^*)$], since we are interested only in the flow field up to and around $y = y^*$. The result of the nondimensionalization is then

$$\begin{cases} \left(\dfrac{y^*}{Ut^*}\right)\dfrac{\partial u'}{\partial t'} = \left[\dfrac{g_x y^*}{U^2} - F''\langle t'\rangle\right] + \dfrac{1}{Uy^*/\nu}\dfrac{\partial^2 u'}{\partial y''^2} & (5.16) \\ p' = x'F''\langle t'\rangle + \left(\dfrac{g_y}{g_x}\right)\left(\dfrac{g_x y^*}{U^2}\right) y'' + f'\langle t'\rangle & (5.17) \\ u'\langle 0; t'\rangle = 1 \qquad t' > 0 & (5.18) \\ u'\langle Y/y^*; t'\rangle = 0 & (5.19) \\ u'\langle y''; 0\rangle = 0 & (5.20) \end{cases}$$

where we have introduced the new symbols

$$y'' = y/y^*$$

$$F''\langle t'\rangle = \frac{y^* F\langle t^* \cdot t'\rangle}{\rho U^2}$$

Equation (5.16) can also be written as

$$\frac{\partial u'}{\partial t'} = \left(\frac{Ut^*}{y^*}\right)\left[\frac{g_x y^*}{U^2} - F''\langle t'\rangle\right] + \frac{\nu t^*}{y^{*2}}\left(\frac{\partial^2 u'}{\partial y''^2}\right) \tag{5.16A}$$

Now, within the framework of (2), we have a very small t^*. Suppose that, for this observational time, we are only interested in the neighborhood of y^* which is

$$y^* \sim \square(\sqrt{\nu t^*})$$

Then

$$\frac{Ut^*}{y^*} \sim \square\left(U\sqrt{\frac{t^*}{\nu}}\right)$$

and

$$\frac{\nu t^*}{y^{*2}} \sim \square(1)$$

The limit $t^* \to 0$ (or, even better, $U\sqrt{t^*/\nu} \to 0$), which implies $y^* \to 0$ (but $\nu t^*/y^{*2}$ stays), yields

$$\begin{cases} \dfrac{\partial u'}{\partial t'} = \dfrac{\nu t^*}{y^{*2}}\left(\dfrac{\partial^2 u'}{\partial y''^2}\right) & \textbf{(5.21)} \\ p' = f'\langle t'\rangle & \textbf{(5.22)} \\ u'\langle 0; t'\rangle = 1 \qquad t' > 0 & \textbf{(5.23a)} \\ u'\langle \infty; t'\rangle = 0 & \textbf{(5.23b)} \\ u'\langle y''; 0\rangle = 0 & \textbf{(5.24)} \end{cases}$$

Equation (5.21) and boundary conditions (5.23a,b), when brought back to their dimensional forms, happen to be exactly the equation and conditions of the Rayleigh problem that was solved in Section 4.1. Physically, this interesting conclusion means that at small values of time, a thin layer of fluid near the moving plate feels only the motion of the plate, and nothing else. In particular, the thin layer does not feel the presence of the fixed plate at all, "thinking" that it is at infinity (Fig. 5.3). The smaller the time, the thinner is this "dumb" layer ($y^* \sim \square(\sqrt{\nu t^*})$). An investigation of the flow beyond this layer is again out of our scope.

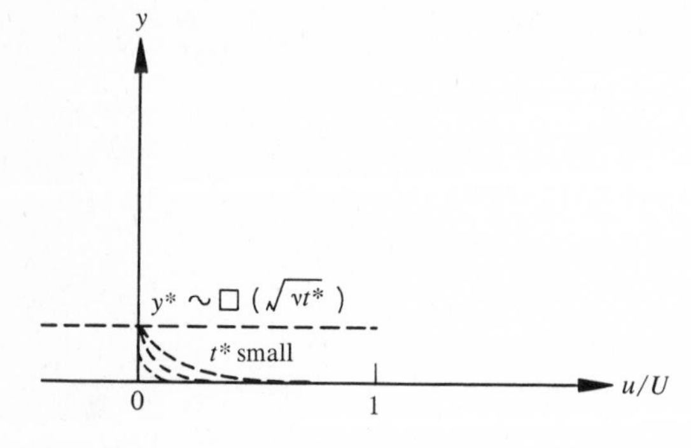

Figure 5.3 Small-time solution in the neighborhood of the moving plate.

It goes without saying that, similarly, a thin layer near the stationary plate will stay stationary for small times.

Before we leave this section, we wish to take this opportunity to discuss also a similar situation.

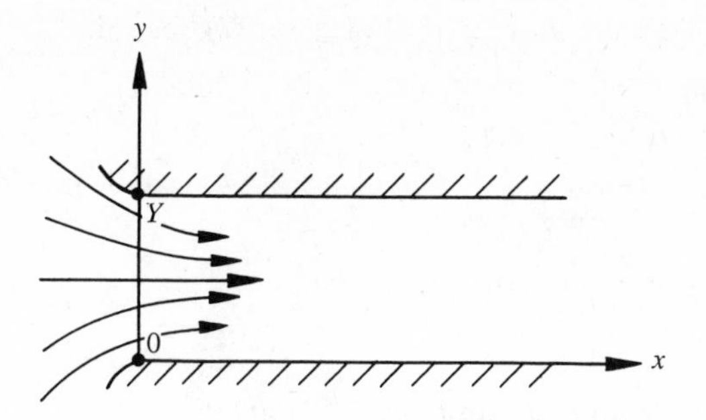

Figure 5.4 Entrance into a long channel.

(3) Entrance into a Long Channel

Here we will consider the steady flow into a long channel $x > 0$ between $y = 0$ and Y (Fig. 5.4). It is known that the flow at the entrance $x = 0$ is such that

$$u = u_0\langle y\rangle$$

$$v = v_0\langle y\rangle$$

The ansatz of $v \equiv 0$ or $\partial \mathbf{q}/\partial x \equiv 0$ in Section 3.2 does not work here, and we have to go back to the original system:

$$\begin{cases} \dfrac{\partial u}{\partial x}+\dfrac{\partial v}{\partial y}=0 & \textbf{(1.9)} \\ u\dfrac{\partial u}{\partial x}+v\dfrac{\partial u}{\partial y}=g_x-\dfrac{1}{\rho}\dfrac{\partial p}{\partial x}+\nu\left(\dfrac{\partial^2 u}{\partial x^2}+\dfrac{\partial^2 u}{\partial y^2}\right) & \textbf{(2.15a)} \\ u\dfrac{\partial v}{\partial x}+v\dfrac{\partial v}{\partial y}=g_y-\dfrac{1}{\rho}\dfrac{\partial p}{\partial y}+\nu\left(\dfrac{\partial^2 v}{\partial x^2}+\dfrac{\partial^2 v}{\partial y^2}\right) & \textbf{(2.15b)} \\ x=0:\quad \left.\begin{matrix} u=u_0\langle y\rangle \\ v=v_0\langle y\rangle \end{matrix}\right\} & \textbf{(5.25)} \\ y=0:\quad u=0,\ v=0 & \textbf{(5.26a)} \\ y=Y:\quad u=0,\ v=0 & \textbf{(5.26b)} \end{cases}$$

In ordering, let us choose

$$\begin{aligned} &v^* = u^* = u_0\langle Y/2\rangle \\ &y^* = Y \\ &x^* = \text{location of observational station} \end{aligned}$$

The nondimensionalization of Eq. (1.9) then yields

$$\left(\frac{Y}{x^*}\right)\frac{\partial u'}{\partial x'}+\frac{\partial v'}{\partial y'}=0 \qquad \textbf{(5.27)}$$

As we extend our attention far downstream, or as $Y/x^* \to 0$, Eq. (5.27) becomes

$$\frac{\partial v'}{\partial y'} = 0$$

that is,

$$v' = v'\langle x' \rangle \text{ only}$$

Since $v' = 0$ on the plate $y' = 0$, we must conclude that[10]

$$v' \text{ or } v = 0 \qquad \textbf{(5.28)}$$

This solution clearly cannot satisfy boundary conditions (5.25) at $x = 0$. Again, this would indicate a damping off of the inlet situation downstream. Please note here that you *cannot* put Eq. (5.28) back into Eq. (5.27) again and deduce that $\partial u'/\partial x' = 0$. On the one hand, Eq. (5.27) has just been used to derive Eq. (5.28). On the other, the first term in Eq. (5.27) was dropped because of small (Y/x^*), not of small or zero $\partial u'/\partial x'$. For all we know, at this stage,

$$\frac{\partial u'}{\partial x'} \sim O(1)$$

that is, $\sim \square(1)$ or smaller. Whether it is vanishingly small remains to be seen.

Substituting Eq. (5.28) into Eq. (2.15b) gives

$$\frac{\partial p}{\partial y} = \frac{\partial \tilde{p}}{\partial y} = \rho g_y \qquad \textbf{(5.29)}$$

where

$$\tilde{p} = p - p\langle 0,0 \rangle \qquad \textbf{(5.30)}$$

is introduced since we are really interested here in the pressure differences. The introduction of $\tilde{p}$ of course does not disturb anything since $p\langle 0,0 \rangle$ (or any other chosen datum value) is a constant. Its use, however, makes the ordering more realistic. Thus, instead of choosing a characteristic pressure, we should choose a characteristic $\tilde{p}$. (We did not bother to do this before, since it would have made no difference.) Let us accordingly choose

$$\tilde{p}^* = Fx^*$$

where F is the capability of the pumping device, expressed in pressure rise per unit channel length.

Equation (2.15a) now becomes, after using Eq. (5.28) and proper nondimensionalization,

$$\left(\frac{Y}{x^*}\right)^2 u' \frac{\partial u'}{\partial x'} = \left(\frac{Y}{x^*}\right)^2 \frac{g_x x^*}{u^{*2}} - \left(\frac{Y}{x^*}\right)^2 \frac{Fx^*}{\rho u^{*2}} \frac{\partial \tilde{p}'}{\partial x'} + \frac{\nu}{u^* x^*} \left[\left(\frac{Y}{x^*}\right)^2 \frac{\partial^2 u'}{\partial x'^2} + \frac{\partial^2 u'}{\partial y'^2}\right]$$

where $\tilde{p}' = \tilde{p}/Fx^*$. There are several x^*'s floating around on the right-hand side. We have to put them into the framework of Y/x^* since we are considering $\lim_{Y/x^* \to 0}$, not $\lim_{x^* \to 0}$!

---

[10] Again, to be exact, we should have written $\partial v'/\partial y' \sim O(Y/x^*)$ and $v' = 0 + y'[O(Y/x^*)]$.

This is easily done:

$$\left(\frac{Y}{x^*}\right)^2 u' \frac{\partial u'}{\partial x'} = \left(\frac{Y}{x^*}\right) \frac{g_x Y}{u^{*2}} - \left(\frac{Y}{x^*}\right) \frac{FY}{\rho u^{*2}} \frac{\partial \tilde{p}'}{\partial x'} + \frac{\nu}{u^* Y} \left[\left(\frac{Y}{x^*}\right)^3 \frac{\partial^2 u'}{\partial x'^2} + \frac{\partial^2 u'}{\partial y'^2} \left(\frac{Y}{x^*}\right) \right]$$

Then, $\lim\limits_{Y/x^* \to 0}$ leads to

$$\frac{\nu}{u^* Y} \frac{\partial^2 u'}{\partial y'^2} + \frac{g_x Y}{u^{*2}} - \frac{FY}{\rho u^{*2}} \frac{\partial \tilde{p}'}{\partial x'} = 0 \tag{5.31}$$

Note again that the x-derivatives drop out because of small Y/x^*, not because of x-independence of u; not at this stage!

Now, starting from Eq. (5.29), one can again go through the steps in Section 3.1 that yielded Eqs. (3.5) to (3.9) and obtain the key equation (in dimensional form)

$$\mu \frac{\partial^2 u}{\partial y^2} = F - \rho g_x \tag{5.32}$$

with the boundary conditions

$$y = 0: \quad u = 0 \tag{5.33a}$$

$$y = Y: \quad u = 0 \tag{5.33b}$$

Integrating Eq. (5.32), we have

$$\mu \frac{\partial u}{\partial y} = (F - \rho g_x) y + f_1\langle x \rangle$$

and

$$\mu u = \tfrac{1}{2}(F - \rho g_x) y^2 + f_1\langle x \rangle y + f_2\langle x \rangle$$

Boundary condition (5.33a) dictates that

$$f_2\langle x \rangle = 0$$

Boundary condition (5.33b) then dictates that

$$f_1\langle x \rangle = -\tfrac{1}{2}(F - \rho g_x) Y$$

With these, the solution becomes identical with Eq. (3.21) in Section 3.2. Note that only now do we know that $u = u\langle y \rangle$ only, and $\partial u'/\partial x' \equiv 0$, and so forth.[11] This solution, of course, cannot satisfy the boundary condition (5.25) at $x = 0$, and the same remark as that after Eq. (5.28) applies here.

Physical Conclusion The inlet situation damps out and thus, far downstream, the fully developed flow (with zero v) of Section 3.2 holds. (This is also true for the general case, in which one plate is moving.)

On the other hand, in Eq. (5.27), if we are interested in the immediate

[11] This really means that $\partial u'/\partial x' \sim o\,(1)$. Reflecting back, we see that $\partial v'/\partial y'$ should be $o\,(Y/x^*)$. Still no contradiction.

neighborhood of the entrance, we would take $\lim_{x^*/Y \to 0}$. This then leads to

$$\frac{\partial u'}{\partial x'} = 0$$

that is,

$$u' = u'\langle y' \rangle \text{ only}$$
$$= u'_0\langle y' \rangle$$

That is, the inlet velocity distribution in the x-direction has no chance to change yet. This is not much information, being physically trivial. And we will not go any further along this line, except to point out that the boundary condition at $y = 0$ and Y can never be satisfied.

If, at the entrance, we investigate only the thin layer near $y = 0$ (or, similarly, near $y = Y$), Y will no longer be the proper characteristic length to use in the ordering process. One should use instead the location of the observational station y^*. If $y^* \sim \square(x^*)$, although both are small, the corresponding dimensionless equation of continuity

$$\left(\frac{y^*}{x^*}\right)\frac{\partial u'}{\partial x'} + \frac{\partial v'}{\partial y''} = 0$$

(where $y'' = y/y^*$) will not yield any simplification. I mention this in order to convey to you the fact that, in treating the asymptotics of a parameter $\alpha_1 \to 0$, one should always pay attention to the orders of magnitude of other parameters, α_2, α_3, and so on, relative to α_1. For example, if $\lim_{\alpha_1 \to 0} (\alpha_2/\alpha_1) \sim \square(1)$, the factor α_2/α_1 stays.

One Last Remark Although ordering and asymptotic investigation usually produce useful approximate information, they constitute, after all, only a special kind of ansatz, namely: With respect to a parameter α,

$$\text{true solution} = \lim_{\alpha \to 0} \text{ in some manner} + \text{error that becomes smaller as } \alpha \to 0$$

The ansatz may work, or it may not work—in the sense that (a) it is not self-consistent, (b) it produces a flow that is unrealizable or impossible, or (c) the error is not as small as expected. When it fails to work (assuming that your ordering is proper), it does not necessarily mean that the problem is not solvable; it may only mean that the solution is not in the form expected. Even if the ansatz works consistently, it may not be any real help to us because (a) it may produce only trivial information, (b) the limiting or asymptotic problem may be as difficult to solve as the original, thus offering no advancement toward the understanding of the problem, or (c) it may have killed the phenomenon you wished to investigate (leaving you holding an empty bag). The technique is potent, but not omnipotent.

5.4 FLOW INDUCED BY AN OSCILLATING PLATE

In the general case of Rayleigh's problem with varying plate speed, Section 4.1, if $U\langle t\rangle$ takes on the form

$$U\langle t\rangle = U_0 \cos \omega t \tag{5.33}$$

that is, if the plate oscillates with (circular) frequency ω, the ansatz of similarity solution of Section 4.1 immediately fails. But this does not mean that we cannot solve this new problem by other routes. (Incidentally, this problem of an oscillating plate is sometimes called Stokes' second problem.[12])

To recapitulate, we wish now to solve the following problem:

$$\begin{cases} \dfrac{\partial u}{\partial t} = \nu \dfrac{\partial^2 u}{\partial y^2} & \textbf{(4.5)} \\ u\langle y; 0\rangle = 0 & \textbf{(4.6a)} \\ u\langle 0; t\rangle = U_0 \cos \omega t & \textbf{(4.6d)} \end{cases}$$

together with proper behavior at $y = \infty$.

To start, let us consider the associated problem for $\mathcal{F}\langle y; t\rangle$

$$\begin{cases} \dfrac{\partial \mathcal{F}}{\partial t} = \nu \dfrac{\partial^2 \mathcal{F}}{\partial y^2} & \textbf{(5.34)} \\ \mathcal{F}\langle y; 0\rangle = 0 & \textbf{(5.35a)} \\ \mathcal{F}\langle 0; t\rangle = U_0 e^{i\omega t} & \textbf{(5.35b)} \end{cases}$$

where $i = \sqrt{-1}$. $\mathcal{F}$ is then a complex-valued function. If we take the real part of both sides of Eq. (5.34) and boundary conditions (5.35a,b), we obtain ($\mathcal{R}$ means "real part of")

$$\begin{cases} \dfrac{\partial \mathcal{R}(\mathcal{F})}{\partial t} = \nu \dfrac{\partial^2 \mathcal{R}(\mathcal{F})}{\partial y^2} & \textbf{(5.36)} \\ \mathcal{R}(\mathcal{F})\,|_{t=0} = 0 & \textbf{(5.37a)} \\ \mathcal{R}(\mathcal{F})\,|_{y=0} = \mathcal{R}(U_0 e^{i\omega t}) & \\ \qquad = \mathcal{R}(U_0 \cos \omega t + iU_0 \sin \omega t) & \\ \qquad = U_0 \cos \omega t & \textbf{(5.37b)} \end{cases}$$

Comparing Eq. (5.36) and boundary conditions (5.37a,b) with Eq. (4.5) and conditions (4.6a,d), we see that

$$u = \mathcal{R}(\mathcal{F}) \tag{5.38}$$

This means that we could solve for $\mathcal{F}$ first, and then obtain u by taking the real part of $\mathcal{F}$.

---

12 Stokes' 1851 paper; see footnote 1 in Chapter 4.

Now, boundary condition (5.35b) suggests an ansatz

$$\mathscr{F} = \xi\langle y \rangle e^{i\omega t} \tag{5.39}$$

Will this work? Substituting ansatz (5.39) into Eq. (5.34) and boundary condition (5.35b), we have

$$\begin{cases} e^{i\omega t} i\omega \xi\langle y \rangle = \nu \dfrac{d^2\xi}{dy^2} e^{i\omega t} & (5.40) \\[2ex] e^{i\omega t} \xi\langle 0 \rangle = U_0 e^{i\omega t} & (5.41) \end{cases}$$

which is easily solvable. To be specific, Eq. (5.40) has the general solution

$$\begin{aligned} \xi &= A \exp\left(\sqrt{\frac{i\omega}{\nu}}\, y\right) + B \exp\left(-\sqrt{\frac{i\omega}{\nu}}\, y\right) \\ &= A \exp\left[\sqrt{\frac{\omega}{\nu}} \left(\frac{1+i}{\sqrt{2}}\right) y\right] + B \exp\left[-\sqrt{\frac{\omega}{\nu}} \left(\frac{1+i}{\sqrt{2}}\right) y\right] \\ &= A \exp\left(\sqrt{\frac{\omega}{2\nu}}\, y\right) \cdot \left(\cos \sqrt{\frac{\omega}{2\nu}}\, y + i \sin \sqrt{\frac{\omega}{2\nu}}\, y\right) \\ &\quad + B \exp\left(-\sqrt{\frac{\omega}{2\nu}}\, y\right) \cdot \left(\cos \sqrt{\frac{\omega}{2\nu}}\, y - i \sin \sqrt{\frac{\omega}{2\nu}}\, y\right) \end{aligned}$$

As $y \to \infty$, the factor $A \exp\,[\sqrt{(\omega/2\nu)}\; y]$ grows without bound, which would make $\lim_{y \to \infty} u$ infinitely large. Therefore, physically acceptable behavior at $y = \infty$ dictates here that

$$A = 0$$

Then, boundary condition (5.41) easily leads to

$$B = U_0$$

and therefore

$$\xi = U_0 \exp\left(-\sqrt{\frac{\omega}{2\nu}}\, y\right) \cdot \left(\cos \sqrt{\frac{\omega}{2\nu}}\, y - i \sin \sqrt{\frac{\omega}{2\nu}}\, y\right)$$

that is,

$$\begin{aligned} \mathscr{F} &= \xi e^{i\omega t} \\ &= U_0 \exp\left(-\sqrt{\frac{\omega}{2\nu}}\, y\right) \cdot \exp\left(-i \sqrt{\frac{\omega}{2\nu}}\, y\right) \cdot \exp\,(i\omega t) \\ &= U_0 \exp\left(-\sqrt{\frac{\omega}{2\nu}}\, y\right) \cdot \exp\left[i\left(\omega t - \sqrt{\frac{\omega}{2\nu}}\, y\right)\right] \\ &= U_0 \exp\left(-\sqrt{\frac{\omega}{2\nu}}\, y\right) \cdot \left[\cos\left(\omega t - \sqrt{\frac{\omega}{2\nu}}\, y\right) + i \sin\left(\omega t - \sqrt{\frac{\omega}{2\nu}}\, y\right)\right] \end{aligned} \tag{5.42}$$

And, finally,

$$u = \mathscr{R}(\mathscr{F}) = U_0 \exp\left(-\sqrt{\frac{\omega}{2\nu}}\, y\right) \cdot \cos\left(\omega t - \sqrt{\frac{\omega}{2\nu}}\, y\right) \tag{5.43}$$

Now, did the ansatz (5.39) work? No! There still remains the initial condition (5.35a) or (4.6a), which is clearly unsatisfiable by Eq. (5.42) or Eq. (5.43). We have thus only obtained a solution of Eq. (4.5) under boundary condition (4.6d), disregarding the initial condition (4.6a)!

However, if we refer to expression (5.43) as $u_{(5.43)}$, and the true unknown solution as u, we can write (or define)

$$u = u_{\text{makeup}} + u_{(5.43)} \tag{5.44}$$

that is, we can make $u_{(5.43)}$ correct by adding to it a makeup term.

Substituting Eq. (5.44) into Eq. (4.5), with conditions (4.6a,d), and remembering that $u_{(5.43)}$ satisfies Eq. (4.5) and boundary condition (4.6d), we have

$$\begin{cases} \dfrac{\partial u_{\text{makeup}}}{\partial t} = \nu \dfrac{\partial^2 u_{\text{makeup}}}{\partial y^2} & (5.45) \\ u_{\text{makeup}}\langle y;0\rangle = -u_{(5.43)}\langle y;0\rangle & \\ \qquad = -U_0 \exp\left(-\sqrt{\dfrac{\omega}{2\nu}}\,y\right)\cdot \cos\sqrt{\dfrac{\omega}{2\nu}}\,y & (5.46a) \\ u_{\text{makeup}}\langle 0;t\rangle = 0 & (5.46b) \end{cases}$$

You must realize that up to now we have not introduced any approximation. (We have not solved the problem either.) Here, we begin to order the problem for u_{makeup}. We will choose

$$u^* = U_0$$

$$y^* = \sqrt{\frac{\nu}{\omega}}$$

$$t^* = \text{observational time}$$

Nondimensionalization then leads to

$$\begin{cases} \left(\dfrac{1/\omega}{t^*}\right)\dfrac{\partial u'_{\text{makeup}}}{\partial t'} = \dfrac{\partial^2 u'_{\text{makeup}}}{\partial y'^2} & (5.47) \\ u'_{\text{makeup}}\langle y';0\rangle = -e^{-y'/\sqrt{2}}\cos\dfrac{y'}{\sqrt{2}} & (5.48a) \\ u'_{\text{makeup}}\langle 0;t'\rangle = 0 & (5.48b) \end{cases}$$

We are now interested in the interval around an observation time that is large compared with $1/\omega$ (the period of oscillation). The limit as $(1/\omega)/t^* \to 0$ yields then

$$\frac{\partial^2 u'_{\text{makeup}}}{\partial y'^2} = 0$$

for Eq. (5.47), which gives

$$\frac{\partial u'_{\text{makeup}}}{\partial y'} = f\langle t'\rangle$$

or

$$u'_{\text{makeup}} = f\langle t'\rangle y' + g\langle t'\rangle$$

at large times.

Boundary condition (5.48b) then demands that $g\langle t'\rangle \equiv 0$, leaving only

$$u'_{\text{makeup}} = f\langle t'\rangle y'$$

But, in order that u'_{makeup} (and hence the true u) be bounded at $y = \infty$, we must have $f\langle t'\rangle \equiv 0$ also. (The initial condition (5.48a) cannot be satisfied.)

$$u'_{\text{makeup}} = 0 \qquad \text{at large times}$$

Equation (5.44) then states

$$u_{(5.43)} = u \qquad \text{at large times}$$

Thus, although our approximate investigation did not give us the true flow u, it did come up with the important information that if we wait until t is very large compared to the period of oscillation, the flow will be in accordance with Eq. (5.43). For this reason, Eq. (5.43) is known as the long-time or sustained solution. (Some authors call it the steady solution. But this is a misnomer since Eq. (5.43) represents a time-dependent, oscillating flow, one that is not steady at all!)

Plots of Eq. (5.43) for a complete cycle of the plate motion are shown in Figure 5.5. It is seen that the motion of the fluid always lags behind the plate,

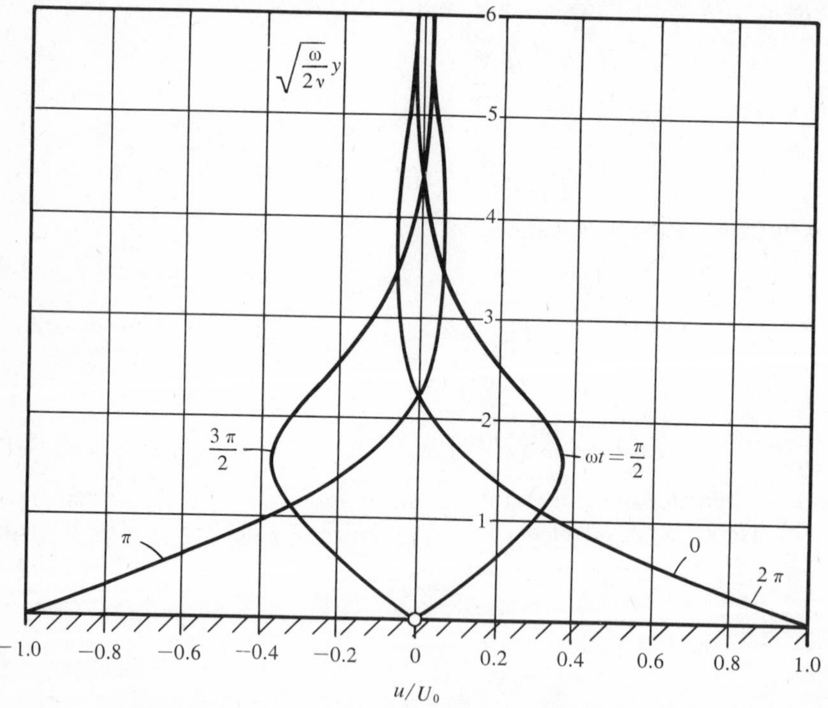

Figure 5.5 Velocity profiles over an oscillating plate.

and the oscillation is damped at locations away from the plate. For practical purposes, the fluid motion for

$$y \geqslant 2\pi\sqrt{2}\cdot\sqrt{\frac{\nu}{\omega}}$$

can be regarded as at rest. The influence of the plate oscillation practically cannot penetrate beyond $y = 2\pi\sqrt{2}\cdot\sqrt{\nu/\omega}$. The layer

$$0 < y < 2\pi\sqrt{2}\cdot\sqrt{\frac{\nu}{\omega}}$$

is therefore called the depth of penetration of the *viscous wave*. Note that the depth of penetration increases with the increasing kinematic viscosity and decreases with increasing frequency.

Going back, let us recall that we have not solved u_{makeup} except to surmise that it vanishes at $t \to \infty$ and that the initial condition (5.46a) must damp out soon. The real solution of u_{makeup} is beyond our scope.

Note that the analysis of this section shows that the correct ansatz to take at the beginning should be

$$u - u_{(5.43)} = \left[\text{lim of } u - u_{(5.43)} \text{ as } \frac{t_{\text{obs}}}{1/\omega} \to \infty\right] + \left[\text{error that vanishes as } \frac{t_{\text{obs}}}{1/\omega} \to \infty\right]$$

where t_{obs} is the time of observation. It would be wrong to try

$$u = \left[\text{lim of } u \text{ as } \frac{t_{\text{obs}}}{1/\omega} \to \infty\right] + \left[\text{error that vanishes as } \frac{t_{\text{obs}}}{1/\omega} \to \infty\right]$$

As a matter of fact, if we had performed the same ordering process on Eq. (4.5), under conditions (4.6a) and (4.6d), and had let $t^*/(1/\omega) \to \infty$, we would have ended up with

$$\begin{cases} \dfrac{\partial^2 u'}{\partial y'^2} = 0 \\ u'\langle y'; 0\rangle = 0 \\ u'\langle 0; t'\rangle = \cos(\infty) \end{cases}$$

This last boundary condition is not definite, being in between -1 and $+1$. (Picturesquely, this means that the fluid "sees" the plate velocity as a blur!) So, the ansatz would not have pulled us through.

5.5 LUBRICATION OF A SLIPPER BEARING

As an introduction to the theory of hydrodynamic lubrication, let us consider a slipper (or slider or rocker) bearing (Fig. 5.6). For convenience, we will fix our coordinate system on the slipper, which moves with uniform velocity U. This will not change the dynamics, since Newton's second law of motion is still valid with respect to a uniformly moving coordinate system.

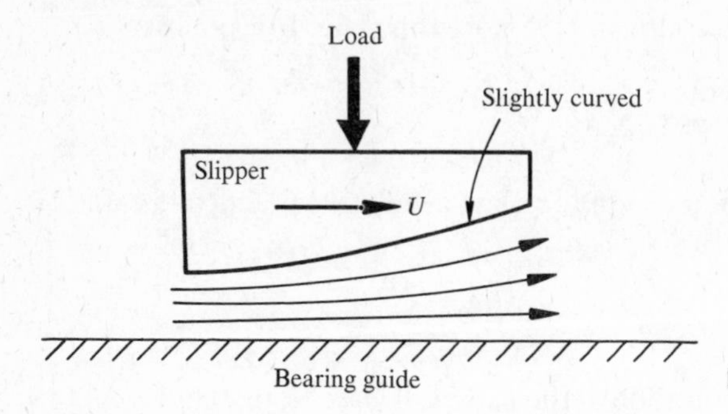

Figure 5.6 Slipper bearing.

To this system the slipper will be stationary, while the guide moves with U in the opposite direction (Fig. 5.7). Neglecting gravity, the (steady) problem to be solved is (using $\tilde{p} = p - p\langle 0,0\rangle$):

$$\begin{cases} \dfrac{\partial u}{\partial x} + \dfrac{\partial v}{\partial y} = 0 & \textbf{(1.9)} \\ u\dfrac{\partial u}{\partial x} + v\dfrac{\partial u}{\partial y} = -\dfrac{1}{\rho}\dfrac{\partial \tilde{p}}{\partial x} + \nu\left(\dfrac{\partial^2 u}{\partial x^2} + \dfrac{\partial^2 u}{\partial y^2}\right) & \textbf{(2.15c)} \\ u\dfrac{\partial v}{\partial x} + v\dfrac{\partial v}{\partial y} = -\dfrac{1}{\rho}\dfrac{\partial \tilde{p}}{\partial y} + \nu\left(\dfrac{\partial^2 v}{\partial x^2} + \dfrac{\partial^2 v}{\partial y^2}\right) & \textbf{(2.15d)} \\ y = 0: \quad u = -U,\, v = 0 & \textbf{(5.49a)} \\ y = h\langle x\rangle: \quad u = 0,\, v = 0,\, 0 < x < L & \textbf{(5.49b)} \\ x = 0: \quad u, v \text{ prescribed for } 0 < y < h_1 & \textbf{(5.49c)} \\ x = L: \quad u, v \text{ prescribed for } 0 < y < h_2 & \textbf{(5.49d)} \end{cases}$$

In practice, the angle of inclination of the slipper face is always small, the gap h is always much smaller than the length L. Taken together, they imply that $h_2 \sim \square(h_1)$ is also much smaller than L. We will try to take advantage of these in introducing a rational approximation.

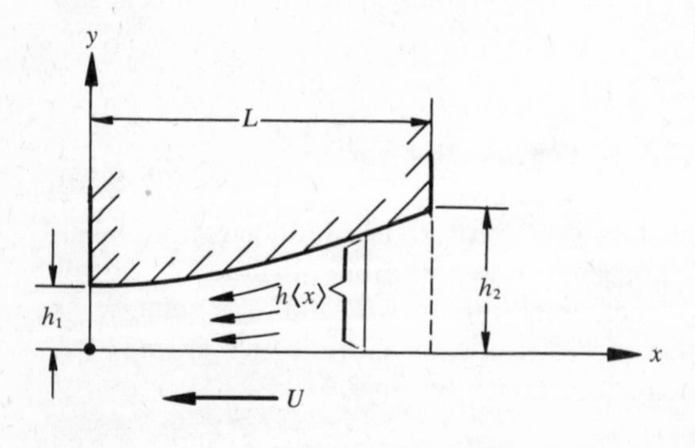

Figure 5.7 Slipper bearing (referred to a coordinate system moving with the slipper).

In ordering, let us use

$$\begin{cases} x^* = L \\ y^* = h_1 \\ v^* = u^* = U \\ \tilde{p}^* = \rho U^2 \end{cases}$$

Nondimensionalization of Eq. (1.9) then yields

$$\left(\frac{h_1}{L}\right)\frac{\partial u'}{\partial x'} + \frac{\partial v'}{\partial y'} = 0$$

As $h_1/L \to 0$, this gives

$$\frac{\partial v'}{\partial y'} = 0$$

that is,

$$v' = v'\langle x'\rangle \text{ only}$$

But $v' = 0$ on $y' = 0$, and thus

$$v' = 0 \tag{5.50}$$

Again, this argument really means that $\partial v'/\partial y' \sim O(h_1/L)$ and

$$v' = 0 + y'[O(h_1/L)]$$

Examination shows that, in order that this approximation scheme be self-consistent, we must require zero[13] v at $x = 0$ and L. If you insist on a perceptible or large v there, the ansatz that "true solution = limit as $h_1/L \to 0$ + error that vanishes as $h_1/L \to 0$" simply does not pull us through. We will assume that this requirement is met in practice.

Substituting Eq. (5.50) into Eq. (2.15d),

$$\frac{\partial \tilde{p}}{\partial y} = 0$$

that is,

$$\tilde{p} = \tilde{p}\langle x\rangle \text{ only} \tag{5.51}$$

Using these, the nondimensionalization of Eq. (2.15c) yields

$$\left(\frac{h_1}{L}\right)^2 u'\frac{\partial u'}{\partial x'} = -\left(\frac{h_1}{L}\right)^2 \frac{d\tilde{p}'}{dx'} + \frac{\nu}{UL}\left[\left(\frac{h_1}{L}\right)^2 \frac{\partial^2 u'}{\partial x'^2} + \frac{\partial^2 u'}{\partial y'^2}\right]$$

If we take $h_1/L \to 0$, this becomes

$$\frac{\partial^2 u'}{\partial y'^2} = 0 \tag{5.52}$$

or

$$u' = f\langle x'\rangle y' + g\langle x'\rangle$$

---

[13] $v \sim O(h_1/L)$ would be enough.

but boundary conditions (5.49a,b) demand now that

$$y' = 0: \quad u' = -1$$

$$y' = h'\langle Lx'\rangle = h''\langle x'\rangle: \quad u' = 0$$

and therefore

$$g\langle x'\rangle = -1$$

$$f\langle x'\rangle = \frac{1}{h''\langle x'\rangle}$$

that is,

$$u' = \frac{y'}{h''\langle x'\rangle} - 1 \tag{5.53}$$

A plot of Eq. (5.53) shows that it represents locally (at every x) the simple shear flow (Fig. 5.8). That is to say, the flow behaves locally as if the slipper surface is parallel to the guide. This approximation is, of course, only compatible when the prescribed u at $x = 0$ and L varies (almost) linearly from 0 to $-U$. (If you force the situation to be otherwise, the ansatz of asymptotics does not work.)

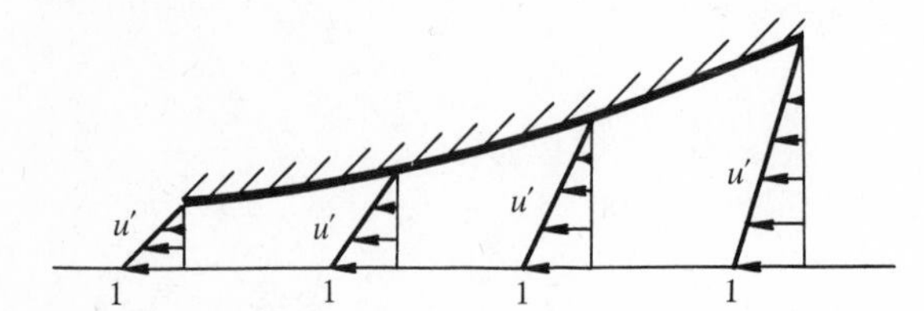

Figure 5.8 "Lubrication" without self-pumping.

Although the foregoing result is interesting in itself, it is not the lubrication problem! In our ordering, we have killed the problem by not realizing one important character of hydrodynamic lubrication; that is, the bearing must be self-pumping, and pumping rather hard (the pressure difference from $x = 0$ to $x = L/2$ must be considerable), in order that the slipper can carry a heavy load. This means that to use $\tilde{p}^* = \rho U^2$ in the ordering is not proper. Actually, the smaller the gap of the bearing, the harder must be the self-pumping. So, $\tilde{p}^*$ must be increased as h_1/L decreases. Let us take a few trials (or ansatz):

1. $\tilde{p}^* = (L/h_1)\rho U^2$. This expression leads right back to Eq. (5.52)!
2. $\tilde{p}^* = (L/h_1)^3\rho U^2$. This expression leads to $d\tilde{p}/dx = 0$; that is,

$$p - p\langle 0, 0\rangle = \text{constant}$$

   In other words, you are pumping so hard that the effect is largely hydrostatic; the hydrodynamic effect is negligible.
3. $\tilde{p}^* = (L/h_1)^2\rho U^2$. This expression leads to

$$\mu\frac{\partial^2 u}{\partial y^2} = \frac{dp}{dx} \tag{5.54}$$

If you are going to succeed at all in getting a solution according to your ansatz of asymptotics, this third route must be the only way.

Before we proceed to solve Eq. (5.54), let us pause and point out that we have just taught you implicitly an advanced technique in ordering—that is, to arrive at the proper ordering with the aid of the governing equation. What we implicitly did here is to inquire, "How should we order $\tilde{p}$ so that pressure term is of the same order as the viscous term?" The necessity of retaining the pressure term can be looked upon from three different angles: (a) The result in the absence of the pressure term does not check with preliminary observations on lubrication. (This is the way we looked at the situation.) (b) Preliminary observations on lubrication show that the pressure force is comparable with (as large as) the viscous force. (c) In simplifying the problem in the asymptotic sense (or any other rational sense) of approximation, one should simplify as little as possible, so that the original problem degenerates as little as possible. This third angle is so important that it is called the principle of least degeneracy (by Van Dyke[14]). Thus Eq. (5.52) is a more degenerate form of Eq. (2.15c) than Eq. (5.54). Of course, we would like to work with Eq. (5.54) if possible[15], since it contains Eq. (5.52) as a special case, anyway. The desirability of such a principle is obvious.

Let us now say that the principle of least degeneracy dictates that Eq. (5.54) be investigated as the satisfactory, approximate, governing equation of hydrodynamic lubrication.

Integrating Eq. (5.54) gives

$$u = \frac{1}{2\mu}\left(\frac{dp}{dx}\right) y^2 + f\langle x\rangle y + g\langle x\rangle$$

Boundary conditions (5.49a,b) easily lead to

$$g\langle x\rangle = -U$$

$$f\langle x\rangle = \frac{U}{h\langle x\rangle} - \frac{h\langle x\rangle}{2\mu}\frac{dp}{dx}$$

Thus

$$u = -U\left(1 - \frac{y}{h}\right) - \frac{h^2}{2\mu}\frac{dp}{dx}\frac{y}{h}\left(1 - \frac{y}{h}\right) \tag{5.55}$$

where $h\langle x\rangle$ is abbreviated to h for ease of writing. To determine dp/dx, let us integrate Eq. (5.55) from $y = 0$ to $y = h$. This is easily seen to be

$$\int_0^{h\langle x\rangle} u\, dy = -\frac{Uh}{2} - \frac{h^3}{12\mu}\frac{dp}{dx}$$

But the left-hand side is just the rate of volume flow of the lubricant through the

---

[14] M. D. Van Dyke, *Perturbation Methods in Fluid Mechanics*, Academic Press, New York, 1964. M. D. Van Dyke is a professor of ae odynamics at Stanford University. See K. O. Friedrichs' remark in a discussion on M. D. Kruskal's "Asymptology," in *Mathematical Models in Physical Sciences*, edited by S. Drobot and P. A. Viebrock, Prentice-Hall, Englewood Cliffs, N.J., 1963: "You want to catch a wild solution. You have to tamper with it. If you tamper too much it tends to something trivial. So you have to make it tame but you have to minimize the tameness."

[15] We would like even more to work with the term $u(\partial u/\partial x)$ retained, but we cannot handle it.

bearing (of unit depth into the paper), which is a constant of the problem easily measured or maintained. We will denote it by Q. Thus

$$\frac{dp}{dx} = -\frac{12\mu}{h^3}\left(\frac{Uh}{2} + Q\right) \tag{5.56}$$

If we wish to integrate Eq. (5.56) to obtain $p\langle x\rangle$, a boundary condition on p must be supplied:

$$p\langle L\rangle - p\langle 0\rangle = -12\mu\left(\frac{U}{2}\int_0^L \frac{dx}{h^2} + Q\cdot\int_0^L \frac{dx}{h^3}\right)$$

Let us say that both ends of the bearing are exposed to a bath of lubricant, or to atmosphere. Then, except for minor modifications due to the flow around the corners of the slipper face, we will have $p\langle L\rangle = p\langle 0\rangle$ as the boundary condition for pressure.[16] This then leads to

$$Q = -\frac{U}{2}\int_0^L \frac{dx}{h^2}\Big/\int_0^L \frac{dx}{h^3} \tag{5.57}$$

Thus the constant Q is not only measurable and maintainable, it is even calculable – purely from the geometry of the gap. For example, if the slipper face is straight – that is, $h = h_1 + (h_2 - h_1)x/L$ – we have

$$Q = -\frac{h_1 h_2}{h_1 + h_2} U$$

But what about the prescribed u at $x = 0$ and L? Well, they are dictated by the solution Eq. (5.55), evaluated at $x = 0$ and L, although a deviation of $O(h_1/L)$ is allowable. If you force any other u-distributions there that differ from those dictated, the ansatz of asymptotics fails. Yet, even if we wish to enforce only the right u-distributions at $x = 0$, L according to Eq. (5.55), how are we going to do it? The answer is easy: Do nothing except to see that $p\langle L\rangle = p\langle 0\rangle$; the flow will adjust itself to conform with Eq. (5.55) to an error $\sim O(h_1/L)$ in the major portion. Near $x = 0$, L, the discrepency may be larger. The influence of this larger deviation from Eq. (5.55) is called the end effect. The end effect on the performance of the entire bearing is small since the bearing is long.

Going back to Eq. (5.56) and substituting the value of Q for a straight slipper face, we have

$$p = p\langle 0\rangle - \frac{6\mu UL}{h^2(h_2^2 - h_1^2)}(h - h_1)(h - h_2) \quad \text{(straight slipper face)} \tag{5.58}$$

A typical plot of Eq. (5.58), see Figure 5.9, shows that p reaches a p_{max} between 0 and L. In the left portion of the bearing, pressure decreases in the flow direction; in the right portion, it increases. A few velocity profiles are also sketched. You should compare them with the plot of Eq. (3.18). The total load that the bearing

---

[16] Do not be fooled by this. It does not mean "no self-pumping." (p_{max} in the bearing) $\gg p\langle 0\rangle$!

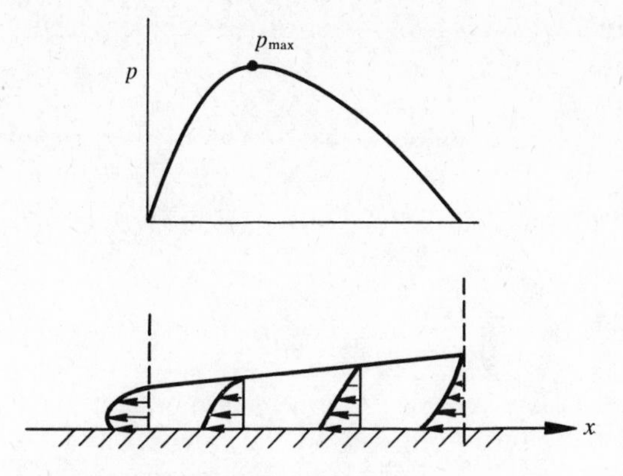

Figure 5.9 Lubrication.

can carry against an ambient pressure of $p\langle 0\rangle$, per unit depth perpendicular to the paper, is just

$$\int_0^L [p - p\langle 0\rangle]\, dx$$

(Heavy load is one object of hydrodynamic lubrication; the other object is small friction.)[17]

[17] Hydrodynamic lubrication was treated theoretically first by Reynolds.

CHAPTER 6

OUTER REGION OF A HIGH-*Re* FLOW FIELD

6.1 INTRODUCTION

In the preceding chapter the asymptotic behavior of a problem was always sought, with the famous Reynolds number, wherever it occurs, assumed to be of $\square(1)$. This is intentional since $Re \to 0$ or ∞ is a troublesome asymptotic process. Its proper handling marks the greatest advancement of fluid dynamics, and is a major gift from the 20th century to the engineering science. Our scope does not allow us a discussion on $Re \to 0$ at this point. From now on, to the end of Part 1, we will devote ourselves exclusively to the study of $Re \to \infty$, or $1/Re \to 0$.

As a key to the understanding of the strange behavior of a fluid field at large Re, let us go back to Section 3.4 and pick up the problem:

$$\begin{cases} \mu \dfrac{d^2u}{dy^2} - \rho v_0 \dfrac{du}{dy} = F - \rho g_x & \textbf{(3.31)} \\ u\langle 0 \rangle = 0 & \textbf{(3.32)} \\ u\langle Y \rangle = U & \textbf{(3.33)} \end{cases}$$

(See also Fig. 3.11.)

To order the problem, we choose

$$y^* = Y$$

$$u^* = U$$

Nondimensionalization then yields

$$\begin{cases} \dfrac{1}{Re}\dfrac{d^2u'}{dy'^2} - v_0{}'\dfrac{du'}{dy'} = C & \textbf{(6.1)} \\ u'\langle 0\rangle = 0 & \textbf{(6.2a)} \\ u'\langle 1\rangle = 1 & \textbf{(6.2b)} \end{cases}$$

where

$$C = \frac{F - \rho g_x}{\rho U^2/Y} \tag{3.33}$$

$$Re = \frac{UY}{\nu} \tag{3.34}$$

and

$$v_0' = \frac{v_0}{U}$$

The exact solution of this problem has been found to be

$$u' = \left(1 + \frac{C}{v_0'}\right)\frac{1 - e^{(Re)v_0'y'}}{1 - e^{(Re)v_0'}} - \frac{Cy'}{v_0'} \tag{3.32}$$

Our task is to find $\lim_{Re \to \infty} u'$ from Eq. (3.32).

In order not to be sidetracked by unimportant trimmings, let us take

$$\begin{cases} v_0 = -U \qquad \text{(that is, } v_0' = -1\text{)} \\ 0 < C < 1 \end{cases}$$

That is, we wish to examine the problem

$$\text{(A)} \quad \begin{cases} \dfrac{1}{Re}\dfrac{d^2u'}{dy'^2} + \dfrac{du'}{dy'} = C < 1, \quad (C > 0) & \textbf{(6.3)} \\ u'\langle 0\rangle = 0 & \textbf{(6.2a)} \\ u'\langle 1\rangle = 1 & \textbf{(6.2b)} \end{cases}$$

with exact solution

$$u' = (1 - C)\frac{1 - e^{-(Re)y'}}{1 - e^{-Re}} + Cy' \tag{6.4}$$

This problem (A) was first investigated with an eye on the limit as $Re \to \infty$ (but as a purely mathematical problem, without the above physical interpretation) by Friedrichs. It is therefore known as the Friedrichs problem.[1]

[1] This name is probably not very well chosen. The problem was apparently common knowledge to the mathematical fluid dynamicists in the 1930s if not already in the 1920s. Furthermore, to honor K. O. Friedrichs with this small problem is just like honoring Kronecker with the "Kronecker delta." However, we do need a short tag here for easy cross reference; hence the name. After all, Friedrichs *did* first put the problem in writing in the 1941 mimeographed notes, *Fluid Dynamics* (with R. von Mises), Brown University, Providence, Rhode Island. K. O. Friedrichs is a professor at the Courant Institute of Mathematical Sciences, New York University.

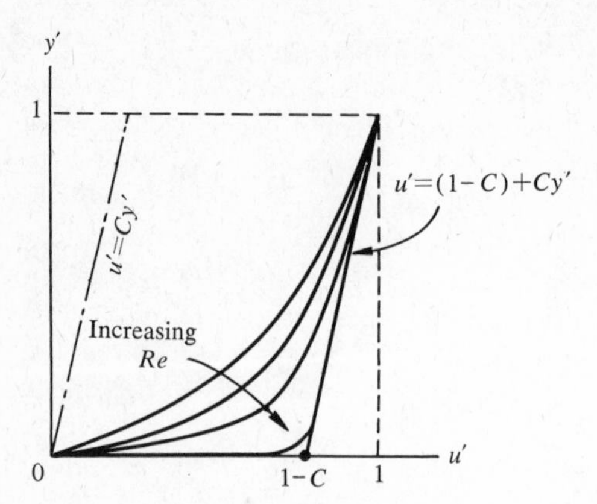

Figure 6.1 Equation (6.4) at large *Re*.

If we plot the exact solution, Eq. (6.4), for large values of Re, we have a sequence of curves as sketched in Figure 6.1. It is seen that for very large Re, u' for the most part follows the straight-line distribution $u' = (1-C)+Cy'$. For the very narrow layer near $y' = 0$, u' changes very rapidly from 0 to $(1-C)$.

Now what is exactly the behavior of the flow at large Re?

If we let $Re \to \infty$ directly in the original system (6.3) and (6.2a,b), we get

$$\begin{cases} \dfrac{du'}{dy'} = C & \textbf{(6.5)} \\ u'\langle 0\rangle = 0 & \textbf{(6.2a)} \\ u'\langle 1\rangle = 1 & \textbf{(6.2b)} \end{cases}$$

We are obviously in trouble. Equation (6.5) is only a first-order ordinary differential equation. Its solution cannot satisfy both boundary conditions (6.2a,b). Whether one uses condition (6.2a) or (6.2b) in solving Eq. (6.5), the result cannot represent the asymptotic behavior of the Friedrichs problem as $Re \to \infty$ *for all values of* y', since it must fail completely in the neighborhood of the boundary where the condition is not satisfied.

The source of trouble here is, of course, the fact that the highest-order derivative d^2u'/dy'^2 happens to be multiplied by $1/Re$ in the original problem. As $Re \to \infty$, this term drops out, thus reducing the order of the ordinary differential equation.

A look at our sketch shows that the solution of Eq. (6.5) with boundary condition (6.2a) satisfied—that is, $u' = Cy'$ (dash-dot line)—is not the approximation of Eq. (6.4) for large Re in any region. However the solution of Eq. (6.5) with boundary condition (6.2b) satisfied—that is, $u' = (1-C)+Cy'$—is the asymptotic limit of Eq. (6.4) as $Re \to \infty$ for the region $y' > 0$. That is, $u' = (1-C)+Cy'$ represents Eq. (6.4) as $Re \to \infty$ everywhere except in the neighborhood of $y' = 0$. It is therefore obvious that the asymptotic limit of Eq. (6.4) must be described differently in the two different regions outside and inside the neighborhood of the boundary $y' = 0$. Again referring back to our sketch, we see that in the neighborhood of $y' = 0$, as $Re \to \infty$, the change of du'/dy' (the slope) is very large. ($y' = 0$ is therefore the *seat of trouble.*) It must change from ∞ down to C in the very narrow neighborhood. So, in this narrow layer, $(1/Re)(d^2u'/dy'^2)$ is not really small and cannot be neglected. This is then the root of our trouble. This narrow neighborhood of $y' = 0$ as $Re \to \infty$ is known as the boundary layer of the flow. The region outside it is known as the outer region of the flow. The solution of Eq. (6.5) with boundary condition (6.2b)—that is, $u' = (1-C)+Cy'$, which is a valid approximation of Eq. (6.4) as $Re \to \infty$ in the outer region—is called the outer flow (the flow outside the boundary layer). Referring back to Eq. (3.31), we see that the outer flow behaves as if the fluid were inviscid. (But it still sticks to the upper plate!) It is really

$$\text{outer flow} = \lim_{\substack{Re\to\infty \\ (y' \text{ fixed})}} \text{ of Eq. (6.4)} \tag{6.6}$$

It satisfies the boundary condition at $y' = 1$ and is valid asymptotically as $Re \to \infty$ *in the outer region.* It cannot satisfy the boundary condition at the seat of trouble. Physically, we say that the no-slip condition at $y' = 0$ is not satisfied.

How can we also get an asymptotic limit, as $Re \to \infty$, that is valid in the boundary layer? To do this, we have to note the fact that $(d^2u'/dy'^2)/Re$ is not small in this region even when $Re \to \infty$. In other words, the proper limiting process here must somehow keep the term $(d^2u'/dy'^2)/Re$ in Eq. (6.3) intact.

To investigate closely the sudden change of du'/dy' over a very thin layer near $y' = 0$, let us *magnify* this layer by introducing a new variable

$$y'' = y'(Re)^N \qquad N > 0 \tag{6.7}$$

As $Re \to \infty$, y' is thus magnified into y''. This will enable us to see more clearly what happens. (In our sketch we cannot see, as $Re \to \infty$, the details of the curve near $y' = 0$ unless we blow up the scale for y' there.) In terms of y'', the original Friedrichs problem in the region $(0,E)$ where E is a small but finite number becomes

$$\begin{cases} (Re)^{N-1}\dfrac{d^2u'}{dy''^2}+\dfrac{du'}{dy''}=\dfrac{C}{(Re)^N} & (6.8) \\ y''=0\text{:}\quad u'\langle y''\rangle = 0 & (6.9\text{a}) \\ y''=E(Re)^N\text{:}\quad u'\langle y''\rangle = A & (6.9\text{b}) \end{cases}$$

where A is the still unknown value of u' at $y' = E$. Now, as $Re \to \infty$, *with* y'' *fixed*, the first term of Eq. (6.8) remains if $N = 1$. We then have, as $Re \to \infty$, with $N = 1$,

$$\begin{cases} \dfrac{d^2u'}{dy''^2}+\dfrac{du'}{dy''}=0 & (6.10) \\ u'\langle 0\rangle = 0 & (6.11\text{a}) \\ u'\langle \infty\rangle = A & (6.11\text{b}) \end{cases}$$

(If $N > 1$, we will obtain

$$\begin{cases} \dfrac{d^2u'}{dy''^2} = 0 & \textbf{(6.12)} \\ u'\langle 0 \rangle = 0 & \textbf{(6.11a)} \\ u'\langle \infty \rangle = A & \textbf{(6.11b)} \end{cases}$$

instead. In view of the principle of least degeneracy, such a drastic simplification is to be avoided.)

Thus, we conclude that the proper saving grace is to magnify y' by Re into

$$y'' = (Re)y' \tag{6.13}$$

In the system, (6.10) and (6.11a,b), we notice three things:

1. The boundary layer experiences, therefore, a negligible effective pressure force.
2. Although the range for the physical distance y' is $(0, E)$, that for the magnified distance y'' is now $(0, \infty)$!
3. A in boundary condition (6.11b) is still unknown. With this unknown A, the solution of Eq. (6.10) under boundary conditions (6.11a,b) is

$$u' = A(1 - e^{-y''}) \tag{6.14}$$

If we did not know the exact solution, Eq. (6.4), of the original Friedrichs problem, we would not know how to go on and determine A. But we do have Eq. (6.4), which, in terms of y'', is

$$u' = (1 - C)\frac{1 - e^{-y''}}{1 - e^{-Re}} + \frac{Cy''}{Re}$$

On taking the corresponding limit, we have

$$\lim_{\substack{Re \to \infty \\ (y'' \text{ fixed})}} u' = (1 - C)(1 - e^{-y''})$$

which checks with Eq. (6.14) if

$$A = 1 - C$$

The proper system to solve in the boundary layer (as $Re \to \infty$) is, therefore,

$$\begin{cases} \dfrac{d^2u'}{dy''^2} + \dfrac{du'}{dy''} = 0 & \textbf{(6.10)} \\ u'\langle 0 \rangle = 0 & \textbf{(6.11a)} \\ u'\langle \infty \rangle = 1 - C & \textbf{(6.11c)} \end{cases}$$

(Note that the effective pressure force C appears in the boundary layer only through a boundary condition.) The problem represented by this system is called the boundary-layer problem and its solution

$$u' = (1-C)(1-e^{-y''}) \tag{6.15}$$

is known as the boundary-layer flow. Figure 6.2 shows a sketch of Eq. (6.15). The same sketch in terms of the unmagnified y' is almost indistinguishable from the u'-axis from $y' = 0$ to $y' = E$ (a very small number); see Figure 6.3. We say that u' changes rapidly from 0 at the wall $y' = 0$ to $1-C$ at the edge of the boundary layer, which is very thin (as $Re \to \infty$).

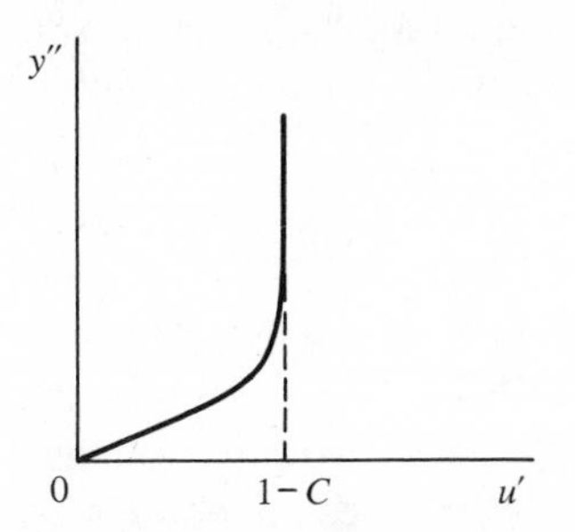

Figure 6.2 Boundary-layer flow.

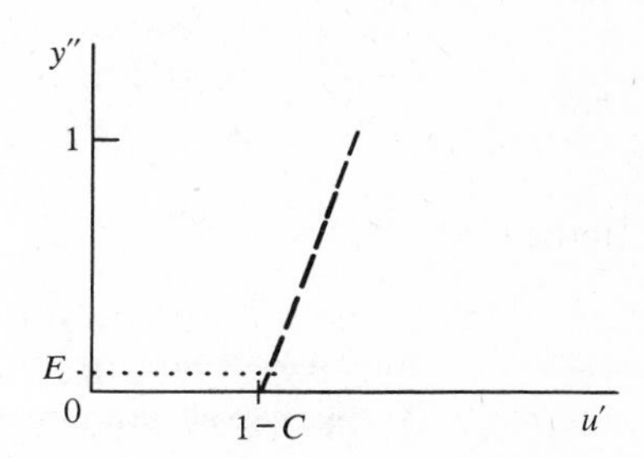

Figure 6.3 Outer flow.

The above is then the correct description of the asymptotic behavior of the flow as $Re \to \infty$. We see, however, that the success lies in our identifying A with $1-C$ (or, physically, identifying the velocity at the edge of the boundary layer with $1-C$). We would like to ask here:

1. What is this magic quantity $1-C$?
2. If we did not know the exact solution, Eq. (6.4), how can we determine A properly?

The answers to these questions are as follows:

1. $1-C$ is just the outer flow evaluated at $y' = 0$. A is actually the boundary-layer flow evaluated at $y'' = \infty$. The identification of A with $1-C$ is, therefore, a matching of the outer flow with the boundary-layer flow through

$$(\text{outer flow})_{y'=0} = (\text{boundary-layer flow})_{y''=\infty} \tag{6.16}$$

2. The previous matching equation (6.16) is actually the heart of the matter. After all, the two asymptotic limits valid in the two regions must match in some fashion to form a complete representation of the entire flow. In the present example, Eq. (6.16) is in reality derived from the exact solution, Eq. (6.4). But it should underlie all similar flow problems with large Re. We will therefore enshrine it by calling it the *matching principle*. If we did not know the exact solution, Eq. (6.4), we could have incurred this matching principle and obtained boundary condition (6.11c). It is important to notice

that the matching principle does *not* require the outer flow and the boundary layer flow to match at $y' = E$, or at the "edge" of the thin boundary layer. This latter event is physically suggestive but mathematically senseless according to our exposition. This must be cautioned against because of the many loose statements on how the boundary-layer flow blends into the outer flow at the "edge" of the layer. These statements can be correct only if Eq. (6.16) is strictly observed. A better description of the matching would be "the boundary-layer flow at the edge of the boundary layer (meaning $y'' = \infty$) is equal to that of the outer flow *at the bottom plate, not* the outer flow at the edge of the boundary layer!"

Before we go on, let us summarize the previous discussion in physical terms.

In the presence of a moderate downward flow, there appears as $Re \to \infty$ a thin boundary layer near the bottom, stationary plate of the channel. Physically speaking, the boundary layer cannot appear on the upper plate because of the downward blowing. The same downward blowing, however, presses the boundary layer onto the bottom plate. The flow inside the boundary layer obeys the no-slip condition at the bottom plate but is not acted upon by the effective pressure force. Outside the boundary layer the flow behaves as if it *were* inviscid, but it still follows the motion of the upper plate. It slips right past the bottom plate, but matches with the (magnified) boundary-layer flow at y'' (magnified distance away from the bottom plate) $= \infty$.

This then is our first encounter with a boundary-layer flow. This simple example is actually the prototype for the more general, and complicated, boundary-layer flow. It is therefore the key to the general boundary-layer theory. It may be desirable to go over the problem once more—this time pretending that we do not know the exact solution!

The original Friedrichs problem is

$$\begin{cases} \dfrac{1}{Re}\dfrac{d^2u'}{dy'^2} + \dfrac{du'}{dy'} = C \\ u'\langle 0 \rangle = 0 \\ u'\langle 1 \rangle = 1 \end{cases}$$

The fact that the highest-order derivative is multiplied by $1/Re$ indicates trouble in investigating the asymptotic behavior as $Re \to \infty$. If we take $\lim Re \to \infty$ with y' fixed, we have

$$\begin{cases} \dfrac{du'}{dy'} = C \\ u'\langle 0 \rangle = 0 \\ u'\langle 1 \rangle = 1 \end{cases}$$

which indicates that the fluid behaves as if it *were* inviscid. Satisfying the boundary condition $u'\langle 1\rangle = 1$, we have $u' = 1-C+Cy'$, leaving the boundary condition at the seat of trouble $y' = 0$ *unsatisfied.* This is the outer flow that slips past the bottom plate and is invalid near this seat of trouble $y' = 0$. For the neighborhood of $y' = 0$, we use a magnified coordinate $y'' = (Re)y'$ so chosen as to (a) keep the highest-order derivative term, and (b) degrade the system as little as possible. We then take the limit $Re \to \infty$ with y'' fixed:

$$\begin{cases} \dfrac{d^2u'}{dy''^2}+\dfrac{du'}{dy''}=0 \\ y''=0: \quad u'=0 \\ y''=\infty: \quad u' = (\text{outer flow})_{y'=0} \\ \qquad\qquad\quad = 1-C \end{cases}$$

where the matching principle is incurred. The solution

$$u' = (1-C)(1-e^{-y''})$$
$$= (1-C)(1-e^{-(Re)y'})$$

is then the boundary-layer flow (see Figs. 6.4 and 6.5). It is important to realize that if $v_0 = 0$ in Eq. (3.31), there will be no boundary layer whatsoever for large Re. The solution is just Eq. (3.18). So it is only *possible* to have a boundary layer at large Re; it is not a necessity! Another thing to watch is that if Re is too large, turbulence sets in.

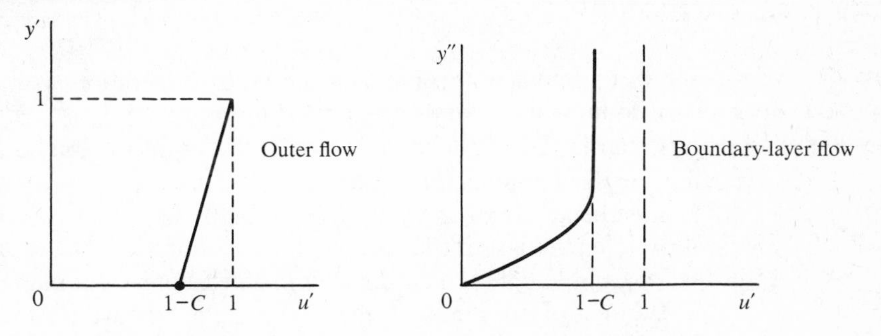

Figure 6.4 Outer and boundary-layer flow.

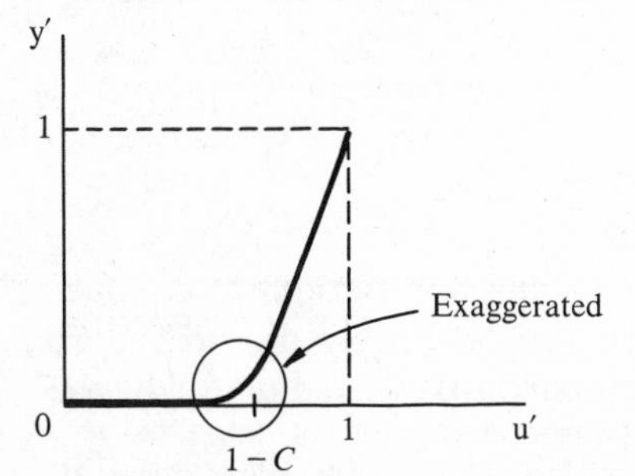

Figure 6.5 The composite flow.

In the absence of the exact solution, how do we know that the seat of trouble is $y' = 0$? Or how do we know that the outer flow should satisfy the condition $u'\langle 1\rangle = 1$, not $u'\langle 0\rangle = 0$? Let us try the alternate route and see. The "outer flow" with $u'\langle 0\rangle = 0$ is just $u' = Cy'$. The seat of trouble here is $y' = 1$. In order to magnify the "boundary layer" near $y' = 1$, let us first introduce $\underset{\sim}{y} = Y - y$; that is, $\underset{\sim}{y}' = 1 - y'$. In terms of $\underset{\sim}{y}'$, the original system is

$$\begin{cases} \dfrac{1}{Re}\dfrac{d^2u'}{d\underset{\sim}{y}'^2} - \dfrac{du'}{d\underset{\sim}{y}'} = C \\ \underset{\sim}{y}' = 1: \quad u' = 0 \\ \underset{\sim}{y}' = 0: \quad u' = 1 \end{cases}$$

Again, we magnify $\underset{\sim}{y}'$ (that is, $1 - y'$) by introducing

$$\underset{\sim}{y}'' = (Re)\underset{\sim}{y}'$$

We have then

$$\begin{cases} \dfrac{d^2u'}{d\underset{\sim}{y}''^2} - \dfrac{du'}{d\underset{\sim}{y}''} = 0 \\ \underset{\sim}{y}'' = 0: \quad u' = 1 \\ \underset{\sim}{y}'' = \infty: \quad u' = (\text{"outer flow"})_{\underset{\sim}{y}' = 0} \\ \qquad\qquad\qquad = C \end{cases}$$

The solution with the boundary condition at $\underset{\sim}{y}'' = 0$ satisfied is

$$u' = (1 - B)e^{\underset{\sim}{y}''} + B$$

But as $\underset{\sim}{y}'' \to \infty$, this blows up and can never satisfy the second condition. We say that there is no solution to this "boundary-layer problem." Please notice that the key here is the positive sign in the exponential index, or the negative sign before $du'/d\underset{\sim}{y}''$. Physically, this means that no boundary layer can form on the upper plate. So the seat of trouble must be at $y' = 0$, not $y' = 1$. You might also look upon the location of the seat of trouble as a part of a large ansatz of asymptotics. The ansatz with $y' = 1$ as the seat of trouble failed to go through in a self-consistent manner, but that with $y' = 0$ as the seat of trouble succeeded. This is then the way of locating the seat of trouble without knowing the exact solution. Actually, once you have some experience with fluid dynamics, you will be able to identify correctly the seat of trouble on physical grounds.

6.2 HIGH-*Re* FLOW PAST A FLAT WALL

With the previous prototype in mind, we can now set out to conquer the general high-Reynolds-number flow.[2] We will follow closely the procedure described at the end of Section 6.1 (where we pretended that we did not know the

[2] We follow here essentially the works of Lagerstrom and his students, especially Kaplun; although their works were already foreshadowed in part by Schmidt and Schroeder (H. Schmidt and

exact solution of the Friedrichs problem). But we will have no exact solution to guide us. Although the heart of the matter is the same as with the Friedrichs problem, we will also encounter some new phenomena which have no counterpart in the Friedrichs problem.

Consider a uniform stream approaching a flat wall ($y = 0$, $x \geqslant 0$) from far left with an angle of inclination α as shown in Figure 6.6. The flat wall is actually the top surface of a body whose lower surface is given as $y = h\langle x\rangle$. We use the general shape $y = h\langle x\rangle$ here only for the practice. Later, we will be treating the special case where $y = h\langle x\rangle$ is a straight line. In other words, our main concern will be a flow past a wedge-shaped body.

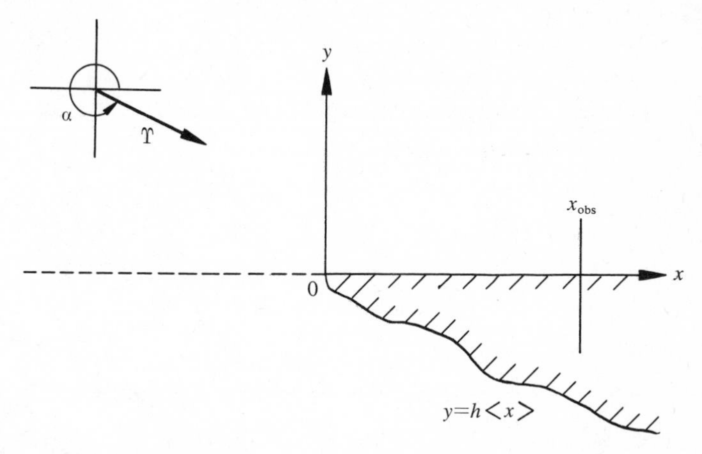

Figure 6.6 A stream approaching a flat wall.

Let us focus our attention on the flow around a location of observation x_{obs}. For proper ordering, we choose

$$y^* = x^* = x_{obs}$$

And, for the characteristic speed, we use the approaching speed Υ (> 0):

$$u^* = v^* = \Upsilon$$

and

$$p^* = \rho\Upsilon^2$$

From now on we will neglect the gravity and treat only steady flows. The governing equations are the dimensionless equation of continuity, and steady Navier-

K. Schroeder, "Laminare Grenzschichten. Ein kritischer Literaturbericht," *Luftfahtforschung*, **19**, 65–97, 1942). We hesitate, however, to give references here because they are rather difficult for beginners. You might try to read Van Dyke's book (see footnote 14 in Chapter 5), which is probably the most accessible. P. A. Lagerstrom is professor of aerodynamics at California Institute of Technology. S. Kaplun (1924–1964) was on Caltech's research staff when he died suddenly.

Stokes equations:

$$\begin{cases} \dfrac{\partial u'}{\partial x'}+\dfrac{\partial v'}{\partial y'}=0 & \textbf{(1.9A)} \\ u'\dfrac{\partial u'}{\partial x'}+v'\dfrac{\partial u'}{\partial y'}=-\dfrac{\partial p'}{\partial x'}+\dfrac{1}{Re}\left(\dfrac{\partial^2 u'}{\partial x'^2}+\dfrac{\partial^2 u'}{\partial y'^2}\right) & \textbf{(6.17a)} \\ u'\dfrac{\partial v'}{\partial x'}+v'\dfrac{\partial v'}{\partial y'}=-\dfrac{\partial p'}{\partial y'}+\dfrac{1}{Re}\left(\dfrac{\partial^2 v'}{\partial x'^2}+\dfrac{\partial^2 v'}{\partial y'^2}\right) & \textbf{(6.17b)} \end{cases}$$

where $Re = \Upsilon x^*/\nu$. Now for fluids with small viscosity at high speed and for a region farther downstream from the origin, Re is very large. We would like to learn what happens as $Re \to \infty$.

The original system of equations is to be solved subject to the following conditions (in the dimensionless form):

$x' > 0, y' = 0$:

$$u' = 0 \quad \textbf{(6.18a)}$$

$$v' = 0 \quad \textbf{(6.18b)}$$

$x' > 0, y' = h'\langle x' \rangle$:

$$q_{\text{tangential to } y'=h'\langle x'\rangle} = 0 \quad \textbf{(6.18c)}$$

$$q_{\text{normal, etc.}} = 0 \quad \textbf{(6.18d)}$$

$\sqrt{x'^2+y'^2} \to \infty$ (excluding the body and its surface):

$$\left.\begin{aligned} u' &\to \cos\alpha \\ v' &\to \sin\alpha \end{aligned}\right\} \quad \textbf{(6.18e)}$$

Boundary condition (6.18e) can be generalized to approach a given flow (not uniform) far away. Boundary conditions (6.18a,c) are the no-slip conditions on the body. Boundary conditions (6.18b,d) indicate that the body is rigid and impermeable. Boundary condition (6.18e) states that the uniform stream of approach must prevail far away from the body.

Now the highest-order derivatives of Eqs. (6.17a,b) are multiplied by $1/Re$. Therefore, as $Re \to \infty$, we should expect the same kind of trouble as in Section 6.1; namely, there will be a boundary layer that must be magnified in order to be seen clearly. Taking the limit of Eqs. (1.9A) and (6.17a,b) as $Re \to \infty$ with y' (distance away from the body) fixed, we have

$$\begin{cases} \dfrac{\partial u'}{\partial x'}+\dfrac{\partial v'}{\partial y'}=0 & \textbf{(1.9A)} \\ u'\dfrac{\partial u'}{\partial x'}+v'\dfrac{\partial u'}{\partial y'}=-\dfrac{\partial p'}{\partial x'} & \textbf{(6.19a)} \\ u'\dfrac{\partial v'}{\partial x'}+v'\dfrac{\partial v'}{\partial y'}=-\dfrac{\partial p'}{\partial y'} & \textbf{(6.19b)} \end{cases}$$

The upshot of this (as in Section 6.1) is that the fluid in the outer region acts as if it were inviscid. (It cannot be truly inviscid, since no real fluid is inviscid. It is only that the viscous force is *negligible* in the outer region.) Equations (6.19a,b) are the famous Euler's equations.[3]

It is physically obvious that such a flow would not satisfy the no-slip conditions (6.18a,c) on the body. So, without further ado, we will identify the seat of trouble with the surface of the body.

The solution of Eqs. (1.9A) and (6.19a,b) with conditions (6.18e) and (6.16) (the matching principle) is then the outer flow. (At this stage, one should not try to enforce boundary conditions (6.18b,d) since we have already expected trouble at the body surface.) A rudimentary discussion of the outer flow will be the subject of the next section. At this point, let us say that we are able to obtain the outer flow and go on from here.

This outer flow is certainly not valid near the body surface where a boundary layer forms across which the terms

$$\frac{\partial^2 u'}{\partial x'^2}, \frac{\partial^2 u'}{\partial y'^2}, \frac{\partial^2 v'}{\partial x'^2}, \frac{\partial^2 v'}{\partial y'^2}$$

are not all $O(1)$. To see clearly in this layer, let us magnify, as in Section 6.1, the y'-coordinate by introducing

$$y'' = (Re)^N y' \qquad N > 0$$

Another way of looking at this is to say that x^* is not proper for ordering y. The proper y^* is a much smaller quantity $x^*/(Re)^N$. Then, $y'' = y/[x^*/(Re)^N] \sim \square(1)$. The object here is to determine N such that (a) at least one of the terms listed above remains after taking lim $Re \to \infty$ with y'' fixed, and (b) the original system is degraded as little as possible (principle of least degeneration).

Now, writing Eq. (1.9A) in terms of y''

$$\frac{\partial u'}{\partial x'} + (Re)^N \left(\frac{\partial v'}{\partial y''}\right) = 0 \tag{6.20}$$

we see that, as $Re \to \infty$ with y'' fixed,

$$\frac{\partial v'}{\partial y''} = 0$$

This is going a little too far! After all, the term $\partial u'/\partial x'$ in (1.9A) is not the trouble-maker. If we start by chopping off a part of the innocent equation of continuity, we would end up with more error (which may be unbearable) in the approximation than necessary. In order to start with the unmutilated equation of continuity, we are *suddenly* led by the principle of least degeneration to the profound realization that

$$\frac{\partial v'}{\partial y''} \textit{ is not } \square(1), \textit{ but } O(Re^{-N})$$

---

[3] Euler's 1755 paper, cited in footnote 2 of Chapter 1. It even appeared in his 1752 paper, "Principles of fluids in motion" (in Latin), but not in clear and definitive form. Solutions of Eqs. (1.9A) and (6.19a,b) are sometimes called Euler limits (of the true solutions) or Euler flows.

such that the two terms in Eq. (6.20) are of the same order of magnitude and both should stay as $Re \to \infty$. This is something new; it has no counterpart in the Friedrichs problem.

Since $y'' \sim \square(1)$ owing to the fact that we have now used the proper $y^* = x^*/(Re)^N$, the above realization is equivalent to

$$v' = v/\Upsilon \textit{ is small of } O(Re^{-N})$$

in the boundary layer. We have not gone very far (we have not even found the value of N); yet, just our desire to have a rational approximation with the least degeneration already led us to a great discovery—that is, in the boundary layer the transverse flow is much smaller than the longitudinal flow. Thus encouraged, we are ready to plunge ahead. But, maybe we should first consolidate our new information by introducing

$$\begin{aligned} v'' &= (Re)^N v' \\ &= \frac{v}{\Upsilon(Re)^{-N}} \end{aligned} \tag{6.21}$$

The second expression means that we found Υ to be too large for the ordering of v in the boundary layer and hence replaced it by $\Upsilon(Re)^{-N}$ to make $v'' \sim O(1)$.

In terms of v'' and y'', Eq. (1.9A) becomes

$$\frac{\partial u'}{\partial x'} + \frac{\partial v''}{\partial y''} = 0 \tag{6.22}$$

which remains the same as $Re \to \infty$ with fixed y''.

Similarly, in terms of v'' and y'', Eq. (6.17a,b) become

$$\begin{cases} u'\dfrac{\partial u'}{\partial x'} + v''\dfrac{\partial u'}{\partial y''} = -\dfrac{\partial p'}{\partial x'} + \dfrac{1}{Re}\left[\dfrac{\partial^2 u'}{\partial x'^2} + (Re)^{2N}\dfrac{\partial^2 u'}{\partial y''^2}\right] & \text{(6.23a)} \\ (Re)^{-N}\left(u'\dfrac{\partial v''}{\partial x'} + v''\dfrac{\partial v''}{\partial y''}\right) = -(Re)^N\dfrac{\partial p'}{\partial y''} + (Re)^{-N-1}\left[\dfrac{\partial^2 v''}{\partial x'^2} + (Re)^{2N}\dfrac{\partial^2 v''}{\partial y''^2}\right] & \text{(6.23b)} \end{cases}$$

As $Re \to \infty$ with y'' fixed, Eq. (6.23b) reduces for any $N > 0$ to

$$\frac{\partial p'}{\partial y''} = 0$$

that is, $p' = p'\langle x' \rangle$ only! So, again, without having determined the value of N we have already learned that the pressure does not change (at each x-value) across the boundary layer.

As to Eq. (6.23a), we see that a viscous term (namely $\partial^2 u'/\partial y''^2$) would remain, without inducing further degeneracy, only when

$$2N - 1 = 0$$

that is,

$$N = \tfrac{1}{2} \tag{6.24}$$

Although this is obtained the same way as in Section 6.1, it is now $\frac{1}{2}$, not 1 as in the

Friedrichs problem. With $N = \frac{1}{2}$, Eq. (6.23a) becomes in the limit $Re \to \infty$ with fixed $y'' = \sqrt{Re}\, y'$

$$u' \frac{\partial u'}{\partial x'} + v'' \frac{\partial u'}{\partial y''} = -\frac{dp'}{dx'} + \frac{\partial^2 u'}{\partial y''^2} \qquad \textbf{(6.25)}$$

We have incorporated the information that "$p' = p'\langle x' \rangle$ only" in Eq. (6.25) by writing dp'/dx'. The approximation represented by Eq. (6.25) is just right—not too much, not too little. Equation (6.25) is then the key to the boundary-layer flow and is known as the Prandtl's boundary-layer equation.[4] It is, of course, to be solved together with Eq. (6.22).

Now, within the present scope, we will concentrate on the *boundary layer adjacent to the upper flat face* $y = 0$ (and $x > 0$). To solve this part of the boundary-layer problem, we need the boundary conditions

$x' > 0, y'' = 0$:

$$u' = 0 \qquad \textbf{(6.18a)}$$

$$v'' = 0 \qquad \textbf{(6.18f)}$$

on the flat wall. As in Section 6.1, it is senseless to enforce boundary conditions (6.18e) far away from the wall. Instead, we have to incur the matching principle:

$$(\text{boundary-layer flow})_{y''=\infty} = (\text{outer flow})_{y'=0}$$

We will denote the flow quantities in the outer region by superscript o, which denotes "outer," and the values at $y' = 0$ by subscript w, which denotes "wall." Strictly speaking, we should have denoted the quantities in the boundary-layer by $(\ \ )_b$. But for the sake of simplicity we shall refrain from doing this. From now on, whether $(\ \ )$ denotes a quantity in a boundary layer or not should be clear from the context. Thus,

$$(\text{boundary-layer flow})_{y''=\infty} = (\ \ )_w{}^o \qquad \textbf{(6.26)}$$

Applying condition (6.26) to the pressure p', we have

$$(p' \text{ in boundary layer})_{y''=\infty} = p'^{o}_{w}\langle x' \rangle \qquad \textbf{(6.27)}$$

[4] Historically, the development of the boundary-layer theory is not at all as we presented. The story went back to Euler whose equations actually took care of a large class of fluid dynamic problems, including the theory of flight. However, in the matter of viscous drags, Euler flows do not yield anything because of their failure to satisfy the no-slip condition. Actual observations on boundary-layer flows started early. Froude in 1872 already reported on "Experiments on the surface-friction experienced by a plane moving through water." But it was Prandtl who, in 1904, first arrived at Eq. (6.25). That today we choose not to follow Prandtl's original argument should not be regarded as belittling Prandtl's genius. It is, on the contrary, a tribute to him in that his physical intuition required 50 years of annotation to be really understood. Solutions of the boundary-layer equation are sometimes termed Prandtl flows or Prandtl limits. Ludwig Prandtl (1875–1953) was professor of mechanics at the University of Göttingen. He was noted for perceiving immediately the heart of the matter of a physical phenomenon, constructing a simple mathematical model for it, and performing crucial experimental tests on the model. The trimmings and details were usually added on later by his students, including von Kármán, Blasius, and Tollmien. At Göttingen, one day Prandtl suddenly decided that it was time for him to get married. So he wrote to Frau Föppl, wife of his teacher at Munich, asking the hand of *either* of her two daughters. The marriage that followed was a successful one; they also had two daughters.

where $p_w'^o\langle x'\rangle$ is the pressure of the outer flow evaluated at the wall ($y' = 0$). Its variation with x' is supposedly known from the solution of the outer-flow problem. Physically, Eq. (6.27) is often stated as "the outer flow impresses its pressure distribution on the boundary layer, undiminished across the boundary layer." But it must be remembered that it is the value of p'^o evaluated at the wall, $y' = 0$, that is impressed on the boundary layer. It is *not* the value of p'^o at a small distance E away from the plate!

With Eq. (6.27), Eq. (6.25) becomes

$$u'\frac{\partial u'}{\partial x'} + v''\frac{\partial u'}{\partial y''} = -\frac{dp_w'^o}{dx'} + \frac{\partial^2 u'}{\partial y''^2} \tag{6.28}$$

The matching of u' is, similarly,

$$(u' \text{ in boundary layer})_{y''=\infty} = u_w'^o\langle x'\rangle \tag{6.29}$$

where $u_w'^o\langle x'\rangle$ is the x-component of the outer (nondimensionalized) velocity evaluated at the wall, and *not* a short distance E away from the wall. $u_w'^o\langle x'\rangle$ is also supposedly known from the outer-flow problem.

In matching up v', we have to recall two things:

1. $v_w'^o$ is still floating around, unknown. It is *not* zero *by way of* boundary condition (6.18b), since the outer flow is not expected to be valid near $y' = 0$.
2. v' (in the boundary layer) is a small quantity, almost zero. To be exact,

$$v' = 0 + O\left(\frac{1}{\sqrt{Re}}\right)$$
$$= 0 + o\,(1)$$

that is, it is zero *with an error* of $O(1/\sqrt{Re})$ or $o\,(1)$. Or, even more exactly,

$$v' = 0 + \frac{v''}{\sqrt{Re}}$$

where $v'' \sim O(1)$.

So a matching, as $Re \to \infty$, leads to

$$v_w'^o = 0 \tag{6.30}$$

which is a (so-far missing) boundary condition for the outer-flow problem! Please notice that

$$x' > 0,\ y' = 0\text{:}\quad v'^o = 0$$

not because of the original boundary condition (6.18b), *but because of the matching principle*. Furthermore, this argument on the matching of v' can also be applied to the lower surface $y' = h'\langle x'\rangle$ with v' replaced by $q'_{\text{normal to the surface}}$ and y replaced by the normal distance from the wall. We then have for the outer flow, the boundary conditions

$$x' > 0, y' = 0: \quad v'^{o} = 0 \tag{6.31a}$$

$$y' = h'\langle x'\rangle: \quad q'^{o}_{\text{normal}} = 0 \tag{6.31b}$$

We have given them new numerals to remind you that they are *not* the same as boundary conditions (6.18b,d).

Notes

1. It is only at this point that the outer-flow problem[5] is complete and ready to be solved.
2. The matching of v' furnishes only boundary conditions for the outer flow. It says nothing at all about v'' in the boundary layer. In other words, there is absolutely *no* condition to be satisfied by v'' at $y'' = \infty$! As a matter of fact, v'' at $y'' = \infty$ will come out as a part of the solution of the boundary-layer flow,[6] which is definitely *not* zero. Physically, this means that there is a vertical flow at the "edge" of the boundary layer contrary to the usual belief that the boundary-layer flow "blends" into one parallel to the wall. This vertical flow is small, $v' \sim \square(1/\sqrt{Re})$, but it is there. (The boundary-layer flow almost "blends" into a parallel flow, but not quite! This imperfection is expected; after all, it is only an approximation.)

Summary

1. As $Re \to \infty$, there *may* be a boundary layer on a wall such that an outer flow describes the situation outside it; and a boundary-layer flow, inside it. (Counterexample: In a fully developed channel flow without suction and blowing there is *no* boundary layer, even at large Re.)
2. The outer flow is the solution of Eqs. (1.9A) and (6.19a,b) subject to conditions (6.18e) and (6.31a,b).
3. The boundary-layer solution is obtained by properly magnifying the perpendicular distance from the wall (for example, $y'' = \sqrt{Re}\, y'$), and by realizing that the flow normal to the wall is very small ($\sim O(1/\sqrt{Re})$) in the boundary layer.
4. The pressure inside the boundary layer does not change across (in the normal direction of) the boundary layer, being equal to the pressure "impressed" by the outer flow (evaluated on the wall). There-

---

[5] Called the Euler problem.

[6] Only the first-order derivatives of v'' appear in the governing equations. So one boundary condition at $y'' = 0$, (6, 18f), is enough.

fore, the forces on the body that originate mainly from the pressure on its surface—for example, lift on an airfoil—are insensitive to the boundary layer near it. The lift on airfoils at large Re calculated from the outer (inviscid) flow (note the term inviscid *flow*, not inviscid *fluid* which is fictitious) checks, therefore, very well with observations.[7] It is only in the calculation of drags that the boundary-layer is needed.

5. The boundary-layer flow can be determined only after the outer flow is found.
6. The boundary-layer flow is the solution of Eqs. (6.22) and (6.28) subject to conditions (6.18a,f) and (6.29).[8]
7. There is a small but not zero ($\sim \square(1/\sqrt{Re})$) "residue" v' at $y'' = \infty$ (the "edge" of the boundary layer).
8. It is interesting to note that the above discussion holds even when there is slight suction or blowing at the wall such that

$$x' > 0, \text{ on the wall:} \quad q'_{\text{normal}} = \frac{a\langle x'\rangle}{\sqrt{Re}}$$

where $a\langle x'\rangle$ is a given function of x', and $a\langle x'\rangle \sim \square(1)$. The only difference is that for the boundary-layer flow, boundary condition (6.18f) must be replaced by

$$x' > 0, y'' = 0: \quad v'' = a\langle x'\rangle \qquad \textbf{(6.18g)}$$

(And similarly for the lower surface, if we are interested.) In particular, the outer-flow problem is exactly the same as before. In other words, the outer flow is not sensitive to a slight blowing or suction at the body surface. The boundary-layer flow is, of course, changed. Boundary-layer flows with slight suction and blowing are important in a number of applications, such as airfoil performance[9] and turbine-blade cooling. (However, if the blowing or suction is not slight, we will have a totally different problem at hand. It will no longer fit in here.)

9. Although we will not strive for generality here, it is interesting to point out that, for a general body shape, y should always read "coordinate normal to the surface"; and v, "velocity component normal to the surface". The discussion then remains the same as before. Furthermore, the surface may actually form the walls of a channel; see Figures 6.7 and 6.8.

---

[7] Thus the theory of flight rests mainly on the Euler problem.

[8] Called the Prandtl problem.

[9] Especially the transition from the laminar to turbulent boundary layer and the separation of the boundary layer from the airfoil.

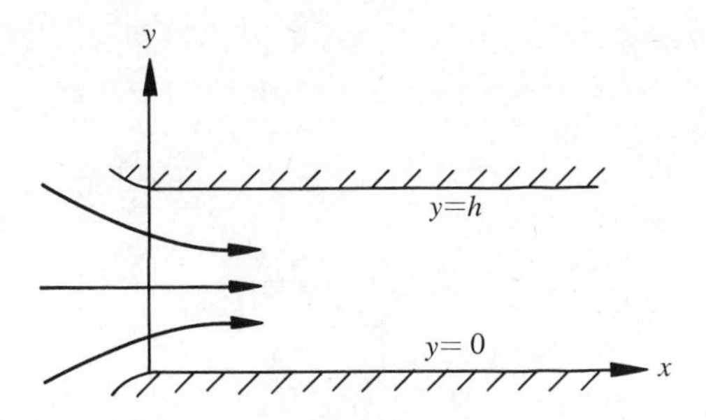

Figure 6.7 Boundary layers inside a channel.

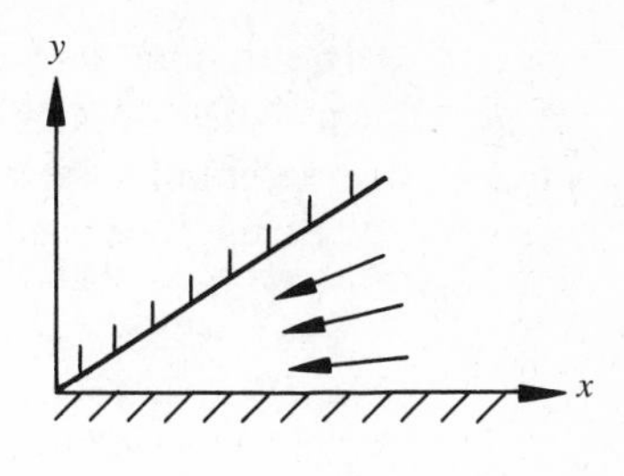

Figure 6.8 Boundary layers inside a wedge.

6.3 THE OUTER FLOW

In this section we will discuss briefly the outer flow:

$$\left\{\begin{array}{ll}
\dfrac{\partial u'^o}{\partial x'}+\dfrac{\partial v'^o}{\partial y'}=0 & \textbf{(1.9A)} \\[2ex]
u'^o\dfrac{\partial u'^o}{\partial x'}+v'^o\dfrac{\partial u'^o}{\partial y'}=-\dfrac{\partial p'^o}{\partial x'} & \textbf{(6.19a)} \\[2ex]
u'^o\dfrac{\partial v'^o}{\partial x'}+v'^o\dfrac{\partial v'^o}{\partial y'}=-\dfrac{\partial p'^o}{\partial y'} & \textbf{(6.19b)} \\[2ex]
\sqrt{x'^2+y'^2}\rightarrow\infty \text{ (excluding the body and its surface):} & \\
\qquad\qquad \left.\begin{array}{l} u'^o\rightarrow\cos\alpha \\ v'^o\rightarrow\sin\alpha \end{array}\right\} & \textbf{(6.18e)} \\
x'>0,\ y'=0: & \\
\qquad\qquad v'^o=0 & \textbf{(6.31a)} \\
\qquad y'=h'\langle x'\rangle: & \\
\qquad\qquad q'^o_{\text{normal}}=0 & \textbf{(6.31b)}
\end{array}\right.$$

where boundary conditions (6.18e) can be generalized to approach a given non-uniform flow far away. It is important to recall that, in the outer flow, the viscous force is negligible compared with the other forces. But the fluid is *not* inviscid. If it were, there would be no boundary layer at all between the wall and the outer region.

We study the above Euler problem because:

1. It is needed for the boundary-layer flow.
2. It is valid everywhere except close to the boundary layer.
3. It leads to valid pressure distribution on the body even in the presence of the boundary layer.

In this section, we will only present enough rudiments of it to enable us to go on with the boundary-layer flow. Details of the Euler problem belong to another course or book (on aerodynamics).

Let us start our study by investigating the Euler equations along a streamline on which (in dimensionless form)

$$\frac{dy'}{dx'} = \frac{v'^o\langle x', y'\rangle}{u'^o\langle x', y'\rangle} \tag{1.20}$$

To do this, we multiply Eq. (6.19a) by dx' and Eq. (6.19b) by dy':

$$u'^o \frac{\partial u'^o}{\partial x'} dx' + \underbrace{\underline{v'^o} \frac{\partial u'^o}{\partial y'} \underline{dx'}}_{u'^o\, dy' \quad \text{[by Eq. (1.20)]}} = -\frac{\partial p'^o}{\partial x'} dx'$$

$$\underbrace{\underline{u'^o} \frac{\partial v'^o}{\partial x'} \underline{dy'}}_{v'^o\, dx' \quad \text{[(by Eq. (1.20)]}} + v'^o \frac{\partial v'^o}{\partial y'} dy' = -\frac{\partial p'^o}{\partial y'} dy$$

$$+)\overline{\qquad d[\tfrac{1}{2}(q'^o)^2] = d[\tfrac{1}{2}(u'^o)^2 + \tfrac{1}{2}(v'^o)^2] = -dp'^o}$$

that is,

$$\tfrac{1}{2}(q'^o)^2 + p'^o = \text{constant along any streamline}^{10} \tag{6.32}$$

For brevity, let us write

$$B' = \tfrac{1}{2}(q'^o)^2 + p'^o$$

and call B' the dimensionless Bernoullian. Equation (6.32) then states that *the Bernoullian B' stays constant along any streamline*. Equation (6.32) is known as the weak Bernouilli's equation[11] – weak because it is valid only along a streamline.

---

[10] Both the conclusion and the derivation appeared in Euler's 1755 paper, cited in footnote 2 of Chapter 1.

[11] Neither Daniel nor John Bernoulli stated the equation in this form. Daniel, in his *Hydrodynamics* (in Latin), only solved a number of specific problems using the energy principle; no general and explicit statement except the energy (or "living-force") principle. John, in his *Hydraulics*, used what Euler called the true and genuine method (namely, balance of force and mass times acceleration for an infinitesimal element) and arrived at a form equivalent to the famous equation, but only for pipe flows. The general form is due entirely to Euler. In 1738, D. Bernoulli published his *Hydrodynamics*, which brought him great fame. This incurred envy from John, his father, who was then 71 years old. Thus, in the very last decade of his life, John wrote his own *Hydraulics*, which was published in 1743 but was deliberately misdated 1732 so as to steal priority from his son. It may appear that the *Hydraulics* must be a plagiarized edition of the *Hydrodynamics*, but it is not. John's work is even more a masterpiece! In it, for the very first time, kinematics and dynamics were separated, the idea of pressure on an imaginary plane cutting through water was conceived, and the above-mentioned "true and genuine method" was introduced. Furthermore, it was John's work that Euler studied for ten years before he, in turn, wrote his great hydrodynamic papers.

In general, B' will change if we go from one streamline to another. But if in a flow field there is a region of uniform B' (for example, a region at rest where $B' = p'^o$ which is constant everywhere, neglecting gravity, or a region of uniform flow where $B' = [\frac{1}{2}(\text{nondimensional uniform speed})^2 + p'^o]$ is constant everywhere), and if every fluid stream either originates from or goes to this region of uniform B', then Eq. (6.32) becomes

$$B' = \tfrac{1}{2}(q'^o)^2 + p'^o = \text{constant everywhere in the entire field} \tag{6.33}$$

A flow field like this is known as homo-Bernoullian. Our current problem of uniform flow past a body is clearly homo-Bernoullian.

On the other hand, in terms of B', it is easy to see that Eqs. (6.19a,b) can also be written as

$$v'^o\left(\frac{\partial v'^o}{\partial x'} - \frac{\partial u'^o}{\partial y'}\right) = \frac{\partial B'}{\partial x'}$$

$$-u'^o\left(\frac{\partial v'^o}{\partial x'} - \frac{\partial u'^o}{\partial y'}\right) = \frac{\partial B'}{\partial y'}$$

For a homo-Bernoullian flow, $\partial B'/\partial x'$ and $\partial B'/\partial y'$ are both zero. So,

$$\frac{\partial v'^o}{\partial x'} - \frac{\partial u'^o}{\partial y'} = 0 \tag{6.34}$$

except at points where both u'^o and v'^o vanish (stagnation points).

Now, the left-hand side of Eq. (6.34) is exactly the local (dimensionless) vorticity of the outer flow; see Eq. (1.25) and the discussion associated with it. So a homo-Bernoullian outer flow has zero vorticity (local average rotation of fluid particle) everywhere except, possibly, at the stagnation points. A flow with zero vorticity everywhere is known as irrotational. Therefore,

$$\text{homo-Bernoullian outer flow} = \text{irrotational flow}$$

except possibly at stagnation points. The possible exception of stagnation points does not bother us since they are isolated points.

Equation (6.34) looks like the equation of continuity (1.9) in form. It therefore brings to mind a possible function similar to the stream function we introduced in Section 1.2–(1), Eqs. (1.10a,b). As a matter of fact, if we can find[12] a function $\phi'\langle x', y'\rangle$ such that

$$\frac{\partial \phi'}{\partial x'} = u'^o \tag{6.35a}$$

$$\frac{\partial \phi'}{\partial y'} = v'^o \tag{6.35b}$$

Equation (6.34) becomes automatically satisfied. So, for a homo-Bernoullian, or irrotational flow, there exists a function ϕ' such that the velocity field can be obtained from it by simple differentiation. ϕ' is known as the (dimensionless) velocity

[12] That we can actually do so follows also from the analogy to the stream function.

potential (function).[13] A flow where such a velocity potential exists is known as potential flow. Therefore,

$$\text{homo-Bernoullian outer flow} = \text{potential flow}$$

We must realize here that Eqs. (6.33) and (6.35a,b) are both consequences of the Euler equations. They may actually be used in lieu of the Euler equations. There is of course still the equation of continuity, (1.9A). Substituting Eqs. (6.35a,b) into Eq. (1.9A), we have

$$\frac{\partial^2 \phi'}{\partial x'^2} + \frac{\partial^2 \phi'}{\partial y'^2} = 0 \tag{6.36}$$

Equation (6.36) is then the key equation[14] for the outer (potential) flow. It is to be solved with the conditions (6.18e) and (6.31a,b) with u'^o replaced by $\partial\phi'/\partial x'$, v'^o by $\partial\phi'/\partial y'$ and q'^o_{normal} by $(\partial\phi'/\partial x')\cos\measuredangle(n,x) + (\partial\phi'/\partial y')\cos\measuredangle(n,y)$, where $\measuredangle(n,x)$ and $\measuredangle(n,y)$ are the angles between the normal direction to the surface and the x- and y-directions, respectively.

Once ϕ' is solved, one obtains the velocities from Eqs. (6.35a,b) and then the pressure distribution from (6.33).

It is not our object here to treat the potential flow in general. For our later use, we only wish to state that

$$\phi' = \frac{C}{n}(x'^2 + y'^2)^{n/2}\cos\left[n\left(\tan^{-1}\frac{y'}{x'}\right)\right] \tag{6.37}$$

where $0 \leq \tan^{-1}(y'/x') < 2\pi$, $n > 0$, and factorial powers always refer to the principal values, is obviously[15] a solution of Eq. (6.36). It represents, therefore, potential functions of possible outer flows past certain bodies or walls. To investigate these possibilities we have to see what conditions are satisfied and decide whether they represent certain physically realizable flow situations.

---

[13] This function made its first appearance in Euler's 1752 paper; see footnote 3. The name velocity potential was introduced by Helmholtz in his 1858 paper, "On integrals of the hydrodynamic equations expressing vortex motion" (in German).

[14] Called the Laplace equation since Laplace studied it extensively in conjunction with the gravitational potential, although Euler already derived it and presented its polynomial solutions in his 1752 paper, cited in footnote 3, when Laplace was only three. It is known that Euler's influence on Laplace was great, although never acknowledged; also that both Laplace and Lagrange tried in their writings to play down Euler in order to play up d'Alembert. Pierre Simon (Marquis) de Laplace (1749–1827) was brilliant, but not above stealing ideas. He was in turn a republican and a Royalist, and thus survived the French Revolution. He was then, again in turn, made a count by Napoleon and a marquis by the returned Bourbons. It might not be an accident that he contributed tremendously to the theory of probability. His 1812 edition of *Analytic Theory of Probability* (in French) was dedicated to Napoleon; but in the 1814 edition, he wrote instead "that the fall of empires which aspired to universal dominion could be predicted with very high probability by one versed in the calculus of chances."

[15] Laplace was fond of the trick of jumping around using the disarming phrase "it is obvious." When pressed for more details about his "obvious" arguments, he was not always able to oblige without many hours of work. However, the use of "obviously" here is honest in the sense that all you have to go through is some differentiation.

Let us first see what velocity field we have here:

$$\mathbf{q}'^o = \begin{cases} u'^o = \dfrac{C}{2}(x'^2+y'^2)^{(n/2)-1}(2x')\cos\left[n\left(\tan^{-1}\dfrac{y'}{x'}\right)\right] \\ \qquad + (x'^2+y'^2)^{n/2}\sin\left[n\left(\tan^{-1}\dfrac{y'}{x'}\right)\right]\dfrac{Cy'/x'^2}{1+(y'^2/x'^2)} \\ v'^o = \dfrac{C}{2}(x'^2+y'^2)^{(n/2)-1}(2y')\cos\left[n\left(\tan^{-1}\dfrac{y'}{x'}\right)\right] \\ \qquad - (x'^2+y'^2)^{n/2}\sin\left[n\left(\tan^{-1}\dfrac{y'}{x'}\right)\right]\dfrac{C/x'}{1+(y'^2/x'^2)} \end{cases}$$

The streamlines are rather complicated. But we notice that $v'^o = 0$ on $y' = 0$ if $x' \geqslant 0$. (On $y' = 0$ and $x' < 0$, $\tan^{-1}(y'/x') = \pi$; and

$$v'^o = -C(x')^n \sin(n\pi)\left(\frac{1}{x'}\right)$$

which is not necessarily zero.) So there is no flow across $y' = 0$ and $x' \geqslant 0$. The positive x'-axis is therefore a streamline. Associated with this, we ask: Are there any other half-lines—that is, constant-(y'/x') lines—that are also streamlines? The answer is yes. Actually, it is easily seen that if

$$\tan^{-1}\frac{y'}{x'} = \frac{2\pi}{n} \tag{6.38}$$

the second terms of v'^o and u'^o drop out, and the first terms form a ratio y'/x'; that is,

$$\frac{v'^o}{u'^o} = \frac{y'}{x'} \qquad \text{if } \tan^{-1}\frac{y'}{x'} = \frac{2\pi}{n} \tag{6.39}$$

Therefore, the half-line with angle of inclination $2\pi/n$ is a streamline (Fig. 6.9). These two streamlines can then be looked upon as two rigid walls of a wedge, or a wedge-shaped cavity (Fig. 6.10). It is obvious from the figure that our practical interest ends with $n = 1$ as the lower bound, which corresponds to a semiinfinite flat plate. Often it is convenient to use the apex angle of the wedge $\vartheta = \beta\pi$ instead

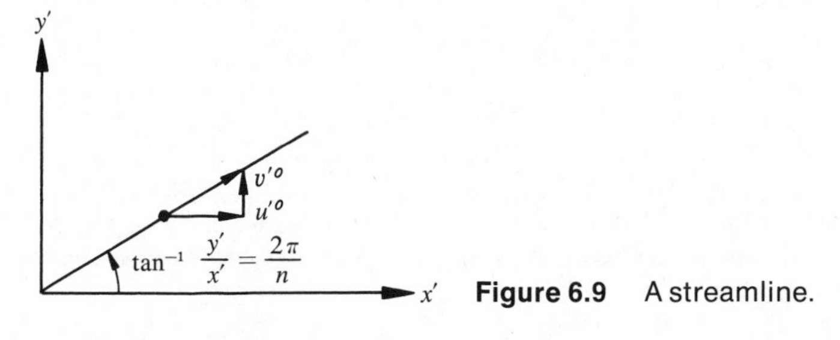

Figure 6.9 A streamline.

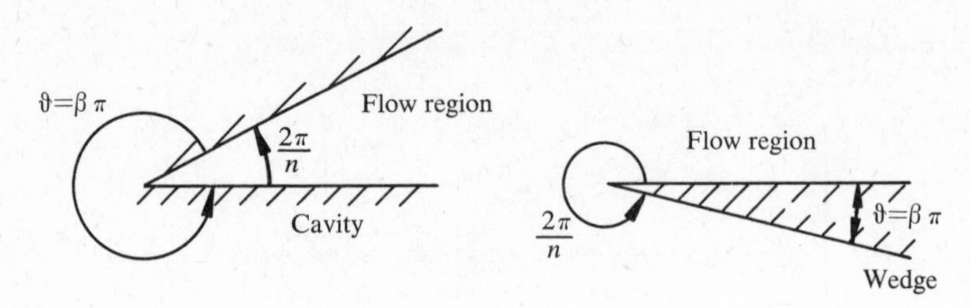

Figure 6.10 Flows inside and outside wedges.

of $2\pi/n$, and use the value of β instead of n to denote different flow situations:[16]

$$\vartheta = \beta\pi = 2\pi - \frac{2\pi}{n} \tag{6.40}$$

Therefore

$$\beta = 2 - \frac{2}{n} \tag{6.41}$$

or

$$n = \frac{2}{2-\beta} \tag{6.42}$$

It is also easy to see that the half-line

$$\tan^{-1}\frac{y'}{x'} = \frac{\pi}{n}$$

is also a streamline, which bisects the wedge or cavity. Let us orient our approaching stream (see Fig. 6.6) such that it is in line with this streamline (Fig. 6.11). That is to say,

$$\alpha = 2\pi - \tfrac{1}{2}\beta\pi \quad \text{or} \quad \pi - \tfrac{1}{2}\beta\pi \tag{6.43}$$

The flow, Eq. (6.37), is thus seen to be the outer potential flow of a certain stream bisecting the wedge or cavity, toward or away from the apex.[17] This potential flow is obviously symmetric about this bisecting stream as shown by some streamlines sketched in the figure. By easy calculation, we can see that

$$\begin{aligned} q'^{o} &= \sqrt{u'^{o2} + v'^{o2}} \\ &= C(x'^2 + y'^2)^{(n-1)/2} \end{aligned}$$

For $n > 1$, we see that $q'^{o} = 0$ at the origin; that is, the apex is a stagnation point of the potential flow. On the other hand, $q'^{o} \to \infty$ as $\sqrt{x'^2 + y'^2} \to \infty$; therefore, far away from the walls the flow is *not uniform*. It is *not* a *uniform* flow past one

---

[16] For $n \geqslant 1$, β lies between 0 and 2.

[17] It was apparently d'Alembert who first attempted a solution of potential flow by complex variables in his 1761 "Remarks on the laws of fluid motion" (in French). But as characteristic of him, he was very confused and doubtful about the usefulness of the whole thing. Lagrange in a letter to d'Alembert (November 13, 1764) corrected one of the errors in the Remarks; and, in this letter, the potential flow in a wedge appeared for the first time.

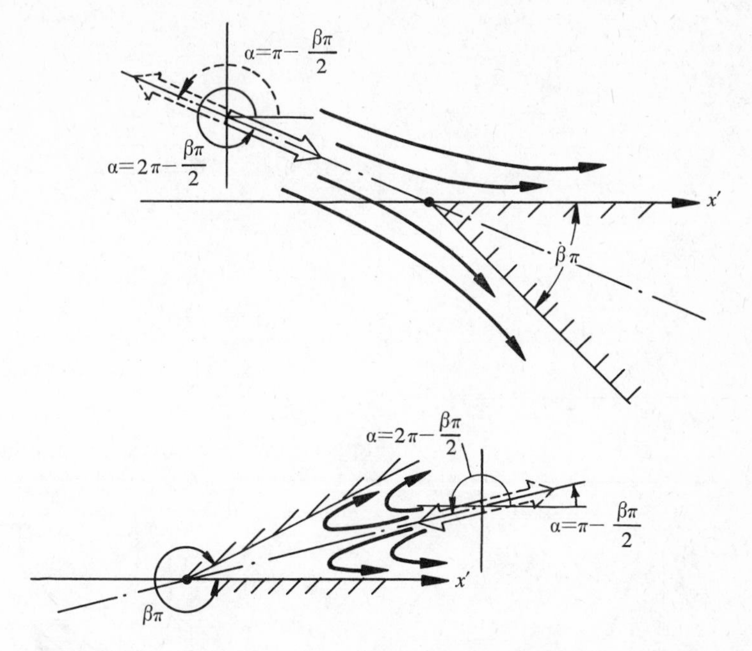

Figure 6.11 Approaching stream bisecting the wedge angle.

wedge as many people misbelieve. As a matter of fact, the infinity here is an exceptional "point" to be avoided in our physical interpretation. All we are interested in is that there exists a certain stream, of certain finite speed, some finite distance away from the wedge. The stream is symmetric about the line bisecting the apex angle. And, if we are only interested in a rather narrow region around this line of symmetry, the stream may even be regarded *approximately* as uniform (Fig. 6.12). This is really what we mean by saying that the flow, Eq. (6.37), can be realized in the laboratory as a uniform stream past a wedge (compare Fig. 4.9).

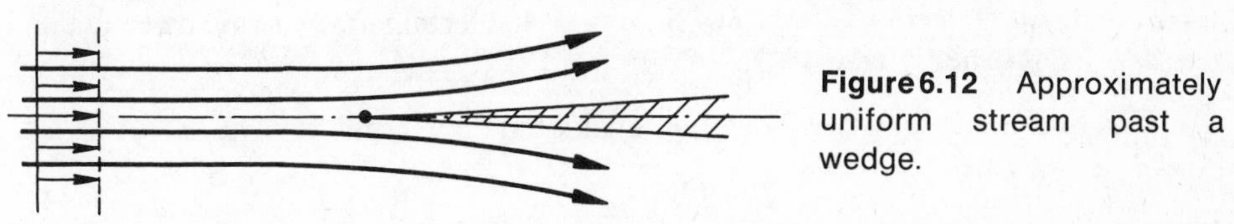

Figure 6.12 Approximately uniform stream past a wedge.

Now that we are clear on this point, let us list a few cases:

1. $n = 1$, $\alpha = 0$ (or 2π): Uniform (*really* uniform) flow toward a semiinfinite flat plate (Fig. 6.13).
2. $n = 1$, $\alpha = \pi$: Uniform flow off a semiinfinite plate (Fig. 6.14).
3. $n = 2$ ($\beta = 1$, $\vartheta = \pi$), $\alpha = 3\pi/2$ or $\pi/2$: Certain flow (approximately uniform, if interest is in the narrow neighborhood of the origin) toward or away from a flat plate (Fig. 6.15). This flow is almost the "stagnation point" flow of Section 4.2, farther away from the plate.

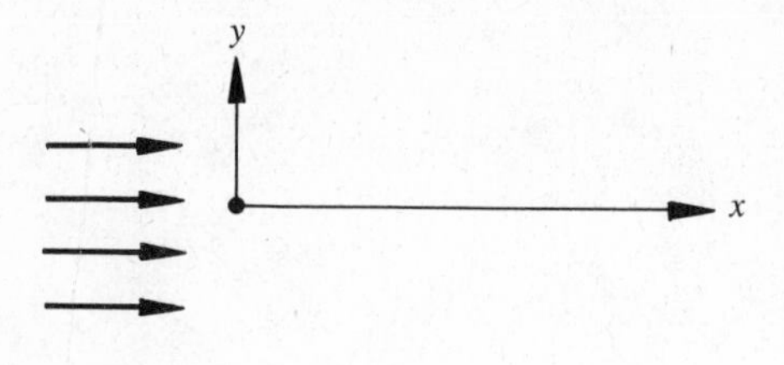

Figure 6.13 Flow toward a semiinfinite plate.

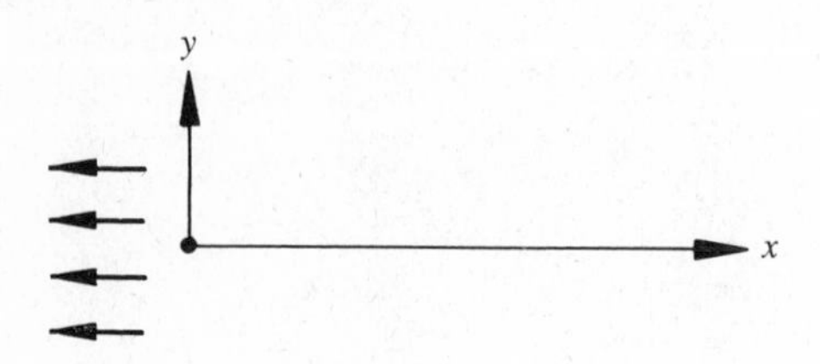

Figure 6.14 Flow off a semiinfinite plate.

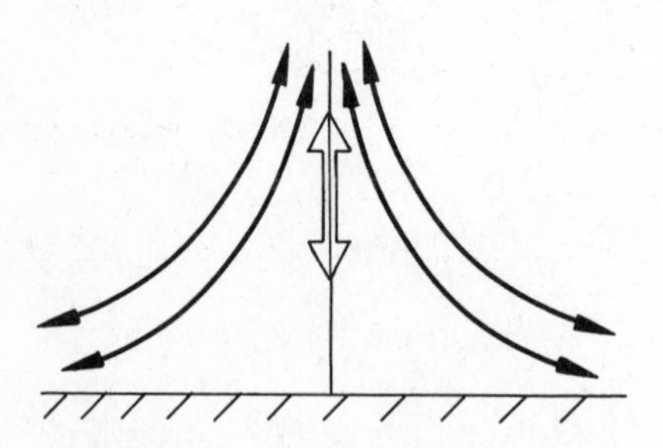

Figure 6.15 Flow toward or away from a flat plate.

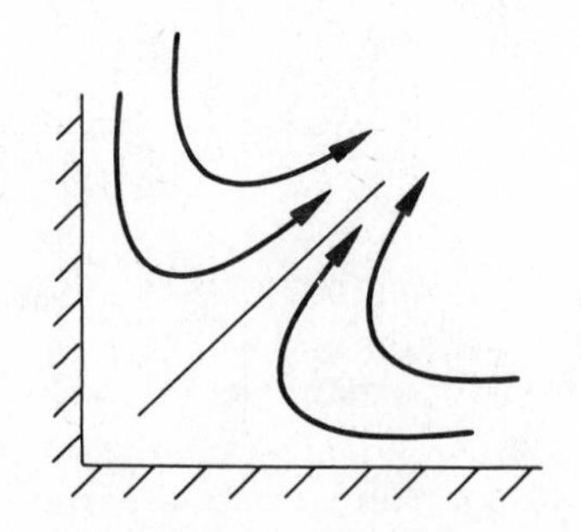

Figure 6.16 Flow in a 90° corner.

(It would be exactly that if the residual term $-b_0$ in Eq. (4.43) is neglected.)

4. $n=4$ ($\beta=\frac{3}{2}$, $\vartheta=3\pi/2$), $\alpha=\pi/4$: Certain flow away from a 90° corner (Fig. 6.16).

For later use, we will need u'^o evaluated on $y'=0$:

$$u'^o_w = Cx'^{n-1} = Cx'^m \tag{6.44}$$

where

$$m = n-1 \tag{6.45}$$

From now on, we will identify C with ± 1. Then, the fact that $|u'^o_w|=1$ at $x'=1$ shows that we have actually used $|u'^o\langle 1,0\rangle|$ as our characteristic speed Υ, $u^o\langle 1,0\rangle$ is positive if $C=1$, negative if $C=-1$.

As for dp'^o_w/dx', we have Eq. (6.19a).

There are, of course, many other outer potential flows of interest. We simply have to stop here. However, before we leave, let us point out that the boundary-layer flow is sensitive only to $u_w'^o\langle x'\rangle$, not to the entire outer flow. This brings to mind three remarks of interest:

1. If two different outer potential flows have the same u-value on the wall $y = 0$, both would have the same boundary-layer flow on $y = 0$.
2. If one can measure the u-values close enough to the wall, and yet not so close as to be within the practical range of the boundary layer, one can have approximately the actual $u_w{}^o\langle x\rangle$ of a given flow (at large Re). The boundary-layer flow can then be computed from the boundary-layer theory using this measured $u_w{}^o\langle x\rangle$.
3. For any prescribed $u_w{}^o\langle x\rangle$, one can compute a corresponding boundary-layer flow. We may not be interested enough to ask what exactly the outer potential flow is that produces such a $u_w{}^o\langle x\rangle$, we may have difficulty in producing such an outer flow in reality, and we may even find it impossible to have such a potential flow in practice or even in theory; however, the corresponding (maybe fictitious) boundary layer can be investigated nevertheless. But, why should one do such a thing? For one thing, maybe the particular $u_w{}^o\langle x\rangle$ reflects a certain character whose influence on the boundary layer we wish to ascertain, at least qualitatively. On the other hand, perhaps its investigation conveys certain information (again qualitative) about the structure of boundary layers in general. For example, Eq. (6.44) with $n < 1$ $(m < 0)$ and $C = 1$; that is,

$$u_w'^o = x'^m \qquad m < 0$$

and

$$\frac{dp_w'^o}{dx} = -u_w'^o \frac{du_w'^o}{dx'}$$

$$= -mx'^{2m-1} \qquad m < 0$$

represents a positive pressure gradient, which acts against the main flow (in the positive-x direction) in the sense that pressure increases in the flow direction. Or, since $du_w'^o/dx' = mx'^{m-1} < 0$, it represents a decelerating main flow at the "edge" of the boundary layer. The corresponding boundary-layer flow is of great qualitative interest since it shows the influence of a decelerating flow (against a positive pressure gradient). The interest becomes tremendous when an investigation shows that under such an unfavorable main flow the boundary layer becomes separated from the wall and a region of back flow forms near the wall. Yet, this flow with $n < 1$ cannot be a realistic potential flow past a wedge. It, of course, can be some other kind of potential flow associated with a wedge—for example, around a wedge. For our purpose, however, it would be easier to visualize it as a potential flow past the flat surface $(y = 0, x > 0)$ with some positive pressure gradient impressed on it. This kind of pressure-impressing is difficult, if not impossible, to realize in a laboratory.[18] But this does not bother us since our object is to learn something qualitative about the separation of boundary layers in such an unfavorable situation. To summarize, in our study, we will allow m to be negative even though the corresponding situation is not easily realizable.

[18] It is, however, qualitatively descriptive of the situation in the back of a circular cylinder.

CHAPTER 7

THE BOUNDARY LAYER OF A HIGH-*Re* FLOW FIELD

7.1 THE FALKNER-SKAN FLOWS

The stage is now all set for the appearance of the boundary layer. Concentrating on what happens near the wall $y' = 0$, $x' > 0$, we wish to solve the following problem:

$$\left\{\begin{array}{ll}
\dfrac{\partial u'}{\partial x'} + \dfrac{\partial v''}{\partial y''} = 0 & \textbf{(6.22)} \\[2ex]
u' \dfrac{\partial u'}{\partial x'} + v'' \dfrac{\partial u'}{\partial y''} = -\dfrac{dp'^{o}_{w}}{dx'} + \dfrac{\partial^2 u'}{\partial y''^2} = u'^{o}_{w}\langle x'\rangle \dfrac{du'^{o}_{w}}{dx'} + \dfrac{\partial^2 u'}{\partial y''^2} & \textbf{(6.28)} \\[2ex]
(v'' = \sqrt{Re}\, v',\ y'' = \sqrt{Re}\, y') & \\[1ex]
x' > 0,\ y'' = 0: \quad u' = 0 & \textbf{(6.18a)} \\[1ex]
\qquad\qquad\qquad\quad v'' = 0 & \textbf{(6.18f)} \\[1ex]
\qquad\quad y'' = \infty: \quad u' = u'^{o}_{w}\langle x'\rangle & \textbf{(6.29)}
\end{array}\right.$$

We will attack this problem by searching for its similarity solutions, as explained in Section 4.1. We start with the ansatz

$$u' = a\langle x'\rangle A \left\langle \frac{y''}{g\langle x'\rangle} \right\rangle \qquad \textbf{(7.1)}$$

We again call $y''/g\langle x'\rangle$ the parameter η for brevity. (But η here is not to be confused with η in Section 4.1, of course.) That is, we set

$$\eta = \frac{y''}{g\langle x'\rangle} \tag{7.2}$$

Then

$$u' = a\langle x'\rangle A\langle \eta\rangle \tag{7.1A}$$

If our ansatz can pull us through at all, it must be able to satisfy the condition (6.29); that is (stipulating that $0 < g\langle x'\rangle < \infty$ for $0 < x' < \infty$),

$$\eta = \infty: \quad a\langle x'\rangle A\langle \eta\rangle = u_w'^{o}\langle x'\rangle$$

This is obviously possible only if

$$a\langle x'\rangle = u_w'^{o}\langle x'\rangle \tag{7.3}$$

since then we have

$$\eta = \infty: \quad A = 1 \tag{7.4}$$

as the corresponding condition for $A\langle \eta\rangle$. Thus, we must have

$$u' = u_w'^{o}\langle x'\rangle A\langle \eta\rangle \tag{7.1B}$$

Next, we substitute Eq. (7.1B) into Eq. (6.22) and integrate to get v''. Because of the integration, it would be more convenient to denote $A\langle \eta\rangle$ by the derivative of another function of η

$$A\langle \eta\rangle = \frac{df\langle \eta\rangle}{d\eta} \tag{7.5}$$

to avoid the need for writing a large number of integration signs. Then,

$$u' = u_w'^{o}\langle x'\rangle \frac{df}{d\eta} \tag{7.1C}$$

and the condition (7.4) becomes

$$\eta = \infty: \quad \frac{df}{d\eta} = 1 \tag{7.4A}$$

All these are a matter of convenience; no generality is lost. Incidentally, the counterpart of boundary condition (6.18a) is

$$\eta = 0: \quad \frac{df}{d\eta} = 0 \tag{7.6a}$$

Substituting Eq. (7.1C) into Eq. (6.22) and integrating with respect to y'', we have

$$v'' = -\int_0^{y''} \left\{ \frac{du_w'^{o}}{dx'}\left(\frac{df}{d\eta}\right) - u_w'^{o}\left(\frac{d^2 f}{d\eta^2}\right)\eta\left(\frac{dg}{dx'}\right)\Big/ g \right\} dy'' + \mathscr{F}\langle x'\rangle$$

$$= u_w'^{o}\left(\frac{dg}{dx'}\right)\int_0^{\eta}\left(\frac{d^2 f}{d\eta^2}\right)\eta \, d\eta - g\left(\frac{du_w'^{o}}{dx'}\right)\int_0^{\eta}\frac{df}{d\eta}\, d\eta + \mathscr{F}\langle x'\rangle$$

$$= u_w'^o\left(\frac{dg}{dx'}\right)\left\{\left(\frac{df}{d\eta}\right)\eta\Big|_0^\eta - \int_0^\eta \frac{df}{d\eta}\,d\eta\right\} - g\left(\frac{du_w'^o}{dx'}\right)\int_0^\eta \frac{df}{d\eta}\,d\eta + \mathscr{F}\langle x'\rangle$$

$$= u_w'^o\left(\frac{dg}{dx'}\right)\eta\left(\frac{df}{d\eta}\right) - \frac{d(gu_w'^o)}{dx'}f + \mathscr{G}\langle x'\rangle$$

But boundary condition (6.18f) states that $v''=0$ at $y''=0$ or $\eta=0$; therefore

$$\mathscr{G}\langle x'\rangle = \frac{d(gu_w'^o)}{dx'}f\langle 0\rangle$$

and

$$v'' = u_w'^o\left(\frac{dg}{dx'}\right)\eta\left(\frac{df}{d\eta}\right) - \frac{d(gu_w'^o)}{dx'}\left[f - f\langle 0\rangle\right]$$

But $f\langle 0\rangle$, being a constant, can naturally be absorbed into the still-to-be-determined function $f\langle\eta\rangle$ without any loss of generality. Let us do exactly this so that

$$v'' = u_w'^o\left(\frac{dg}{dx'}\right)\eta\left(\frac{df}{d\eta}\right) - \frac{d(gu_w'^o)}{dx'}f \tag{7.7}$$

Boundary condition (6.18f) then becomes (unless $gu_w'^o=$ constant)

$$\eta = 0:\quad f = 0 \tag{7.6b}$$

Now substitute Eqs. (7.1C) and (7.7) into Eq. (6.28):

$$u_w'^o\left(\frac{df}{d\eta}\right)\left\{\left(\frac{du_w'^o}{dx'}\right)\frac{df}{d\eta} + u_w'^o\left(\frac{d^2f}{d\eta^2}\right)\left[-\eta\left(\frac{dg}{dx'}\right)\Big/ g\right]\right\}$$

$$+\left\{u_w'^o\left(\frac{dg}{dx'}\right)\eta\left(\frac{df}{d\eta}\right) - \frac{d(gu_w'^o)}{dx'}f\right\}u_w'^o\left(\frac{d^2f}{d\eta^2}\right)\Big/ g$$

$$= u_w'^o\left(\frac{du_w'^o}{dx'}\right) + u_w'^o\left(\frac{d^3f}{d\eta^3}\right)\Big/ g^2$$

or

$$\frac{d^3f}{d\eta^3} + \gamma f\left(\frac{d^2f}{d\eta^2}\right) + \lambda\left[1 - \left(\frac{df}{d\eta}\right)^2\right] = 0 \tag{7.8}$$

where

$$\gamma = g\left[\frac{d(gu_w'^o)}{dx'}\right] \tag{7.9a}$$

$$\lambda = g^2\left(\frac{du_w'^o}{dx'}\right) \tag{7.9b}$$

The success of our ansatz now lies in whether we can solve Eq. (7.8). We can probably solve it if it is an ordinary differential equation of f with respect to η; that is, if γ and λ are both constants:

$$g\left[\frac{d(gu_w'^o)}{dx'}\right] = \gamma, \text{ a constant}$$

$$g^2\left(\frac{du_w'^o}{dx'}\right) = \lambda, \text{ a constant}$$

Then,

$$\begin{aligned}\frac{d(g^2u_w'^o)}{dx'} &= 2g\left(\frac{dg}{dx'}\right)u_w'^o + g^2\left(\frac{du_w'^o}{dx'}\right)\\ &= 2g\left[\frac{d(gu_w'^o)}{dx'}\right] - 2g^2\left(\frac{du_w'^o}{dx'}\right) + g^2\left(\frac{du_w'^o}{dx'}\right)\\ &= 2\gamma - \lambda, \text{ a constant}\end{aligned}$$

or

$$g^2u_w'^o = (2\gamma-\lambda)x' + \text{constant}$$

The constant of integration here is only a shift of the origin of the coordinate system. We will not be bothered by this, since no generality is lost by any chosen value for this constant. Let us take it to be zero:

$$g^2u_w'^o = (2\gamma-\lambda)x' \tag{7.10}$$

Together with Eq. (7.9b), this leads to

$$\begin{aligned}\frac{du_w'^o}{dx'}\Big/u_w'^o &= \left(\frac{\lambda}{2\gamma-\lambda}\right)\Big/x' \qquad 2\gamma-\lambda \neq 0\\ &= \frac{M}{x'} \qquad M \neq \infty\end{aligned}$$

where

$$M = \frac{\lambda}{2\gamma-\lambda} \tag{7.11}$$

that is,

$$\ln|u_w'^o| = \ln|x'|^M + \text{constant}$$

or

$$|u_w'^o| \propto |x'|^M \tag{7.12}$$

Since our ansatz is

$$u' = u_w'^o\langle x'\rangle\frac{df}{d\eta}$$

whatever constant of proportionality involved in relation (7.12) can be immediately absorbed into the still-to-be-determined $f\langle\eta\rangle$ without any loss of generality. So, why cannot we simply conclude that

$$|u_w'^o| = |x'|^M$$

Or, since $x' > 0$,

$$|u_w'^o| = x'^M \tag{7.13}$$

that is,

$$u_w'^o = \begin{cases} x'^M, \text{ where } u_w'^o > 0 \\ -x'^M, \text{ where } u_w'^o < 0 \end{cases}$$

(If you had forgotten that $\int dx/x = \ln |x|$, not $\ln x$, you would have lost one half of it!)

The conclusion then is this: A similarity solution exists if the outer potential flow is such that its velocity on the wall $y' = 0$, $x' > 0$, is a power of distance along the wall from the leading edge $x' = 0$. (The characteristic speed used in the nondimensionalization is $|u_w{}^o|$ at $x' = 1$, or at $x = x_{\text{obs}}$.) We are here overjoyed to see that the potential flows treated in Section 6.3 – that is, the wedge flows – are among this category. For good measure, we will also throw in the cases where $M < 0$, as discussed at the end of Section 6.3.

(1) Wedge Flows

Here, we must have from Eq. (6.44)

$$M = m = n - 1 = \frac{2}{2-\beta} - 1 = \frac{\beta}{2-\beta}$$

where $0 \leqslant \beta < 2$. Then, Eq. (7.11) yields $\lambda/\gamma = \beta$. (You of course recall that $\beta\pi$ is the wedge angle.)

(2) $M < 0, \neq -1$

No identification is called for here. However, let us write m for M, and β for λ/γ, so this class becomes *formally* the same as the wedge flows.

For all these, Eq. (7.10) dictates that

$$g^2 = \frac{(2\gamma - \lambda)x'}{u_w'^o}$$

be always positive. Its square root

$$g = \pm \sqrt{\frac{(2\gamma - \lambda)x'}{u_w'^o}}$$

is then always real. This is needed for the similarity parameter η in Eq. (7.2). The + sign before the square root is to be chosen as stipulated earlier. (Actually, the − sign will not yield a different solution.) So,

$$\eta = \frac{y''}{\sqrt{(2\gamma - \lambda)x'/u_w'^o}}$$

But, $\lambda/\gamma = \beta$ for both cases. Therefore

$$g = \sqrt{\gamma(2-\beta)x'/u_w'^o} \tag{7.14}$$

Similarly to what we said before, the constant factor γ is immaterial and can be chosen at random, without loss of generality, as long as

$$\frac{\gamma(2-\beta)x'}{u_w'^o} > 0 \tag{7.15}$$

For $\beta < 2$ and $x' > 0$, we can choose conveniently (you may wish to choose differently, as many authors do; but the final result will be equivalent):

$$\gamma = 1 \text{ if } u_w'^o > 0 \text{ (forward flow toward the leading edge)}$$
$$= -1 \text{ if } u_w'^o < 0 \text{ (backward flow from the leading edge)}$$

For both backward and forward flows, we have

$$\eta = \frac{y''}{\sqrt{2-\beta}\, x'^{(1-m)/2}} \tag{7.16}$$

The difference is in the equation for f:

$$\frac{d^3f}{d\eta^3} \pm f\left(\frac{d^2f}{d\eta^2}\right) \pm \beta\left[1-\left(\frac{df}{d\eta}\right)^2\right] = 0 \tag{7.17}$$

where

$$\begin{cases} +: & \text{forward flow} \\ -: & \text{backward flow} \end{cases}$$

Equation (7.17) is subject to the conditions

$$\eta = \infty: \quad \frac{df}{d\eta} = 1 \tag{7.4A}$$

$$\eta = 0: \quad \frac{df}{d\eta} = 0 \tag{7.6a}$$

$$f = 0 \tag{7.6b}$$

Before we go on, let us state that the analysis for large y at the end of Section 4.2—that is, the derivation of relation (4.46)—applied to Eq. (7.17) with boundary condition (7.4A) shows that there cannot be a solution in the case of the backflow.

What does this mean? Either of the following explanations is possible:

1. The solution is not in the conjectured form.
2. No boundary layer forms (that is, the large *ansatz* of asymptotics described from Section 6.2 on fails to pull us through here).

We will therefore exclude this case and consider only the forward flows. Equation (7.17) for the forward flow (plus signs) is known as the Falkner-Skan[1] (or F-S) equation. We will call the flows described by it and the conditions (7.4A), (7.6a,b) the F-S flows.

---

[1] V. M. Falkner and S. W. Skan, "Some approximate solutions of the boundary layer equations," *Phil. Mag.* (7), **12**, 865–869, 1931.

Summary For *F-S flows* $(x' > 0)$,

$$u' = x'^m \left(\frac{df}{d\eta}\right) \tag{7.18}$$

$$v'' = -\sqrt{2-\beta}\, x'^{(m-1)/2} \left\{\frac{m-1}{2}\eta\left(\frac{df}{d\eta}\right) + \frac{m+1}{2} f\right\} \tag{7.19}$$

$$\eta = \frac{y''}{\sqrt{2-\beta}\, x'^{(1-m)/2}} \tag{7.2A}$$

$$\frac{d^3 f}{d\eta^3} + f\left(\frac{d^2 f}{d\eta^2}\right) + \beta\left[1 - \left(\frac{df}{d\eta}\right)^2\right] = 0 \tag{7.17A}$$

$$\eta = \infty: \quad \frac{df}{d\eta} = 1 \tag{7.4A}$$

$$\eta = 0: \quad \frac{df}{d\eta} \quad \text{and} \quad f = 0 \tag{7.6a,b}$$

$$\beta = \frac{2m}{m+1} \qquad m = \frac{\beta}{2-\beta}$$

$$\left(0 \leqslant \beta < 2; \qquad \vartheta = \beta\pi = \frac{2m\pi}{m+1}; \qquad \alpha = 2\pi - \frac{1}{2}\beta\pi\right)$$

7.2 BLASIUS FLOW

The special case of the flows of Section 7.1 with $m = 0$ [$\beta = 0$, $\vartheta = 0$, $\alpha = 2\pi$ (or 0)] is the famous Blasius flow[2] in a boundary layer on a semiinfinite flat plate in an outer flow that is uniform (really!), parallel to the plate, and toward the leading edge (Fig. 7.1). The boundary-layer theory, being established around large x_{obs} (so that $Re = \Upsilon x_{obs}/\nu$ is large), will not hold in the region near the leading edge,

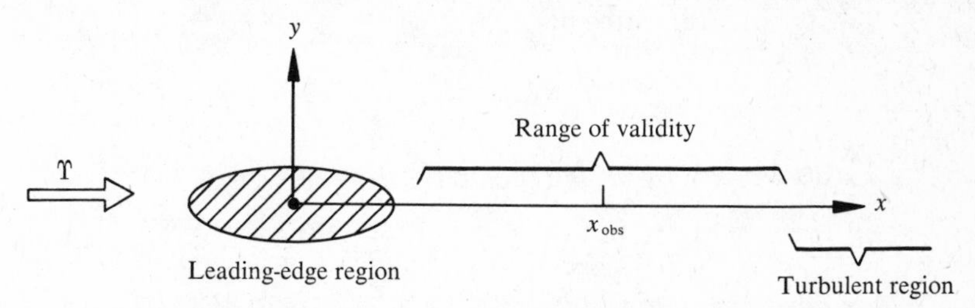

Figure 7.1 Regions of Blasius and non-Blasius flows.

[2] (Paul Richard) Heinrich Blasius (1883–) of Berlin, a student of Prandtl, predicted this in 1908. Blasius was equally famous for the *Blasius theorem* in the theory of flight (plane flows). There is also a Blasius formula for the turbulent flow through smooth pipes. After 1913 he suddenly disappeared from the scene of active hydrodynamic creativity.

and, being based on the laminar assumption, will also not hold in the turbulent region where $\Upsilon x/\nu$ exceeds roughly 200,000.

In its region of validity, the Blassius flow is the solution of

$$\frac{d^3f}{d\eta^3}+f\left(\frac{d^2f}{d\eta^2}\right)=0 \tag{7.20}$$

subject to the boundary conditions

$$\eta=\infty: \quad \frac{df}{d\eta}=1$$

$$\eta=0: \quad f, \frac{df}{d\eta}=0$$

where

$$\eta=\frac{y''}{\sqrt{2x'}}$$

$$u'=\frac{df}{d\eta} \tag{7.21}$$

$$v''=\frac{1}{\sqrt{2x'}}\left[\eta\left(\frac{df}{d\eta}\right)-f\right] \tag{7.22}$$

($u'^{o}=1$ throughout, leaving zero pressure gradient everywhere. Blasius flow is then a boundary-layer flow without pressure gradient.)

Before we present the solution, let us notice the following points:

1. The leading edge $x=0$ is a possible exceptional point where the solution may not behave as well as elsewhere, but our boundary-layer approximation is not valid near $x=0$ anyway. (Near $x=0$, x' is no longer $\sim\Box(1)$, but o (1).) So, we should not be alarmed when something strange happens in our solution at $x'=0$.
2. The initial profile of $u'\langle y''\rangle$ at $x'=0$ is inherently fixed in our similarity solution, since in

$$u'=\frac{df}{d\eta}$$

where $\eta=y''/\sqrt{2x'}$ we see that

$$x'\to 0: \quad \eta\to\infty \qquad \text{if } y''\neq 0$$

$$\eta\to 0 \qquad \text{if } y''=0$$

that is,

$$x'\to 0: \quad u'=1 \qquad \text{if } y''\neq 0$$

$$u'=0 \qquad \text{if } y''=0$$

from the conditions (Fig. 7.2). This means that in order to realize our similarity solution we must have an initial boundary-layer flow maintained right at $x'=0$, which is uniform except at the leading

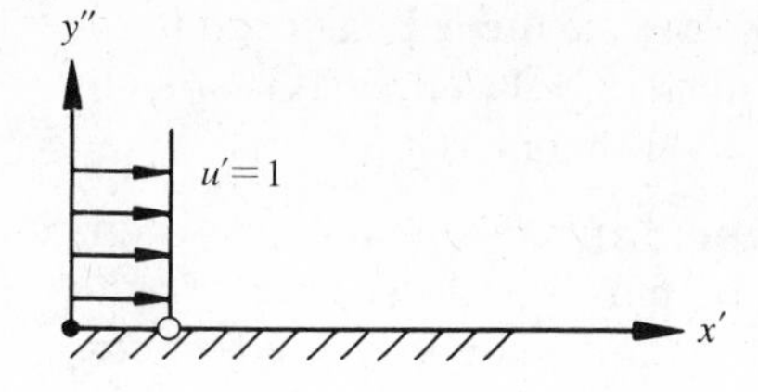

Figure 7.2 Initial profile of the Blasius flow.

edge, where it vanishes. This inherent requirement of an initial profile has a number of practical repercussions:

(a) It is not so easy to maintain even approximately such a discontinuous profile.
(b) To be precise, the Blasius flow is, therefore, *not* the boundary-layer flow with a uniform flow far upstream, as usually believed, since any flow far upstream must produce a *continuous* profile at $x' = 0$. It goes without saying that any "incorrect" initial profile will not give us exactly the Blasius flow—some other boundary-layer flow, but not the Blasius flow.
(c) Fortunately, the boundary-layer equation has such an inherent structure that the influence of the initial profile on its solution dies down quickly as x' increases. Therefore, if we have a uniform flow far upstream, the boundary-layer flow will be approximately that of Blasius except in the region close to the leading edge. This fact then saves our similarity solution of the boundary-layer equation from being unrealizable.

3. It is also an inherent character of the boundary-layer equation that its solution at an x'-value depends only on what happens before this, not after. This is very important. For example, although we know that the flow will eventually become turbulent, this turbulent region will not influence the laminar region before it, where the Blasius flow exists. As another example, suppose that the plate has a rather long but finite length. Then certainly the flow after the trailing edge will be non-Blasius; however, this will not prevent the Blasius flow from appearing before the trailing edge (Fig. 7.3).

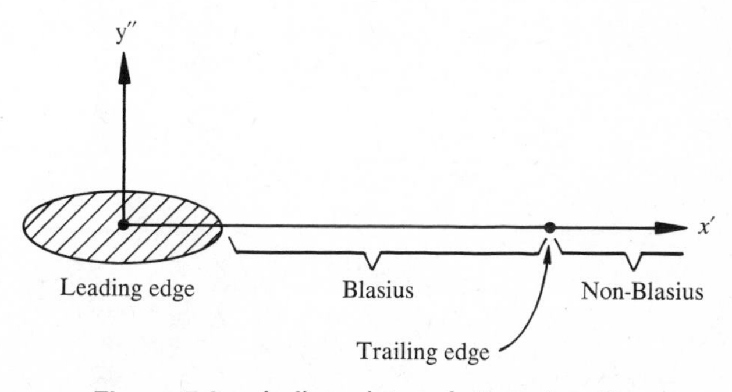

Figure 7.3 A flat plate of finite length.

4. An exact investigation of the influence of the leading edge in the neighborhood of it is beyond the scope of this book. (It is actually beyond the scope of the boundary-layer theory.)

Now, the solution. Equation (7.20) is unfortunately a nonlinear ordinary differential equation whose solution cannot be put in a closed form. In this book we can only state that its solution, subject to the conditions, was first constructed by Blasius and is plotted in Figure 7.4.

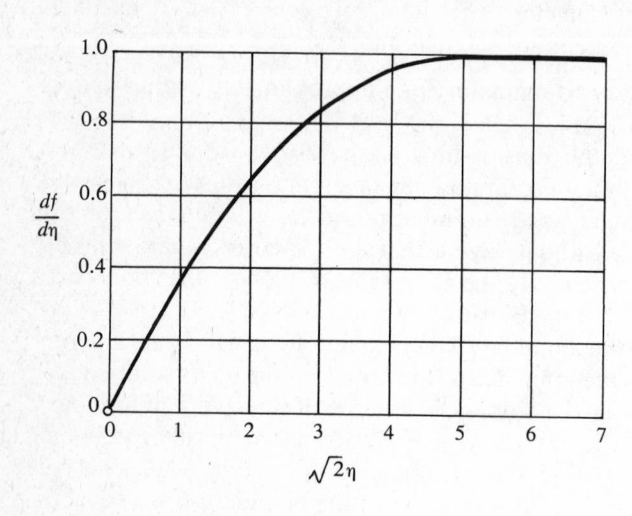

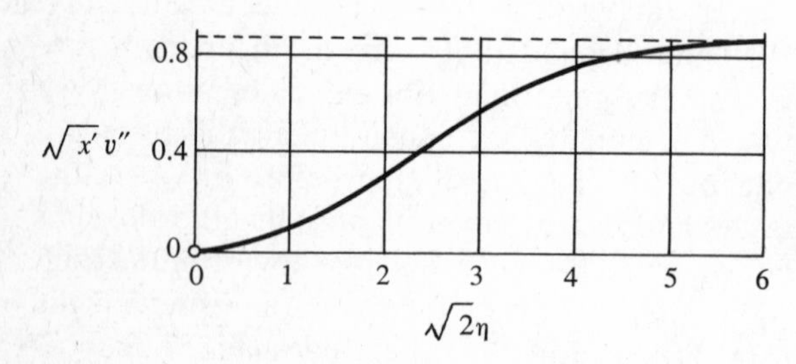

Figure 7.4 Blasius velocity profiles.

It is interesting to see that

$$v''\langle x',\infty\rangle = \frac{0.8604}{\sqrt{x'}} \tag{7.23}$$

$$\neq 0$$

or

$$v'\langle x',\infty\rangle = \frac{0.8604}{\sqrt{Re\, x'}}$$

$$\sim \square(1/\sqrt{Re})$$

That is, at the "edge" of the boundary layer there is a very small but nonzero $v' \sim \square(1/\sqrt{Re})$, as predicted before. So, the boundary-layer flow does not exactly "blend" into the uniform outer flow; there is a little residue of $\square(1/\sqrt{Re})$ left.

Up to now, we have often mentioned in a figurative manner the "edge" of the boundary layer to imply that the boundary layer has a thickness. In reality, there is of course no finite thickness for a boundary layer. However, this does not prevent us from introducing some *measure* of the boundary-layer thickness. For example, we may compare the distances y' (not y'') at which u' reaches 0.9 or 0.99, or 0.999 (that is, u reaches 90 percent of $u_w{}^o$, or 99 percent, or 99.9 percent) for different boundary layers and call them, respectively, the 90 percent thickness, the 99 percent thickness, and the 99.9 percent thickness. The idea of boundary-layer thickness is of course inherently imprecise and somewhat vague. But, it is of some practical value in that it offers some comparison for the behavior of different boundary-layer flows and also conveys a vivid, picturesque, and suggestive mental image of the boundary-layer flow.

To this end, it is much better to utilize the *displacement thickness* δ_1' defined as (Fig. 7.5)

$$\sqrt{Re}\,\delta_1' = \left\{\int_0^\infty (u_w'^{o} - u')\,dy''\right\} \Big/ u_w'^{o} = \int_0^\infty \left(1 - \frac{u'}{u_w'^{o}}\right) dy'' \qquad \textbf{(7.24)}$$

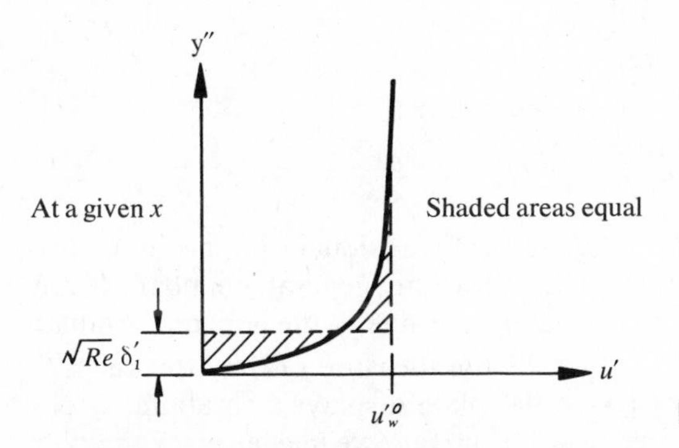

Figure 7.5 Displacement thickness.

A physical explanation of the term "displacement" can be introduced by supposing that the above curve ends at a finite but large distance Δ'' (Fig. 7.6). Then, because of the boundary layer, the volume flow in the fluid slab $0 \leqslant y'' \leqslant \Delta''$ is $\sqrt{Re}\,\delta_1'$ smaller than that in the absence of the boundary layer. If originally (in the absence of the boundary layer) the outer flow is only of height Δ'', then, because of the boundary layer, the approaching stream must go to the height $\Delta'' + \sqrt{Re}\,\delta_1'$ (*displaced* by an amount $\sqrt{Re}\,\delta_1'$) in order to carry the same amount of volume flow as before. Thus the name *displacement* thickness.

In the Blasius flow, we have

$$\delta_1' = \frac{\sqrt{2x'}}{\sqrt{Re}} \int_0^\infty \left(1 - \frac{df}{d\eta}\right) d\eta = \frac{\sqrt{2x'}}{\sqrt{Re}} \lim_{\eta \to \infty} (\eta - f)$$

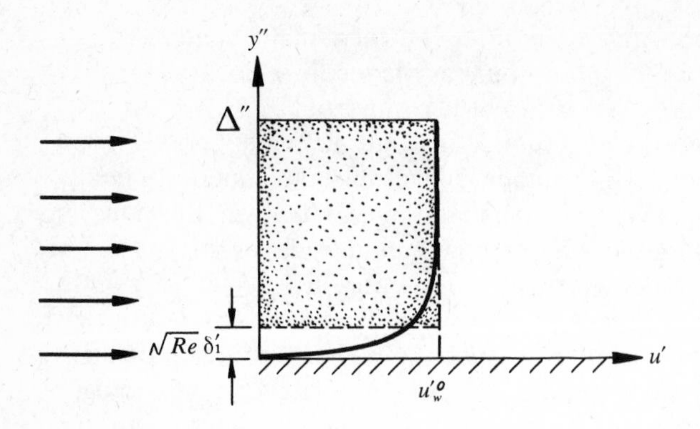

Figure 7.6 Physical explanation of displacement thickness.

Since $\lim_{\eta\to\infty} df/d\eta = 1$, this can also be written as

$$\delta_1' = 2\sqrt{\frac{x'}{Re}}\lim_{\eta\to\infty}\left\{\frac{1}{\sqrt{2}}\left(\eta\frac{df}{d\eta}-f\right)\right\}$$

$$= 2\sqrt{\frac{x'}{Re}}\,v''\langle x',\infty\rangle\sqrt{x'}$$

$$= \frac{2x'}{\sqrt{Re}}\,v''\langle x',\infty\rangle \tag{7.25}$$

$$= 1.7208\sqrt{\frac{x'}{Re}}$$

We see that (a) $\delta_1' \sim \square(1/\sqrt{Re})$, (b) δ_1' is related to the residual v'' at the "edge" of the boundary layer, (c) $\delta_1' \propto \sqrt{x'}$ ($\delta_1' = \delta_1'\langle x'\rangle$ for the general boundary-layer flow); or (a) the thickness is small, (b) the thickness is due to the nonzero residual transverse velocity at the "edge," (c) with δ_1' as the measure of the "thickness," the boundary layer of a large-*Re* flow past a flat plate displays a parabolic, ever-thickening profile (Fig. 7.7). With reference to Figure 7.7, we may also say that the transverse velocity at the "edge" is nonzero because the boundary layer thickens and pushes (or displaces) the main flow out. It is then obvious that, if we wish to improve on our boundary-layer theory, we should first improve the outer flow so that it sees the thickened parabolic body (crosshatched in the figure) instead of the flat plate. However, these improvements[3] must be small of $\square(1/\sqrt{Re})$ compared to the Blasius solution, which is $\square(1)$.

[3] To improve on the boundary-layer theory is definitely beyond the scope of this book. Furthermore, the above nice outlook on how to improve the Blasius flow is not always really suitable for other boundary-layer flows; the improvement may not be possible along the line just described. However, this need not concern us if only the boundary-layer theory itself is considered. (Interested students may read, for example, Van Dyke's book, cited in footnote 14 of Chapter 5.)

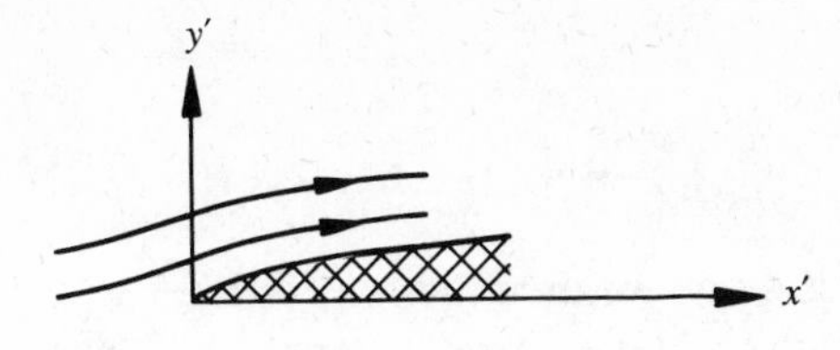

Figure 7.7 Boundary-layer-thickened flat plate.

7.3 FALKNER-SKAN FLOWS AND SEPARATION OF THE BOUNDARY LAYER

The nonlinear ordinary differential equation (7.17A) has also been solved with the associated conditions (and the requirements that $\lim_{\eta \to \infty} (df/d\eta)$ approach zero faster than any power of $1/\eta$, and that $0 \leq df/d\eta \leq 1$) for

$$-0.1988 \leq \beta < 2 \qquad (-0.0904 \leq m < \infty)$$

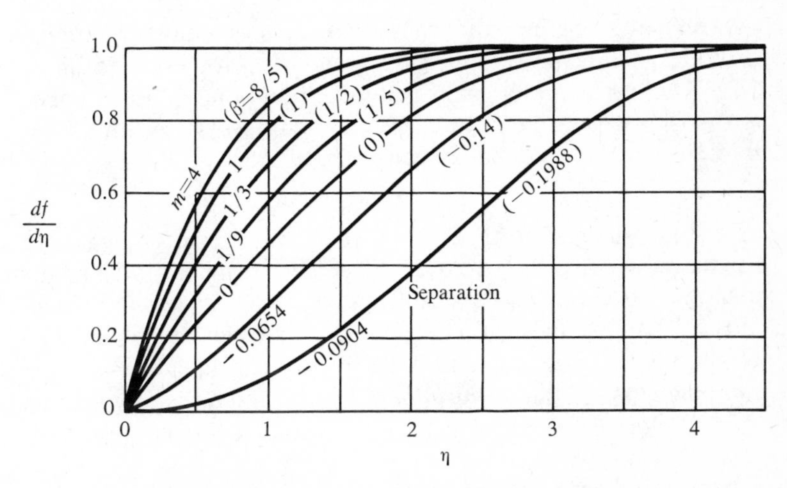

Figure 7.8 Falkner-Skan velocity profiles.

Figure 7.8 shows several cases of which the Blasius flow is a member ($\beta = 0$, $m = 0$). We are especially interested in the following cases:

(1) $m = 1 \quad (\beta = 1)$

The wedge angle ϑ here is π, and the angle of inclination of the approaching stream α is $3\pi/2$—that is, downward. Furthermore,

$$u' = x' \frac{df}{d\eta}$$

$$v'' = -f$$

$$\eta = y''$$

Equation (7.17A) becomes

$$\frac{d^3f}{dy''^3}+f\left(\frac{d^2f}{dy''^2}\right)+1-\left(\frac{df}{dy''}\right)^2=0$$

The problem is actually almost the same as the one we dealt with in Section 4.2. The difference is that here we are treating a boundary-layer flow (an approximation) at large Re, but in Section 4.2 there is no restriction on Re and the solution is exact.

The two problems are completely identical if the outer flow has an *additional* transverse velocity equal to $0.6479/\sqrt{Re}$, which just accommodates the one showing up at the "edge" of the boundary layer. Under such a contrived situation, the boundary layer would be so good as to be identical with the exact solution—that is, valid all the way. (Incidentally, δ_1' here is just $0.6479/\sqrt{Re}$, which ties in with the required additional transverse velocity.)

(2) $m=-0.0654$ $(\beta=-0.14)$

As indicated before, this case represents qualitatively a decelerating flow against certain unfavorable pressure gradients. However, nothing unusual happens here. The curve simply shows an inflection point and the slope $d^2f/d\eta^2$ at $\eta=0$ is smaller than that for the positive-m cases. At any rate, unfavorable pressure gradients are still too small to cause any mishap.

(3) $m=-0.0904$ $(\beta=-0.1988)$

Here the slope $d^2f/d\eta^2$ at the wall $\eta=0$ is zero. The velocity profile $u'\langle x',y''\rangle$ has therefore zero slope at every x' (Fig. 7.9). This is then an indication of the beginning of trouble, because a still lower value of m would induce a back-flow region ($u'<0$) near the wall (Fig. 7.10). When this latter happens, we say that the boundary layer is *separated* from the wall. A value of $m=-0.0904$ then indicates the beginning of this *separation*, beyond which the flow can no longer negotiate the unfavorable pressure gradient. (Note that the unfavorable pressure gradient with $m=-0.0904$ is not very large.)

It is the inherent character of our similarity (F-S) solutions that the separation occurs along the entire wall, when it occurs at all. In other, more general (nonsimilarity) boundary-layer flows, the separation occurs at one x' value where $\partial u'/\partial y''=0$, beyond which there is backflow (Fig. 7.11).

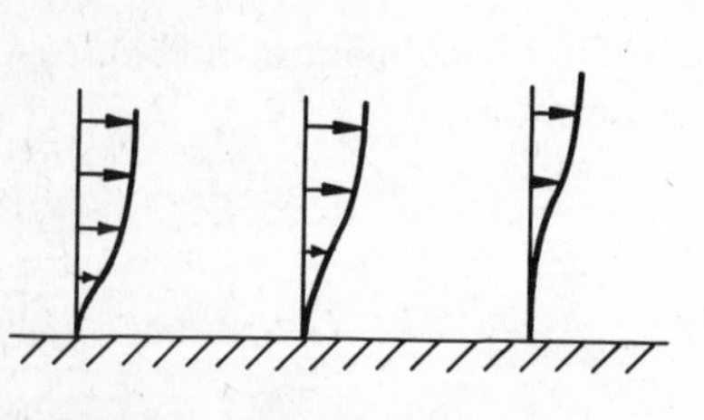

Figure 7.9 Separation of boundary layer.

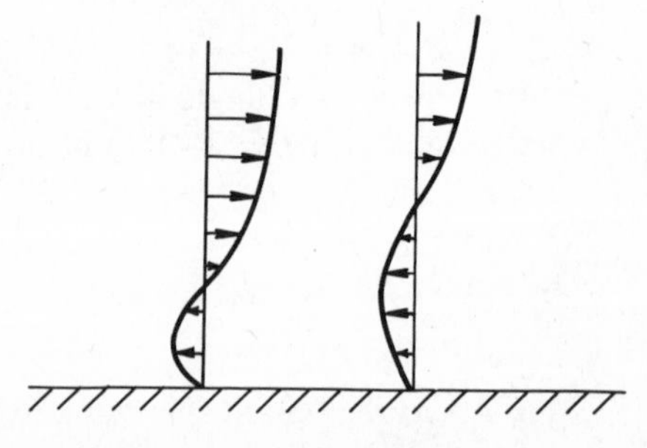

Figure 7.10 Back-flow region.

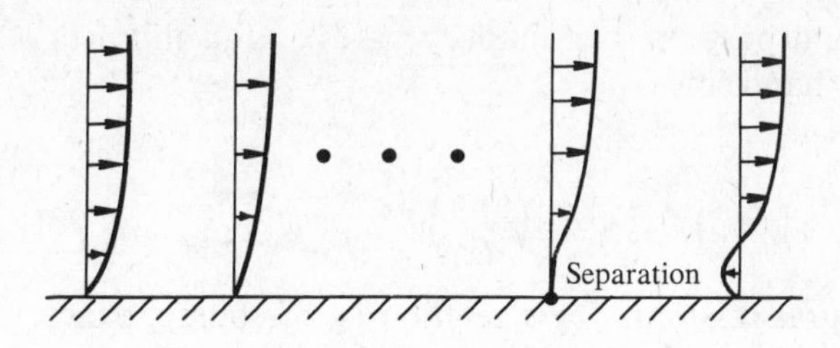

Figure 7.11 Separation farther downstream.

7.4 KÁRMÁN-POHLHAUSEN TECHNIQUE

In this section we will discuss briefly an approximate technique for obtaining boundary-layer flows. (It is an approximation within an approximation, then.) This is known as the Kármán-Pohlhausen (K-P), or integral, technique.

Consider the boundary-layer problem posed at the beginning of Section 7.1. Multiply the equation of continuity (6.22) by $u'^{o}_{w} - u'$ and then subtract from it the boundary-layer equation (6.28). A little rearrangement yields, finally,

$$-\frac{\partial^2 u'}{\partial y''^2} = \frac{\partial}{\partial x'}\,(u'^{o}_{w}u' - u'^2) + (u'^{o}_{w} - u')\,\frac{du'^{o}_{w}}{dx'} - \frac{\partial}{\partial y''}\,(u'v'') + u'^{o}_{w}\,\frac{\partial v''}{\partial y''}$$

Integrating this equation with respect to y'' from 0 to ∞ gives

$$\left.\frac{\partial u'}{\partial y''}\right|_{y''=0} - \left.\frac{\partial u'}{\partial y''}\right|_{y''=\infty} = \frac{d}{dx'}\int_0^\infty (u'^{o}_{w}u' - u'^2)\,dy'' + \left(\frac{du'^{o}_{w}}{dx'}\right)\int_0^\infty (u'^{o}_{w} - u')\,dy''$$
$$-(u'v'')_{y''=\infty} + (u'v'')_{y''=0} + u'^{o}_{w}[v''(\infty) - v''(0)]$$

The fact that u' approaches a function of x' as $y'' \to \infty$ implies that $\partial u'/\partial y'' \to 0$ as $y'' \to \infty$; see relation (4.37). Furthermore,

$$\int_0^\infty (u'^{o}_{w} - u')\,dy'' = \sqrt{Re}\,\delta'_1 u'^{o}_{w}$$
$$= \delta''_1 u'^{o}_{w} \qquad \textbf{(7.24A)}$$

and

$$\int_0^\infty (u'^{o}_{w}u' - u'^2)\,dy'' = (u'^{o}_{w})^2 \int_0^\infty \left[\frac{u'}{u'^{o}_{w}} - \frac{u'^2}{(u'^{o}_{w})^2}\right] dy''$$
$$= (u'^{o}_{w})^2 \sqrt{Re}\,\delta'_2$$
$$= (u'^{o}_{w})^2 \delta''_2 \qquad \textbf{(7.26)}$$

where

$$\delta'_2 = \int_0^\infty \left[\frac{u'}{u'^{o}_{w}} - \frac{u'^2}{(u'^{o}_{w})^2}\right] dy'' \Big/ \sqrt{Re} \qquad \textbf{(7.27)}$$

is called the (dimensionless) momentum thickness. The physical explanation of δ'_2 is quite similar to δ'_1, except that δ'_1 is based on volume flow whereas δ'_2 is based on momentum flux. For our present purpose, we can simply regard δ_2' as an abbreviation of the integration on the right-hand side.

So, finally, we have

$$\left.\frac{\partial u'}{\partial y''}\right|_{y''=0} = \frac{d}{dx'}\left[(u_w'^o)^2\delta_2''\right] + u_w'^o\left(\frac{du_w'^o}{dx'}\right)\delta_1'' \tag{7.28}$$

This is known as Kármán's integral relation.[4] This relation is, of course, exact. But, in integrating the original equations, some local information is lost. The integral relation is an exact overall, or average, relation. It certainly contains far less information than the original differential equations.

The K-P technique is an attempt, then, to get as much local information back as possible using this Kármán relation. The technique consists of choosing a family of curves, usually with one parameter, as trial curves of u' and pick up the one that satisfies Eq. (7.28). It is required that the family of curves satisfy as many conditions originally set for u' as possible, including the hidden ones:

$$y''=0:\quad \frac{\partial^2 u'}{\partial y''^2} = -u_w'^o\frac{du_w'^o}{dx'}\quad \text{(from Eq. (6.28))}$$

$$\frac{\partial^3 u'}{\partial y''^3} = 0\ \text{(from the } y'' \text{ differentiation of Eqs. (6.28), and (6.22))}$$

$$\frac{\partial^4 u'}{\partial y''^4} = \frac{\partial u'}{\partial y''}\frac{\partial^2 u'}{\partial x'\,\partial y''}\quad \text{(differentiating Eq. (6.28) with respect to } y'' \text{ once more)}$$

$$\cdots\cdots$$

$$y''=\infty:\quad u' = u_w'^o\langle x'\rangle$$

$$\frac{\partial u'}{\partial y''}, \frac{\partial^2 u'}{\partial y''}, \ldots = 0\quad \text{(see Eq. (4.37))}$$

In choosing the conditions to be satisfied, we of course start with the explicit ones and move on to the hidden ones, for derivatives of higher and higher order.

To be more exact, we will follow Pohlhausen[5] in choosing

$$u' = u_w'^o\langle x'\rangle F\langle\eta\rangle \tag{7.29}$$

---

[4] Theodore von Kármán (1881–1963), the most illustrious student of Prandtl, was born in Hungary. He migrated to the United States in the 1930s and was director of Caltech's Guggenheim Aeronautics Laboratory for more than 20 years, covering the most exciting period of aeronautics (namely that of high-speed flight and jet propulsion). He "inherited" from his teacher the ability of cutting immediately to the heart of the matter of a complicated physical phenomenon, by simple but to-the-point analysis and simple but decisive experiments. Because of his cosmopolitan contacts, he was able to spread the teaching of the Prandtl-Kármán school on both continents. A basic character of this school is the invention of many *ad hoc* theories, equations, formulas, methods, and techniques. Although today's trend in engineering science is toward a more general, fundamental, and rational approach, the Prandtl-Kármán central teaching that one should first look for the heart of the matter, temporarily disregarding the trimmings, is still with us. As to the specific gifts from Kármán, this integral relation is but the beginning.

[5] K. Pohlhausen, "Zur näherungsweisen Integration der Differentialgleichung der laminaren Reibungsschicht," *ZAMM*, **1**, 252–268, 1921.

where

$$\eta = \frac{y''}{\delta''\langle x' \rangle} \tag{7.30}$$

and $\delta''\langle x' \rangle$ is a parametric function of x'. (F, η, δ'' here are suggestive of A, η, and g of Section 7.1; but they are *not* the same.) Our purpose is to determine $\delta''\langle x' \rangle$ so that Eq. (7.28) is satisfied. The form of F is arbitrary except that it must satisfy, as far as possible, the following conditions:

$\eta = 0$:

$$F = 0$$

$$\frac{d^2F}{d\eta^2} = -\Lambda$$

$$\frac{d^3F}{d\eta^3} = 0$$

$$\frac{d^4F}{d\eta^4} = \delta''^3 \left(\left. \frac{dF}{d\eta} \right|_{\eta=0} \right) \frac{d}{dx'} \left\{ \frac{u_w'^o}{\delta''} \left(\left. \frac{dF}{d\eta} \right|_{\eta=0} \right) \right\}$$

$$\ldots\ldots$$

$\eta = \infty$:

$$F = 1$$

$$\frac{dF}{d\eta}, \frac{d^2F}{d\eta^2}, \ldots = 0$$

where

$$\Lambda = (\delta''^2) \frac{du_w'^o}{dx'} \tag{7.31}$$

With F thus properly chosen, Eq. (7.28) becomes an ordinary differential equation of the first order for $\delta''\langle x' \rangle$ which can be solved.

EXAMPLE Blasius Flow

Here $u_w'^o = 1$, $\Lambda = 0$. And, with $u' = F\langle y''/\delta''\langle x' \rangle \rangle = F\langle \eta \rangle$, we have

$$\delta_2'' = \int_0^\infty (u' - u'^2)\, dy''$$

$$= \delta'' \left[\int_0^\infty F(1-F)\, d\eta \right]$$

$$\left. \frac{\partial^2 u'}{\partial y''^2} \right|_{y''=0} = \frac{1}{\delta''} \left(\left. \frac{dF}{d\eta} \right|_{\eta=0} \right)$$

Therefore, Eq. (7.28) requires that

$$\left. \frac{dF}{d\eta} \right|_{\eta=0} \Big/ \delta'' = \frac{d\delta_2''}{dx'} = \frac{d\delta''}{dx'} \left[\int_0^\infty F(1-F)\, d\eta \right]$$

or

$$\tfrac{1}{2}[\delta''\langle x'\rangle]^2 = C_1 x' + C_2$$

where

$$C_1 = \left(\frac{dF}{d\eta}\bigg|_{\eta=0}\right) \bigg/ \int_0^\infty F(1-F)\, d\eta$$

At the leading edge $x' = 0$, u' must be 1 unless $y'' = 0$ (remembering that our object is to approximate the Blasius solution). So we require that

$$\lim_{\substack{x'\to 0\\ y''\neq 0}} \frac{y''}{\delta''\langle x'\rangle} = \infty$$

or $\delta''\langle 0\rangle = 0$. Then, at $x' = 0$,

$$u' = F\langle\infty\rangle = 1$$

From this condition, the integration constant C_2 must vanish, and

$$\delta''\langle x'\rangle = \sqrt{2C_1 x'}$$

If we choose

$$(1)\quad F\langle\eta\rangle = \begin{cases}\eta & \eta \leqslant 1\\ 1 & \eta \geqslant 1\end{cases}$$

where $F\langle 0\rangle = 0$, $F\langle\infty\rangle = 1$, $dF/d\eta|_{\eta=\infty}, \cdots = 0$, and F is continuous at $\eta = 1$, we have

$$C_1 = 1 \bigg/ \int_0^1 \eta(1-\eta)\, d\eta = 6$$

and

$$\delta''\langle x'\rangle = \sqrt{12x'}$$

$$u' = \begin{cases} y''/\sqrt{12x'} & y'' \leqslant \sqrt{12x'}\\ 1 & y'' \geqslant \sqrt{12x'}\end{cases}$$

Compared to the Blasius solution, this approximation may look very bad. But it is qualitatively good in the sense that it shows the *effective* boundary-layer thickness (δ') as being proportional to $\sqrt{x'}$. If we calculate the drag coefficient, we have

$$C_D = \frac{1.155}{\sqrt{Re_L}}$$

$Re_L = \Upsilon L/\nu$, where L is the length of plate over which drag is summed up; C_D is based on one face only. Compared to the correct $1.328/\sqrt{Re_L}$, it is surprisingly good for such a bad guess. It shows that the K-P technique would yield very good results near the wall.

Let us try a few more curves:

$$(2)\quad F\langle\eta\rangle = \sin\frac{\pi}{2}\eta \qquad \eta \leqslant 1$$
$$= 1 \qquad \eta \geqslant 1$$

where $F\langle 0\rangle$, $d^2F/d\eta^2|_{\eta=0} = 0$; $F\langle\infty\rangle = 1$, $dF/d\eta|_{\eta=\infty}, \cdots = 0$ and F, $dF/d\eta$ are continuous at $\eta = 1$.

$$\delta'' = 4.79\sqrt{x'}$$

$$C_D = \frac{1.310}{\sqrt{Re_L}}$$

This result is considered excellent.

$$(3) \quad F\langle\eta\rangle = \tfrac{3}{2}\eta - \tfrac{1}{2}\eta^2 \qquad \eta \leqslant 1$$
$$= 1 \qquad \eta \geqslant 1$$

where $F\langle 0\rangle$, $d^2F/d\eta^2|_{\eta=0} = 0$; $F\langle\infty\rangle = 1$, $dF/d\eta|_{\eta=\infty}, \cdots = 0$ and F, $dF/d\eta$ are continuous at $\eta = 1$.

$$\delta'' = \sqrt{\frac{560}{39}}\,(\sqrt{x'})$$

$$C_D = \frac{1.292}{\sqrt{Re_L}}$$

$$(4) \quad F\langle\eta\rangle = 2\eta - 2\eta^3 + \eta^4 \qquad \eta \leqslant 1$$
$$= 1 \qquad \eta \geqslant 1$$

where $F\langle 0\rangle$, $d^2F/d\eta^2|_{\eta=0} = 0$, $F\langle\infty\rangle = 1$, $dF/d\eta|_{\eta=\infty}, \cdots = 0$ and F, $dF/d\eta$, $d^2F/d\eta^2$ are continuous at $\eta = 1$.

$$\delta'' = \sqrt{\frac{630}{37}}\,(\sqrt{x'})$$

$$C_D = \frac{1.372}{\sqrt{Re_L}}$$

(5) $F\langle\eta\rangle = \tanh\eta$, for all η.

$$\delta'' = 2.55\sqrt{x'}$$

$$C_D = \frac{1.568}{\sqrt{Re_L}}$$

The result is not as good as the others.

The form that Pohlhausen recommends for all boundary-layer flows is the quartic form

$$F\langle\eta\rangle = a\eta + b\eta^2 + c\eta^3 + d\eta^4 \qquad \eta \leqslant 1$$
$$= 1 \qquad \eta \geqslant 1$$

together with

$$\begin{cases} \left.\dfrac{d^2F}{d\eta^2}\right|_{\eta=0} = -\Lambda = -\delta''^2\left(\dfrac{du_w'^o}{dx'}\right) \\ F\langle 1\rangle = 1 \\ \left.\dfrac{dF}{d\eta}\right|_{\eta=1}, \left.\dfrac{d^2F}{d\eta^2}\right|_{\eta=1} = 0 \end{cases}$$

Note that $F\langle 0\rangle = 0$ is already satisfied. It is easy to show that

$$a = 2+\frac{\Lambda}{6} \qquad b = -\frac{\Lambda}{2} \qquad c = -2+\frac{\Lambda}{2} \qquad d = 1-\frac{\Lambda}{6}$$

which yields, for the Blasius flow, $a = 2$, $b = 0$, $c = -2$, $d = 1$, as used in (4).

Observations

MERITS

1. It yields good approximations to known exact solutions of the boundary-layer problem with little effort.
2. It leads quickly to practical results in cases where exact solutions are unknown.
3. It makes complicated investigations such as on separation manageable.
4. It is applicable to all problems with a rapid boundary-layer-like decay. The results are expected to be qualitatively good, and even quantitatively good near the wall.

DEMERITS In general, we notice that this method as it stands is a makeshift. It has no rational background. It is the laughingstock of the mathematicians. Its only justification lies in that, if done by an experience-guided expert, it yields results that compare favorably with the exact solution of either the problem under investigation, or a similar problem of the same structure. Now let us go into more detail.

1. It is inherently average in nature. It cannot be expected to yield detailed local information correctly.
2. From the original system, an infinite number of integral relations can be formulated—for example, the Leibenson relation[6] obtained in one of the Problems. A similar technique can then be established for every one of them.
3. In the absence of exact solutions, there is no way of finding out the errors involved. There is also no way of comparing the results obtained from different choices of the form of F.
4. The procedure is not iterative in nature. Thus, there is no way to get better and better results that approach the exact solution in some manner.
5. The result is at the mercy of the trial curves you choose. If you happen to try the correct family, you might even get the exact solution. But if you do not know the general behavior of the exact solution of the problem (or, at least, a similar problem), the technique might give you some outlandish results. For example, in the

[6] L. S. Leibenson, "Energy form of the integral condition in the boundary layer theory" (in Russian), CAHI (Central Aero. & Hydro. Institute) Report No. 240, 1935; as quoted in L. G. Loitsianskii, "Integral methods in the theory of the boundary layer," translated as NACA TM No. 1070, 1944.

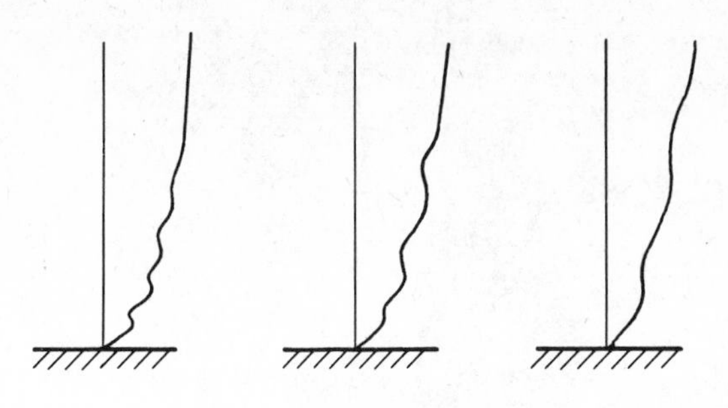

Figure 7.12 A family of wiggly curves.

boundary-layer flow, if you choose a family of curves that wiggle as y'' increases, the technique is going to pick up one for you (Fig. 7.12). But, we all know that the boundary-layer velocity profiles do not wiggle!

6. In certain problems it is conceivable that some important term might get lost in the integral relation. Then, whatever trial curve you choose, the result is not going to be near.

But, in all fairness, it should be added that there are attempts being made to modify the technique so that it will have a more rational basis; moreover, the technique has served engineers well for many years, and will continue to do so for many more years to come. (See, for example, D. E. Abbott and H. E. Bethel, "Application of the Galerkin-Kantorovich-Dorodnitsyn method of integral relations to the solution of steady laminar boundary layers," *Ingenieur-Archiv*, **37**, 110–124, 1968.)

7.5 JETS AND WAKES

The main purpose of this last section of Part 1 is to show that boundary layers do not of necessity appear only adjacent to rigid boundaries. They can also form where two different streams come together side by side in some fashion.

Take first the narrow jet issuing from a small slit into the same fluid at rest (Fig. 7.13) as our example. This jet will induce two streams that come together along the axis of the jet. Referring to the station of observation x_{obs} downstream from the slit (Fig. 7.13), we will render the problem dimensionless by choosing

$$y^* = x^* = x_{\text{obs}}$$
$$u^* = v^* = \Upsilon$$
$$p^* = \rho\Upsilon^2$$

where the characteristic speed Υ will be chosen later for convenience. The dimensionless equations governing u', v', and p' are then the same as Eqs. (1.9A) and (6.17a,b). The boundary conditions are obviously (1) $u', v' \to 0$ far away from the jet; (2) $v' = 0$ on $y' = 0$ for $x' > 0$; and (3) $\partial u'/\partial y' = 0$ on $y' = 0$ for $x' > 0$. Condition (2) comes from the physical fact that v is antisymmetric about the axis $y = 0$, whereas condition (3) follows the fact that u is symmetric about $y = 0$, and differ-

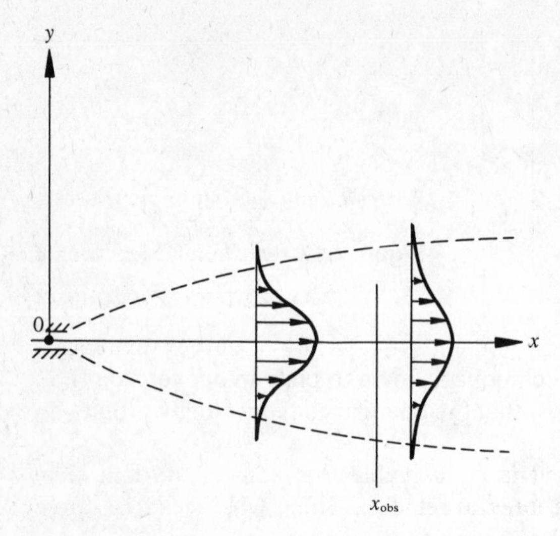

Figure 7.13 A thin jet.

entiable there. Both taken together indicate that the flow in the lower half-plane is just the mirror image of that in the upper half-plane. In addition to these three conditions, we also require that $u\langle x, 0\rangle \neq 0$ so that there is really a jet.

In this problem, we again let $Re \to \infty$ and ask for the corresponding asymptotic behavior. (But we can afford to be rather brief this second time around.) Equations (6.17a,b) again yield the Euler equations as $Re \to \infty$ knocks off the second-order derivatives. Without further ado, we will point out that the fluid at rest, that is, with

$$\begin{cases} u' = 0 \\ v' = 0 \\ p' = \text{constant} \end{cases}$$

obviously satisfies the Euler equations and Eq. (1.9A), together with all the conditions except our additional requirement that $u'\langle x', 0\rangle \neq 0$. In other words, the fluid at rest is the true description of the situation except near the x-axis where it fails to account for the jet. We thus recognize that $y' = 0$ $(x' > 0)$ is the seat of trouble, and that the source of trouble is in the knocking-off of all viscous terms in the limiting process $Re \to \infty$ with y' fixed. The fluid at rest is then the outer flow, which is valid everywhere except inside a thin boundary layer near $y' = 0$, downstream from the slit.

To see the boundary layer clearly, we must stretch the neighborhood of $y' = 0$ by introducing

$$y'' = \sqrt{Re}\, y'$$

with $p' = p'\langle x'\rangle$ and $v' \sim O(1/\sqrt{Re})$ as before (Section 6.2). After we introduce $v'' = \sqrt{Re}\, v'$, the matching principle yields the following facts:

1. The constant pressure in the outer flow is impressed on the boundary layer, so that

$$\frac{dp'}{dx'} = 0$$

2. The boundary-layer flow "blends" into the outer fluid at rest in the sense that

$$y'' = \infty; \quad u' = 0$$

but the value $v''\langle x', \infty\rangle$ is to be left open.

To summarize, the boundary-layer flow in a narrow jet, with $Re \to \infty$, emitted into the same fluid at rest is governed by the following system:

$$\begin{cases} \dfrac{\partial u'}{\partial x'} + \dfrac{\partial v''}{\partial y''} = 0 & \textbf{(6.22)} \\[2ex] u' \dfrac{\partial u'}{\partial x'} + v'' \dfrac{\partial u'}{\partial y''} = \dfrac{\partial^2 u'}{\partial y''^2} & \textbf{(7.32)} \\[2ex] x' > 0, y'' = 0: \quad \dfrac{\partial u'}{\partial y''} = 0 & \textbf{(7.33)} \\[2ex] \qquad\qquad\qquad v'' = 0 & \textbf{(6.18f)} \\[2ex] y'' = \infty: \quad u' = 0 & \textbf{(7.34)} \end{cases}$$

Here, again, we must append the additional requirement that

$$u'\langle x', 0\rangle \not\equiv 0 \qquad \textbf{(7.35)}$$

which excludes the trivial solution $u' = 0, v'' = 0$ from further consideration.

To solve this boundary-layer system, we could follow the established procedure in the same manner as in Section 7.1, while keeping in mind two things: (a) Boundary condition (7.34) under ansatz (7.1) simply demands that

$$\eta = \infty: \quad \frac{df}{d\eta} = 0$$

without yielding any information on $a\langle x'\rangle$; and thus, in paraphrasing the rest of the derivation in Section 7.1, $u_w'^o\langle x'\rangle$ must be replaced by $a\langle x'\rangle$. (b) There is only a viscous term and no pressure-gradient term, on the right-hand side of Eq. (7.32); the counterpart of Eq. (7.8) will thus have only three terms on the left-hand side, the term λ being absent.

However, instead of repeating Section 7.1 in essence here, we will follow a shortcut by starting with a narrower form of the ansatz (7.1), thus:

$$u'\langle x', y''\rangle = x'^c \left(\frac{df}{d\eta}\right) \qquad \textbf{(7.36)}$$

where

$$\eta = \frac{y''}{x'^b} \qquad \textbf{(7.37)}$$

We introduce this narrower ansatz[7] not only to avoid repetition and to sustain interest but also to familiarize you with the most often used ansatz in the search

[7] H. Schlichting, "Laminare Strahlausbreitung," *ZAMM*, **13**, 260–263, 1933. H. Schlichting is a professor at Braunschweig Technical University, Braunschweig, Germany.

for similarity solutions. Its relative simplicity makes it very appealing as a first trial; if it does not work, or if more generality is called for, one can always go back to ansatz (7.1).

Now, the first things we notice under this new ansatz are boundary conditions (7.34) and (7.33), which become:

$$\eta = \infty: \quad \frac{df}{d\eta} = 0 \tag{7.38}$$

$$\eta = 0: \quad \frac{d^2 f}{d\eta^2} = 0 \tag{7.39a}$$

Next, just as in Section 7.1, the equation of continuity, Eq. (6.22), yields the counterpart of Eq. (7.7):

$$v'' = x'^{c+b-1}\left[b\eta\left(\frac{df}{d\eta}\right) - (c+b)f\right] \tag{7.40}$$

with the associated boundary conditions

$$\eta = 0: \quad f = 0 \tag{7.39b}$$

Substituting into Eq. (7.32), we have

$$x'^{c}\left(\frac{df}{d\eta}\right)\left[cx'^{c-1}\left(\frac{df}{d\eta}\right) - bx'^{c-1}\eta\left(\frac{d^2 f}{d\eta^2}\right)\right]$$
$$+ x'^{c+b-1}\left[b\eta\frac{df}{d\eta} - (c+b)f\right]x'^{c-b}\left(\frac{d^2 f}{d\eta^2}\right) = x'^{c-2b}\frac{d^3 f}{d\eta^3}$$

that is,

$$\frac{d^3 f}{d\eta^3} + x'^{c+2b-1}\left[(c+b)f\left(\frac{d^2 f}{d\eta^2}\right) - c\left(\frac{df}{d\eta}\right)^2\right] = 0$$

which becomes an ordinary differential equation

$$\frac{d^3 f}{d\eta^3} + (c+b)f\left(\frac{d^2 f}{d\eta^2}\right) - c\left(\frac{df}{d\eta}\right)^2 = 0 \tag{7.41}$$

if

$$c + 2b - 1 = 0 \tag{7.42}$$

We see thus that this procedure yields here only one relationship for the two indices c and b. For another relationship, let us integrate Eq. (7.32) with respect to y'' from 0 to ∞:

$$\int_0^\infty \left(u'\frac{\partial u'}{\partial x'} + v''\frac{\partial u'}{\partial y''}\right) dy'' = \left.\frac{\partial u'}{\partial y''}\right|_{y''=\infty} - \left.\frac{\partial u'}{\partial y''}\right|_{y''=0}$$
$$= 0$$

where boundary conditions (7.33) and (7.34) have been used. (Boundary condition

(7.34) implies that $\partial u'/\partial y'' \to 0$ as $y'' \to \infty$.) With the aid of Eq. (6.22), we have

$$\int_0^\infty \left(u' \frac{\partial u'}{\partial x'} + \frac{\partial u'v''}{\partial y''} - u' \frac{\partial v''}{\partial y''}\right) dy'' = \int_0^\infty \left(u' \frac{\partial u'}{\partial x'} + u' \frac{\partial u'}{\partial x'}\right) dy'' + (u'v'')\Big|_{y''=0}^{y''=\infty}$$

$$= \frac{d}{dx'} \int_0^\infty u'^2 \, dy''$$

$$= 0$$

where boundary conditions (6.18f) and (7.34) are employed. Thus, the jet has an overall property

$$\int_0^\infty u'^2 \, dy'' = \text{constant} \qquad \textbf{(7.43)}$$

or

$$x'^{2c+b} \int_0^\infty \left(\frac{df}{d\eta}\right)^2 d\eta = \text{constant}$$

Now this is possible only if

$$2c + b = 0 \qquad \textbf{(7.44)}$$

(unless $u' \equiv 0$). This is then the second relationship needed for the determination of c and b. Equations (7.42) and (7.44) then yield

$$\begin{cases} c = -\frac{1}{3} & \textbf{(7.45a)} \\ b = \frac{2}{3} & \textbf{(7.45b)} \end{cases}$$

Equation (7.43) has a very simple physical explanation, namely, the total momentum of the jet in the x-direction

$$\rho \int_{-\infty}^{\infty} u^2 \, dy = \frac{2\rho \Upsilon^2 x^*}{\sqrt{Re}} \int_0^\infty u'^2 \, dy''$$

is a constant along the jet axis. This constant is then a measure of the strength of the jet. We will call it $J(\neq 0)$; that is,

$$J = \rho \int_{-\infty}^{\infty} u^2 \, dy = \text{constant} \qquad \textbf{(7.46)}$$

To summarize, we have found a similarity solution for the thin jet *if it carries* an appreciable amount of momentum:

$$u' = \frac{df/d\eta}{x'^{1/3}} \qquad \textbf{(7.47)}$$

$$v'' = \frac{2\eta(df/d\eta) - f}{3x'^{2/3}} \qquad \textbf{(7.48)}$$

$$\eta = \frac{y''}{x'^{2/3}} \qquad \textbf{(7.49)}$$

where $f\langle\eta\rangle$ satisfies the nonlinear ordinary differential equation

$$3\frac{d^3f}{d\eta^3}+f\left(\frac{d^2f}{d\eta^2}\right)+\left(\frac{df}{d\eta}\right)^2=0 \tag{7.50}$$

with conditions

$$\eta=0:\quad f,\frac{d^2f}{d\eta^2}=0 \tag{7.51a,b}$$

$$\eta=\infty:\quad \frac{df}{d\eta}=0 \tag{7.52}$$

and

$$\int_0^\infty\left(\frac{df}{d\eta}\right)^2 d\eta=\frac{\sqrt{Re}\,J}{2\rho\Upsilon^2x^*}\neq 0 \tag{7.53}$$

At this point, it becomes convenient to choose the characteristic speed

$$\Upsilon=\left(\frac{J^2}{\rho\mu x^*}\right)^{1/3} \tag{7.54}$$

so that (remembering that $Re=\rho\Upsilon x^*/\mu$)

$$\frac{\sqrt{Re}\,J}{\rho\Upsilon^2x^*}=1$$

and Eq. (7.53) becomes

$$\int_0^\infty\left(\frac{df}{d\eta}\right)^2 d\eta=\frac{1}{2} \tag{7.53A}$$

Equation (7.50) is surprisingly simple to solve.[8] All we have to notice is that

$$\frac{d}{d\eta}\left[f\left(\frac{df}{d\eta}\right)\right]=f\left(\frac{d^2f}{d\eta^2}\right)+\left(\frac{df}{d\eta}\right)^2$$

Therefore we can integrate once and get

$$3\frac{d^2f}{d\eta^2}+f\left(\frac{df}{d\eta}\right)=C_1$$

At $\eta=0$, boundary conditions (7.51a,b) yield

$$C_1=0$$

Therefore

$$3\frac{d^2f}{d\eta^2}+f\left(\frac{df}{d\eta}\right)=0$$

This is easily integrated to yield

$$3\frac{df}{d\eta}+\frac{1}{2}f^2=C_2$$

[8] W. Bickley, "The plane jet," *Phil. Mag.* (7), **23**, 727–731, 1939.

At $\eta = \infty$, boundary condition (7.52) assures us that

$$C_2 = \tfrac{1}{2}(f\langle\infty\rangle)^2 \geqslant 0$$

So for convenience we will write

$$3\frac{df}{d\eta}+\frac{1}{2}f^2=\frac{1}{2}\alpha^2$$

or

$$\frac{6\,df}{\alpha^2-f^2}=d\eta$$

that is,

$$\frac{6}{\alpha}\tanh^{-1}\left(\frac{f}{\alpha}\right)=\eta+C_3$$

Boundary condition (7.51a) then demands that $C_3 = 0$. Therefore

$$f=\alpha\tanh\left(\frac{\alpha\eta}{6}\right)$$

This solution satisfies all the boundary conditions (7.51a,b) and (7.52) for any α. To determine α, we have to incur the condition (7.53A):

$$\frac{\alpha^4}{36}\int_0^\infty \operatorname{sech}^4\left(\frac{\alpha\eta}{6}\right)d\eta=\frac{\alpha^3}{9}=\frac{1}{2}$$

therefore

$$\alpha=\left(\frac{9}{2}\right)^{1/3}=1.6510 \tag{7.55}$$

Our solution is now complete:

$$f=1.6510\tanh(0.2752\eta)$$

$$u'=\frac{0.4543\operatorname{sech}^2(0.2752\eta)}{x'^{1/3}} \tag{7.56}$$

$$v''=\frac{0.5503\,[0.5503\eta\operatorname{sech}^2(0.2752\eta)-\tanh(0.2752\eta)]}{x'^{2/3}} \tag{7.57}$$

It is again noticed that the transverse velocity at the "edge" of the jet (or boundary layer) is

$$v''\langle x',\infty\rangle=\frac{-0.5503}{x'^{2/3}} \tag{7.58}$$

$$\neq 0$$

That is, there is a residue transverse flow feeding into the jet at the "edge."

In this lucky problem where we can obtain the solution to the governing *nonlinear* equation in closed form, you should have no difficulty in visualizing the result by sketching a few velocity profiles and by figuring out that the "spread" of the jet (or the "thickness" of the boundary layer) goes like $x'^{2/3}$. It is more important to ask what we should do in order to produce such a jet in a laboratory. The first thing we must do is, of course, to see that

$$Re = \left(\frac{\rho J x_{obs}}{\mu^2}\right)^{2/3}$$

is large. The slit must, of course, be very narrow. But how should we blow at the slit? To answer this question, we note that

$$x' \to 0: \quad u' = 0 \qquad y'' \neq 0$$
$$= \infty \qquad y'' = 0$$

This seems to mean that we should blow hard through a narrow slit. However, if we calculate the mass flux across a line perpendicular to the jet axis at x, we have

$$Q = 2\rho \int_0^\infty u \, dy$$
$$= \frac{2\rho \Upsilon x^*}{\sqrt{Re}} \int_0^\infty u' \, dy''$$
$$= 3.3019 \left(\frac{\rho \Upsilon x^*}{\sqrt{Re}}\right) x'^{1/3} \tag{7.59}$$

which shows that Q at the slit should be exactly zero; no mass flow through the slit! Are we in trouble? Have we in fact solved a fictitious problem? No!

This flat "no" can be explained in the following manner: Let us say, in practice, the slit is of width δ and the speed at the slit is U. Then

$$J = \rho U^2 \delta$$
$$Q_{slit} = \rho U \delta$$

In order that the foregoing similarity jet flow be realizable, all we have to do is to make δ so small that $\rho U \delta$ is negligibly small (although U is large), and to make U so large that $\rho U^2 \delta$ is a moderate number (which is a measure of the strength of the jet), and to observe far enough downstream that

$$Re = \left(\frac{\rho J x_{obs}}{\mu^2}\right)^{2/3}$$

is large. We may say that such a jet carries appreciable momentum but negligible mass. The nonzero Q at $x \neq 0$ is due to the fluid dragged into the jet from the surroundings; it is induced, not carried, by the jet; see Eq. (7.58).

As another example where the boundary layer forms in the neighborhood of a nonrigid surface, we will consider the laminar *wake* behind a flat plate of finite length (Fig. 7.14). This is actually a continuation of our previous discussion in Section 7.2, point 3. When the length of the plate is long enough that the Reynolds number based on this length is large (but still within the laminar regime), the trailing edge will display a Blasius flow. But what happens after the trailing edge needs a fresh investigation.

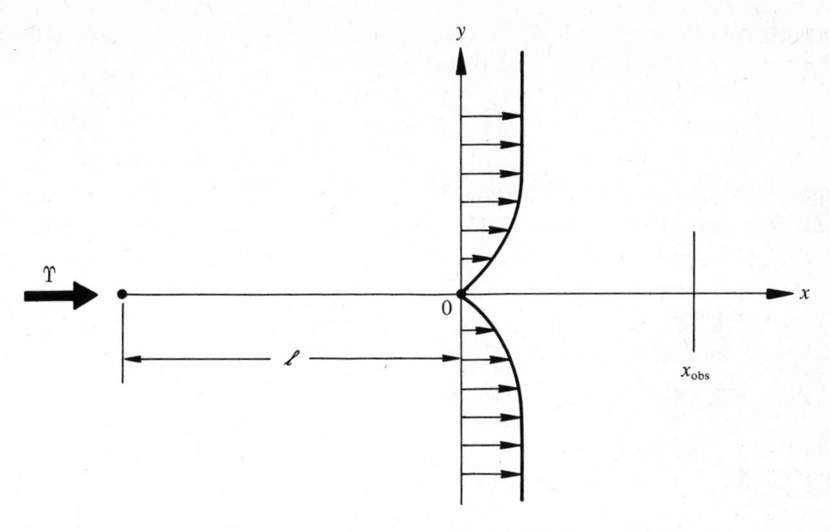

Figure 7.14 Wake after a flat plate.

For convenience, we have denoted the trailing edge by the origin 0 in Figure 7.14. Thus the wake appears in the region $x > 0$. Again, just as in Section 7.2, an investigation of the asymptotic behavior as $Re = \rho\Upsilon x_{obs}/\mu \to \infty$ yields first an outer flow

$$\begin{cases} u^o = \Upsilon \\ v^o = 0 \\ p^o = \text{constant} \end{cases}$$

which satisfies the Euler equations and equation of continuity, but cannot account for the effect of the Blasius flow at 0. In other words, this outer flow does not satisfy the additional (wake) requirement that

$$u'\langle x', 0\rangle \neq 1 \tag{7.60}$$

To look more closely into the trouble near $y = 0$, we use the same stretching of y' into y'' as before and obtain the following boundary-layer problem:

$$\begin{cases} \dfrac{\partial u'}{\partial x'} + \dfrac{\partial v''}{\partial y''} = 0 & \textbf{(6.22)} \\[2ex] u'\dfrac{\partial u'}{\partial x'} + v''\dfrac{\partial u'}{\partial y''} = \dfrac{\partial^2 u'}{\partial y''^2} & \textbf{(7.32)} \\[2ex] x' > 0,\ y'' = 0\text{:} \quad \dfrac{\partial u'}{\partial y''} = 0 & \textbf{(7.33)} \\[2ex] \qquad\qquad\qquad v'' = 0 & \textbf{(6.18f)} \\[1ex] y'' = \infty\text{:} \quad u' = 1 & \textbf{(7.61)} \end{cases}$$

This problem is the same as the thin jet except for boundary condition (7.61). Furthermore, while for the thin jet we can hold back the condition at $x' = 0$ until the end of our search for the similarity solution, here we must state the condition that

$$\text{at } x = 0: \quad \text{the flow must be Blasius, developed over the length } \ell \tag{7.62}$$

Otherwise, it would not be the wake after a flat plate of length ℓ.

The above problem is easier to set up than to solve. A discussion of the solution close to the trailing edge[9] is out of the scope of this book. The solution far downstream[10] is rather easy to obtain; unfortunately, however, its practical value is very doubtful since the flow there is most certainly no longer laminar.

With this remark, we now cut short the wake problem and pick up, instead, the problem of the wall jet as our last topic in Part 1. You will see that this is another "fortunate" problem that we can carry to the bitter end.

Consider[11] a thin jet of fluid issuing from a narrow slit, parallel to a flat wall, into the same fluid at rest; or a narrow jet striking a flat wall while immersed in the same fluid at rest (Fig. 7.15). Again, letting Υ be the characteristic speed, to

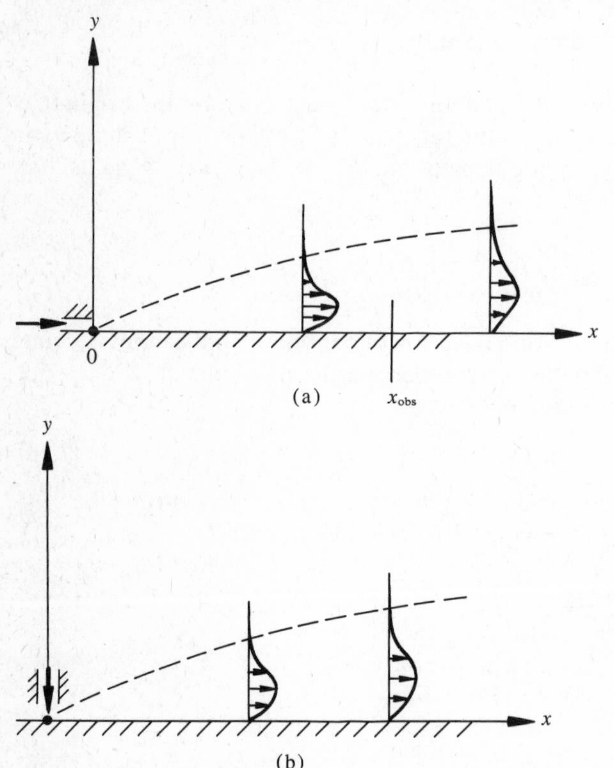

Figure 7.15 Two ways to realize a thin wall jet.

[9] S. Goldstein, "Concerning some solutions of the boundary layer equations in hydrodynamics," *Proc. Camb. Phil. Soc.*, **26**, 1–30, 1930. S. Goldstein is Gordon McKay Professor Emeritus at Harvard University.

[10] W. Tollmien, "Grenzschichttheorie," *Handbuch der Experimentalphysik*, **IV/1**, Akademische Verlag, Leipzig, 1931. W. Tollmien is a professor at Göttingen University.

[11] M. B. Glauert, "The wall jet," *J. Fluid Mech.*, **1**, 625–643, 1956.

be identified later, we choose x_{obs} such that $Re = \rho \Upsilon x_{obs}/\mu$ is very large. As $Re \to \infty$ with y' fixed, we find the outer "flow" again:

$$\begin{cases} u^o = 0 \\ v^o = 0 \\ p^o = \text{constant} \end{cases}$$

which even satisfies the boundary condition on the wall; that is,

$$x > 0, y = 0: \quad u, v = 0$$

As a matter of fact, this outer flow would be the solution for the entire flow field if not for the fact that it does not reflect the presence of the jet at all. The jet actually poses an additional condition on the problem:

$$u \not\equiv 0$$

The failure of the outer flow to satisfy this additional condition occurs near the wall. The neighborhood of the wall must therefore be magnified through the stretching:

$$y'' = \sqrt{Re}\, y'$$

The resulting boundary-layer problem, as $Re \to \infty$ with y'' fixed, is

$$\begin{cases} \dfrac{\partial u'}{\partial x'} + \dfrac{\partial v''}{\partial y''} = 0 & \textbf{(6.22)} \\ u' \dfrac{\partial u'}{\partial x'} + v'' \dfrac{\partial u'}{\partial y''} = \dfrac{\partial^2 u'}{\partial y''^2} & \textbf{(7.32)} \\ x' > 0, y'' = 0: \quad u' = 0 & \textbf{(6.18a)} \\ \qquad\qquad\qquad v'' = 0 & \textbf{(6.18f)} \\ \qquad y'' = \infty: \quad u' = 0 & \textbf{(7.34)} \end{cases}$$

together with

$$u' \not\equiv 0 \qquad \textbf{(7.63)}$$

To search for the similarity solution, we start with the same ansatz (7.36) plus (7.37). We then have boundary conditions (7.38) and (7.39b), while boundary condition (7.39a) is replaced by

$$\eta = 0: \quad \frac{df}{d\eta} = 0 \qquad \textbf{(7.39c)}$$

Our search then leads us to the same equation, (7.41), with the same relation, (7.42).

To get another relation between c and b, we must modify the route that led to Eq. (7.44), since here

$$\left.\frac{\partial u'}{\partial y''}\right|_{y''=0} \neq 0$$

and

$$J \neq \text{constant}$$

Instead, we will integrate Eq. (7.32) with respect to y'' from $y''=y''$ to $y''=\infty$, and obtain

$$\frac{\partial}{\partial x'}\int_{y''}^{\infty} u'^2\,dy'' - u'v'' = -\frac{\partial u'}{\partial y''}$$

We then multiply this new relation by u' before integrating again with respect to y'' from 0 to ∞:

$$\int_0^{\infty} u'\left(\frac{\partial}{\partial x'}\int_{y''}^{\infty} u'^2\,dy''\right)dy'' - \int_0^{\infty} u'^2 v''\,dy'' = -\frac{1}{2}u'^2\Big|_{y''=0}^{y''=\infty} = 0$$

where the first term on the left-hand side can be written as

$$\begin{aligned}
&\frac{d}{dx'}\int_0^{\infty} u'\left(\int_{y''}^{\infty} u'^2\,dy''\right)dy'' - \int_0^{\infty}\frac{\partial u'}{\partial x'}\left(\int_{y''}^{\infty} u'^2\,dy''\right)dy''\\
&\qquad = \frac{d}{dx'}\int_0^{\infty} u'\left(\int_{y''}^{\infty} u'^2\,dy''\right)dy'' + \int_0^{\infty}\frac{\partial v''}{\partial y''}\left(\int_{y''}^{\infty} u'^2\,dy''\right)dy''\\
&\qquad = \frac{d}{dx'}\int_0^{\infty} u'\left(\int_{y''}^{\infty} u'^2\,dy''\right)dy'' + v''\int_{y''}^{\infty} u'^2\,dy''\Big|_{y''=0}^{y''=\infty}\\
&\qquad\quad + \int_0^{\infty} u'^2 v''\,dy''\\
&\qquad = \frac{d}{dx'}\int_0^{\infty} u'\left(\int_{y''}^{\infty} u'^2\,dy''\right)dy'' + \int_0^{\infty} u'^2 v''\,dy''
\end{aligned}$$

where Eq. (6.22), integration by parts, and boundary condition (6.18f) have been used in turn. Therefore

$$\frac{d}{dx'}\int_0^{\infty} u'\left(\int_{y''}^{\infty} u'^2\,dy''\right)dy'' = 0$$

or

$$\int_0^{\infty} u'\left(\int_{y''}^{\infty} u'^2\,dy''\right)dy'' = \text{constant} \qquad \textbf{(7.64)}$$

that is,

$$x'^{3c+2b}\int_0^{\infty}\frac{df}{d\eta}\left[\int_{\eta}^{\infty}\left(\frac{df}{d\eta}\right)^2 d\eta\right]d\eta = \text{constant}$$

This is only possible, since $df/d\eta \not\equiv 0$, if

$$3c+2b=0 \qquad \textbf{(7.65)}$$

From Eqs. (7.42) and (7.65), we have

$$\begin{cases} c=-\frac{1}{2} & \textbf{(7.66a)}\\ b=\frac{3}{4} & \textbf{(7.66b)}\end{cases}$$

Summary

$$u' = \frac{df/d\eta}{x'^{1/2}} \tag{7.67}$$

$$v'' = \frac{3\eta(df/d\eta) - f}{4x'^{3/4}} \tag{7.68}$$

$$\eta = \frac{y''}{x'^{3/4}} \tag{7.69}$$

$$4\frac{d^3f}{d\eta^3} + f\left(\frac{d^2f}{d\eta^2}\right) + 2\left(\frac{df}{d\eta}\right)^2 = 0 \tag{7.70}$$

$$\eta = 0: \quad f, \frac{df}{d\eta} = 0 \tag{7.71a,b}$$

$$\eta = \infty: \quad \frac{df}{d\eta} = 0 \tag{7.72}$$

and

$$\int_0^\infty \frac{df}{d\eta}\left[\int_\eta^\infty \left(\frac{df}{d\eta}\right)^2 d\eta\right] d\eta \neq 0 \tag{7.73}$$

Before we solve the problem, we can make the following observations:

1. The spread of the jet is proportional to $x'^{3/4}$.
2. $x' \to 0$: $u' = 0 \quad y'' \neq 0$
   $\qquad\qquad = 0 \quad y'' = 0$

However, for $n > \frac{3}{4}$,

$$\lim_{\substack{x' \to 0 \\ y'' \sim \square(x'^n)}} u' \sim \frac{0 + A\eta}{x'^{1/2}}$$

$$\sim Ay''/x'^{5/4}$$

$$\sim x'^{(n-5/4)}$$

$$= \infty \quad \text{if } n < \tfrac{5}{4}$$

Also, if $n = \frac{3}{4}$, then $0 < \eta < \infty$, $df/d\eta \neq 0$; and

$$\lim_{x' \to 0} u' = \infty$$

In practice this calls for a very large speed U at the slit, somewhere near the wall.

3. $Q = \rho \int_0^\infty u \, dy$

$$= \frac{\rho \Upsilon x^*}{\sqrt{Re}} \int_0^\infty u' \, dy''$$

$$\propto x'^{(-1/2)+(3/4)}$$

$$\propto x'^{1/4}$$

This indicates that the slit must be very narrow so as to carry negligible mass; that is, $\rho U\delta$ must be very small. (Q increases downstream as the jet drags in more fluid.)

$$4. \quad J = \rho \int_0^\infty u^2\, dy$$
$$\propto \int_0^\infty u'^2\, dy''$$
$$\propto x'^{-1+(3/4)}$$
$$\propto x'^{-1/4}$$

This indicates that the momentum carried by the jet at the slit must be very large; that is, $\rho U^2\delta$ must be very large. (J decreases downstream because of the friction force on the wall.)

$$5. \quad \rho^2 \int_0^\infty u \left(\int_y^\infty u^2\, dy \right) dy = \text{constant}$$
$$= \rho^2 \int_0^\delta U \left(\int_y^\delta U^2\, dy \right) dy$$
$$= \tfrac{1}{2}\rho^2 U^3 \delta^2$$
$$= \tfrac{1}{2}(JQ)_{\text{slit}}$$
$$= \tfrac{1}{2}K$$

This indicates that, in practice, the product JQ at the slit must be a moderate number (whereas J is large and Q small). This number is then a measure of the strength of the jet. In terms of K, we will have

$$\frac{\rho^2 \Upsilon^3 x^{*2}}{Re} \int_0^\infty \left(\frac{df}{d\eta}\right) \left[\int_\eta^\infty \left(\frac{df}{d\eta}\right)^2 d\eta \right] d\eta = \frac{1}{2}K$$

or

$$\int_0^\infty \left(\frac{df}{d\eta}\right) \left[\int_\eta^\infty \left(\frac{df}{d\eta}\right)^2 d\eta \right] d\eta = \frac{K}{2\rho\mu x^* \Upsilon^2}$$
$$\neq 0$$

Here, it is convenient to choose

$$\Upsilon = \left(\frac{K}{\rho\mu x^*}\right)^{1/2} \tag{7.74}$$

such that

$$\int_0^\infty \left(\frac{df}{d\eta}\right) \left[\int_\eta^\infty \left(\frac{df}{d\eta}\right)^2 d\eta \right] d\eta = \frac{1}{2} \tag{7.75}$$

To solve for f, we first notice that

$$f = 2\beta F\langle \beta\eta/2 \rangle$$

for any constant β, if $F\langle\xi\rangle$ is *a* solution of the ordinary differential equation

$$\frac{d^3F}{d\xi^3}+F\left(\frac{d^2F}{d\xi^2}\right)+2\left(\frac{dF}{d\xi}\right)^2=0 \tag{7.76}$$

with boundary conditions

$$\xi=0:\quad F,\frac{dF}{d\xi}=0 \tag{7.77a,b}$$

$$\xi=\infty:\quad \frac{dF}{d\xi}=0 \tag{7.78}$$

Condition (7.75) then fixes the arbitrary constant β, thus:

$$\beta^4\int_0^\infty\left(\frac{dF}{d\xi}\right)\left[\int_\xi^\infty\left(\frac{dF}{d\xi}\right)^2 d\xi\right]d\xi=\frac{1}{8} \tag{7.75A}$$

To obtain F, let us multiply Eq. (7.76) by F:

$$\begin{aligned}F\left(\frac{d^3F}{d\xi^3}\right)+F^2\left(\frac{d^2F}{d\xi^2}\right)+2F\left(\frac{dF}{d\xi}\right)^2&=F\left(\frac{d^3F}{d\xi^3}\right)+\frac{d}{d\xi}\left[F^2\left(\frac{dF}{d\xi}\right)\right]\\&=\frac{d}{d\xi}\left[F\left(\frac{d^2F}{d\xi^2}\right)\right]-\left(\frac{dF}{d\xi}\right)\left(\frac{d^2F}{d\xi^2}\right)+\frac{d}{d\xi}\left[F^2\left(\frac{dF}{d\xi}\right)\right]\\&=0\end{aligned}$$

This can be integrated directly to yield

$$F\left(\frac{d^2F}{d\xi^2}\right)-\frac{1}{2}\left(\frac{dF}{d\xi}\right)^2+F^2\left(\frac{dF}{d\xi}\right)=\text{constant} \tag{7.79}$$

Because of boundary condition (7.78), which also implies that

$$\xi=\infty:\quad \frac{d^2F}{d\xi^2}=0$$

the constant of integration must be zero. Multiplying this new equation (with the right-hand side = 0) by $F^{-3/2}$ and integrating, we have

$$\frac{dF/d\xi}{F^{1/2}}+\frac{2}{3}F^{3/2}=\text{constant}$$

where the first term comes from the first two terms of Eq. (7.79). This constant of integration is arbitrary; it is not fixed by any of the boundary conditions. For convenience, we choose for it the value $\frac{2}{3}$. (There is absolutely nothing lost in this choice, since $F\langle\xi\rangle$ is only required as "*a* solution" in the relation $f=2\beta F\langle\beta\eta/2\rangle$. Another choice will, of course, yield a different $F\langle\xi\rangle$; but β will also be different through Eq. (7.75A), thus yielding the same f as before because of the inherently convenient structure of Eq. (7.70).) Setting

$$g=F^{1/2}$$

or

$$g^2=F$$

we have finally

$$\frac{dg}{d\xi} = \frac{1}{3}(1 - g^3)$$

since

$$\frac{dF}{d\xi} = 2g\frac{dg}{d\xi}$$

The ordinary differential equation for g is easily integrated under the condition $g\langle 0\rangle = 0$:

$$\xi = \ln\frac{\sqrt{1+g+g^2}}{1-g} + \sqrt{3}\tan^{-1}\left(\frac{\sqrt{3}g}{2+g}\right) \tag{7.80}$$

Figure 7.16 shows the variation of $F(= g^2)$ and $dF/d\xi$ $(= 2g\, dg/d\xi)$ vs. ξ according to Eq. (7.80). A numerical integration shows that

$$\int_0^\infty \left(\frac{dF}{d\xi}\right)\left[\int_\xi^\infty \left(\frac{dF}{d\xi}\right)^2 d\xi\right] d\xi = \frac{1}{10}$$

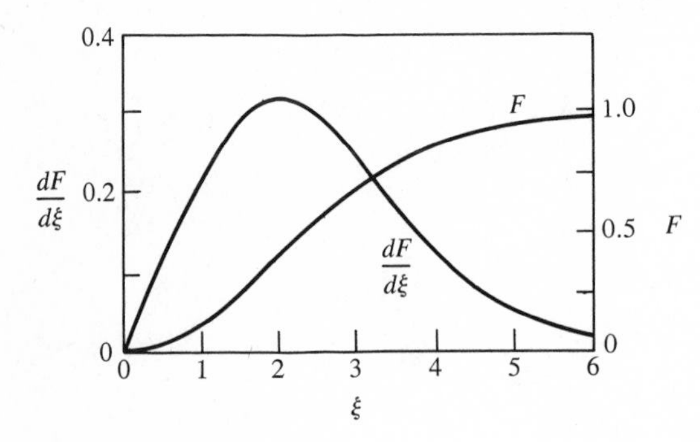

Figure 7.16 Solution of the wall-jet problem.

Therefore

$$\beta = \left(\frac{5}{4}\right)^{1/4} \tag{7.81}$$

and

$$f\langle\eta\rangle = (20)^{1/4} F\langle (5/64)^{1/4}\eta\rangle \tag{7.82}$$

It is at this point that we have solved the problem of the wall jet!

PART 2

FLOWS OF INCOMPRESSIBLE FLUIDS—2

CHAPTER 8

PLANE FLOWS IN POLAR COORDINATES

8.1 INTRODUCTION

It is our object here in Part 2 to lead you gradually out of the limitations we posed in Part 1 for didactic reasons. To be specific we will (1) adopt non-Cartesian coordinate systems (for example, polar coordinates) since they are more convenient to use in many problems (with circular boundaries); (2) eventually investigate nonplane flow fields since many practically interesting problems are not plane (such as flow inside a round pipe); and (3) stop shying away from the problem when it begins to get more complicated. In this manner we will be filling in those gaps left in Part 1, as well as broadening our outlook. The assumption of incompressibility, however, will still be with us all through Part 2. For the effects of compressibility, you have to wait for Part 3.

To begin with, let us try to put the key equations of Part 1 into a framework where the polar coordinates (R, θ) are used instead of the rectangular Cartesian coordinates (x, y); see Figure 8.1.

The relationship between (R, θ) and (x, y) is, of course,

$$\begin{cases} R = \sqrt{x^2 + y^2} & \textbf{(8.1a)} \\ \theta = \tan^{-1}\dfrac{y}{x} & \textbf{(8.1b)} \end{cases}$$

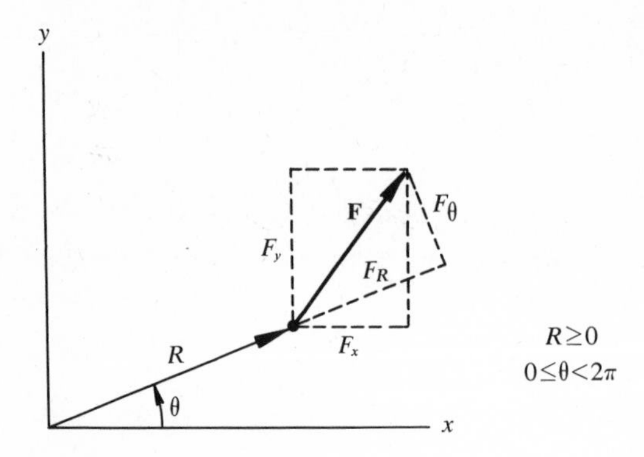

Figure 8.1 Polar coordinates.

where

$$0 \leqslant \theta < 2\pi \quad \text{and} \quad R \geqslant 0$$

or

$$\begin{cases} x = R \cos \theta & \textbf{(8.1c)} \\ y = R \sin \theta & \textbf{(8.1d)} \end{cases}$$

If **F** denotes any vector (Fig. 8.1), it is also necessary to relate its components in the R- and θ-directions to those in the x- and y-directions:

$$\begin{cases} F_R = F_x \cos \theta + F_y \sin \theta & \textbf{(8.2a)} \\ F_\theta = -F_x \sin \theta + F_y \cos \theta & \textbf{(8.2b)} \end{cases}$$

or

$$\begin{cases} F_x = F_R \cos \theta - F_\theta \sin \theta & \textbf{(8.2c)} \\ F_y = F_R \sin \theta + F_\theta \cos \theta & \textbf{(8.2d)} \end{cases}$$

For the velocity **q**, we have always been writing (u, v) for (q_x, q_y). Then,

$$\begin{cases} q_R = u \cos \theta + v \sin \theta & \textbf{(8.3a)} \\ q_\theta = -u \sin \theta + v \cos \theta & \textbf{(8.3b)} \end{cases}$$

or

$$\begin{cases} u = q_R \cos \theta - q_\theta \sin \theta & \textbf{(8.3c)} \\ v = q_R \sin \theta + q_\theta \cos \theta & \textbf{(8.3d)} \end{cases}$$

In general,

$$\mathbf{F} = F_R \hat{\mathbf{R}} + F_\theta \hat{\boldsymbol{\theta}} \qquad \textbf{(8.4a)}$$

$$= F_x \hat{\mathbf{x}} + F_y \hat{\mathbf{y}} \qquad \textbf{(8.4b)}$$

where $\mathbf{R}$ and $\hat{\boldsymbol{\theta}}$ are unit vectors in the R- and θ-directions, respectively. But, you must notice that whereas $\hat{\mathbf{x}}$ and $\hat{\mathbf{y}}$ are fixed for any given rectangular Cartesian system, $\hat{\mathbf{R}}$ and $\hat{\boldsymbol{\theta}}$ *change their directions* from point to point. As a matter of fact,

$$\begin{cases} \hat{\mathbf{R}} = \hat{\mathbf{x}} \cos \theta + \hat{\mathbf{y}} \sin \theta & \textbf{(8.5a)} \\ \hat{\boldsymbol{\theta}} = -\hat{\mathbf{x}} \sin \theta + \hat{\mathbf{y}} \cos \theta & \textbf{(8.5b)} \end{cases}$$

and, conversely,

$$\begin{cases} \hat{\mathbf{x}} = \hat{\mathbf{R}} \cos\theta - \hat{\boldsymbol{\theta}} \sin\theta & \textbf{(8.5c)} \\ \hat{\mathbf{y}} = \hat{\mathbf{R}} \sin\theta + \hat{\boldsymbol{\theta}} \cos\theta & \textbf{(8.5d)} \end{cases}$$

We will now start our long journey of transformation from (x, y) to (R, θ) with the formula for the contact force per unit area acting on an area element whose unit normal vector is $\hat{\mathbf{n}}$; that is, from Section 2.1,

$$\mathbf{f}\langle\hat{\mathbf{n}}\rangle = \mathbf{f}\langle\hat{\mathbf{x}}\rangle \cos \measuredangle(n, x) + \mathbf{f}\langle\hat{\mathbf{y}}\rangle \cos \measuredangle(n, y) \qquad \textbf{(2.3)}$$

Applying Eq. (2.3) to two local area elements with $\hat{\mathbf{n}} = \hat{\mathbf{R}}$ and $\hat{\boldsymbol{\theta}}$ in turn, we have (Fig. 8.2):

$$\begin{aligned} \mathbf{f}\langle\hat{\mathbf{R}}\rangle &= \mathbf{f}\langle\hat{\mathbf{x}}\rangle \cos \measuredangle(R, x) + \mathbf{f}\langle\hat{\mathbf{y}}\rangle \cos \measuredangle(R, y) \\ &= \mathbf{f}\langle\hat{\mathbf{x}}\rangle \cos\theta + \mathbf{f}\langle\hat{\mathbf{y}}\rangle \sin\theta \qquad \textbf{(8.6a)} \\ \mathbf{f}\langle\hat{\boldsymbol{\theta}}\rangle &= \mathbf{f}\langle\hat{\mathbf{x}}\rangle \cos \measuredangle(\theta, x) + \mathbf{f}\langle\hat{\mathbf{y}}\rangle \cos \measuredangle(\theta, y) \\ &= -\mathbf{f}\langle\hat{\mathbf{x}}\rangle \sin\theta + \mathbf{f}\langle\hat{\mathbf{y}}\rangle \cos\theta \qquad \textbf{(8.6b)} \end{aligned}$$

where $\measuredangle(\theta, x)$, and so on, stand for the angles between the θ- and the x-directions, and so on. From Eqs. (8.6a,b), we can also write

$$\begin{cases} \mathbf{f}\langle\hat{\mathbf{x}}\rangle = \mathbf{f}\langle\hat{\mathbf{R}}\rangle \cos\theta - \mathbf{f}\langle\hat{\boldsymbol{\theta}}\rangle \sin\theta & \textbf{(8.6c)} \\ \mathbf{f}\langle\hat{\mathbf{y}}\rangle = \mathbf{f}\langle\hat{\mathbf{R}}\rangle \sin\theta + \mathbf{f}\langle\hat{\boldsymbol{\theta}}\rangle \cos\theta & \textbf{(8.6d)} \end{cases}$$

Furthermore, using the same symbolism as in Eqs. (2.4a,b),

$$\begin{cases} \mathbf{f}\langle\hat{\mathbf{R}}\rangle = \sigma_{RR}\hat{\mathbf{R}} + \sigma_{R\theta}\hat{\boldsymbol{\theta}} & \textbf{(8.7a)} \\ \mathbf{f}\langle\hat{\boldsymbol{\theta}}\rangle = \sigma_{\theta R}\hat{\mathbf{R}} + \sigma_{\theta\theta}\hat{\boldsymbol{\theta}} & \textbf{(8.7b)} \end{cases}$$

where σ_{RR} is the R-component of the contact force intensity (normal stress) acting on a plane whose normal is in the R-direction, and so forth.

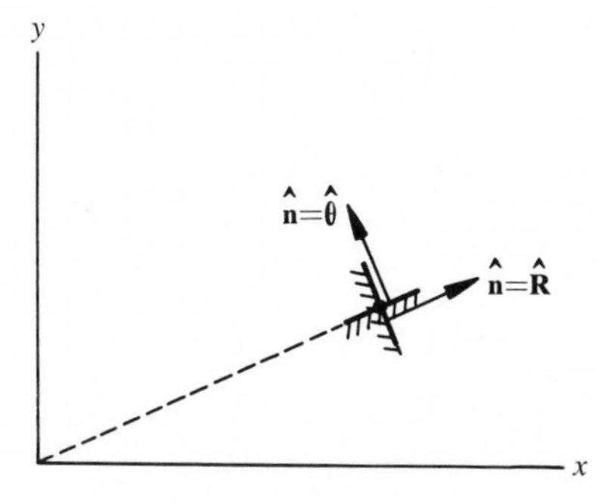

Figure 8.2 Area elements in the R- and θ-directions.

In view of Eqs. (8.5a,b), Eqs. (8.7a,b) can be rewritten as

$$\begin{cases} \mathbf{f}\langle\hat{\mathbf{R}}\rangle = (\sigma_{RR} \cos\theta - \sigma_{R\theta} \sin\theta)\hat{\mathbf{x}} + (\sigma_{RR} \sin\theta + \sigma_{R\theta} \cos\theta)\hat{\mathbf{y}} & \textbf{(8.8a)} \\ \mathbf{f}\langle\hat{\boldsymbol{\theta}}\rangle = (\sigma_{\theta R} \cos\theta - \sigma_{\theta\theta} \sin\theta)\hat{\mathbf{x}} + (\sigma_{\theta R} \sin\theta + \sigma_{\theta\theta} \cos\theta)\hat{\mathbf{y}} & \textbf{(8.8b)} \end{cases}$$

Substituting Eqs. (8.8a,b) and Eqs. (2.4a,b) into Eqs. (8.6c,d), we have

$$\sigma_{xx}\hat{\mathbf{x}}+\sigma_{xy}\hat{\mathbf{y}} = [(\sigma_{RR}\cos\theta-\sigma_{R\theta}\sin\theta)\cos\theta-(\sigma_{\theta R}\cos\theta-\sigma_{\theta\theta}\sin\theta)\sin\theta]\hat{\mathbf{x}} + [(\sigma_{RR}\sin\theta+\sigma_{R\theta}\cos\theta)\cos\theta-(\sigma_{\theta R}\sin\theta+\sigma_{\theta\theta}\cos\theta)\sin\theta]\hat{\mathbf{y}} \tag{8.9a}$$

$$\sigma_{yx}\hat{\mathbf{x}}+\sigma_{yy}\hat{\mathbf{y}} = [(\sigma_{RR}\cos\theta-\sigma_{R\theta}\sin\theta)\sin\theta+(\sigma_{\theta R}\cos\theta-\sigma_{\theta\theta}\sin\theta)\cos\theta]\hat{\mathbf{x}} + [(\sigma_{RR}\sin\theta+\sigma_{R\theta}\cos\theta)\sin\theta+(\sigma_{\theta R}\sin\theta+\sigma_{\theta\theta}\cos\theta)\cos\theta]\hat{\mathbf{y}} \tag{8.9b}$$

that is,

$$\sigma_{xx} = \sigma_{RR}\cos^2\theta+\sigma_{\theta\theta}\sin^2\theta-(\sigma_{R\theta}+\sigma_{\theta R})\sin\theta\cos\theta \tag{8.10a}$$

$$\sigma_{xy} = (\sigma_{RR}-\sigma_{\theta\theta})\sin\theta\cos\theta+\sigma_{R\theta}\cos^2\theta-\sigma_{\theta R}\sin^2\theta \tag{8.10b}$$

$$\sigma_{yx} = (\sigma_{RR}-\sigma_{\theta\theta})\sin\theta\cos\theta-\sigma_{R\theta}\sin^2\theta+\sigma_{\theta R}\cos^2\theta \tag{8.10c}$$

$$\sigma_{yy} = \sigma_{RR}\sin^2\theta+\sigma_{\theta\theta}\cos^2\theta+(\sigma_{R\theta}+\sigma_{\theta R})\sin\theta\cos\theta \tag{8.10d}$$

But

$$\sigma_{yx} = \sigma_{xy} \tag{2.10}$$

Therefore, Eqs. (8.10b,c) yield

$$\sigma_{\theta R} = \sigma_{R\theta} \tag{8.11}$$

and

$$\sigma_{xy} = (\sigma_{RR}-\sigma_{\theta\theta})\sin\theta\cos\theta+\sigma_{R\theta}(\cos^2\theta-\sin^2\theta) \tag{8.12a}$$

Equations (8.10a,d) then become

$$\sigma_{xx} = \sigma_{RR}\cos^2\theta+\sigma_{\theta\theta}\sin^2\theta-2\sigma_{R\theta}\sin\theta\cos\theta \tag{8.12b}$$

$$\sigma_{yy} = \sigma_{RR}\sin^2\theta+\sigma_{\theta\theta}\cos^2\theta+2\sigma_{R\theta}\sin\theta\cos\theta \tag{8.12c}$$

Conversely, σ_{RR}, $\sigma_{\theta\theta}$, and $\sigma_{R\theta}$ can be solved from Eqs. (8.12a,b,c):

$$\sigma_{R\theta} = -(\sigma_{xx}-\sigma_{yy})\sin\theta\cos\theta+\sigma_{xy}(\cos^2\theta-\sin^2\theta) \tag{8.12d}$$

$$\sigma_{RR} = \sigma_{xx}\cos^2\theta+\sigma_{yy}\sin^2\theta+2\sigma_{xy}\sin\theta\cos\theta \tag{8.12e}$$

$$\sigma_{\theta\theta} = \sigma_{xx}\sin^2\theta+\sigma_{yy}\cos^2\theta-2\sigma_{xy}\sin\theta\cos\theta \tag{8.12f}$$

Next, by the chain rule of differentiation, we have

$$\frac{\partial(\ )}{\partial x} = \frac{\partial(\ )}{\partial R}\frac{\partial R}{\partial x}+\frac{\partial(\ )}{\partial\theta}\frac{\partial\theta}{\partial x} = \cos\theta\frac{\partial(\ )}{\partial R}-\frac{\sin\theta}{R}\frac{\partial(\ )}{\partial\theta} \tag{8.13a}$$

$$\frac{\partial(\ )}{\partial y} = \frac{\partial(\ )}{\partial R}\frac{\partial R}{\partial y}+\frac{\partial(\ )}{\partial\theta}\frac{\partial\theta}{\partial y} = \sin\theta\frac{\partial(\ )}{\partial R}+\frac{\cos\theta}{R}\frac{\partial(\ )}{\partial\theta} \tag{8.13b}$$

because of Eqs. (8.1a,b). Applying relations (8.13a,b) to Eqs. (8.3c,d), we obtain

$$\frac{\partial u}{\partial x}=\left(\cos\theta\frac{\partial}{\partial R}-\frac{\sin\theta}{R}\frac{\partial}{\partial\theta}\right)(q_R\cos\theta-q_\theta\sin\theta)$$

$$=\cos^2\theta\frac{\partial q_R}{\partial R}+\sin^2\theta\left(\frac{1}{R}\frac{\partial q_\theta}{\partial\theta}+\frac{q_R}{R}\right)$$

$$-\cos\theta\sin\theta\left(\frac{\partial q_\theta}{\partial R}+\frac{1}{R}\frac{\partial q_R}{\partial\theta}-\frac{q_\theta}{R}\right) \qquad \textbf{(8.14a)}$$

$$\frac{\partial u}{\partial y}=\left(\sin\theta\frac{\partial}{\partial R}+\frac{\cos\theta}{R}\frac{\partial}{\partial\theta}\right)(q_R\cos\theta-q_\theta\sin\theta)$$

$$=\cos^2\theta\left(\frac{1}{R}\frac{\partial q_R}{\partial\theta}-\frac{q_\theta}{R}\right)-\sin^2\theta\frac{\partial q_\theta}{\partial R}$$

$$+\cos\theta\sin\theta\left(\frac{\partial q_R}{\partial R}-\frac{1}{R}\frac{\partial q_\theta}{\partial\theta}-\frac{q_R}{R}\right) \qquad \textbf{(8.14b)}$$

$$\frac{\partial v}{\partial y}=\sin^2\theta\frac{\partial q_R}{\partial R}+\cos^2\theta\left(\frac{1}{R}\frac{\partial q_\theta}{\partial\theta}+\frac{q_R}{R}\right)$$

$$+\cos\theta\sin\theta\left(\frac{\partial q_\theta}{\partial R}+\frac{1}{R}\frac{\partial q_R}{\partial\theta}-\frac{q_\theta}{R}\right) \qquad \textbf{(8.14c)}$$

$$\frac{\partial v}{\partial x}=-\sin^2\theta\left(\frac{1}{R}\frac{\partial q_R}{\partial\theta}-\frac{q_\theta}{R}\right)+\cos^2\theta\frac{\partial q_\theta}{\partial R}$$

$$+\cos\theta\sin\theta\left(\frac{\partial q_R}{\partial R}-\frac{1}{R}\frac{\partial q_\theta}{\partial\theta}-\frac{q_R}{R}\right) \qquad \textbf{(8.14d)}$$

Now, substituting Eqs. (2.14a,b,c) into Eq. (8.12d), we have

$$\sigma_{R\theta}=-2\mu\left(\frac{\partial u}{\partial x}-\frac{\partial v}{\partial y}\right)\sin\theta\cos\theta+\mu\left(\frac{\partial v}{\partial x}+\frac{\partial u}{\partial y}\right)(\cos^2\theta-\sin^2\theta)$$

$$=\mu\left(\frac{1}{R}\frac{\partial q_R}{\partial\theta}+\frac{\partial q_\theta}{\partial R}-\frac{q_\theta}{R}\right) \qquad \textbf{(8.15a)}$$

because of Eqs. (8.14a,b,c,d). Similarly, you can convince yourself that Eqs. (8.12e,f) yield

$$\sigma_{RR}=-p+2\mu\frac{\partial q_R}{\partial R} \qquad \textbf{(8.15b)}$$

$$\sigma_{\theta\theta}=-p+2\mu\left(\frac{1}{R}\frac{\partial q_\theta}{\partial\theta}+\frac{q_R}{R}\right) \qquad \textbf{(8.15c)}$$

The equation of continuity, Eq. (1.9), is very easy to handle. Just an appeal to Eqs. (8.14a,c) will enable us to write it down in the polar coordinates:

$$\frac{\partial q_R}{\partial R}+\frac{q_R}{R}+\frac{1}{R}\frac{\partial q_\theta}{\partial \theta}=0 \tag{8.16}$$

which can also be written in the following more convenient form

$$\frac{\partial (Rq_R)}{\partial R}+\frac{\partial q_\theta}{\partial \theta}=0 \tag{8.16A}$$

Your attention is here directed to two things:

1. Equations (8.15b,c) and (8.16) yield

$$-p=\tfrac{1}{2}(\sigma_{RR}+\sigma_{\theta\theta})$$

2. Equation (8.16A) implies right away the existence of a stream function $\psi\langle R, \theta; t\rangle$ such that

$$Rq_R=\frac{\partial \psi}{\partial \theta}$$

or

$$q_R=\frac{1}{R}\frac{\partial \psi}{\partial \theta} \tag{8.17a}$$

and

$$q_\theta=-\frac{\partial \psi}{\partial R} \tag{8.17b}$$

This stream function is the same as the one introduced in Section 1.2, except that (x, y) is now transformed into (R, θ).

We now come to the equations of motion, Eqs. (2.11a,b):

$$\begin{cases} \rho\left(\dfrac{\partial u}{\partial t}+u\dfrac{\partial u}{\partial x}+v\dfrac{\partial u}{\partial y}\right)=\rho g_x+\left(\dfrac{\partial \sigma_{xx}}{\partial x}+\dfrac{\partial \sigma_{xy}}{\partial y}\right) & \text{(2.11aA)} \\ \rho\left(\dfrac{\partial v}{\partial t}+u\dfrac{\partial v}{\partial x}+v\dfrac{\partial v}{\partial y}\right)=\rho g_y+\left(\dfrac{\partial \sigma_{xy}}{\partial x}+\dfrac{\partial \sigma_{yy}}{\partial y}\right) & \text{(2.11bA)} \end{cases}$$

To transform these two important equations into polar coordinates, we had better remember that they are actually the (x, y)-component equations of the following vectorial equation:

$$\begin{aligned} &\rho\left[\left(\frac{\partial u}{\partial t}+u\frac{\partial u}{\partial x}+v\frac{\partial u}{\partial y}\right)\hat{\mathbf{x}}+\left(\frac{\partial v}{\partial t}+u\frac{\partial v}{\partial x}+v\frac{\partial v}{\partial y}\right)\mathbf{y}\right] \\ &\qquad =\rho[g_x\hat{\mathbf{x}}+g_y\hat{\mathbf{y}}]+\left[\left(\frac{\partial \sigma_{xx}}{\partial x}+\frac{\partial \sigma_{xy}}{\partial y}\right)\hat{\mathbf{x}}+\left(\frac{\partial \sigma_{xy}}{\partial x}+\frac{\partial \sigma_{yy}}{\partial y}\right)\hat{\mathbf{y}}\right] \end{aligned} \tag{8.18}$$

In Eq. (8.18),

$$\begin{aligned} \rho[g_x\hat{\mathbf{x}}+g_y\hat{\mathbf{y}}] &= \rho\mathbf{g} \\ &= \rho[g_R\hat{\mathbf{R}}+g_\theta\hat{\boldsymbol{\theta}}] \end{aligned} \tag{8.19}$$

Thus, ρg_R and ρg_θ (*not constants* even for constant **g**) will appear as the body-force terms in the (R, θ)-component equations. Similarly, from Eq. (1.2), the left-hand side of Eq. (8.18) is equal to

$$\rho \mathbf{a} = \rho(a_R \hat{\mathbf{R}} + a_\theta \hat{\boldsymbol{\theta}})$$

Now, we know that

$$\begin{aligned}
a_x &= \frac{\partial u}{\partial t} + u\frac{\partial u}{\partial x} + v\frac{\partial u}{\partial y} \\
&= \frac{\partial}{\partial t}(q_R \cos\theta - q_\theta \sin\theta) \\
&\qquad + (q_R \cos\theta - q_\theta \sin\theta)\left(\cos\theta \frac{\partial}{\partial R} - \frac{\sin\theta}{R}\frac{\partial}{\partial\theta}\right)(q_R\cos\theta - q_\theta\sin\theta) \\
&\qquad\qquad + (q_R \sin\theta + q_\theta\cos\theta)\left(\sin\theta\frac{\partial}{\partial R} + \frac{\cos\theta}{R}\frac{\partial}{\partial\theta}\right)(q_R\cos\theta - q_\theta\sin\theta) \\
&= \cos\theta\left(\frac{\partial q_R}{\partial t} + q_R\frac{\partial q_R}{\partial R} + \frac{q_\theta}{R}\frac{\partial q_R}{\partial\theta} - \frac{q_\theta{}^2}{R}\right) \\
&\qquad - \sin\theta\left(\frac{\partial q_\theta}{\partial t} + q_R\frac{\partial q_\theta}{\partial R} + \frac{q_\theta}{R}\frac{\partial q_\theta}{\partial\theta} + \frac{q_R q_\theta}{R}\right)
\end{aligned}$$

using Eqs. (8.3c,d) and (8.13a,b). On the other hand, Eq. (8.2c) dictates that

$$a_x = a_R\cos\theta - a_\theta\sin\theta$$

Comparing this with the previous formula (both being valid for any θ), we must have

$$\left\{\begin{aligned}
a_R &= \frac{\partial q_R}{\partial t} + q_R\frac{\partial q_R}{\partial R} + \frac{q_\theta}{R}\frac{\partial q_R}{\partial\theta} - \frac{q_\theta{}^2}{R} && \textbf{(8.20a)} \\
a_\theta &= \frac{\partial q_\theta}{\partial t} + q_R\frac{\partial q_\theta}{\partial R} + \frac{q_\theta}{R}\frac{\partial q_\theta}{\partial\theta} + \frac{q_R q_\theta}{R} && \textbf{(8.20b)}
\end{aligned}\right.$$

Thus ρa_R and ρa_θ will appear on the left-hand side of the (R, θ)-component equations. Finally, we will denote the remaining group of terms in Eq. (8.18) by some vectorial symbol:

$$\text{Second bracket on the right-hand side of Eq. (8.18)} = \mathscr{F}_R\hat{\mathbf{R}} + \mathscr{F}_\theta\hat{\boldsymbol{\theta}}$$

Again, we already know that

$$\begin{aligned}
\mathscr{F}_x &= \frac{\partial\sigma_{xx}}{\partial x} + \frac{\partial\sigma_{xy}}{\partial y} \\
&= \left(\cos\theta\frac{\partial}{\partial R} - \frac{\sin\theta}{R}\frac{\partial}{\partial\theta}\right)(\sigma_{RR}\cos^2\theta + \sigma_{\theta\theta}\sin^2\theta - 2\sigma_{R\theta}\sin\theta\cos\theta) \\
&\qquad + \left(\sin\theta\frac{\partial}{\partial R} + \frac{\cos\theta}{R}\frac{\partial}{\partial\theta}\right)[(\sigma_{RR} - \sigma_{\theta\theta})\sin\theta\cos\theta + \sigma_{R\theta}(\cos^2\theta - \sin^2\theta)] \\
&= \left(\frac{\partial\sigma_{RR}}{\partial R} + \frac{1}{R}\frac{\partial\sigma_{R\theta}}{\partial\theta} + \frac{\sigma_{RR} - \sigma_{\theta\theta}}{R}\right)\cos\theta - \left(\frac{1}{R}\frac{\partial\sigma_{\theta\theta}}{\partial\theta} + \frac{\partial\sigma_{R\theta}}{\partial R} + \frac{2\sigma_{R\theta}}{R}\right)\sin\theta
\end{aligned}$$

Thus

$$\begin{cases} \mathscr{F}_R = \dfrac{\partial \sigma_{RR}}{\partial R} + \dfrac{1}{R}\dfrac{\partial \sigma_{R\theta}}{\partial \theta} + \dfrac{\sigma_{RR}-\sigma_{\theta\theta}}{R} & \textbf{(8.21a)} \\[2ex] \mathscr{F}_\theta = \dfrac{1}{R}\dfrac{\partial \sigma_{\theta\theta}}{\partial \theta} + \dfrac{\partial \sigma_{R\theta}}{\partial R} + \dfrac{2\sigma_{R\theta}}{R} & \textbf{(8.21b)} \end{cases}$$

Therefore, the (R,θ)-components of Eq. (8.18), or the equations of motion in polar coordinates, turn out to be

$$\begin{cases} \rho a_R = \rho g_R + \mathscr{F}_R & \textbf{(8.22a)} \\ \rho a_\theta = \rho g_\theta + \mathscr{F}_\theta & \textbf{(8.22b)} \end{cases}$$

Substituting Eqs. (8.15a,b,c) into Eqs. (8.21a,b), we have (with constant viscosity),

$$\begin{aligned} \mathscr{F}_R = & -\frac{\partial p}{\partial R} + 2\mu \frac{\partial^2 q_R}{\partial R^2} + \mu\left(\frac{1}{R^2}\frac{\partial^2 q_R}{\partial \theta^2} + \frac{1}{R}\frac{\partial^2 q_\theta}{\partial R\,\partial \theta} - \frac{1}{R^2}\frac{\partial q_\theta}{\partial \theta}\right) \\ & + 2\mu\left(\frac{1}{R}\frac{\partial q_R}{\partial R} - \frac{1}{R^2}\frac{\partial q_\theta}{\partial \theta} - \frac{q_R}{R^2}\right) \\ = & -\frac{\partial p}{\partial R} + \mu\left(\frac{\partial^2 q_R}{\partial R^2} + \frac{1}{R}\frac{\partial q_R}{\partial R} - \frac{q_R}{R^2} + \frac{1}{R^2}\frac{\partial^2 q_R}{\partial \theta^2} - \frac{2}{R^2}\frac{\partial q_\theta}{\partial \theta}\right) \\ & + \mu\left(\frac{\partial^2 q_R}{\partial R^2} + \frac{1}{R}\frac{\partial^2 q_\theta}{\partial R\,\partial \theta} - \frac{1}{R^2}\frac{\partial q_\theta}{\partial \theta} + \frac{1}{R}\frac{\partial q_R}{\partial R} - \frac{q_R}{R^2}\right) \\ = & -\frac{\partial p}{\partial R} + \mu\left(\frac{\partial^2 q_R}{\partial R^2} + \frac{1}{R}\frac{\partial q_R}{\partial R} - \frac{q_R}{R^2} + \frac{1}{R^2}\frac{\partial^2 q_R}{\partial \theta^2} - \frac{2}{R^2}\frac{\partial q_\theta}{\partial \theta}\right) \\ & + \mu\frac{\partial}{\partial R}\left(\frac{\partial q_R}{\partial R} + \frac{q_R}{R} + \frac{1}{R}\frac{\partial q_\theta}{\partial \theta}\right) \\ = & -\frac{\partial p}{\partial R} + \mu\left(\frac{\partial^2 q_R}{\partial R^2} + \frac{1}{R}\frac{\partial q_R}{\partial R} - \frac{q_R}{R^2} + \frac{1}{R^2}\frac{\partial^2 q_R}{\partial \theta^2} - \frac{2}{R^2}\frac{\partial q_\theta}{\partial \theta}\right) \qquad \textbf{(8.23a)} \end{aligned}$$

where the equation of continuity, Eq. (8.16), has been used; and similarly,

$$\mathscr{F}_\theta = -\frac{1}{R}\frac{\partial p}{\partial \theta} + \mu\left(\frac{\partial^2 q_\theta}{\partial R^2} + \frac{1}{R}\frac{\partial q_\theta}{\partial R} - \frac{q_\theta}{R^2} + \frac{1}{R^2}\frac{\partial^2 q_\theta}{\partial \theta^2} + \frac{2}{R^2}\frac{\partial q_R}{\partial \theta}\right) \qquad \textbf{(8.23b)}$$

The corresponding Navier-Stokes equations in polar coordinates[1] are

---

[1] The procedure adopted in this section, though straightforward, is lengthy and somewhat messy. We hope that it *has* overtaxed your patience to such an extent that you are beginning to ask whether there is a better way of handling this. There *is* a better way through the use of dyadics or tensors; but you must first be motivated enough before you can start a study of these more abstract concepts without wavering. Dyadics will be employed in Part 3; and it will be shown there that transformation from one coordinate system to another can be handled in a very economical fashion – a lazybones' paradise! At the present, there is nothing else you can do but to proceed by brute force.

thus

$$\frac{\partial q_R}{\partial t}+q_R\frac{\partial q_R}{\partial R}+\frac{q_\theta}{R}\frac{\partial q_R}{\partial \theta}-\frac{q_\theta^2}{R}=g_R-\frac{1}{\rho}\frac{\partial p}{\partial R}$$

$$+\nu\left(\frac{\partial^2 q_R}{\partial R^2}+\frac{1}{R}\frac{\partial q_R}{\partial R}-\frac{q_R}{R^2}+\frac{1}{R^2}\frac{\partial^2 q_R}{\partial \theta^2}-\frac{2}{R^2}\frac{\partial q_\theta}{\partial \theta}\right) \quad \textbf{(8.24a)}$$

$$\frac{\partial q_\theta}{\partial t}+q_R\frac{\partial q_\theta}{\partial R}+\frac{q_\theta}{R}\frac{\partial q_\theta}{\partial \theta}+\frac{q_R q_\theta}{R}=g_\theta-\frac{1}{\rho}\left(\frac{1}{R}\right)\frac{\partial p}{\partial \theta}$$

$$+\nu\left(\frac{\partial^2 q_\theta}{\partial R^2}+\frac{1}{R}\frac{\partial q_\theta}{\partial R}-\frac{q_\theta}{R^2}+\frac{1}{R^2}\frac{\partial^2 q_\theta}{\partial \theta^2}+\frac{2}{R^2}\frac{\partial q_R}{\partial \theta}\right) \quad \textbf{(8.24b)}$$

Equations (8.24a,b) and (8.16), or (8.16A), are then the key equations for the (plane) flows of viscous (incompressible) Newtonian fluids in polar coordinates.

For the (non-Newtonian) Reiner-Rivlin fluids (Section 2.4), Eqs. (8.16) or (8.16A), (8.20a,b), (8.21a,b) and (8.22a,b) still hold. In addition, we must list the (R,θ) counterpart of Eqs. (2.18a,b,c,d) and (2.17):

$$\sigma_{RR}=-p+2\mu\langle \mathrm{II}\rangle\frac{\partial q_R}{\partial R}+\frac{1}{6}\eta\langle \mathrm{II}\rangle \mathrm{II} \quad \textbf{(8.25a)}$$

$$\sigma_{\theta\theta}=-p+2\mu\langle \mathrm{II}\rangle\left(\frac{1}{R}\frac{\partial q_\theta}{\partial \theta}+\frac{q_R}{R}\right)+\tfrac{1}{6}\eta\langle \mathrm{II}\rangle \mathrm{II} \quad \textbf{(8.25b)}$$

$$\sigma_{R\theta}=\mu\langle \mathrm{II}\rangle\left(\frac{1}{R}\frac{\partial q_R}{\partial \theta}+\frac{\partial q_\theta}{\partial R}-\frac{q_\theta}{R}\right) \quad \textbf{(8.25c)}$$

$$\sigma_{zz}=-p-\tfrac{1}{3}\eta\langle \mathrm{II}\rangle \mathrm{II} \quad \textbf{(8.25d)}$$

where[2]

$$\mathrm{II}=2\left[\left(\frac{\partial q_R}{\partial R}\right)^2+\frac{1}{4}\left(\frac{1}{R}\frac{\partial q_R}{\partial \theta}+\frac{\partial q_\theta}{\partial R}-\frac{q_\theta}{R}\right)^2\right] \quad \textbf{(8.26)}$$

8.2 COUETTE FLOW[3]

Consider two very long concentric cylinders, represented by two circles in Figure 8.3, which are rotating at different angular speeds. The fluid in between the cylinders must move in such a fashion as to satisfy Eqs. (8.16A) and

---

[2] It must first be shown that the values of II in (x, y) and (R, θ) coordinate systems are the same in order that $\mu\langle \mathrm{II}\rangle$ and $\eta\langle \mathrm{II}\rangle$ stay the same. We do not recommend that you try deriving these equations. If you are curious, you should go on and study Part 3.

[3] M. Couette first discussed the flow between two concentric cylinders, one stationary and one rotating, in his 1890 paper, "Studies on the friction of liquids" (in French). In this book we will follow

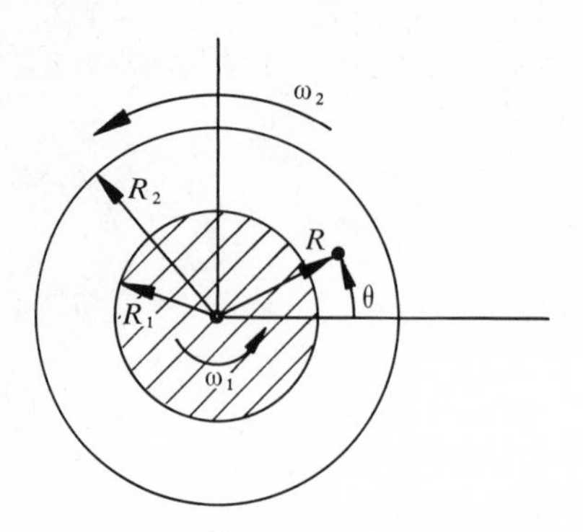

Figure 8.3 Two concentric cylinders.

(8.24a,b), under the boundary conditions

$$\begin{cases} R = R_1: \quad q_R = 0, \, q_\theta = \omega_1 R_1 & \textbf{(8.27a)} \\ R = R_2: \quad q_R = 0, \, q_\theta = \omega_2 R_2 & \textbf{(8.27b)} \end{cases}$$

The boundary conditions suggest that the ansatz $q_R = 0$ (or $q_\theta = q_\theta\langle R\rangle$ only, which is equivalent) be tried. Equation (8.16A) then yields

$$\frac{\partial q_\theta}{\partial \theta} = 0$$

that is,

$$q_\theta = q_\theta\langle R\rangle \text{ only} \tag{8.28}$$

Equations (8.24a,b) then become,[4] for a steady flow,

$$\frac{q_\theta^2}{R} = \frac{1}{\rho}\frac{\partial p}{\partial R} \tag{8.29}$$

$$\frac{1}{\rho R}\frac{\partial p}{\partial \theta} = \nu\left(\frac{d^2 q_\theta}{dR^2} + \frac{1}{R}\frac{dq_\theta}{dR} - \frac{q_\theta}{R^2}\right) \tag{8.30}$$

that is,

$$\frac{\partial p}{\partial \theta} = \rho R C\langle R\rangle$$

or

$$p = (\rho R C\langle R\rangle)\theta + F\langle R\rangle \tag{8.31}$$

But, since the gap in between the two cylinders is a full annulus, p at any R must

---

the usage in hydrodynamic stability and call the flow between any two rotating, concentric, long cylinders by the name Couette. In the literature, the flows in between two parallel plates (that is, two cylinders of infinite radii), one stationary and one moving, is sometimes also called the Couette flow, but this is historically unfounded. Some authors are cautious and call this the plane Couette flow. But we can not follow this practice, because the (circular) Couette flow is also a plane flow according to our definition. We will use instead the term *flat* Couette flow to denote the fact that the bounding walls are flat, not circular. A flat Couette flow is more general than the simple shear flow of Section 3.2, Special Case 1, if the fluid is non-Newtonian, being pumped, with varying viscosity, or compressible.

[4] From now on we again neglect the effects of gravity, unless otherwise stated.

be the same at $\theta = 0$ as at $\theta \to 2\pi$. This physical requirement is fulfilled only if $C\langle R\rangle \equiv 0$. Therefore, we must have

$$p = F\langle R\rangle \text{ only} \tag{8.32}$$

and

$$\frac{d^2 q_\theta}{dR^2} + \frac{1}{R}\frac{dq_\theta}{dR} - \frac{q_\theta}{R^2} = 0 \tag{8.33}$$

Equation (8.29) then yields

$$\frac{1}{\rho}\frac{dp}{dR} = \frac{q_\theta^2}{R}$$

or

$$\frac{p}{\rho} = \int_{R_1}^{R} \frac{q_\theta^2}{R}\, dR + C_1 \tag{8.34}$$

where $C_1 = p\langle R_1\rangle/\rho$ is a known or measurable quantity.

You should find no difficulty in solving Eq. (8.33), which yields, under the boundary conditions (8.27a,b),

$$q_\theta = \frac{1}{R_2^2 - R_1^2}\left[R(\omega_2 R_2^2 - \omega_1 R_1^2) - \frac{R_1^2 R_2^2}{R}(\omega_2 - \omega_1)\right] \tag{8.35}$$

A simple integration then gives

$$\begin{aligned} p = p\langle R_1\rangle + \frac{\rho}{(R_2^2 - R_1^2)^2}\Bigg[&(\omega_2 R_2^2 - \omega_1 R_1^2)^2\left(\frac{R^2 - R_1^2}{2}\right) \\ &- 2R_1^2 R_2^2(\omega_2 - \omega_1)(\omega_2^2 R_2^2 - \omega_1^2 R_1^2)\ln\frac{R}{R_1} \\ &- \frac{1}{2}R_1^4 R_2^4(\omega_2 - \omega_1)^2\left(\frac{1}{R^2} - \frac{1}{R_1^2}\right)\Bigg] \end{aligned} \tag{8.36}$$

It is interesting to see that the *steady* flow inside a rotating cylinder of radius R_2 is, with R_1 set to zero,

$$\begin{cases} q_\theta = R\omega_2 \\ p = p\langle 0\rangle + \dfrac{\rho}{2}\omega_2^2 R^2 \end{cases}$$

that is, a rigid-body-like rotation. (Physically, this would be the case if a long enough time has elapsed after the start of the rotation. As to the transient stage following the start of the motion, it is definitely not rigid-body-like.) On the other hand, letting $R_2 \to \infty$ and $\omega_2 \to 0$, we have the steady flow due to one rotating cylinder with a radius R_1 in an otherwise tranquil fluid:

$$\begin{cases} q_\theta = \dfrac{R_1^2 \omega_1}{R} \\ p = p\langle R_1\rangle + \dfrac{\rho}{2} R_1^2 \omega_1^2\left(1 - \dfrac{R_1^2}{R^2}\right) \end{cases}$$

We see that the motion diminishes as R increases with the infinity remaining tranquil. Referring back to the discussion of "free"-vortex flow in Section 1.3-(1), Example 2, we see that this special one-cylinder case happens to be *irrotational.* Usually, an irrotational flow cannot satisfy the no-slip condition; but in special cases, such as here, it may just happen to satisfy all governing equations and conditions of a viscous fluid, and thus represent a viscous flow. (If the no-slip condition were not enforced, the "solution" would be $q_\theta \equiv 0$!)

A look at Eq. (8.35) shows that the velocity distribution does not depend on the viscosity at all, just as in the simple shear flow; the torques exerted on the cylinders by the fluid, of course, still depend on the viscosity. By the same token, the flow will not exhibit any change in its inherent structure as Re (chosen as, for example, $\omega_2 R_2^2/\nu$) increases. In other words, this is one of the cases where no boundary layer forms as Re increases. Equation (8.35) can also be written as

$$q_\theta = AR + \frac{B}{R} \tag{8.35A}$$

with easily identifiable constants A and B. It is then seen that a general Couette flow is just a linear combination or superposition of the wheel flow and the irrotational vortex, with the constants A and B so adjusted as to satisfy boundary conditions (8.27a,b). This comes about only because the governing equation, Eq. (8.33), happens to be linear. The principle of superposition, which is in effect for all linear phenomena, therefore applies.

At this point, it may be worthwhile to inquire into the Couette flow of a Reiner-Rivlin fluid, and to compare results.

For a Reiner-Rivlin fluid, the equation of continuity and the boundary conditions are still the same as before. So, the ansatz $q_R = 0$ leads again to Eq. (8.28). Equation (8.26) then becomes

$$\mathrm{II} = \frac{1}{2}\left(\frac{dq_\theta}{dR} - \frac{q_\theta}{R}\right)^2$$

and Eqs. (8.25a,b,c,d) yield

$$\begin{cases} \sigma_{RR} = -p + \frac{1}{6}\eta \mathrm{II} \\ \sigma_{\theta\theta} = -p + \frac{1}{6}\eta \mathrm{II} \\ \sigma_{R\theta} = \mu\left(\dfrac{dq_\theta}{dR} - \dfrac{q_\theta}{R}\right) \\ \sigma_{zz} = -p - \frac{1}{3}\eta \mathrm{II} \end{cases}$$

where μ and η are functions of II. We then have, from Eqs. (8.21b) and (8.22b),

$$\frac{1}{R}\frac{\partial p}{\partial \theta} = \frac{d}{dR}\left[\mu\left(\frac{dq_\theta}{dR} - \frac{q_\theta}{R}\right)\right] + \frac{2\mu}{R}\left(\frac{dq_\theta}{dR} - \frac{q_\theta}{R}\right)$$

Exactly the same argument as before will lead us to Eq. (8.32) again, together with

$$\frac{d}{dR}\left[\mu\left(\frac{dq_\theta}{dR} - \frac{q_\theta}{R}\right)\right] + \frac{2\mu}{R}\left(\frac{dq_\theta}{dR} - \frac{q_\theta}{R}\right) = 0 \tag{8.36}$$

Equation (8.36) is easily solved, yielding

$$\mu\left(\frac{dq_\theta}{dR}-\frac{q_\theta}{R}\right)=C_1R^{-2}$$

or

$$\mu\left[R\frac{d}{dR}\left(\frac{q_\theta}{R}\right)\right]=C_1R^{-2} \tag{8.37}$$

where μ is a function of

$$\text{II}=\frac{1}{2}\left[R\frac{d}{dR}\left(\frac{q_\theta}{R}\right)\right]^2$$

or of

$$\left|R\frac{d}{dR}\left(\frac{q_\theta}{R}\right)\right|$$

To continue the investigation, we must be given the particular form of the function

$$\mu\left\langle\left|R\frac{d}{dR}\left(\frac{q_\theta}{R}\right)\right|\right\rangle$$

for the fluid considered. If μ can be regarded as constant, Eq. (8.37) can be further integrated:

$$\frac{\mu q_\theta}{R}=-\frac{C_1}{2}R^{-2}+C_2$$

That is,

$$q_\theta=AR+\frac{B}{R}$$

where A and B are fixed by the boundary conditions, as before. As another example, consider the case where

$$\mu\propto\left[R\frac{d}{dR}\left(\frac{q_\theta}{R}\right)\right]^{2m}$$

Equation (8.37) then yields

$$R\frac{d}{dR}\left(\frac{q_\theta}{R}\right)=C_3R^{-2/(2m+1)}$$

that is,

$$\frac{q_\theta}{R}=C_4R^{-2/(2m+1)}+C_5$$

or

$$q_\theta=C_4R^{(2m-1)/(2m+1)}+C_5R$$

and C_4 and C_5 are to be determined from the boundary conditions.

In any event, Eq. (8.37) is solvable. So let us move on and consider Eqs. (8.21a) and (8.22a); that is,

$$-\frac{\rho q_\theta^2}{R}=\frac{d\sigma_{RR}}{dR}$$

where $\sigma_{RR} = -p + \frac{1}{6}\eta\langle \mathrm{II}\rangle \mathrm{II}$ is a function of R only. We thus have

$$\sigma_{RR}\langle R\rangle - \sigma_{RR}\langle R_1\rangle = -\rho \int_{R_1}^{R} \left(\frac{q_\theta^2}{R}\right) dR$$

8.3 JEFFERY-HAMEL FLOW[5]

By Jeffery-Hamel flow, we mean a plane flow in between two inclined flat plates (Fig. 8.4). The plates are very long, both radially and in the direction perpendicular to the representative plane, so they appear as two radial lines in Figure 8.4. We also assume that they are fixed and impermeable; thus, we have the boundary conditions.

$$\theta = \Theta_1, \Theta_2:\ q_R, q_\theta = 0$$

Let us introduce here the ansatz

$$q_\theta \equiv 0$$

Equation (8.16A) then yields

$$Rq_R = f\langle\theta\rangle$$

or

$$q_R = \frac{f\langle\theta\rangle}{R} \tag{8.38}$$

Before we check whether this is compatible with Eqs. (8.24a,b), let us first see if it can possibly pass the boundary conditions. In this respect, we are glad to see that the requirement that $q_R = 0$ at $\theta = \Theta_1$ and Θ_2 only demands that

$$\theta = \Theta_1, \Theta_2:\ f = 0 \tag{8.39a,b}$$

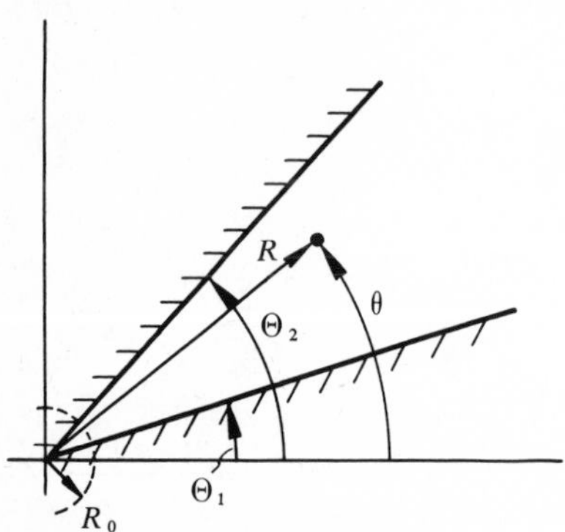

Figure 8.4 Jeffery-Hamel flow.

[5] G. B. Jeffery, "The two-dimensional steady motion of a viscous fluid," *Phil. Mag.* (6), **29**, 455–465, 1915. G. Hamel, "Spiralförmige Bewegungen zaher Flüssigkeiten," *Jahresbericht d. deutschen Mathematiker-Vereinigung*, **25**, 34–60, 1917 (translated as NACA TM No. 1342, 1953). It is suggested here that you turn to the last paragraph of Section 8.5 and read the eight motivating points before going on. Georg Hamel (1877–1954) contributed significantly to the foundations of mechanics for both particles and continua.

Furthermore, as $R \to \infty$, $q_R \to 0$, which is physically proper. Thus, our ansatz is admissible as far as the boundary conditions are concerned. But Eq. (8.38) does give an unbounded q_R as $R \to 0$ with $\Theta_1 < \theta < \Theta_2$. Is this physically proper? Certainly not. Yet suppose we equip the inclined-plate channel with an opening at $R = R_0$ (Fig. 8.4), and consider only the region $R > R_0$ ($\Theta_1 < \theta < \Theta_2$) as the physical domain of the flow, the singular point $R = 0$ (where q_R grows unbounded) will then be harmlessly outside the region of physical interest.

With this saving grace, we can now continue to investigate our ansatz, knowing that it is physically realizable for $R > R_0$. There is, however, still the small matter of providing the dictated distribution of q_R at the inlet $R = R_1$ in between Θ_1 and Θ_2. Here, the ansatz dictate that

$$q_R\langle R_1, \theta\rangle = \frac{f\langle\theta\rangle}{R_1}$$

This, being a certain function of θ, is not at all easy to provide physically. But the difficulty is lessened in practice by one fortunate fact: The influence of the detailed character of $q_R\langle R_1, \theta\rangle$ damps down very fast in the direction of the flow, and is not felt upstream.[6] Any velocity distribution at R_1 should induce very little error, as long as the mass flow rate across $R = R_1$ matches that across any R downstream. With this assurance, let us now go on and test our ansatz against the N-S equations, Eqs. (8.24a,b).

The ansatz and Eq. (8.38) actually reduce the N-S equations to

$$\begin{cases} -\dfrac{f^2}{R^3} = -\dfrac{1}{\rho}\dfrac{\partial p}{\partial R} + \dfrac{\nu}{R^3}\left(\dfrac{d^2 f}{d\theta^2}\right) \\[2ex] 0 = -\dfrac{1}{\rho R}\dfrac{\partial p}{\partial \theta} + \dfrac{2\nu}{R^3}\left(\dfrac{df}{d\theta}\right) \end{cases}$$

The second of these can be easily integrated to yield

$$\frac{p}{\rho} = \left(\frac{2\nu}{R^2}\right) f + F\langle R\rangle$$

and, therefore,

$$\frac{1}{\rho}\left(\frac{\partial p}{\partial R}\right) = -\left(\frac{4\nu}{R^3}\right) f + \frac{dF}{dR}$$

Thus, the first of the previous pair of equations becomes

$$-f^2 = 4\nu f - R^3\left(\frac{dF}{dR}\right) + \nu\left(\frac{d^2 f}{d\theta^2}\right)$$

or

$$\nu\left(\frac{d^2 f}{d\theta^2}\right) + 4\nu f + f^2 = R^3\left(\frac{dF}{dR}\right)$$

---

[6] This has something to do with the inherent mathematical structure of the problem. Here you have to regard it as a qualitative, observational fact. (It is of course true that the damping might not be fast enough should the initial deviation be *too* large.)

Now the left-hand side of this equation is a function of θ only, whereas the right-hand side is a function of R only. Therefore

$$R^3\left(\frac{dF}{dR}\right) = \text{constant, } A$$

that is,

$$F\langle R\rangle = -\frac{A}{2R^2} + B \tag{8.40}$$

and

$$\nu\left(\frac{d^2f}{d\theta^2}\right) + 4\nu f + f^2 = A \tag{8.41}$$

This last ordinary differential equation is to be solved under boundary conditions (8.39a,b). The constants A and B are determinate when we are given (a) the value of p at one point in the flow field, and (b) the (constant) volume flow[7] across any R = constant, that is,

$$Q = \int_{\Theta_1}^{\Theta_2} q_R R\, d\theta$$

$$= \int_{\Theta_1}^{\Theta_2} f\, d\theta \tag{8.42}$$

To summarize, the success of our ansatz now hinges on the solvability of Eq. (8.41) under boundary conditions (8.39a,b). Once f is solved, the pressure can be calculated from the following pressure formula:

$$p = \left(\frac{2\rho\nu}{R^2}\right)f - \frac{A\rho}{2R^2} + B\rho$$

To demonstrate the solvability, we notice that, if we multiply Eq. (8.41) by $df/d\theta$ and integrate the result once, the following intermediate formula is obtained:

$$\frac{1}{3}f^3 + 2\nu f^2 + \frac{\nu}{2}\left(\frac{df}{d\theta}\right)^2 = Af + C_2$$

or

$$\left(\frac{df}{d\theta}\right)^2 = -\frac{2}{3\nu}f^3 - 4f^2 + \frac{2Af}{\nu} + C_1$$

$$= -\frac{2}{3\nu}G\langle f\rangle \tag{8.43}$$

where

$$G\langle f\rangle = f^3 + 6\nu f^2 - 3Af + C \tag{8.44}$$

[7] Given volume flow actually does not quite pin the problem down. For example, zero volume flow may mean no flow at all, or a flow that changes direction once or several times as θ goes from Θ_1 to Θ_2, yielding a zero *net* volume flow. This is to be supplemented by what happens at $R = R_1$, as mentioned before. More details on this will come to light when we present the final solution.

Therefore, Jeffery-Hamel flow is solvable in the form of

$$\begin{cases} q_\theta = 0 \\ q_R = \dfrac{f\langle\theta\rangle}{R} \end{cases}$$

where $f\langle\theta\rangle$ is given implicitly by

$$\theta = \pm \int \frac{df}{\sqrt{-(2/3\nu)G\langle f\rangle}} \tag{8.45}$$

The integral (8.45), which is the integration of Eq. (8.43) square-rooted, is known as an elliptical integral and its numerical values are tabulated in many books on higher transcendental functions. The three constants, namely A, C, and the one implied by the indefinite integral in Eq. (8.45), are to be determined by boundary conditions (8.39a,b) and the given volume flow.[7]

However, even without the help of numerical calculations, we can already learn a great deal about the qualitative nature of the result. To this end it is convenient to start with the general solution, Eq. (8.45), with the three constants unspecified. We will then look for the admissible arrangements of the channel wall (that is, the values of Θ_1 and Θ_2) and find the corresponding volume flow. In other words, we will work on the problem backward.

The cubic function $G\langle f\rangle$ in Eq. (8.45) can be written, for any given A and C, as

$$G\langle f\rangle = (f-e_1)(f-e_2)(f-e_3) \tag{8.44A}$$

The coefficients of f^2 in Eqs. (8.44) and (8.44A) must be the same; that is,

$$e_1 + e_2 + e_3 = -6\nu \tag{8.46}$$

So, only two of the three e's are really independent.

To sketch the curve representing $G\langle f\rangle$ versus f, we notice that (a) the curve must cross the f-axis at three places, $f = e_1,\ e_2,\ e_3$; (b) $G\langle f\rangle \to \pm\infty$ as $f \to \pm\infty$; (c) the integration in Eq. (8.45) implies that $G < 0$ for any real θ. All these are incorporated in Figure 8.5, where the portions with $G > 0$ are crossed

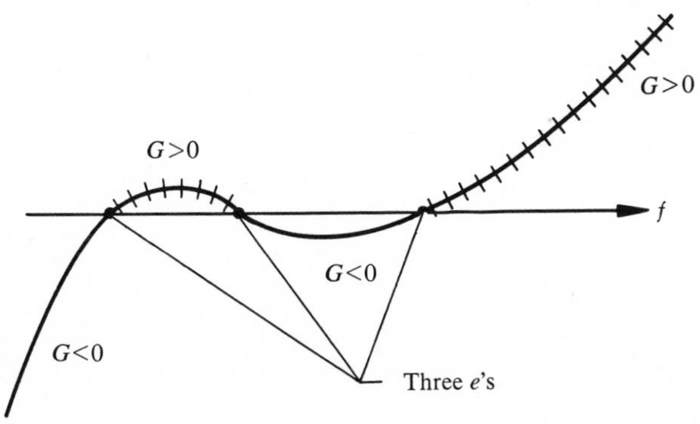

Figure 8.5 Variation of $G\langle f\rangle$.

out as of no physical significance. Furthermore, of the two portions with $G < 0$, only one has anything to do with a real channel flow; this is the one that contains the origin $f = 0$, since f must vanish on the channel walls (no-slip condition) and must be contained in the interval of integration in Eq. (8.45).

We can distinguish two cases here.

CASE 1. *A and C are such that e_1, e_2, e_3 are all real and distinct.* Let us now number the three e's such that

$$e_1 > e_2 > e_3$$

Since $e_1 + e_2 + e_3 = -6\nu < 0$, at least one of the e's must be negative. Then, Figure 8.5 shows that the only physically realizable case is where $e_2, e_3 < 0$ (that is, there is only one positive e). (See Fig. 8.6.) Incidentally, the left end of the curve (to the left of e_3) has nothing to do with the channel-flow problem since it does not embrace $f = 0$. Also, if $e_1 = e_2$, there will be no physically realizable solution to the problem.

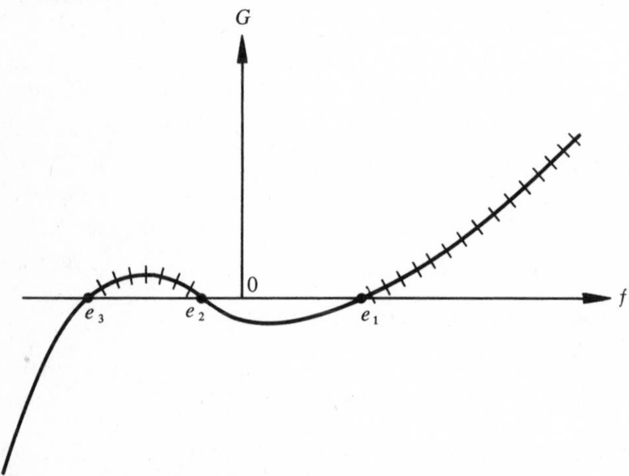

Figure 8.6 e_1, e_2, e_3 real and distinct.

CASE 2. *There is only one real e.*

The only real root, say e_1, may be either positive or negative. But, only the case with $e_1 > 0$ is physically significant (Fig. 8.7).

Finally, the case where $e_1 = e_2 = e_3$ $(= e)$ is physically unrealizable since, then, e must be negative, Eq. (8.46), and there is no negative portion of the G curve that contains the point $f = 0$ (Fig. 8.8).

The corresponding portions of the G curve in Cases 1 and 2 can now be used to construct the function

$$\frac{d\theta}{df} = \pm \frac{1}{\sqrt{-(2/3\nu)G\langle f \rangle}} \tag{8.43A}$$

The results for both cases are sketched in Figure 8.9 as curves[8] of $d\theta/df$ versus f.

[8] Only the two curves associated with the plus sign in Eq. (8.46) are shown; those associated with the minus sign are obtained by a reflection about the f-axis.

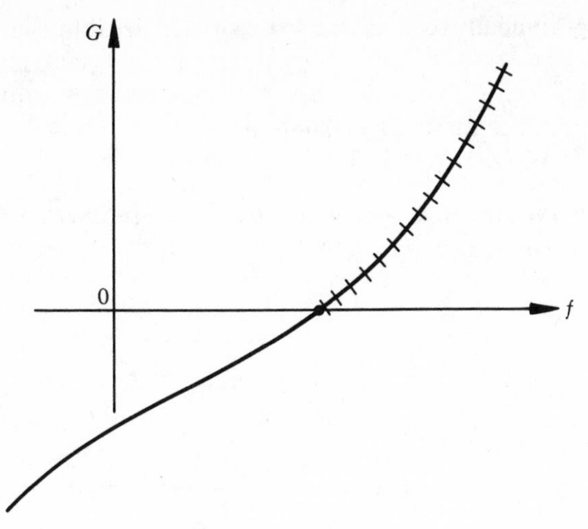

Figure 8.7 Only one real e.

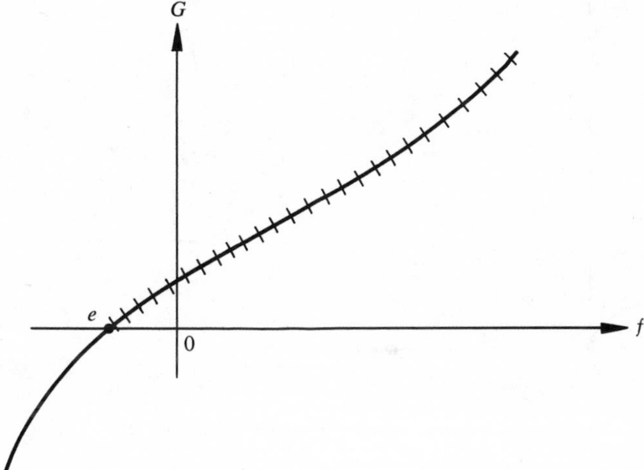

Figure 8.8 $e_1 = e_2 = e_3$.

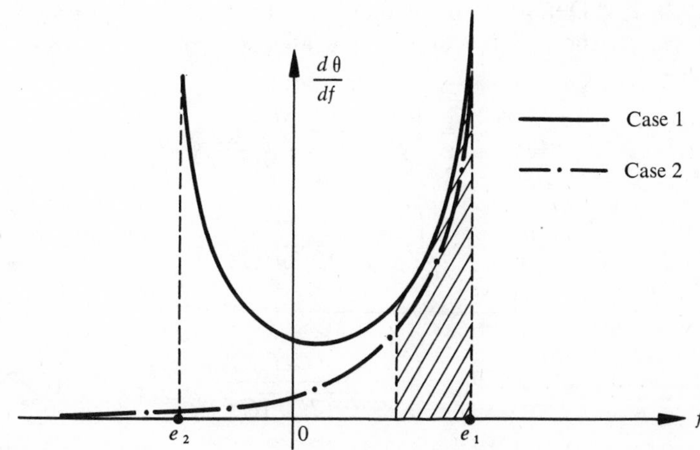

Figure 8.9 $d\theta/df$ versus f.

This can then be integrated (say, graphically as shown by the shaded area[8] in Fig. 8.9) to yield

$$\theta = \pm \int_f^{e_1} \frac{df}{\sqrt{-(2/3\nu)G\langle f\rangle}} + \text{constant} \tag{8.47}$$

Since the integrating constant only fixes the line from which θ is measured, we can choose it to be zero without any loss of generality. Then[8] (Fig. 8.10), $\theta = 0$, where $f = e_1$.

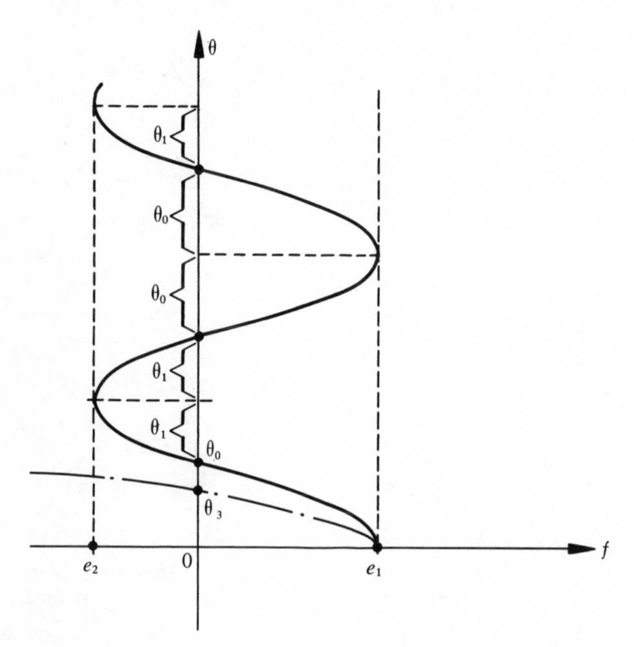

Figure 8.10 θ versus f.

CASE 1. Denoting the values of θ at $f = 0$, e_2 by $\pm\theta_0$, $\pm(\theta_0 + \theta_1)$, respectively, where the $\pm$ sign is again associated with that in Eq. (8.47), we have

$$\theta_0 = \int_0^{e_1} \frac{df}{\sqrt{-(2/3\nu)G\langle f\rangle}} \tag{8.48}$$

$$\theta_0 + \theta_1 = \int_{e_2}^{e_1} \frac{df}{\sqrt{-(2/3\nu)G\langle f\rangle}}$$

Therefore

$$\theta_1 = \int_{e_2}^{0} \frac{df}{\sqrt{-(2/3\nu)G\langle f\rangle}} \tag{8.49}$$

All these are indicated[8] in Figure 8.10. Furthermore, the $f \sim \theta$ curve is periodic[9] with period $2(\theta_0 + \theta_1)$ as shown in the figure.

CASE 2. The $f \sim \theta$ curve in this case is not periodic. It cuts the θ-axis at $\pm\theta_3$, where

$$\theta_3 = \int_0^{e_1} \frac{df}{\sqrt{-(2/3\nu)G\langle f\rangle}} \tag{8.50}$$

In both Cases 1 and 2 it is obvious that, in order to have $f = 0$ on the two walls, they must be situated at the dots[8] shown in Figure 8.10 with a distance between them smaller than 2π. For Case 2, there is only one possibility—namely, a channel in which $\Theta_1 = -\theta_3$ and $\Theta_2 = \theta_3$ (Fig. 8.11), with a pure outflow (that is, $q_R = f/R > 0$) symmetric about the centerline $\theta = 0$. For Case 1, the possibilities are

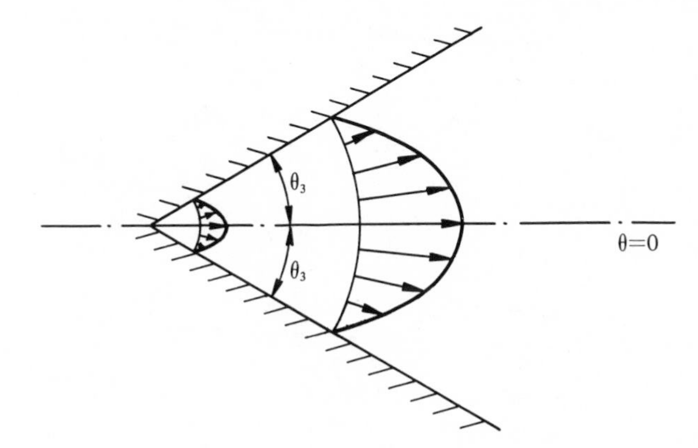

Figure 8.11 A channel of pure outflow (Case 2).

many. Some of these are sketched in Figure 8.12. Whether these possible flow patterns will occur or not in reality will depend on how closely you conform to the requirements. For example, in order to have pattern (a) in Figure 8.12, you must suck out fluid backward near the lower wall. Otherwise, this asymmetric flow with a backflow region simply will not occur. This is exactly what we meant before when we stated that the flow rate across $R = R_1$ must match that across any $R = R$. It is also to be noted that several patterns of flow may yield the same overall flow rate. Then which pattern will occur if the rate across $R = R_1$ matches only this overall value, not the detailed ins and outs? It is probably safe to say that, if a situation admits many patterns, the simplest pattern that is symmetric about the centerline of the channel will actually occur, unless particular care is exercised to

[9] To prove this periodicity, one should go into the complex domain. We will have to omit it here. Compare, however, with

$$\int_{-1}^{f(\leq 1)} \frac{df}{\sqrt{1-f^2}} = \sin^{-1} f \Big|_{-1}^{f}$$

which also exhibits a periodicity.

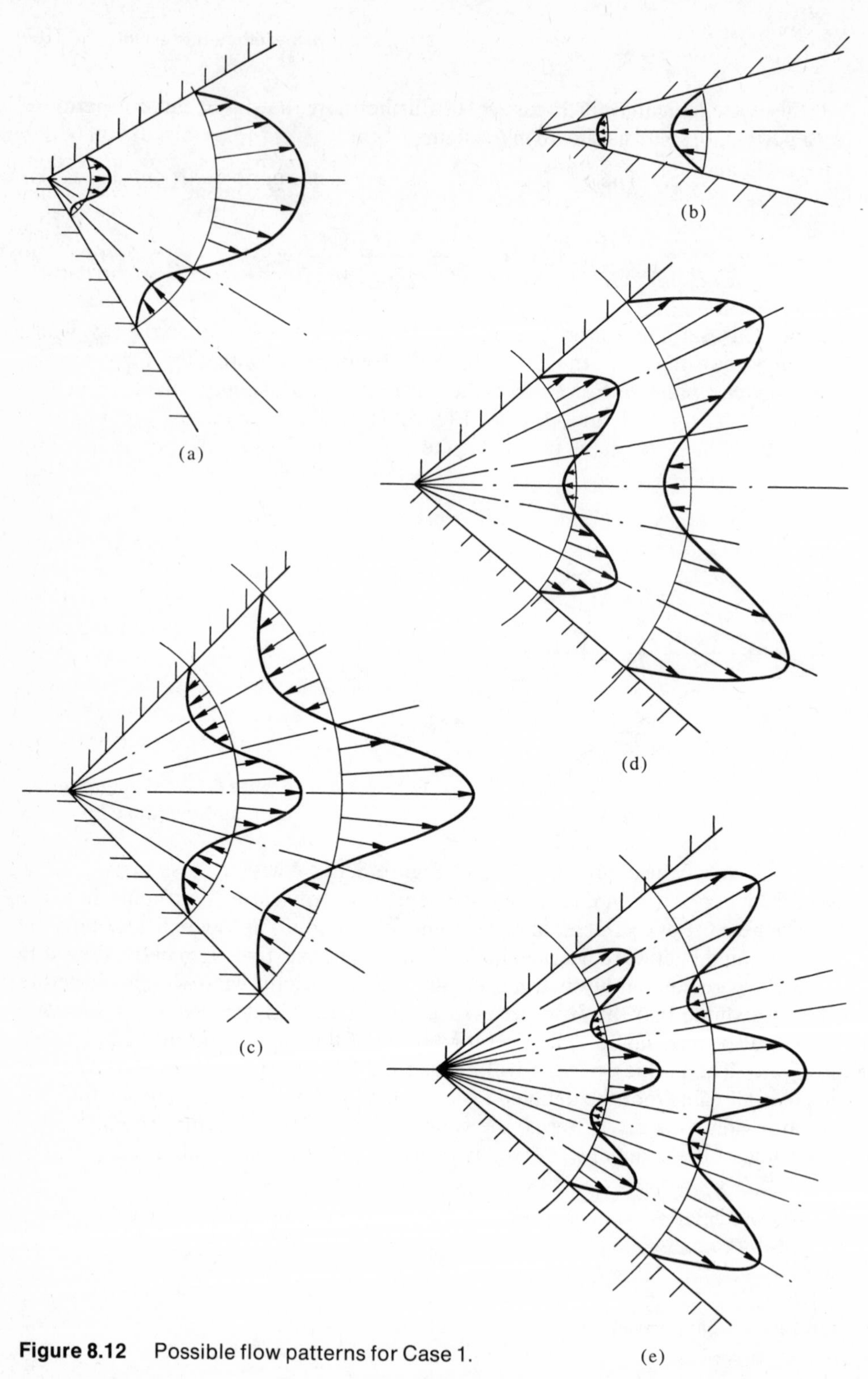

Figure 8.12 Possible flow patterns for Case 1.

induce a specific pattern. This is so probably because the simple pattern is physically most stable and thus most likely to show up. In our further discussion of the Jeffery-Hamel flow in the next section, we will present more taboos on certain flow patterns. But here let us be satisfied with just the following intermediate summary.

Intermediate Summary The flow pattern now hinges on the nature of the cubic function

$$G\langle f\rangle = f^3 + 6\nu f^2 - 3Af + C$$

The values of A and C must be such that:

CASE 1. G has three real roots of which one and only one is positive.

CASE 2. G has only one real (and positive) root.

Otherwise, the ansatz breaks down. Furthermore, we see that for Case 2 (one positive root), f (and hence q_R) must be positive and symmetric about the centerline of the channel. That is, $Q > 0$. For Case 1 (three real roots), it is possible to have either pure outflow (that is, in the R-direction) with f symmetric about the centerline, or mixed inflow and outflow with either symmetric or asymmetric f. In Case 1, it is also possible to have a symmetric, pure inflow ($Q < 0$) if the channel walls are situated at $\theta = \theta_0$, $(\theta_0 + 2\theta_1)$. The important thing in this last situation is that the walls form an angle $2\theta_1$ between them (for example, the two walls may be at $\theta = \pm\theta_1$), since the f-axis (where we start measuring θ) can be shifted to any convenient position without disturbing anything.

8.4 JEFFERY-HAMEL FLOW (Continued)

After gaining a general knowledge about all possible patterns of Jeffery-Hamel flow, we are going to concentrate in this section on the flow through a *given* channel in between $\theta = \pm\Theta$ (Fig. 8.13) and inquire as to what happens when $|Q|$ increases? We have postponed raising this question until now since we need first some general knowledge on what to expect.

To start with, we wish to introduce here a Reynolds number. We will

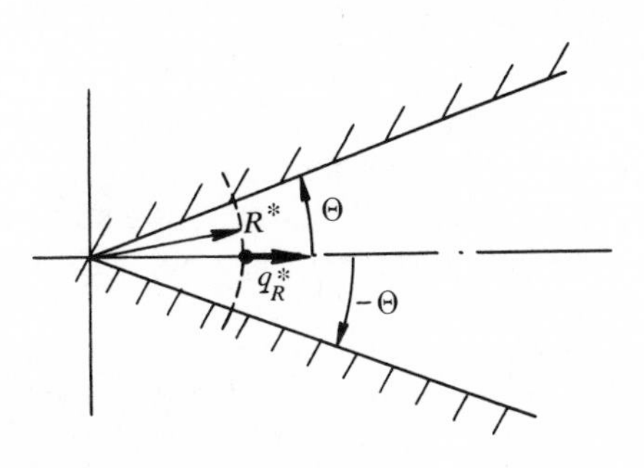

Figure 8.13 A given channel.

pick up *any* radius R^* as our characteristic length, and use the *velocity* (not speed) on the centerline at R^* as our characteristic velocity. Then[10]

$$Re = \frac{q_R^* R^*}{\nu} = \frac{f_{\text{center}}}{\nu} \tag{8.51}$$

Let us also keep the setup carefully symmetric about $\theta = 0$ so that the asymmetric patterns will never occur.

Referring back to Figure 8.10, we see that

$$f_{\text{center}} = \begin{Bmatrix} e_1 \\ e_2 \end{Bmatrix} \quad \text{if } q_R^* \begin{Bmatrix} > 0 \\ < 0 \end{Bmatrix} \tag{8.52a,b}$$

and $Re = e_{1,2}/\nu$ is a constant of the flow.

Next, we will nondimensionalize the results of Section 8.4 except the pressure formula:[11]

$$f' = \frac{f}{f_{\text{center}}}$$

$$e' = \frac{e}{f_{\text{center}}} \quad \text{(that is, } e_1' \text{ or } e_2' = 1\text{)}$$

$$R' = \frac{R}{R^*}$$

$$q_R' = \frac{q_R}{q_R^*} = \frac{f'}{R'}$$

$$Re = \frac{f_{\text{center}}}{\nu} = \frac{e_{1,2}}{\nu}$$

$$\begin{cases} \dfrac{d^2 f'}{d\theta^2} + 4f' + Re(f')^2 = \dfrac{A}{\nu f_{\text{center}}} = A' & \text{(8.41A)} \\ f'\langle \pm\Theta \rangle = 0 & \text{(8.39aA,bA)} \end{cases}$$

or

$$\theta = \pm \int \frac{df'}{\sqrt{-(2/3)Re(f' - e_1')(f' - e_2')(f' - e_3')}} \tag{8.45A}$$

For small values of Re, considerable information can be obtained by looking at the ordinary differential equation (8.41A) from another angle.

In Chapter 5, especially in the last paragraphs of Section 5.3, we advocated the ordering and asymptotics as a special kind of ansatz. Here, we will go

---

[10] For convenience in presenting data, q_R^* or f_{center} will be regarded as given (or actually measured). It then takes the place of the flow rate Q, which in turn becomes a calculated quantity after the solution is obtained.

[11] The calculation of pressure presents no problem, once f is found.

one step further and introduce the following two-term *asymptotic* form for the solution of Eq. (8.41A), again as an ansatz:

$$f'\langle\theta\rangle = f_0'\langle\theta\rangle + f_1'\langle\theta\rangle Re + O[(Re)^2] \tag{8.53}$$

where the last term represents the error that vanishes at least as fast as $(Re)^2$ for small Re, if not faster. One way to proceed is to write first

$$f'\langle\theta\rangle = f_0'\langle\theta\rangle + O(Re)$$

and substitute it into Eq. (8.41A) and boundary conditions (8.39aA,bA). Next, let $Re \to 0$. This leaves you with only $f_0'\langle\theta\rangle$ to solve, which is a much simpler job than the original one of solving for $f'\langle\theta\rangle$. We then take this newly solved $f_0'\langle\theta\rangle$ and insert it into Eq. (8.53). Substituting Eq. (8.53) again into Eq. (8.41A) and boundary conditions (8.39aA,bA), dividing through by Re, and letting $Re \to 0$, we have a problem involving $f_1'\langle\theta\rangle$ only, which we solve. Another equivalent but simpler way to proceed is to substitute Eq. (8.53) at once into Eq. (8.41A) and boundary conditions (8.39aA,bA). We then group all the terms that are multiplied by $(Re)^0$ (that is, 1), those by $(Re)^1$ (that is, Re), and so on, respectively, and equate each group separately to zero. This will yield at once two problems with f_0' and f_1', which we solve. In the following, we will adopt this second procedure. Incidentally, both procedures are known as the *perturbation method* in the literature. The small parameter Re involved is called the perturbation parameter. Equation (8.53) can of course be generalized to three or more terms.

Before we really put the second procedure to work, we must notice also the following:

$$Q' = \frac{Q}{q_R^* R^*} = \frac{Q}{f_{\text{center}}}$$

$$= \int_{-\Theta}^{\Theta} f'd\theta = \int_{-\Theta}^{\Theta} f_0'd\theta + \left(\int_{-\Theta}^{\Theta} f_1'd\theta\right) Re + O[(Re)^2] \tag{8.54}$$

and, since the still-to-be-determined parameter A' is related to Q',

$$A' = A_0' + A_1' Re + O[(Re)^2] \tag{8.55}$$

Both Q' and A' are parameters *to be calculated.*

Substituting, finally, Eqs. (8.53) and (8.55) into Eq. (8.41A) and boundary conditions (8.39aA,bA), we have

$$O[(Re)^0]: \begin{cases} \dfrac{d^2 f_0'}{d\theta^2} + 4f_0' = A_0' & \textbf{(8.56)} \\ f_0'\langle\pm\Theta\rangle = 0 & \textbf{(8.57a,b)} \end{cases}$$

$$O[(Re)^1]: \begin{cases} \dfrac{d^2 f_1'}{d\theta^2} + 4f_1' = -(f_0')^2 + A_1' & \textbf{(8.58)} \\ f_1'\langle\pm\Theta\rangle = 0 & \textbf{(8.59a,b)} \end{cases}$$

The solution of Eq. (8.56) under boundary conditions (8.57a,b) is easily seen to be

$$f'_0 = \frac{A'_0}{4}\left(1 - \frac{\cos 2\theta}{\cos 2\Theta}\right) \tag{8.60}$$

Now

$$\begin{aligned} f' &= \frac{f}{f_{\text{center}}} \\ &= \frac{f_0}{f_{\text{center}}} + \left(\frac{f_1}{f_{\text{center}}}\right) Re + O[(Re)^2] \end{aligned}$$

must be equal to $1[+0 \cdot Re + 0 \cdot (Re)^2 + \cdots]$ on the centerline $\theta = 0$ by definition.[12] Therefore, we must have

$$f'_0\langle 0\rangle = 1 \tag{8.61a}$$

$$f'_1\langle 0\rangle = 0 \tag{8.61b}$$

$$\cdots\cdots$$

Using Eq. (8.61a), we have

$$\frac{A'_0}{4} = 1 \bigg/ \left(1 - \frac{1}{\cos 2\Theta}\right)$$

and

$$\begin{aligned} f'_0 &= \left(1 - \frac{\cos 2\theta}{\cos 2\Theta}\right) \bigg/ \left(1 - \frac{1}{\cos 2\Theta}\right) \\ &= 1 - \left(\frac{\sin\theta}{\sin\Theta}\right)^2 \end{aligned} \tag{8.62}$$

Equation (8.54) then gives

$$\begin{aligned} Q' &= \frac{Q}{f_{\text{center}}} \\ &= \int_{-\Theta}^{\Theta} f'_0 \, d\theta + O(Re) \\ &= \frac{2\Theta \cos 2\Theta - \sin 2\Theta}{\cos 2\Theta - 1} + O(Re) \end{aligned} \tag{8.63}$$

Incidentally,

$$A'_0 = \frac{4Q'}{2\Theta - \tan 2\Theta} + O(Re)$$

is related to Q'.

It is interesting to see that f'_0 is symmetric about the centerline $\theta = 0$, and is always positive. Therefore, to the order $O(Re)$, the flow is symmetric, and

[12] In other words, $f'\langle 0\rangle = 1$ for any Re.

in one direction without any backflow. The sign of the actual flow velocity

$$q_R = \frac{f' q_R^*}{R'} = \frac{f' f_{\text{center}}}{R}$$

$$= \frac{[f_0' + O(Re)] f_{\text{center}}}{R}$$

follows that of f_{center} (or q_R^*) for small Re. That is, the flow for small Re will be pure inflow or outflow according to whether f_{center} (or q_R^*) is smaller or greater than zero.

If Θ is small, θ must also be small in Eq. (8.62). Then

$$f_0' \cong 1 - \left(\frac{\theta}{\Theta}\right)^2$$

assumes a parabolic profile. This is easily seen to be equivalent to the plane Poiseuille flow of Section 3.2, Special Case 2.

If we are not satisfied with the error $O(Re)$ involved here, we can go on to improve the situation by solving for f_1' from Eq. (8.58), which now reads

$$\frac{d^2 f_1'}{d\theta^2} + 4f_1' = -a^2(b - \cos 2\theta)^2 + A_1' \tag{8.64}$$

where

$$a = \frac{A_0'}{4} \Big/ \cos 2\Theta \tag{8.65a}$$

$$b = \cos 2\Theta \tag{8.65b}$$

subject to the conditions (8.59a,b) and (8.61b). The result is

$$f_1' = C_1(\cos 2\theta - 1) + \frac{a^2 b}{2}\theta \sin 2\theta + \frac{a^2}{24}(\cos 4\theta - 1) \tag{8.66}$$

where

$$C_1 = -\left[\frac{a^2 b}{2}\Theta \sin 2\Theta + \frac{a^2}{24}(\cos 4\Theta - 1)\right] \Big/ (\cos 2\Theta - 1) \tag{8.67}$$

Figure 8.14 sketches both $f_0'\langle\theta\rangle$ and $f_1'\langle\theta\rangle$ (for rather small values of Θ). Here an interesting thing happened. Remembering that

$$f' = f_0' + f_1' Re + O[(Re)^2]$$

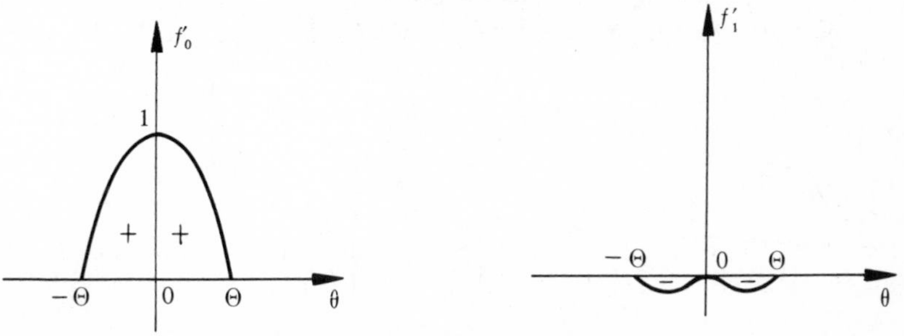

Figure 8.14 f_0' and t_1' for small Θ.

we see that the contribution of the second term on the right-hand side for $Re > 0$ (outflow) is in the opposite direction from that for $Re < 0$ (inflow). This is shown schematically in Figure 8.15 for f', and in Figure 8.16 for f. (Note that the solid curves in Figure 8.16 represent $Re \cong 0$.) From this fact, we expect the following drastic difference in behavior between the inflow and the outflow.

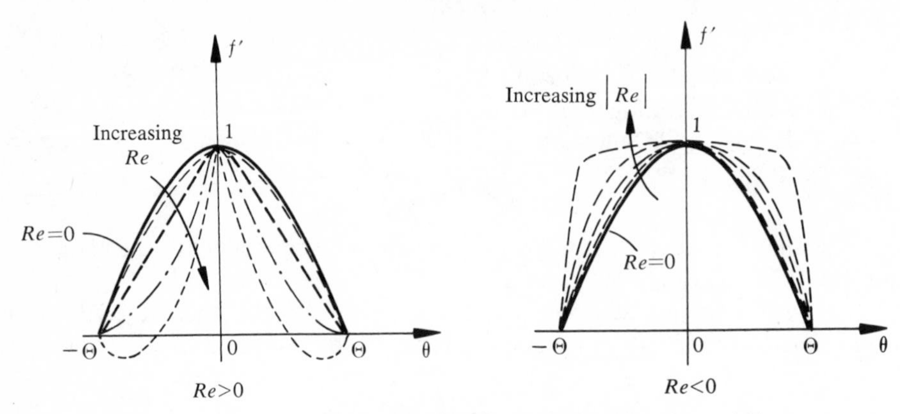

Figure 8.15 Speculative sketches of f'.

Inflow As $|Re|$ increases, the velocity profiles as represented by $f\langle\theta\rangle$ flatten down in the center, but vary sharply near the two walls. Boundary layers thus seem to form on the walls. Moreover, the flow does not change direction.

Outflow As Re increases, the velocity profiles sharpen up in the center with accompanying decrease in slope at the walls. It is therefore expected that, for a certain value of Re, the profiles show a zero slope (physically, no shear stress) at the walls. This is indicated by the dash-dot curves in Figures 8.15 and 8.16. See the discussion of the phenomenon of separation in Section 7.3, Case 3. For still larger values of Re, we expect to see backflow regions near the walls (fine dashed curves in the figures); see Figure 7.11.

The previous speculations[13] will now be substantiated. In the rest of this section, we will discuss the case of the outflow when Re is not small. The inflow case will be taken up in the next section.

When Re is not small, let us go back for a moment to Eq. (8.43):

$$\left(\frac{df}{d\theta}\right)^2 = -\frac{2}{3\nu}(f^3 + 6\nu f^2 - 3Af + C) \tag{8.43}$$

Since $f\langle\pm\Theta\rangle = 0$, we have

$$C = -\frac{3\nu}{2}\left[\left.\frac{df}{d\theta}\right|_{\theta=\pm\Theta}\right]^2 \leqslant 0$$

[13] They are speculations because we have extrapolated small-Re solutions to large values of Re.

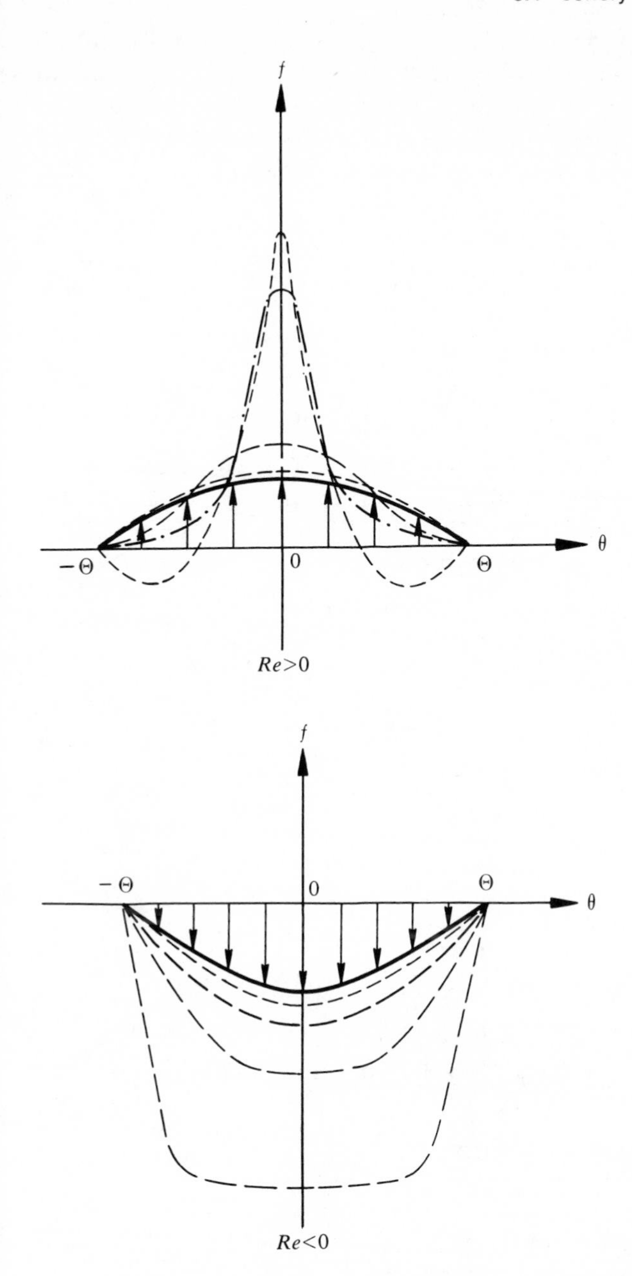

Figure 8.16 Speculative sketches of *f*.

We are glad to see that $C = 0$ is exactly the critical situation of zero shear stress on the wall when backflow is about to appear. Furthermore,

$$\theta = 0: \begin{cases} f = f_{\text{center}} = e_1 & \text{(Fig. 8.10)} \\ \dfrac{df}{d\theta} = 0 & \text{(symmetry about } \theta = 0) \end{cases}$$

Therefore Eq. (8.43) applied to the centerline yields

$$e_1^3 + 6\nu e_1^2 - 3A e_1 + C = 0$$

Solving here for $3A$ and substituting back into Eq. (8.43), we have

$$\left(\frac{df}{d\theta}\right)^2 = \frac{2}{3\nu}(e_1 - f)\left[f^2 + f(6\nu + e_1) - \frac{C}{e_1}\right] \quad \text{(for symmetric flows)}$$

Then,

$$\theta = \pm \int_f^{e_1} \frac{df}{\sqrt{(df/d\theta)^2}} \tag{8.68}$$

since $f\langle 0\rangle = e_1$.

Referring once more back to Figure 8.10, we see that the velocity profiles must be represented by the portion of the $f \sim \theta$ curve with $f \geqslant 0$ before backflow sets in. This portion is reproduced in Figure 8.17 for clarity. Thus, $|\theta|$ for $f = 0$ has the unambiguous value Θ, before back flow appears. (With backflow

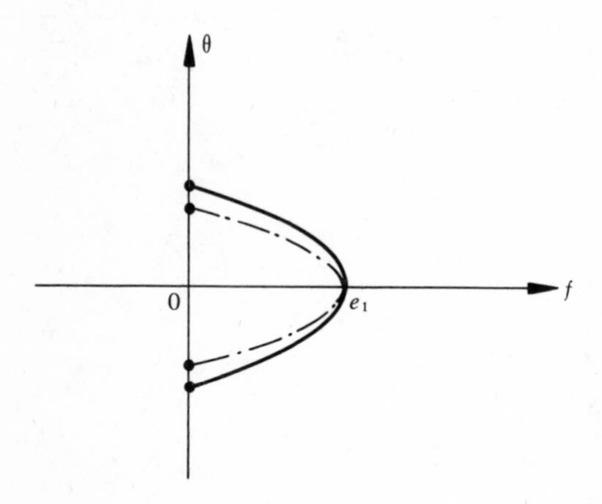

Figure 8.17 θ vs. f (without back flow).

regions, there may be several $|\theta|$ for which $f = 0$.) Thus, Eq. (8.68) can be evaluated at $f = 0$ to yield

$$\Theta = \int_0^{e_1} \frac{df}{\sqrt{(2/3\nu)(e_1 - f)[f^2 + f(6\nu + e_1) - C/e_1]}}$$

without any ambiguity, *before backflow appears*. As stated before, $C = 0$ marks the critical threshold of backflow; the corresponding

$$\Theta_{\text{crit}} = \int_0^{e_1} \frac{df}{\sqrt{(2/3\nu)(e_1 - f)[f^2 + f(6\nu + e_1)]}}$$

$$= \sqrt{\frac{3}{2Re}} \int_0^1 \frac{df'}{\sqrt{(1 - f')[f'^2 + f'(1 + 6/Re)]}}$$

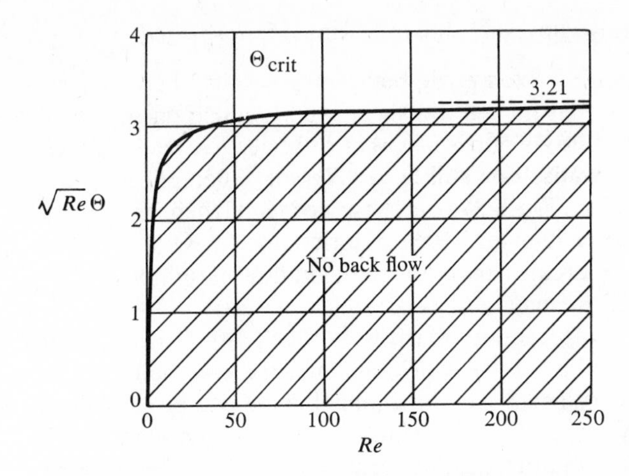

Figure 8.18 Critical value of Θ for pure outflow.

then marks, for a given Re, the largest half-angle that a channel can have and still maintain pure outflow. The definite integration involved can be performed numerically; the result is plotted in Figure 8.18 in the form of $\sqrt{Re}\,\Theta_{\text{crit}}$ versus Re. The shaded region below this curve shows the combinations of Θ and Re that guarantee pure outflow. For large Re, it is seen that $\Theta_{\text{crit}} \approx 3.2/\sqrt{Re}$; that is,

$$\lim_{Re\to\infty} \Theta_{\text{crit}} = 0$$

It is interesting to note that, for small Re,

$$\begin{aligned}\Theta_{\text{crit}} &= \sqrt{\frac{3}{2Re}}\int_0^1 \frac{df'}{\sqrt{6/Re}\cdot\sqrt{f'(1-f')}}\\ &= \frac{1}{2}\int_0^1 \frac{df'}{\sqrt{f'-f'^2}}\\ &= \frac{\pi}{2}\end{aligned}$$

This finding is sketched in Figure 8.19.

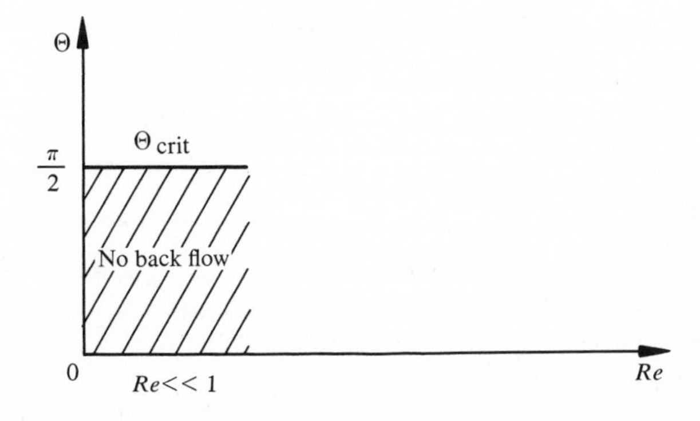

Figure 8.19 Critical Θ for small *Re*.

To end this section, we now offer the following *physical interpretation*:

1. For a channel with a fixed angle between the walls, there is a certain value of Re that makes the shear stress vanish on the walls. For smaller Re, there is pure outflow. For larger Re, backflow will appear near the walls. For still larger Re, more complicated flow (as indicated in Section 8.3) will show up. (But, then, the flow is probably no longer laminar.) Please notice that for $\Theta > \pi/2$ (that is, channel angle greater than π), pure outflow is impossible no matter how small Re may be.
2. For a fixed Re, only channels narrow enough (with $\Theta \leqslant \Theta_{crit}$) can produce pure outflow. In a wider channel, there is bound to be backflow. For very small Re, the channel angle can be as wide as π. But for very large Re, the channel angle must be very nearly zero in order to maintain a pure outflow.

8.5 BOUNDARY-LAYER FLOW IN A CONVERGENT CHANNEL

For a Jeffery-Hamel inflow, $Re < 0$, the situation is entirely different. Referring back to Figure 8.10, we see that here we should reorient the f-axis so that $f\langle 0\rangle = e_2$ (Fig. 8.20). Then

$$f_{\text{center}} = e_2 < 0$$

and there must be three real roots (Case 1, Section 8.3) e_1, e_2, e_3, with $e_1 > 0$, $e_{2,3} < 0$ and $e_2 > e_3$ (Fig. 8.20). Furthermore,

$$Re = \frac{f_{\text{center}}}{\nu} = \frac{e_2}{\nu} < 0$$

Therefore

$$Re = -|Re|$$

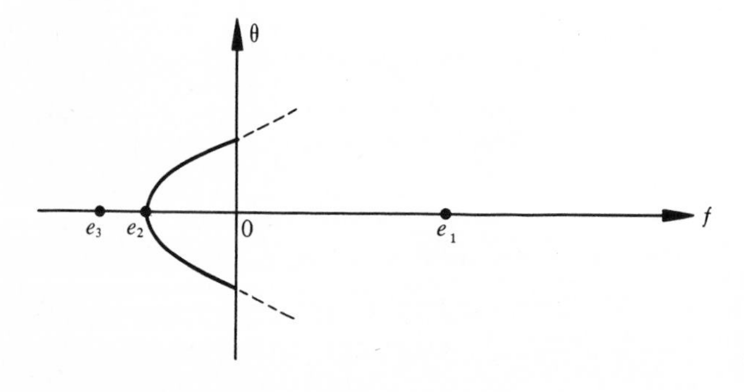

Figure 8.20 θ versus f (pure inflow) with f-axis reoriented.

In terms of dimensionless quantities, the three roots are

$$\begin{cases} e_3' = \dfrac{e_3}{e_2} > 0 \\ e_1' = \dfrac{e_1}{e_2} < 0 \\ e_2' = \dfrac{e_2}{e_2} = 1 \end{cases}$$

and the relation $e_1 + e_2 + e_3 = -6\nu$ becomes

$$\begin{aligned} e_1' + 1 + e_3' &= \frac{-6}{Re} \\ &= \frac{6}{|Re|} \end{aligned} \tag{8.69}$$

The solution is, as before,

$$\theta = \pm \int_{f'}^{1} \frac{df'}{\sqrt{-(2/3)Re(f'-e_1')(f'-1)(f'-e_3')}}$$

or

$$\sqrt{|Re|}\,\theta = \pm\sqrt{\frac{3}{2}} \int_{f'}^{1} \frac{df'}{\sqrt{(f'-e_1')(f'-1)(f'-e_3')}} \tag{8.70}$$

A similar clarification to that put forward in Section 8.4, immediately after Eq. (8.68), gives us

$$\sqrt{|Re|}\,\Theta = \sqrt{\frac{3}{2}} \int_{0}^{1} \frac{df'}{\sqrt{(f'-e_1')(f'-1)(f'-e_3')}} \tag{8.70A}$$

before backflow appears. Whereas the corresponding definite integral in Section 8.4 leads to numerical values without any mishap, the integral here becomes unbounded as $e_3' \to 1$. To see this, we first notice that the definite integral is improper at the upper limit $f' = 1$. This in itself may or may not lead to an unbounded result, depending on the behavior of the integrand near $f' = 1$. We have to distinguish between the cases where $e_3' \neq 1$ and where $e_3' = 1$. (Note that e_1', being negative, can never be close to 1.) If $e_3' \neq 1$, the integrand behaves near $f' = 1$ like

$$\frac{1}{\sqrt{(1-e_1')(1-e_3')}}\left(\frac{1}{\sqrt{f'-1}}\right)$$

which does not grow fast enough, as $f' \to 1$, to cause unboundedness of the integral. However, if $e_3' = 1$, the integrand would grow as

$$\frac{1}{\sqrt{1-e_1'}}\left(\frac{1}{1-f'}\right)$$

near $f' = 1$, which is too fast. In mathematical language, we say that the integral has a logarithmic singularity at $e_3' = 1$. Since $\sqrt{|Re|}\ \Theta$ is not bounded at $e_3' = 1$, its value grows and grows as we let $e_3' \to 1$.

Now $|Re|$ *and* Θ determine e_1' *and* e_3', and vice versa. So if *either* $|Re|$ *or* Θ is given, e_3' can be arranged to assume any value at all.[14] We can then, for any $|Re|$ *or* Θ, make $\sqrt{|Re|}\ \Theta$ as large as we wish by letting e_3' move toward 1. This means that (a) for given $|Re|$, Θ can be made as large as one wishes while still maintaining pure inflow and (b) for given Θ, $|Re|$ can be made as large as one wishes without inducing backflow (outflow).

But, are there not situations where it is impossible to move e_3' toward 1? Yes. Consider the requirement that

$$e_1' + e_3' = \frac{6}{|Re|} - 1 \tag{8.69A}$$

Setting $e_3' = 1$, we have

$$e_1' = \frac{6}{|Re|} - 2$$

But $e_1' < 0$; so e_3' can be equal to 1 only when

$$|Re| > 3$$

For $|Re| < 3$, the best one can do is to make $e_1' = 0$; then

$$e_3' = \frac{6}{|Re|} - 1$$

becomes the lowest possible value for e_3'. For these cases with $|Re| < 3$, $\sqrt{|Re|}\ \Theta$ will have an upper bound $\sqrt{|Re|}\ \Theta_{crit}$, beyond which backflow appears. This Θ_{crit} is shown in Figure 8.21. (We will omit here the detailed analytical procedure of obtaining this curve, dismissing it as not being particularly worthwhile or enlightening.[15]) From the figure, we see that there is always pure inflow when $\Theta \leqslant \pi/2$. For $\Theta > \pi/2$, there is pure inflow when $|Re| \geqslant 3$ but not when $|Re| < 3$. Since the range $\Theta > \pi/2$ and $|Re| < 3$ represents a small, *practically* insignificant portion of the flow situation,[16] we may *roughly* conclude that there will "never" be any back flow in a symmetric Jeffery-Hamel flow toward the apex of the channel, and the two conclusions drawn at the end of the preceding paragraph stand largely valid.

---

[14] It is more truthful to say "any value that is beyond any inherent structural taboos." (See next paragraph.) Essentially, this means that each pair $(\Theta, |Re|)$ is equivalent to a pair (e_1', e_3'); the region with $0 \leqslant \Theta < \pi$ and $0 \leqslant |Re| < \infty$ is equivalent to a certain region in the $e_1'e_3'$-plane; and a line $\Theta =$ constant or $|Re| =$ constant will be equivalent to a certain curve in the $e_1'e_3'$-plane. On this curve, e_3' can assume any value within the region of equivalence.

[15] See L. Rosenhead, "The steady two-dimensional radial flow of viscous fluid between two inclined plane walls," *Proc. Roy. Soc.*, **A175**, 436–467, 1940.

[16] That is why we did not bother to enter a detailed investigation of this range.

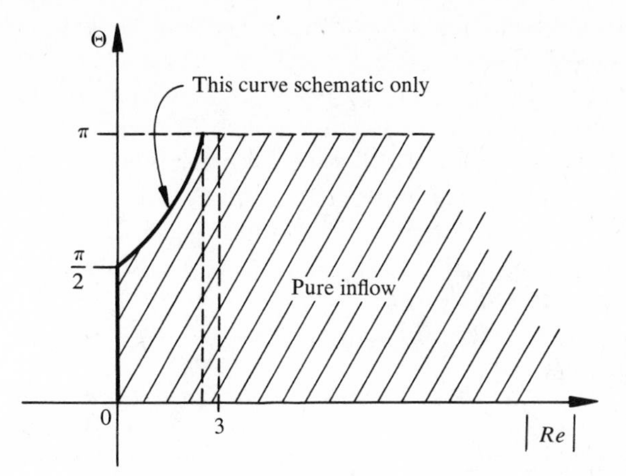

Figure 8.21 Critical Θ for pure inflow.

In the following, we will refer to the pure inflow as the flow in a convergent channel (and the pure outflow as that in a divergent channel). We will want to see what happens to the flow in a convergent channel as $|Re| \to \infty$. We especially wish to substantiate our previous conjecture that a boundary layer would form on the walls.

We start by noticing that, the flow being symmetric, one half of the channel $0 \leqslant \theta \leqslant \Theta$ is all that we need. In this half, $df'/d\theta$ is clearly negative, reaching zero only at $\theta = 0$. Therefore, the negative sign must be used in Eq. (8.43A):

$$-\frac{df'}{d\theta} = \sqrt{\tfrac{2}{3}|Re|} \cdot \sqrt{(f'-e_1')(1-f')(e_3'-f')} \tag{8.71}$$

The three factors $(f'-e_1')$, $(1-f')$ and $(e_3'-f')$ are all positive in the flow through a convergent channel, since $e_1' < 0$ and $0 \leqslant f' \leqslant 1 \leqslant e_3'$. Furthermore,

$$(f'-e_1') \geqslant (0-e_1')$$

$$(e_3'-f') \geqslant (e_3'-1)$$

Therefore

$$-\frac{df'}{d\theta} > \sqrt{\tfrac{2}{3}|Re|} \cdot \sqrt{(-e_1')(1-f')(e_3'-1)}$$

For $|Re| > 6$, we also have from Eq. (8.69A)

$$e_3'+1 = -e_1' + \frac{6}{|Re|}$$

$$< -e_1' + 1$$

or

$$-e_1' > e_3'$$

Therefore

$$-\frac{df'}{d\theta} > \sqrt{\tfrac{2}{3}|Re|} \cdot \sqrt{e_3'(e_3'-1)}\,\sqrt{1-f'} \qquad |Re| > 6$$

Writing

$$a_1 = \sqrt{\tfrac{2}{3}|Re|e_3'(e_3'-1)}$$

the above becomes

$$-\frac{d\theta}{df'} < \frac{1/a_1}{\sqrt{1-f'}}$$

and

$$\begin{aligned}\Theta &= \int_0^\Theta d\theta = \int_\Theta^0 (-d\theta)\\ &= \int_0^1 \left(-\frac{d\theta}{df'}\right) df'\\ &< \frac{1}{a_1}\int_0^1 \frac{df'}{\sqrt{1-f'}} = \frac{2}{a_1}\end{aligned}$$

or

$$\Theta < \frac{2}{\sqrt{\tfrac{2}{3}|Re|e_3'(e_3'-1)}} \qquad |Re| > 6$$

From this conclusion we see that it is necessary to have

$$\lim_{|Re|\to\infty} e_3' = 1$$

for Θ to be fixed as $|Re| \to \infty$. (Otherwise, the apex angle of the channel will be forced to zero as $|Re| \to \infty$, which is physically absurd.) This requirement, coupled with Eq. (8.69A), demands that

$$\lim_{|Re|\to\infty} e_1' = -2$$

So, as $|Re| \to \infty$, the physics of the problem would force the solution (8.70) to become

$$\lim_{|Re|\to\infty} \theta = \sqrt{\frac{3}{2|Re|}} \int_{f'}^1 \frac{df'}{\sqrt{(f'+2)(1-f')^2}} \tag{8.72}$$

The above formula looks indeterminate as it stands, since the integral diverges, owing to the factor $1/(1-f')$, as mentioned before. However, we notice that

$$\begin{aligned}\frac{d\theta}{df'} &= -\sqrt{\frac{3}{2|Re|}} \Big/ \sqrt{(f'-e_1')(1-f')(e_3'-f')}\\ &\to 0 \text{ as } |Re| \to \infty\end{aligned}$$

as long as $f' < 1$. If $f' = 1$, the slope $df'/d\theta = 0$ for any $|Re|$. Therefore, Eq. (8.72) must actually show a broken line, horizontal at $\theta = \Theta$, vertical for $0 \leqslant \theta < \Theta$ (Fig. 8.22). Now, for any large but finite $|Re|$ (10^4, for example), the $f' \sim \theta$ curve must change its slope continuously, because there is no physical mechanism that

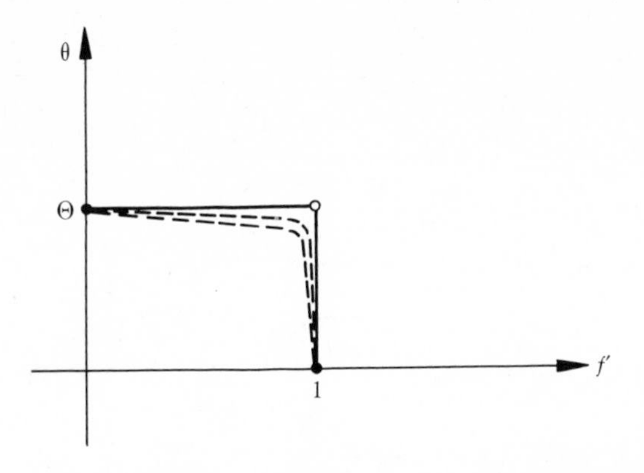

Figure 8.22 Boundary-layer structure in a convergent channel flow.

would support a sharp change. The slope must change rapidly, but not abruptly, as shown by the dashed curves in Figure 8.22.

In our previous terminology of Chapters 6 and 7, $f' = 1$ represents the Euler limit, or outer flow, which is valid in the center region of the channel as $|Re| \to \infty$. To gain a detailed information on the boundary-layer flow in the neighborhood of the wall, we must investigate the Prandtl limit. To do this, let us introduce (Fig. 8.23):

$$\underset{\sim}{\theta} = \Theta - \theta \tag{8.73}$$

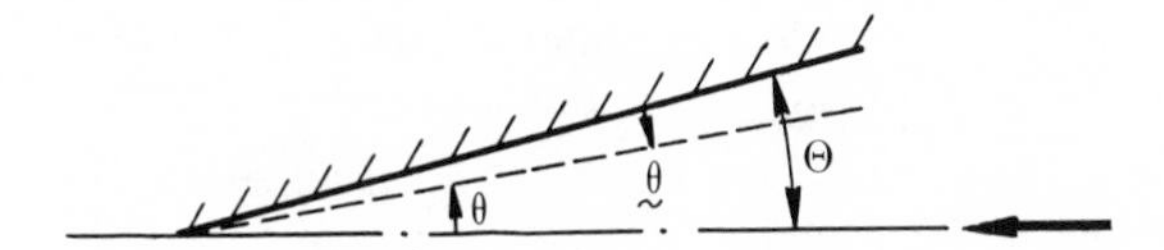

Figure 8.23 Definition of $\underset{\sim}{\theta}$.

Then Eqs. (8.70) and (8.70A) yield

$$\underset{\sim}{\theta} = \sqrt{\frac{3}{2|Re|}} \int_0^{f'} \frac{df'}{\sqrt{(f'-e_1')(1-f')(e_3'-f')}} \tag{8.74}$$

where $0 \leqslant \underset{\sim}{\theta} \leqslant \Theta$. Now, if we stretch $\underset{\sim}{\theta}$ by introducing

$$\underset{\sim}{\theta}'' = \sqrt{|Re|}\, \underset{\sim}{\theta} \tag{8.75}$$

we then have

$$\underset{\sim}{\theta}'' = \sqrt{\frac{3}{2}} \int_0^{f'} \frac{df'}{\sqrt{(f'-e_1')(1-f')(e_3'-f')}} \tag{8.76}$$

where f' stands for $f'\langle \underset{\sim}{\theta}'' \rangle$. As explained in Chapters 6 and 7, Eq. (8.76) with $|Re| \to \infty$ is the Prandtl limit representing the boundary-layer flow:

$$\lim_{|Re| \to \infty} \underset{\sim}{\theta}'' = \sqrt{\frac{3}{2}} \int_0^{f'} \frac{df'}{(1-f')\sqrt{f'+2}}$$

This integral is no longer improper, since the range of integration no longer contains $f' = 1$. As a matter of fact, it is even possible to express it in simple, closed form:

$$\lim_{|Re|\to\infty} \underset{\sim}{\theta}'' = \sqrt{2}\left[\tanh^{-1}\sqrt{\frac{f'+2}{3}}\right]_0^{f'}$$

$$= \sqrt{2}\left[\tanh^{-1}\sqrt{\frac{f'+2}{3}} - 1.146\right]$$

or[17]

$$\lim_{|Re|\to\infty} f' = 3\tanh^2\left(\frac{\underset{\sim}{\theta}''}{\sqrt{2}} + 1.146\right) - 2 \tag{8.77}$$

Equation (8.77) is then the boundary-layer solution of the flow in a convergent channel at large $|Re|$. Please note that

$$f'\langle 0\rangle = 0, f'\langle\infty\rangle = 1$$

In other words, it satisfies the boundary condition on the wall and matches with the outer flow in the central core according to the matching principle (Section 6.1).

The foregoing shows that, from the exact solution, one can demonstrate the existence of an outer flow and a boundary-layer flow matched nicely together according to the matching principle. This is analogous to the first half of Section 6.1. It would be didactically desirable to go through also the program of the latter half of Section 6.1—namely, to obtain the Euler and Prandtl limits, pretending that we do not know the exact solution, Eq. (8.45A).

We start with the system

$$\begin{cases} \dfrac{d^2f'}{d\theta^2} + 4f' + Re(f')^2 = A' & \text{(8.41A)} \\ f'\langle\Theta\rangle = 0 & \text{(8.39aA)} \\ f'\langle 0\rangle = 1 & \text{(8.78)} \end{cases}$$

To avoid the complexity of A' being a parameter to be calculated, let us differentiate Eq. (8.41A) once:

$$\frac{d^3f'}{d\theta^3} + 4\frac{df'}{d\theta} + 2Re\,f'\left(\frac{df'}{d\theta}\right) = 0 \tag{8.41B}$$

Then, the limit as $Re \to \pm\infty$ yields

$$f'^0\left(\frac{df'^0}{d\theta}\right) = 0$$

---

[17] Incidentally, this satisfies the equation

$$\frac{d^2f'_i}{d\underset{\sim}{\theta}''^2} - (f')^2 = -1$$

with $f'\langle 0\rangle = 0$ (and $df'/d\underset{\sim}{\theta}''$, $d^2f'/d\underset{\sim}{\theta}''^2 \to 0$ as $\underset{\sim}{\theta}'' \to \infty$).

Since $f'^o \neq 0$, we must have

$$\frac{df'^o}{d\theta} = 0$$

or

$$f'^o = 1$$

where boundary condition (8.78) is satisfied. For the Prandtl limit with $Re < 0$, we have

$$Re = -|Re|$$

and

$$\underset{\sim}{\theta}'' = \sqrt{|Re|}\,(\Theta - \theta)$$

Equation (8.41B) then becomes, as $|Re| \to \infty$,

$$\frac{d^3 f'}{d\underset{\sim}{\theta}''^3} - 2f'\left(\frac{df'}{d\underset{\sim}{\theta}''}\right) = 0$$

or

$$\frac{d^2 f'}{d\underset{\sim}{\theta}''^2} - (f')^2 = \text{constant}$$

The matching principle states that

$$f'\langle\infty\rangle = f'^o|_{\underset{\sim}{\theta}=0} = 1$$

Therefore

$$\frac{d^2 f'}{d\underset{\sim}{\theta}''^2} - (f')^2 = -1 \tag{8.79}$$

since $d^2f'/d\underset{\sim}{\theta}''^2$ must vanish as $\underset{\sim}{\theta}'' \to \infty$ in keeping with the requirement $f'\langle\infty\rangle = 1$ (Fig. 4.5). Finally, referring to footnote 17, we have again Eq. (8.77) as the solution of Eq. (8.79) under boundary condition (8.39aA).

On the other hand, for $Re > 0$ (that is, the flow in a divergent channel), although the Euler limit is still $f'^o = 1$, the Prandtl limit becomes

$$\frac{d^2 f'}{d\underset{\sim}{\theta}''^2} + (f')^2 = 1 \tag{8.80}$$

where the matching principle $f'\langle\infty\rangle = 1$ is again incurred. Equation (8.80) differs from Eq. (8.79) by one sign. But this sign difference is all that is needed to upset the apple cart. As a matter of fact,

$$f' = 1 + h\langle\underset{\sim}{\theta}''\rangle$$

where

$$\lim_{\underset{\sim}{\theta}'' \to \infty} h = 0$$

Substituting into Eq. (8.80) and taking the limit as $\underset{\sim}{\theta}'' \to \infty$, we have

$$\frac{d^2 h}{d\underset{\sim}{\theta}''^2} + 2h = 0 \qquad \text{for large } \underset{\sim}{\theta}''$$

from which

$$h = D_1 \sin \sqrt{2}\underset{\sim}{\theta}'' + D_2 \cos \sqrt{2}\underset{\sim}{\theta}''$$

or $f' = 1 + D_1 \sin \sqrt{2}\underset{\sim}{\theta}'' + D_2 \cos \sqrt{2}\underset{\sim}{\theta}''$, for large $\underset{\sim}{\theta}''$. But this result clearly oscillates as θ'' increases, and never approaches the value 1 as speculated. Thus, for the flow in a divergent channel, the boundary-layer structure is absent, even if $Re \to \infty$.

In closing, let us explain why we think so highly of the Jeffery-Hamel flow as to devote three sections to it.

1. It is of some practical value.
2. It is one of a limited number of exact solutions of the Navier-Stokes equations.
3. Exact solutions are hard to come by; when we have one we should squeeze it dry.
4. It is complicated enough to encompass a number of interesting phenomena, and thus yields considerable qualitative information on these phenomena.
5. It exhibits the existence and structure of boundary layers beautifully, thereby serving as a more sophisticated prototype of the boundary-layer theory than the Friedrichs problem of Section 6.1.
6. Since it is an exact solution, it can serve as a test case for various schemes of perturbation and approximation.
7. It demonstrates that much can be learned, even from a complicated problem, without detailed, brute-force computation.[18]
8. It fits into so many things discussed previously that it provides a unifying, complementing, and consolidating touch.

8.6 SMALL-*Re* FLOW PAST A CIRCULAR CYLINDER

Figure 8.24 shows a circular cylinder of radius R_0 fixed in a steady stream with velocity Υ in the direction $\theta = 0$. Using Υ and R_0 as the characteristic speed and length, respectively, we form the Reynolds number $Re = R_0\Upsilon/\nu$. We will investigate here the asymptotic behavior of the flow as $Re \to 0$. As mentioned once before, small Re may also be a troublemaker. Its proper handling can be as difficult as that of large Re. Fortunately, the background against which we view the limits of $Re \to 0$ and $Re \to \infty$ stays the same; and we can rely on our experience with boundary-layer flows to get a number of subtle points settled.

Similarly to Section 5.5, we will use $\tilde{p} = p - p\langle \infty, \pi \rangle$ for convenience. Also as in Section 5.5, we are not very sure about the order of magnitude of $\tilde{p}$ and thus do not quite know what to use for $\tilde{p}^*$. To begin, we will try the obvious choice

$$\tilde{p}^* = \rho \Upsilon^2$$

and check it against the principle of least degeneracy.

[18] We may even state that one can hardly learn anything by brute-force computation *without* first exploring the problem for qualitative information and guidelines.

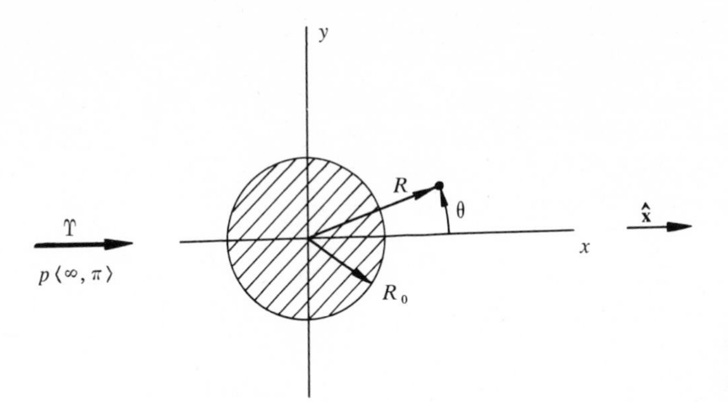

Figure 8.24 Flow past a circular cylinder.

The dimensionless forms of Eqs. (8.24a,b) (for a steady flow and with negligible gravitational effect) are:

$$\begin{cases} q'_R \dfrac{\partial q'_R}{\partial R'} + \dfrac{q'_\theta}{R'}\dfrac{\partial q'_R}{\partial \theta} - \dfrac{q'^2_\theta}{R'} = -\dfrac{\partial \tilde{p}'}{\partial R'} + \dfrac{1}{Re}\left(\dfrac{\partial^2 q'_R}{\partial R'^2} + \dfrac{1}{R'}\dfrac{\partial q'_R}{\partial R'} - \dfrac{q'_R}{R'^2} + \dfrac{1}{R'^2}\dfrac{\partial^2 q'_R}{\partial \theta^2} - \dfrac{2}{R'^2}\dfrac{\partial q'_\theta}{\partial \theta}\right) \\ q'_R \dfrac{\partial q'_\theta}{\partial R'} + \dfrac{q'_\theta}{R'}\dfrac{\partial q'_\theta}{\partial \theta} + \dfrac{q'_R q'_\theta}{R'} = -\dfrac{1}{R'}\dfrac{\partial \tilde{p}'}{\partial \theta} + \dfrac{1}{Re}\left(\dfrac{\partial^2 q'_\theta}{\partial R'^2} + \dfrac{1}{R'}\dfrac{\partial q'_\theta}{\partial R'} - \dfrac{q'_\theta}{R'^2} + \dfrac{1}{R'^2}\dfrac{\partial^2 q'_\theta}{\partial \theta^2} + \dfrac{2}{R'^2}\dfrac{\partial q'_R}{\partial \theta}\right) \end{cases}$$

If we let $Re \to 0$, the groups of terms in the two pairs of parentheses will be equal to zero. These are two partial differential equations for the two unknowns q_R and q_θ. But this would be degenerating the original problem too much, since there is a third equation, namely, the equation of continuity

$$\frac{\partial (R' q'_R)}{\partial R'} + \frac{\partial q'_\theta}{\partial \theta} = 0 \qquad \textbf{(8.16B)}$$

governing the same two unknowns. As it stands, therefore, the limiting process overdetermines the problem. The ordering is thus seen to be improper as far as $\tilde{p}$ is concerned. The remedy lies very clearly in keeping the pressure terms as $Re \to 0$; we will then have three equations for the three unknowns, q_R, q_θ, and $\tilde{p}$, and the degeneracy will be less than before. Furthermore, it is easy to see that physically this means nothing but the hard fact that

$$\tilde{p}' = \frac{\tilde{p}}{\rho \Upsilon^2} \sim O\left(\frac{1}{Re}\right)$$

which was misjudged to be $O(1)$. Then

$$\tilde{p}'' = Re\left(\frac{\tilde{p}}{\rho \Upsilon^2}\right) \sim O(1) \qquad \textbf{(8.81)}$$

that is to say, the proper $\tilde{p}^*$ to use in ordering should have been $\rho\Upsilon^2/Re$. So, again, before we really started on anything, the principle of least degeneracy already revealed one prime fact; namely, the pressure difference in the small-Re flow is going to be quite large compared to $\rho\Upsilon^2$.

In terms of $\tilde{p}''$, the dimensionless forms of Eqs. (8.24a,b) yields, as $Re \to 0$:

$$\begin{cases} \dfrac{\partial^2 q_R'}{\partial R'^2} + \dfrac{1}{R'}\dfrac{\partial q_R'}{\partial R'} - \dfrac{q_R'}{R'^2} + \dfrac{1}{R'^2}\dfrac{\partial^2 q_R'}{\partial \theta^2} - \dfrac{2}{R'^2}\dfrac{\partial q_\theta'}{\partial \theta} = \dfrac{\partial \tilde{p}''}{\partial R'} & \textbf{(8.82a)} \\ \dfrac{\partial^2 q_\theta'}{\partial R'^2} + \dfrac{1}{R'}\dfrac{\partial q_\theta'}{\partial R'} - \dfrac{q_\theta'}{R'^2} + \dfrac{1}{R'^2}\dfrac{\partial^2 q_\theta'}{\partial \theta^2} + \dfrac{2}{R'^2}\dfrac{\partial q_R'}{\partial \theta} = \dfrac{1}{R'}\dfrac{\partial \tilde{p}''}{\partial \theta} & \textbf{(8.82b)} \end{cases}$$

Equations (8.82a,b) are to be solved together with Eq. (8.16B), subject to as many of the following boundary conditions as feasible:

$$R' = 1: \quad q_R', q_\theta' = 0 \qquad \textbf{(8.83a,b)}$$

$$R' \to \infty: \quad \mathbf{q}' \to \hat{\mathbf{x}} \qquad \textbf{(8.83c)}$$

Now, with respect to the boundary conditions, why did we say "as many as feasible"? The highest-order derivatives of Eqs. (8.24a,b) did not get lost as $Re \to 0$. Are we unable to satisfy all the boundary conditions? Well, surprisingly enough, there is trouble below the surface of the matter. True, the trouble is not very obvious, but it is there. We can get some wind of it by noticing that all the nonlinear terms (on the left-hand side) of Eqs. (8.24a,b) are omitted as we arrive at Eqs. (8.82a,b) as $Re \to 0$. Now in the neighborhood of the cylinder, $\mathbf{q}$ is small because of boundary conditions (8.83a,b). The nonlinear terms, being products of q_R, q_θ, and their derivatives, are expected to be negligible in this neighborhood. Equations (8.82a,b) are then valid near the cylinder. However, as one gets away from the cylinder, $\mathbf{q} \to \Upsilon\hat{\mathbf{x}}$; and the nonlinear terms may no longer remain negligible. We therefore expect the solution of Eqs. (8.82a,b) and (8.16B), with boundary conditions (8.83a,b) satisfied, to break down far away from the cylinder; or, in other words, it probably will not satisfy boundary condition (8.83c).

More exactly, what happens may be described as follows. Whenever a problem is posed in a region of infinite extent, one must ask for proper behavior at infinity. But since this proper behavior is quite different in nature from the conditions prescribed on boundaries at finite distances, it is not to be completely forced onto the solution. It is, at least to some extent, not as flexible and as accommodating as an ordinary boundary condition. In our particular problem here, it so happens that the solution to Eqs. (8.82a,b) and (8.16B), subject to boundary conditions (8.83a,b), does not behave properly at infinity.

The above discussion not only indicates the nature of the expected trouble, it also identifies the *seat of trouble* as $R' = \infty$! The limit as $Re \to 0$ with R' fixed is known as the Stokes limit, and the corresponding flow represented by it as the Stokes flow.[19] The Stokes limit or Stokes flow is not expected to satisfy boundary condition (8.83c). In analogy to Chapter 6, we see that Stokes limit is an outer limit (that is, a limit with the unstretched, natural, physical coordinate fixed); is the counterpart of the Euler limit; and, like all outer limits, is valid everywhere

[19] Stokes first investigated this kind of flow in his 1851 paper, cited in footnote 1 of Chapter 4. The interpretation through matched inner and outer limits is from Kaplun, Lagerstrom, Proudman, and Pearson. See footnote 24.

except in the neighborhood of the seat of trouble—that is, $R' = \infty$. Stokes flow is, then, an outer flow valid in the outer region, with the word "outer" referring to "outside of *the neighborhood of* $R' = \infty$."

Once the shock of this "inverted" terminology is over, we can proceed to exploit what we learned from Chapter 6. The Stokes flow must satisfy Eqs. (8.82a,b) and (8.16B), with boundary conditions (8.83a,b). In place of boundary condition (8.83c), a matching principle must be in force to connect the Stokes limit with the corresponding inner limit, which is valid in the neighborhood of the seat of trouble, $R' = \infty$. The inner limit is the result of a "magnified" investigation on the "boundary layer" (or inner region) near the seat of trouble. The "magnification" is done by stretching the coordinate R' near $R' = \infty$ in a proper manner into R'' such that, in taking the limit as $Re \to 0$ with R'' fixed, at least one nonlinear term on the left-hand side of Eqs. (8.24a,b) remains, while degrading the problem as little as feasible. This inner limit will satisfy boundary condition (8.83c) but not (8.83a,b); instead of (8.83a,b), it must satisfy the matching principle. The inner limit[20] here is then the counterpart of the Prandtl limit. It is especially disturbing to the uninitiated to see that the inner flow must be the counterpart of the boundary-layer flow, since there is absolutely no physical resemblance between the inner region near $R' = \infty$ and the boundary layer near a solid wall. In this respect one must always keep in mind that the analogy is mathematical, not physical; it lies in the existence of a seat of trouble, in the stretching of coordinates near the seat of trouble, and in the distinction of two limiting processes. (See Fig. 8.25.)

Once this switch of mental frame of reference is achieved, we can move on to the details and ask exactly how R' is to be stretched so that we can have a closer view of the neighborhood of $R' = \infty$. This is simply done by stretching $1/R'$ around $1/R' = 0$ (that is, $R' = \infty$), into

$$\frac{1}{R''} = \frac{1}{Re}\frac{1}{R'}$$

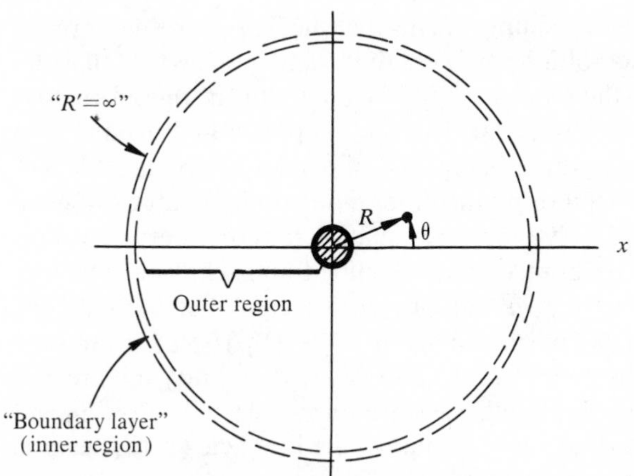

Figure 8.25 "Boundary layer" at infinity.

[20] This inner limit shall remain nameless in this book.

which is equivalent to

$$R'' = Re\,R' \tag{8.84}$$

The stretching factor $1/Re$ is proper since, in terms of R'' and $\tilde{p}'$ (*not* $\tilde{p}''$), Eqs. (8.16A) and (8.24a,b) become

$$\frac{\partial(R''q'_R)}{\partial R''} + \frac{\partial q'_\theta}{\partial \theta} = 0 \tag{8.85}$$

$$q'_R\frac{\partial q'_R}{\partial R''} + \frac{q'_\theta}{R''}\frac{\partial q'_R}{\partial \theta} - \frac{q'^2_\theta}{R''} = -\frac{\partial \tilde{p}'}{\partial R''} + \left(\frac{\partial^2 q'_R}{\partial R''^2} + \frac{1}{R''}\frac{\partial q'_R}{\partial R''} - \frac{q'_R}{R''^2} + \frac{1}{R''^2}\frac{\partial^2 q'_R}{\partial \theta^2} - \frac{2}{R''^2}\frac{\partial q'_\theta}{\partial \theta}\right) \tag{8.86a}$$

$$q'_R\frac{\partial q'_\theta}{\partial R''} + \frac{q'_\theta}{R''}\frac{\partial q'_\theta}{\partial \theta} + \frac{q'_R q'_\theta}{R''} = -\frac{1}{R''}\frac{\partial \tilde{p}'}{\partial \theta} + \left(\frac{\partial^2 q'_\theta}{\partial R''^2} + \frac{1}{R''}\frac{\partial q'_\theta}{\partial R''} - \frac{q'_\theta}{R''^2} + \frac{1}{R''^2}\frac{\partial^2 q'_\theta}{\partial \theta^2} + \frac{2}{R''^2}\frac{\partial q'_R}{\partial \theta}\right) \tag{8.86b}$$

which remain unaltered as $Re \to 0$ with R'' fixed. The inner limit has thus retained the nonlinear terms on the left-hand side of Eqs. (8.24a,b) and has the least degeneracy. As a matter of fact, Eqs. (8.85) and (8.86a,b) are exactly equivalent to the original Eqs. (8.16A) and (8.24a,b); the degeneracy only comes in through the suppression of boundary conditions (8.83a,b). In other words, Eqs. (8.85) and (8.86a,b) are to be solved subject to the boundary condition

$$R'' \to \infty \;\left(\text{or } \frac{1}{R''} \to 0\right):\; \mathbf{q}' \to \hat{\mathbf{x}} \tag{8.83cA}$$

and the requirement that it be matched properly with the outer limit. At this important juncture, we wish also to make the following explanatory remarks:

1. It is interesting to notice that, in terms of $(1/R'')$, the cylinder $R' = 1$ (that is, $R'' = Re$) is just $(1/R'') = \infty$ (that is, $R'' = 0$) in the inner limit; or, in terms of R'', the cylinder shrinks to $R'' = 0$ in the limit. Physically, this simply means that the flow in the inner region (the neighborhood of infinity) *sees* the cylinder as a very small dot, all but shrunk to the point $R'' = 0$. The inner flow is, therefore, to be solved without any boundary conditions prescribed on the cylinder, or without anything being said about its behavior at $R'' = 0$.
2. If we used p'' instead of $\tilde{p}'$, the inner limit would exhibit two equations involving $\tilde{p}''$ only, and one equation (of continuity) involving $\mathbf{q}'$. This would be both underdetermined and with too much degeneracy. The fact that $\tilde{p}'$ must be used means that $\tilde{p}$ is $O(\rho\Upsilon^2)$ in the inner region as contrasted against $\tilde{p} \sim O(\rho\Upsilon^2/Re)$ in the outer region. Physically, this is also quite clear: At infinity the pressure difference would be small, or be comparable to $\rho\Upsilon^2$ at most,[21] because there is hardly any change in the flow; closer to the cylinder the pressure is expected to change very much (comparable to

[21] We shall soon see that this latter is not the case; the pressure hardly changes in the inner region.

$\rho\Upsilon^2/Re$) because, loosely speaking, the *large* viscosity will tend to drag the fluid to a halt by the presence of a fixed solid wall, and thus change the approaching flow drastically.

3. It is not difficult to solve Eqs. (8.85) and (8.86a,b) subject to boundary condition (8.83cA), with (8.83a,b) ignored. As a matter of fact, contrary to the large-*Re* case, here the inner flow is easy to obtain; it is the outer (Stokes) flow that is difficult. To this end, it is better to transform Eqs. (8.85) and (8.86a,b) back to the Cartesian coordinates:

$$\begin{cases} \dfrac{\partial u'}{\partial x''}+\dfrac{\partial v'}{\partial y''}=0 & \textbf{(8.85A)} \\[2ex] u'\dfrac{\partial u'}{\partial x''}+v'\dfrac{\partial u'}{\partial y''}=-\dfrac{\partial \tilde{p}'}{\partial x''}+\left(\dfrac{\partial^2 u'}{\partial x''^2}+\dfrac{\partial^2 u'}{\partial y''^2}\right) & \textbf{(8.86aA)} \\[2ex] u'\dfrac{\partial v'}{\partial x''}+v'\dfrac{\partial v'}{\partial y''}=-\dfrac{\partial \tilde{p}'}{\partial y''}+\left(\dfrac{\partial^2 v'}{\partial x''^2}+\dfrac{\partial^2 v'}{\partial y''^2}\right) & \textbf{(8.86bA)} \end{cases}$$

where $x''=R''\cos\theta$, $y''=R''\sin\theta$. It is then obvious that

$$\tilde{p}'=0,\ u'=1,\ v'=0$$

satisfy Eqs. (8.85A) and (8.86aA,bA) as well as boundary condition (8.83cA). So, the inner flow turns out to be just the uniform approaching flow, which definitely does not satisfy boundary conditions (8.83a,b). In polar coordinates this uniform inner flow, from Eqs. (8.3a,b), is

$$\tilde{p}'=0,\ q_R'=\cos\theta,\ q_\theta'=-\sin\theta \qquad \textbf{(8.87a,b,c)}$$

The task of finding a solution of Eqs. (8.85A) and (8.86aA,bA), subject to boundary conditions (8.83cA) *and* (8.83a,b), is, of course, not easy; it is exactly the difficulty of such a task that forced us into an investigation of the limit $Re \to 0$. We also see that there is no pressure change at all in the inner flow; in other words, $\tilde{p}'$ is actually $o(\rho\Upsilon^2)$, not $\Box(\rho\Upsilon^2)$.

Before we turn to obtain the Stokes flow, we have to clear up one small matter with respect to the name *Oseen.* Although it is out of our set scope to speculate on ways to improve the outer and inner limits, we might mention that the ansatz

$$\mathbf{q}'=\mathbf{q}_0'+\mathbf{q}_1'f_1\langle Re\rangle+O[f_2\langle Re\rangle]$$

$$\tilde{p}'=\tilde{p}_0'+\tilde{p}_1'f_1\langle Re\rangle+O[f_2\langle Re\rangle]$$

where f_1 and f_2 are two properly chosen functions of Re such that

$$\lim_{Re\to 0} f_1\langle Re\rangle=0$$

$$\lim_{Re\to 0}\frac{f_2\langle Re\rangle}{f_1\langle Re\rangle}=0$$

leads to our inner limit again with error $O[f_1\langle Re\rangle]$; that is,

$$\mathbf{q}'_0 = \hat{\mathbf{x}}$$

$$\tilde{p}'_0 = 0$$

To an error of $O[f_2\langle Re\rangle]$, we can substitute

$$\mathbf{q}' = \hat{\mathbf{x}} + \mathbf{q}'_1 f_1\langle Re\rangle$$

$$\tilde{p}' = \tilde{p}'_1 f_1\langle Re\rangle$$

into the original system of governing equations and let $Re \rightarrow 0$. The result turns out to be equivalent to the N-S equations for $\mathbf{q}_1$ with the left-hand side replaced by (in the Cartesian coordinates)

$$\begin{cases} \Upsilon \dfrac{\partial u_1}{\partial x} & \textbf{(a)} \\ \Upsilon \dfrac{\partial v_1}{\partial x} & \textbf{(b)} \end{cases}$$

the equation of continuity for $\mathbf{q}_1$ being of the same form as before. Then $\mathbf{q}_1$ is to be solved subject to the boundary condition

$$R \rightarrow \infty: \ \mathbf{q}_1 = 0$$

and a certain matching condition, not boundary conditions (8.83a,b), provided by a similar investigation in the outer region.

Oseen was the first to obtain $\mathbf{q}_1$ in this way for the small-Re flow past a sphere.[22] It is therefore called the Oseen flow. The corresponding problem of finding the Oseen flow is called the Oseen problem. It is important to notice that Oseen flow is actually the inner limit of $[1/f_1\langle Re\rangle](\mathbf{q}' - \hat{\mathbf{x}})$, while $\hat{\mathbf{x}}$ is the inner limit of $\mathbf{q}'$. Oseen originally proposed the scheme as an improvement over the Stokes scheme, which essentially omitted the left-hand side altogether from the N-S equations; see Eqs. (8.82a,b). He argued that the flow velocity farther away from the cylinder (or sphere, in his case), being $\Upsilon\hat{\mathbf{x}}$, should be partially reflected in the acceleration through the terms (a) and (b). This argument comes to be known as the Oseen linearization because the nonlinear acceleration is represented by the linear terms (a) and (b). (Similarly, dropping the acceleration altogether is called the Stokes linearization.) The solution of the Oseen problem is interesting but difficult. We will not pursue it in this book.

Going back to the Stokes flow, we have to solve now Eqs. (8.16B) and (8.82a,b), subject to boundary conditions (8.83a,b), and to be matched in some fashion with the inner flow, Eqs. (8.87a,b,c). In keeping with the convention of Section 6.2 we will denote the flow quantities in the outer region by superscript o. The equation of continuity, Eq. (8.16B), then suggests a stream function $\psi'^o\langle R', \theta\rangle$

[22] C. W. Oseen, "Über die Stokes'sche Formel, und über eine verwandte Aufgabe in der Hydrodynamik," *Ark. Math. Astronom. Fys.*, **6**, No. 29, 1910.

such that

$$q_R'^o = \frac{1}{R'}\frac{\partial \psi'^o}{\partial \theta} \tag{8.17aA}$$

$$q_\theta'^o = -\frac{\partial \psi'^o}{\partial R'} \tag{8.17bA}$$

Differentiating Eqs. (8.82a,b) with respect to θ and R', respectively, eliminating $\partial^2 \tilde{p}''^o / \partial R' \partial \theta$ by subtracting, and substituting Eqs. (8.17aA,bA) into the result, we have finally

$$\left(\frac{\partial^2}{\partial R'^2} + \frac{1}{R'}\frac{\partial}{\partial R'} + \frac{1}{R'^2}\frac{\partial^2}{\partial \theta^2}\right)^2 \psi'^o = 0 \tag{8.88}$$

Boundary conditions (8.83a,b) become

$$R' = 1: \quad \frac{\partial \psi'^o}{\partial \theta}, \frac{\partial \psi'^o}{\partial R'} = 0 \tag{8.89a,b}$$

Other conditions are to be imposed such that $(q_R'^o, q_\theta'^o)$ or

$$\left(\frac{\partial \psi'^o / \partial \theta}{R'}, -\frac{\partial \psi'^o}{\partial R'}\right)$$

matches in some way with the inner limit $(\cos\theta, -\sin\theta)$. This contemplated matching suggests then a simple ansatz for ψ'^o:

$$\psi'^o = f\langle R' \rangle \sin\theta \tag{8.90}$$

Substituting ansatz (8.90) into Eq. (8.88) and boundary conditions (8.89a,b), we do get a certain linear ordinary differential equation for $f\langle R' \rangle$ subject to the boundary conditions

$$R' = 1: \quad f, \frac{df}{dR'} = 0 \tag{8.91a,b}$$

The solution is easily seen to be

$$f\langle R' \rangle = A_1\left[(1+2A_2)R' \ln R' - \tfrac{1}{2}R' + \frac{1+A_2}{2}\frac{1}{R'} - \frac{A_2}{2}R'^3\right] \tag{8.92}$$

where A_1 and A_2 are two constants of integration to be determined by matching.

The matching principle, Eq. (6.16), of Section 6.1 is too specially trimmed for large-*Re* flows to be of any use here. We have to trace Eq. (6.16) back to some more subtle, more general, and at the same time more fundamental matching principle. It must include Eq. (6.16) as a special case; it must hold for all known exact solutions that exhibit inner-outer limits; it must have some rational background so that future mathematicians may hopefully substantiate or modify it. One such candidate is[23]

$$\text{inner limit of (outer limit)} = \text{outer limit of (inner limit)} \tag{8.93}$$

[23] M. D. Van Dyke, cited in footnote 14 of Chapter 5.

Applying this matching principle to q_R, we have, for the right-hand side,

$$\lim_{\substack{Re\to 0\\(R'\text{ fixed})}} \cos\theta = \cos\theta \tag{8.94}$$

while the left-hand side is

$$\lim_{\substack{Re\to 0\\(R''\text{ fixed})}} \left(\frac{1}{R'}\frac{\partial\psi'^o}{\partial\theta}\right) = \lim_{\substack{Re\to 0\\(R''\text{ fixed})}} \left(\frac{f(R')}{R'}\cos\theta\right)$$

$$= \lim_{\substack{Re\to 0\\(R''\text{ fixed})}} A_1\Bigg[(1+2A_2)(\ln R''-\ln Re)$$

$$-\frac{1}{2}+\frac{1+A_2}{2}\frac{(Re)^2}{R''^2}-\frac{A_2}{2}\frac{R''^2}{(Re)^2}\Bigg]\cos\theta \tag{8.95}$$

$$\sim\begin{cases}-A_1(\ln Re)\cos\theta & \text{if } A_2=0\\ A_3R'^2\cos\theta & \text{if } A_2\neq 0\end{cases}$$

where $A_3=-A_1A_2/2$. The matching is thus possible only if $A_2=0$ together with $A_1=-1/\ln Re$. In other words, the matching principle yields

$$\psi'^o=\frac{1}{\ln Re}\left(\frac{1}{2}R'-R'\ln R'-\frac{1}{2R'}\right)\sin\theta \tag{8.96}$$

Here you should not be disturbed by the fact that the constant A_1 turns out to be dependent on the parameter Re; after all, it is meant to be a constant only as far as R and θ are concerned. All it is trying to tell us is the fact that ψ' is a small quantity of $\Box(1/\ln Re)$ in the outer region, and is approximated by ψ'^o there with an error o $(1/\ln Re)$. (Incidentally, a matching of q_θ would yield the same result.)

Stokes in his 1851 paper[19] did formulate the problem for ψ'^o, arguing that the acceleration terms in the N-S equations should be negligible near the wall since $\mathbf{q}$ is nearly zero there (Stokes linearization). He also realized that, as $R'\to\infty$, ψ'^o will not approach $R'\sin\theta$. Since he believed that ψ'^o must behave this way, he all but abandoned his scheme in the plane flows, and went to work on the small-Re flow past a sphere, which he solved without any mishap.[24]

From Eq. (8.96), one can calculate the drag on the cylinder. The result is (per unit length of cylinder)

$$\text{drag}=(\rho\Upsilon^2R_0)\left(\frac{4\pi}{-Re\ln Re}\right) \tag{8.97}$$

[24] For investigations following our line of thinking and beyond, see S. Kaplun, "Low Reynolds number flow past a circular cylinder," *J. Math. Mech.*, **6**, 595–603, 1957; S. Kaplun and P. A. Lagerstrom, "Asymptotic expansions of Navier-Stokes solution for small Reynolds numbers," *J. Math. Mech.*, **6**, 585–593, 1957; I. Proudman and J. R. A. Pearson, "Expansions at small Reynolds numbers for the flow past a sphere and a circular cylinder," *J. Fluid Mech.*, **2**, 237–262, 1957. The most accurate form of Eq. (8.97) is, as of today,

$$\frac{\text{drag}}{\rho\Upsilon^2R_0}=\frac{4\pi}{Re}[\Delta-0.87\Delta^3+O(\Delta^4)]$$

where $\Delta=1/\ln(3.703/Re)$.

In the literature, you might see this drag formula quoted as

$$\text{drag} = (\rho \Upsilon^2 R_0) \left[\frac{4\pi}{Re \ln (3.703/Re)} \right]$$

But, as $Re \to 0$, these two are the same, since $\ln Re$ will be predominant over $\ln 3.703$. Within our asymptotic framework, it is impossible to obtain this latter, improved formula.[24]

CHAPTER 9

NONPLANE AND AXISYMMETRIC FLOWS

9.1 INTRODUCTION

In going from the plane to the general three-dimensional flows, we will use the previous eight chapters to the greatest advantage. We will go through them once more, and indicate how the essential concepts and important problems can be generalized and/or modified to fit the three-dimensional picture. With Chapters 1 through 8 as the background, we will be able to save a lot of time and effort. You will find that the generalization is usually obvious, and the modification seldom drastic. As a rule, we will not try to substantiate formally these generalizations and modifications. Any dissatisfaction on our part should be interpreted as an inner urge to go on and study Part 3, where everything is taken up at once with unrelenting rigor.

As our starting point, we will go back to Eqs. (1.1a,b,c) where the velocity is described with respect to a rectangular Cartesian coordinate system. We then come to the acceleration, Eq. (1.2). The obvious generalization here for a nonplane flow is

$$\mathbf{a} = \left(\frac{Du}{Dt}, \frac{Dv}{Dt}, \frac{Dw}{Dt}\right) \tag{9.1}$$

where

$$\frac{D}{Dt} = \frac{\partial}{\partial t} + u\frac{\partial}{\partial x} + v\frac{\partial}{\partial y} + w\frac{\partial}{\partial z} \tag{9.2}$$

The equation of continuity for an incompressible fluid in steady *or* transient flow, Eq. (1.9), obviously has a counterpart here

$$\frac{\partial u}{\partial x}+\frac{\partial v}{\partial y}+\frac{\partial w}{\partial z}=0 \tag{9.3}$$

for three-dimensional flows. But the stream function is out, in general.[1] Two exceptions are the plane and the axisymmetric flows. The streamlines defined by the formula

$$(dx, dy, dz)\|(u, v, w)$$

or

$$\frac{dx}{u\langle x, y, z; t\rangle}=\frac{dy}{v\langle x, y, z; t\rangle}=\frac{dz}{w\langle x, y, z; t\rangle} \tag{9.4}$$

are still there as space curves parallel to the local **q**. These streamlines will make up space surfaces known as the stream surfaces. For example, draw any curve in space. Then, passing through every point on the curve, there is a streamline; all these lines form a stream surface (Fig. 9.1). If the curve we draw is a closed one, the stream surface will be tubelike; this surface is called a stream tube (Fig. 9.1).

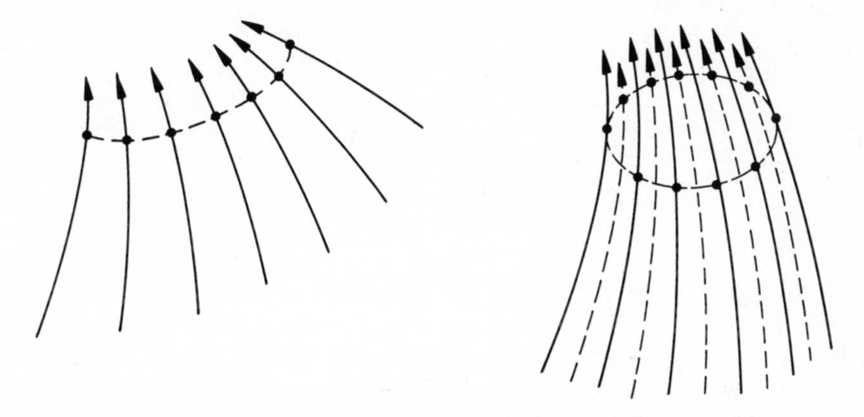

Figure 9.1 Stream surface and stream tube.

As in Section 1.3, a fluid particle in the neighborhood of another particle P in a three-dimensional flow will have a relative velocity

$$\begin{cases} \left.\frac{\partial u}{\partial x}\right|_P dx+\left.\frac{\partial u}{\partial y}\right|_P dy+\left.\frac{\partial u}{\partial z}\right|_P dz & \text{(9.5a)} \\ \left.\frac{\partial v}{\partial x}\right|_P dx+\left.\frac{\partial v}{\partial y}\right|_P dy+\left.\frac{\partial v}{\partial z}\right|_P dz & \text{(9.5b)} \\ \left.\frac{\partial w}{\partial x}\right|_P dx+\left.\frac{\partial w}{\partial y}\right|_P dy+\left.\frac{\partial w}{\partial z}\right|_P dz & \text{(9.5c)} \end{cases}$$

[1] *Two* stream functions are needed at once to yield (u, v, w). Since this duo of stream functions is not particularly useful, we will adopt the attitude that no (single) stream function exists.

The corresponding split-up, as in Eqs. (1.22a,b), can also be easily executed. The first half, or the rotational part, of the split-up will involve three quantities

$$\frac{1}{2}\left(\frac{\partial w}{\partial y}-\frac{\partial v}{\partial z}\right), \frac{1}{2}\left(\frac{\partial u}{\partial z}-\frac{\partial w}{\partial x}\right), \frac{1}{2}\left(\frac{\partial v}{\partial x}-\frac{\partial u}{\partial y}\right)$$

which constitute now a vector representing the average angular velocity of all line segments $\overline{PQ}$ within an infinitesimal sphere of center P. Twice this mean angular velocity is known as the vorticity vector

$$\mathbf{\Omega}\langle x, y, z; t\rangle = \left(\frac{\partial w}{\partial y}-\frac{\partial v}{\partial z}, \frac{\partial u}{\partial z}-\frac{\partial w}{\partial x}, \frac{\partial v}{\partial x}-\frac{\partial u}{\partial y}\right) \tag{9.6}$$

The concept of irrotationality is the same as before.

The second half, or the deformational part, of the split-up now involves six quantities

$$\epsilon_{xx} = \frac{\partial u}{\partial x} \tag{9.7a}$$

$$\epsilon_{yy} = \frac{\partial v}{\partial y} \tag{9.7b}$$

$$\epsilon_{zz} = \frac{\partial w}{\partial z} \tag{9.7c}$$

$$\epsilon_{xy} = \frac{1}{2}\left(\frac{\partial v}{\partial x}+\frac{\partial u}{\partial y}\right) \tag{9.7d}$$

$$\epsilon_{yz} = \frac{1}{2}\left(\frac{\partial w}{\partial y}+\frac{\partial v}{\partial z}\right) \tag{9.7e}$$

$$\epsilon_{xz} = \frac{1}{2}\left(\frac{\partial w}{\partial x}+\frac{\partial u}{\partial z}\right) \tag{9.7f}$$

where ϵ_{xx}, ϵ_{yy}, ϵ_{zz} are three rates of normal strain in the x-, y-, and z-directions, respectively; and ϵ_{xy}, ϵ_{yz}, ϵ_{xz} are the rates of shearing strain by which squares in the xy-, yz-, and xz-planes, respectively, are sheared into rhombi. The physical summary at the end of Section 1.3 also holds.

In Section 2.1 we can accept everything up to Eq. (2.1), if we (a) regard all vectors as three-dimensional, (b) replace ℓ by A, where A is a plane generally inclined in space with normal unit vector $\hat{\mathbf{n}}$, and (c) replace (x, y) by (x, y, z). The equivalence of Eq. (2.5) is clearly

$$\mathbf{f}\langle\hat{\mathbf{n}}\rangle = \begin{cases} \sigma_{xx}\cos\measuredangle(n, x)+\sigma_{yx}\cos\measuredangle(n, y)+\sigma_{zx}\cos\measuredangle(n, z) & \text{(9.8a)} \\ \sigma_{xy}\cos\measuredangle(n, x)+\sigma_{yy}\cos\measuredangle(n, y)+\sigma_{zy}\cos\measuredangle(n, z) & \text{(9.8b)} \\ \sigma_{xz}\cos\measuredangle(n, x)+\sigma_{yz}\cos\measuredangle(n, y)+\sigma_{zz}\cos\measuredangle(n, z) & \text{(9.8c)} \end{cases}$$

where σ_{zx}, and so on, here have *similar* meanings to those of σ_{xy}, and so on, in Section 2.1. In deriving this formula, a three-dimensional element as shown in Figure 9.2 must be used. Equations (2.6a) to (2.8d) are still valid even when the subscript z is introduced

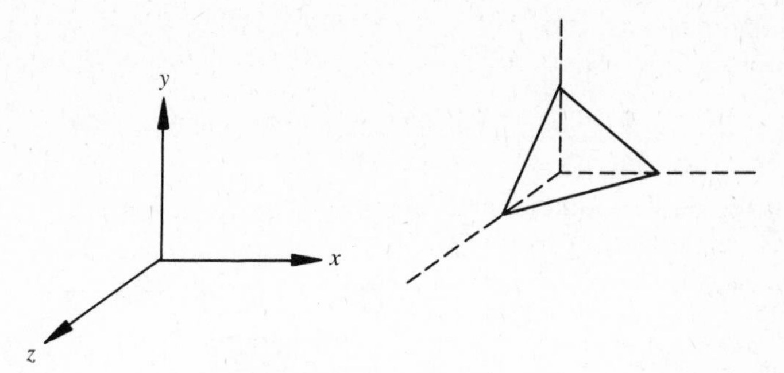

Figure 9.2 A tetrahedron.

The extension of Section 2.2 to the three-dimensional case is too obvious to require any preparation. The generalized results should read

$$\begin{cases} \sigma_{yx} = \sigma_{xy} & \textbf{(9.9a)} \\ \sigma_{zx} = \sigma_{xz} & \textbf{(9.9b)} \\ \sigma_{zy} = \sigma_{yz} & \textbf{(9.9c)} \end{cases}$$

and

$$\begin{cases} \rho \frac{Du}{Dt} = \rho g_x + \left(\frac{\partial \sigma_{xx}}{\partial x} + \frac{\partial \sigma_{xy}}{\partial y} + \frac{\partial \sigma_{xz}}{\partial z} \right) & \textbf{(9.10a)} \\ \rho \frac{Dv}{Dt} = \rho g_y + \left(\frac{\partial \sigma_{xy}}{\partial x} + \frac{\partial \sigma_{yy}}{\partial y} + \frac{\partial \sigma_{yz}}{\partial z} \right) & \textbf{(9.10b)} \\ \rho \frac{Dw}{Dt} = \rho g_z + \left(\frac{\partial \sigma_{xz}}{\partial x} + \frac{\partial \sigma_{yz}}{\partial y} + \frac{\partial \sigma_{zz}}{\partial z} \right) & \textbf{(9.10c)} \end{cases}$$

In treating the Newtonian fluid (Section 2.3), the first required modification is about the average pressure:

$$p = -\tfrac{1}{3}(\sigma_{xx} + \sigma_{yy} + \sigma_{zz}) \qquad \textbf{(9.11)}$$

for any orientation of the x, y, z coordinates. The stress-rate-of-strain relationship, Eqs. (2.14a,b,c), should take on the following generalized form:

$$\sigma_{xx} = -p + 2\mu \frac{\partial u}{\partial x} \qquad \textbf{(9.12a)}$$

$$\sigma_{yy} = -p + 2\mu \frac{\partial v}{\partial y} \qquad \textbf{(9.12b)}$$

$$\sigma_{zz} = -p + 2\mu \frac{\partial w}{\partial z} \qquad \textbf{(9.12c)}$$

$$\sigma_{xy} = \mu \left(\frac{\partial v}{\partial x} + \frac{\partial u}{\partial y} \right) \qquad \textbf{(9.12d)}$$

$$\sigma_{yz} = \mu\left(\frac{\partial w}{\partial y} + \frac{\partial v}{\partial z}\right) \tag{9.12e}$$

$$\sigma_{xz} = \mu\left(\frac{\partial w}{\partial x} + \frac{\partial u}{\partial z}\right) \tag{9.12f}$$

And, the Navier-Stokes equations become

$$\frac{\partial u}{\partial t} + u\frac{\partial u}{\partial x} + v\frac{\partial u}{\partial y} + w\frac{\partial u}{\partial z} = g_x - \frac{1}{\rho}\frac{\partial p}{\partial x} + \nu\left(\frac{\partial^2 u}{\partial x^2} + \frac{\partial^2 u}{\partial y^2} + \frac{\partial^2 u}{\partial z^2}\right) \tag{9.13a}$$

$$\frac{\partial v}{\partial t} + u\frac{\partial v}{\partial x} + v\frac{\partial v}{\partial y} + w\frac{\partial v}{\partial z} = g_y - \frac{1}{\rho}\frac{\partial p}{\partial y} + \nu\left(\frac{\partial^2 v}{\partial x^2} + \frac{\partial^2 v}{\partial y^2} + \frac{\partial^2 v}{\partial z^2}\right) \tag{9.13b}$$

$$\frac{\partial w}{\partial t} + u\frac{\partial w}{\partial x} + v\frac{\partial w}{\partial y} + w\frac{\partial w}{\partial z} = g_z - \frac{1}{\rho}\frac{\partial p}{\partial z} + \nu\left(\frac{\partial^2 w}{\partial x^2} + \frac{\partial^2 w}{\partial y^2} + \frac{\partial^2 w}{\partial z^2}\right) \tag{9.13c}$$

Next, we come to the Reiner-Rivlin fluid (Section 2.4). For a general three-dimensional flow, the stress-rate-of-strain relationship is too messy to quote here. We prefer to quote its specialized forms when needed.

9.2 PSEUDOPLANE, AXISYMMETRIC, AND PSEUDOAXISYMMETRIC FLOWS

There are four important classes of special flows:

1. *Plane flows*
2. *Pseudoplane flows*
3. *Axisymmetric flows*
4. *Pseudoaxisymmetric flows*

As a special case of the three-dimensional flow, the plane flow can be seen to be characterized by the following:[2]

$$\begin{cases} w \equiv 0 \\ \dfrac{\partial(\ )}{\partial z} \equiv 0 \end{cases}$$

In the same vein, we can define the pseudoplane flow as one with

$$\begin{cases} w \equiv 0 \\ \dfrac{\partial(\ )}{\partial z} \equiv 0 \end{cases}$$

As an example, consider the flow past a long, nontapering but swept-back airfoil (Fig. 9.3). In the midspan portion, if we consider planes perpendicular to the

[2] For incompressible fluids this can be liberalized somewhat by allowing p to vary with z. From Eq. (9.13c), this variation of p can only be hydrostatic anyway.

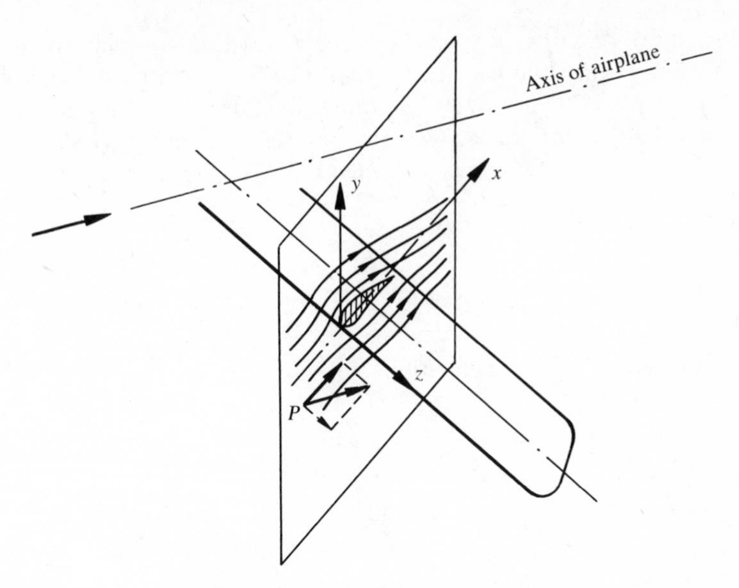

Figure 9.3 A swept-back airfoil.

leading edge, the flow situation would be the same from one plane to another. If one plane is chosen as the representative plane, the spatial flow is obtained if we shift the representative plane perpendicularly to itself. The only difference between this and the plane flow past an airfoil without any sweepback (Fig. 1.6) lies in the flow component out of the representative plane (see point P in Fig. 9.3). This is then a pseudoplane flow, or a plane flow with a nonzero distribution of $w\langle x, y; t\rangle$.

An axisymmetric flow can be defined as one where a representative plane can be recognized such that there is no flow out of this plane, and that the spatial flow is produced by simply rotating this plane about a line in the plane (Fig. 9.4). This line will be called the axis of the flow. Thus, a point on the representative plane is actually a circle around the axis; a fluid particle, a thin-doughnut element; a curve, a surface of revolution around the axis. To pin down the concept of axisymmetry, we will use the axis of the flow as the z-axis, and transform the other

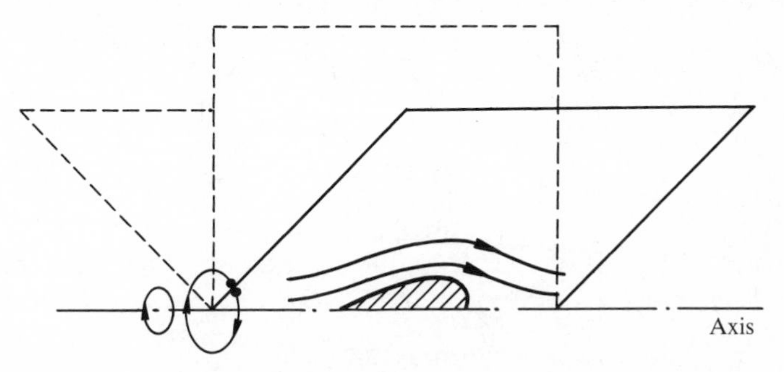

Figure 9.4 Axisymmetric flow.

two Cartesian coordinates (x, y) into the corresponding polar coordinates (R, θ), with $(u, v) \to (q_R, q_\theta)$. Together with z and w, this constitutes a transformation from the rectangular Cartesian to the circular cylindrical coordinates. The details of this transformation *parallel closely* the one that brought the polar coordinates to plane flows (Section 8.1). We will only list the key results here:

(a) Continuity equation

$$\frac{\partial q_R}{\partial R} + \frac{q_R}{R} + \frac{1}{R}\frac{\partial q_\theta}{\partial \theta} + \frac{\partial w}{\partial z} = 0 \tag{9.14}$$

(b) Navier-Stokes equations

$$\frac{\partial q_R}{\partial t} + q_R\frac{\partial q_R}{\partial R} + \frac{q_\theta}{R}\frac{\partial q_R}{\partial \theta} - \frac{q_\theta^2}{R} + w\frac{\partial q_R}{\partial z} = g_R - \frac{1}{\rho}\frac{\partial p}{\partial R} + \nu\left(\frac{\partial^2 q_R}{\partial R^2} + \frac{1}{R}\frac{\partial q_R}{\partial R} - \frac{q_R}{R^2} + \frac{1}{R^2}\frac{\partial^2 q_R}{\partial \theta^2} - \frac{2}{R^2}\frac{\partial q_\theta}{\partial \theta} + \frac{\partial^2 q_R}{\partial z^2}\right) \tag{9.15a}$$

$$\frac{\partial q_\theta}{\partial t} + q_R\frac{\partial q_\theta}{\partial R} + \frac{q_\theta}{R}\frac{\partial q_\theta}{\partial \theta} + \frac{q_R q_\theta}{R} + w\frac{\partial q_\theta}{\partial z} = g_\theta - \frac{1}{\rho R}\frac{\partial p}{\partial \theta} + \nu\left(\frac{\partial^2 q_\theta}{\partial R^2} + \frac{1}{R}\frac{\partial q_\theta}{\partial R} - \frac{q_\theta}{R^2} + \frac{1}{R^2}\frac{\partial^2 q_\theta}{\partial \theta^2} + \frac{2}{R^2}\frac{\partial q_R}{\partial \theta} + \frac{\partial^2 q_\theta}{\partial z^2}\right) \tag{9.15b}$$

$$\frac{\partial w}{\partial t} + q_R\frac{\partial w}{\partial R} + \frac{q_\theta}{R}\frac{\partial w}{\partial \theta} + w\frac{\partial w}{\partial z} = g_z - \frac{1}{\rho}\frac{\partial p}{\partial z} + \nu\left(\frac{\partial^2 w}{\partial R^2} + \frac{1}{R}\frac{\partial w}{\partial R} + \frac{1}{R^2}\frac{\partial^2 w}{\partial \theta^2} + \frac{\partial^2 w}{\partial z^2}\right) \tag{9.15c}$$

(c) Stress-rate-of-strain relationship (Newtonian)

$$\sigma_{RR} = -p + 2\mu\frac{\partial q_R}{\partial R} \tag{9.16a}$$

$$\sigma_{\theta\theta} = -p + 2\mu\left(\frac{1}{R}\frac{\partial q_\theta}{\partial \theta} + \frac{q_R}{R}\right) \tag{9.16b}$$

$$\sigma_{zz} = -p + 2\mu\frac{\partial w}{\partial z} \tag{9.16c}$$

$$\sigma_{R\theta} = \mu\left(\frac{1}{R}\frac{\partial q_R}{\partial \theta} + \frac{\partial q_\theta}{\partial R} - \frac{q_\theta}{R}\right) \tag{9.16d}$$

$$\sigma_{\theta z} = \mu\left(\frac{\partial q_\theta}{\partial z} + \frac{1}{R}\frac{\partial w}{\partial \theta}\right) \tag{9.16e}$$

$$\sigma_{Rz} = \mu\left(\frac{\partial q_R}{\partial z} + \frac{\partial w}{\partial R}\right) \tag{9.16f}$$

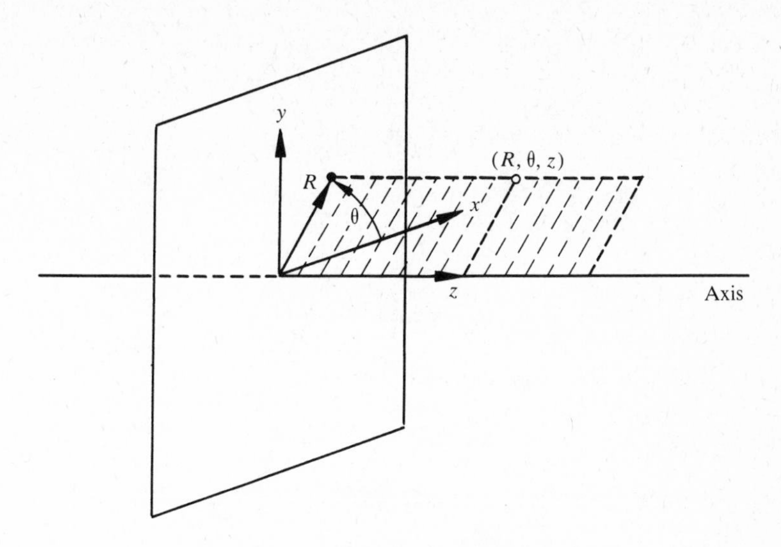

Figure 9.5 Cylindrical coordinates.

An axisymmetric flow is now seen to be characterized by

$$\begin{cases} q_\theta \equiv 0 \\ \dfrac{\partial(\quad)}{\partial\theta} \equiv 0 \end{cases}$$

(*Note*: $\partial g_{\theta,R}/\partial\theta = 0$ is not realistic *unless* $g_{\theta,R} = 0$.)

Referring to a representative plane (Fig. 9.5), the key equations for an axisymmetric flow are:

(a) Equation of continuity

$$\frac{\partial q_R}{\partial R}+\frac{q_R}{R}+\frac{\partial w}{\partial z}=0 \tag{9.17}$$

(b) Navier-Stokes equations

$$\frac{\partial q_R}{\partial t}+q_R\frac{\partial q_R}{\partial R}+w\frac{\partial q_R}{\partial z}=-\frac{1}{\rho}\frac{\partial p}{\partial R}+\nu\left(\frac{\partial^2 q_R}{\partial R^2}+\frac{1}{R}\frac{\partial q_R}{\partial R}-\frac{q_R}{R^2}+\frac{\partial^2 q_R}{\partial z^2}\right) \tag{9.18a}$$

$$\frac{\partial w}{\partial t}+q_R\frac{\partial w}{\partial R}+w\frac{\partial w}{\partial z}=g_z-\frac{1}{\rho}\frac{\partial p}{\partial z}+\nu\left(\frac{\partial^2 w}{\partial R^2}+\frac{1}{R}\frac{\partial w}{\partial R}+\frac{\partial^2 w}{\partial z^2}\right) \tag{9.18b}$$

(c) Stress-rate-of-strain relationship (Newtonian)

$$\sigma_{RR}=-p+2\mu\frac{\partial q_R}{\partial R} \tag{9.19a}$$

$$\sigma_{\theta\theta} = -p + 2\mu\left(\frac{q_R}{R}\right) \quad \textbf{(9.19b)}^3$$

$$\sigma_{zz} = -p + 2\mu\frac{\partial w}{\partial z} \quad \textbf{(9.19c)}$$

$$\sigma_{R\theta} = 0 \quad \textbf{(9.19d)}$$

$$\sigma_{\theta z} = 0 \quad \textbf{(9.19e)}$$

$$\sigma_{Rz} = \mu\left(\frac{\partial q_R}{\partial z} + \frac{\partial w}{\partial R}\right) \quad \textbf{(9.19f)}$$

Equation (9.17) can also be written as

$$\frac{\partial(Rw)}{\partial z} + \frac{\partial(Rq_R)}{\partial R} = 0 \quad \textbf{(9.17A)}$$

which right away suggests the existence of a function $\widehat{\psi}\langle z, R; t\rangle$ such that

$$\begin{cases} Rw = \dfrac{\partial\widehat{\psi}}{\partial R} & \textbf{(9.20a)} \\ Rq_R = -\dfrac{\partial\widehat{\psi}}{\partial z} & \textbf{(9.20b)} \end{cases}$$

or

$$\begin{cases} w = \dfrac{1}{R}\dfrac{\partial\widehat{\psi}}{\partial R} & \textbf{(9.21a)} \\ q_R = -\dfrac{1}{R}\dfrac{\partial\widehat{\psi}}{\partial z} & \textbf{(9.21b)} \end{cases}$$

Equations (9.20a,b) are equivalent to Eq. (1.10a,b) of Section 1.2–(1). The rest of Section 1.2–(1) can also be followed through, replacing u by Rw, v by Rq_R, x by z, y by R, and ψ by $\widehat{\psi}$. For the physical interpretation of $\widehat{\psi}$, a similar journey through Section 1.2–(2) yields the equivalence of Eq. (1.15):

$$\int_{A(z_0,R_0)}^{P(z,R)} (Rw\,dR - Rq_R\,dz) = \frac{1}{2\pi}\int_{A(z_0,R_0)}^{P(z,R)} 2\pi R(w\,dR - q_R\,dz) = \widehat{\psi}\langle z, R; t\rangle \quad \textbf{(9.22)}$$

for any curve connecting A to P in the representative plane (Fig. 9.6). This curve, when rotated about the z-axis by 2π radians, yields a surface of revolution as shown. Obviously, Eq. (9.22) states that $\widehat{\psi}\langle x, y; t\rangle$ equals $(1/2\pi)$ times (the volume flow rate past any surface of revolution whose trace on the representative plane connects the point (z, R) to a chosen datum point (z_0, R_0)); $\widehat{\psi}$ is positive if the

[3] Note that $\sigma_{\theta\theta} \neq -p$, while $\sigma_{zz} = -p$ in a plane flow. Thus, in a plane flow, one can measure the average pressure by *one* gauge with the opening in the representative plane (we did not advertise this practice in Section 2.3 for fear of sidetracking you); but this would not do in an axisymmetric flow. Furthermore, $p = -\frac{1}{2}(\sigma_{xx} + \sigma_{yy})$ in a plane flow; yet, $p = -\frac{1}{3}(\sigma_{RR} + \sigma_{\theta\theta} + \sigma_{zz}) \neq \frac{1}{2}(\sigma_{RR} + \sigma_{zz})$ in an axisymmetric flow. So, ideally, three gauges are needed in measuring p in axisymmetric as well as general three-dimensional flows.

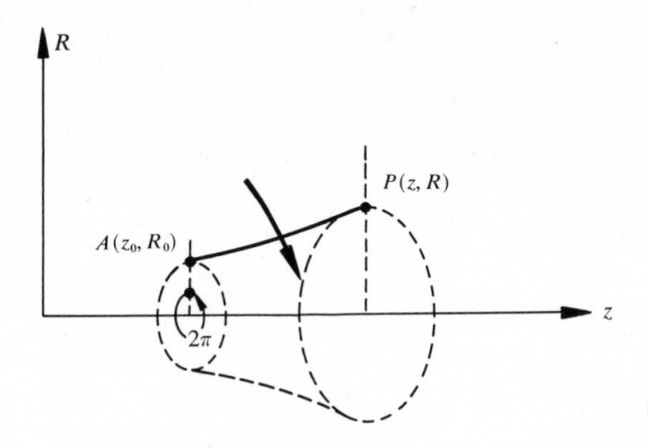

Figure 9.6 A surface of revolution.

general flow direction across the trace on the representative plane is from left to right, when viewed from (z_0, R_0) looking in the general direction of (z, R). Equi-$\widehat{\psi}$ lines in the representative plane are again streamlines, since

$$d\widehat{\psi} = R(w\,dR - q_R\,dz) = 0 \quad \text{leads to} \quad \frac{dR}{dz} = \frac{q_R}{w}$$

In the representative plane, it is sometimes more convenient to use a polar coordinate system (r, φ) in place of the rectangular Cartesian system (z, R); see Figure 9.7. The transformation from (z, R) to (r, φ) is exactly the same as that from (x, y) to (R, θ) in Section 8.1. We have only to replace x by z, y by R,

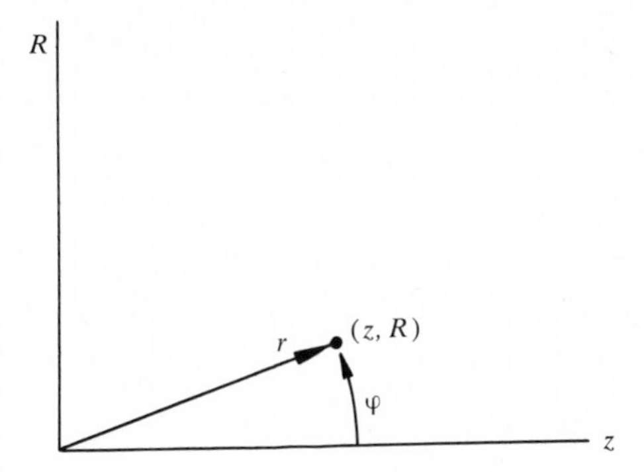

Figure 9.7 Spherical coordinates.

R of Section 8.1 by r, and θ by φ (and u by w, v by q_R). The only minor modification is that $0 \leqslant \varphi \leqslant \pi$ and $R \geqslant 0$ here, while $0 \leqslant \theta < 2\pi$ and $-\infty < y < \infty$ in Section 8.1. However, we might add that the foregoing remark carries us only up to Eqs. (8.14a,b,c,d). We will then be able to work on Eqs. (9.17) through (9.19f) and transform them into (r, φ) form:[4]

(a) Equation of continuity

$$\frac{\partial q_r}{\partial r} + \frac{q_r}{r} + \frac{1}{r}\frac{\partial q_\varphi}{\partial \varphi} + \frac{1}{r \sin\varphi}(q_r \sin\varphi + q_\varphi \cos\varphi) = 0$$

[4] The following is in effect a formulation in the spherical coordinates (r, φ, θ), specialized to axisymmetric flows.

where the first three terms on left-hand side come from the terms $(\partial q_R/\partial R)+(\partial w/\partial z)$ in Eq. (9.17). Rearranging, we have

$$\frac{\partial q_r}{\partial r}+\frac{2q_r}{r}+\frac{1}{r^2\sin\varphi}\frac{\partial(rq_\varphi\sin\varphi)}{\partial\varphi}=0 \quad \textbf{(9.23)}$$

or

$$\frac{\partial(r^2q_r\sin\varphi)}{\partial r}+\frac{\partial(rq_\varphi\sin\varphi)}{\partial\varphi}=0 \quad \textbf{(9.23A)}$$

From Eq. (9.23A), we see that

$$\begin{cases} q_r=\dfrac{1}{r^2\sin\varphi}\dfrac{\partial\widehat{\psi}}{\partial\varphi} & \textbf{(9.24a)}\\[2ex] q_\varphi=-\dfrac{1}{r\sin\varphi}\dfrac{\partial\widehat{\psi}}{\partial r} & \textbf{(9.24b)}\end{cases}$$

where $\widehat{\psi}$ is the same (axisymmetric) stream function as before, only expressed in (r,φ).

(b) Navier-Stokes equations

$$\begin{cases} \dfrac{\partial q_r}{\partial t}+q_r\dfrac{\partial q_r}{\partial r}+\dfrac{q_\varphi}{r}\dfrac{\partial q_r}{\partial\varphi}-\dfrac{q_\varphi{}^2}{r}=g_r-\dfrac{1}{\rho}\dfrac{\partial p}{\partial r}\\ \qquad +\nu\left[\dfrac{1}{r^2}\dfrac{\partial}{\partial r}\left(r^2\dfrac{\partial q_r}{\partial r}\right)+\dfrac{1}{r^2\sin\varphi}\dfrac{\partial}{\partial\varphi}\left(\sin\varphi\dfrac{\partial q_r}{\partial\varphi}\right)\right.\\ \qquad\qquad \left.-\dfrac{2q_r}{r^2}-\dfrac{2}{r^2}\dfrac{\partial q_\varphi}{\partial\varphi}-\dfrac{2q_\varphi\cot\varphi}{r^2}\right] & \textbf{(9.25a)}\\[2ex] \dfrac{\partial q_\varphi}{\partial t}+q_r\dfrac{\partial q_\varphi}{\partial r}+\dfrac{q_\varphi}{r}\dfrac{\partial q_\varphi}{\partial\varphi}+\dfrac{q_rq_\varphi}{r}=g_\varphi-\dfrac{1}{\rho r}\dfrac{\partial p}{\partial\varphi}\\ \qquad +\nu\left[\dfrac{1}{r^2}\dfrac{\partial}{\partial r}\left(r^2\dfrac{\partial q_\varphi}{\partial r}\right)+\dfrac{1}{r^2\sin\varphi}\dfrac{\partial}{\partial\varphi}\left(\sin\varphi\dfrac{\partial q_\varphi}{\partial\varphi}\right)\right.\\ \qquad\qquad \left.+\dfrac{2}{r^2}\dfrac{\partial q_r}{\partial\varphi}-\dfrac{q_\varphi}{r^2\sin^2\varphi}\right] & \textbf{(9.25b)}\end{cases}$$

(c) Stress-rate-of-strain relationship (Newtonian)

$$\begin{cases} \sigma_{rr}=-p+2\mu\dfrac{\partial q_r}{\partial r} & \textbf{(9.26a)}\\[2ex] \sigma_{\varphi\varphi}=-p+2\mu\left(\dfrac{1}{r}\dfrac{\partial q_\varphi}{\partial\varphi}+\dfrac{q_r}{r}\right) & \textbf{(9.26b)}\\[2ex] \sigma_{\theta\theta}=-p+2\mu\left(\dfrac{q_r}{r}+\dfrac{q_\varphi\cot\varphi}{r}\right) & \textbf{(9.26c)}\\[2ex] \sigma_{r\varphi}=\mu\left[r\dfrac{\partial}{\partial r}\left(\dfrac{q_\varphi}{r}\right)+\dfrac{1}{r}\dfrac{\partial q_r}{\partial\varphi}\right] & \textbf{(9.26d)}\\[2ex] \sigma_{\varphi\theta}=0 & \textbf{(9.26e)}\\[1ex] \sigma_{r\theta}=0 & \textbf{(9.26f)}\end{cases}$$

Finally, we define a pseudoaxisymmetric flow by

$$\begin{cases} q_\theta \not\equiv 0 \\ \dfrac{\partial(\ )}{\partial\theta} \equiv 0 \end{cases}$$

It is obviously just an axisymmetric flow plus a flow $q_\theta\langle z,R;t\rangle$ out of the representative plane. The key equations for a pseudoaxisymmetric flow are:

(a) Equation of continuity

$$\frac{\partial q_R}{\partial R}+\frac{q_R}{R}+\frac{\partial w}{\partial z}=0 \tag{9.27}$$

(b) Navier-Stokes equations

$$\frac{\partial q_R}{\partial t}+q_R\frac{\partial q_R}{\partial R}+w\frac{\partial q_R}{\partial z}-\frac{q_\theta^{\,2}}{R}=-\frac{1}{\rho}\frac{\partial p}{\partial R}+\nu\left(\frac{\partial^2 q_R}{\partial R^2}+\frac{1}{R}\frac{\partial q_R}{\partial R}-\frac{q_R}{R^2}+\frac{\partial^2 q_R}{\partial z^2}\right) \tag{9.28a}$$

$$\frac{\partial q_\theta}{\partial t}+q_R\frac{\partial q_\theta}{\partial R}+w\frac{\partial q_\theta}{\partial z}+\frac{q_R q_\theta}{R}=\nu\left(\frac{\partial^2 q_\theta}{\partial R^2}+\frac{1}{R}\frac{\partial q_\theta}{\partial R}-\frac{q_\theta}{R^2}+\frac{\partial^2 q_\theta}{\partial z^2}\right) \tag{9.28b}$$

$$\frac{\partial w}{\partial t}+q_R\frac{\partial w}{\partial R}+w\frac{\partial w}{\partial z}=g_z-\frac{1}{\rho}\frac{\partial p}{\partial z}+\nu\left(\frac{\partial^2 w}{\partial R^2}+\frac{1}{R}\frac{\partial w}{\partial R}+\frac{\partial^2 w}{\partial z^2}\right) \tag{9.28c}$$

(c) Stress-rate-of-strain relationship (Newtonian)

$$\sigma_{RR}=-p+2\mu\frac{\partial q_R}{\partial R} \tag{9.29a}$$

$$\sigma_{\theta\theta}=-p+2\mu\left(\frac{q_R}{R}\right) \tag{9.29b}$$

$$\sigma_{zz}=-p+2\mu\frac{\partial w}{\partial z} \tag{9.29c}$$

$$\sigma_{R\theta}=\mu R\frac{\partial}{\partial R}\left(\frac{q_\theta}{R}\right) \tag{9.29d}$$

$$\sigma_{\theta z}=\mu\frac{\partial q_\theta}{\partial z} \tag{9.29e}$$

$$\sigma_{Rz}=\mu\left(\frac{\partial q_R}{\partial z}+\frac{\partial w}{\partial R}\right) \tag{9.29f}$$

We note that Eq. (9.27) is identical with Eq. (9.17). This makes the stream function $\widehat{\psi}\langle z, R; t\rangle$ here mathematically and physically the same as that for the axisymmetric flows. The reason behind this is the fact that q_θ represents a flow *around* a surface of revolution and does not enter the definition of $\widehat{\psi}$. The streamlines, however, are now space curves spiraling around the axis, and are not directly related to the equi-$\widehat{\psi}$ lines in the representative plane.

9.3 POISEUILLE FLOW[5]

Consider the steady flow through a long, straight pipe with circular cross section (Fig. 9.8). Gravitational effects are regarded as negligible. The boundary condition that $\mathbf{q}\langle R_0, \theta, z\rangle = 0$ for any θ and z suggests the ansatz

$$\begin{cases} \dfrac{\partial(\ )}{\partial\theta} = 0 \text{ with } q_\theta \equiv 0 \text{ (axisymmetric)} \\ \\ \dfrac{\partial \mathbf{q}}{\partial z} = 0 \text{ (fully developed)} \end{cases}$$

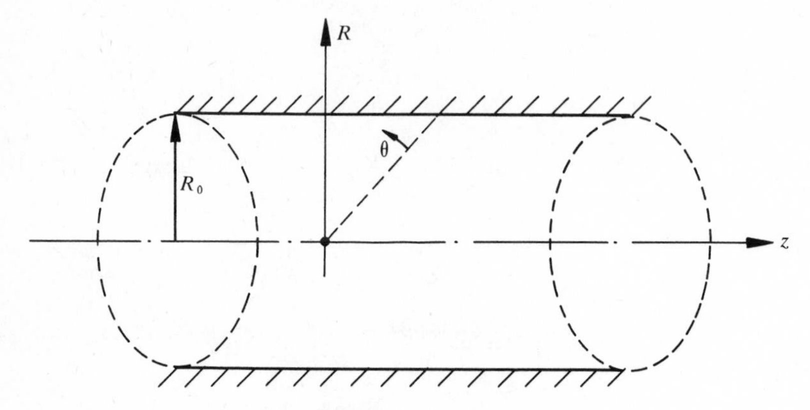

Figure 9.8 A circular pipe.

Equation (9.17A) then demands that

$$\frac{d\,(Rq_R)}{dR} = 0$$

that is,

$$Rq_R = \text{constant}$$

But $q_R = 0$ at $R = R_0$; so the above constant must be zero, leaving

$$q_R = 0 \quad \text{(everywhere)}$$

Equation (9.18a) then yields

$$\frac{\partial p}{\partial R} = 0 \quad \text{or } p = p\langle z\rangle \text{ only}$$

and Eq. (9.18b) reads

$$\frac{dp}{dz} = \mu\left(\frac{d^2w}{dR^2} + \frac{1}{R}\frac{dw}{dR}\right)$$

---

[5] See footnote 4 in Chapter 3.

which leads to the unmistakable conclusion that

$$\frac{dp}{dz} = F = \frac{\Delta p}{L}, \text{ a constant}$$

where $\Delta p = p_2 - p_1$ is the pressure difference across two measuring stations with $\Delta z = z_2 - z_1 = L$; see Eq. (3.7). Therefore,

$$\frac{d^2 w}{dR^2} + \frac{1}{R}\frac{dw}{dR} = \frac{F}{\mu}$$

This ordinary differential equation is to be solved under the boundary condition

$$R = R_0 \colon w = 0$$

The result is

$$w = -\frac{F}{4\mu}(R_0^2 - R^2) \tag{9.30}$$

A simple integration shows that the volume flow rate through the pipe is

$$Q = \frac{\pi R_0^4}{8\mu}(-F) = \frac{\pi R_0^4}{8\mu}\left(\frac{p_1 - p_2}{L}\right) \tag{9.31}$$

Equation (9.31) is the famous Hagen-Poiseuille efflux formula from which we can calculate the average velocity

$$\bar{w} = \frac{Q}{\pi R_0^2} = \frac{R_0^2}{8\mu}\left(-\frac{\Delta p}{L}\right) \tag{9.32}$$

If we define the Euler number (with $D = 2R_0$ denoting the pipe diameter) as

$$Eu = \left(-\frac{\Delta p}{L}\right)\frac{D}{\rho \bar{w}^2}$$

we have

$$Eu = \frac{32}{Re} \tag{9.33}$$

where $Re = \rho \bar{w} D/\mu$ is the Reynolds number.

For the sake of comparison, we will treat in the following the same kind of flow for a Reiner-Rivlin fluid.[6]

We start again with the same ansatz of axisymmetry and fully developed flow, which also leads to $q_R \equiv 0$ since the equation of continuity, Eq. (9.17A), holds also for a non-Newtonian fluid.

In a flow field with

$$\mathbf{q} = [0, 0, w\langle R \rangle]$$

[6] J. B. Serrin, "Poiseuille and Couette flow of non-Newtonian fluids," *ZAMM*, **39**, 295–299, 1959.

the stress-rate-of-strain relationship becomes[7]

$$\begin{cases} \sigma_{RR} = -p + \frac{1}{6}\eta \text{II} & \textbf{(9.33a)} \\ \sigma_{\theta\theta} = -p - \frac{1}{3}\eta \text{II} & \textbf{(9.33b)} \\ \sigma_{zz} = -p + \frac{1}{6}\eta \text{II} & \textbf{(9.33c)} \\ \sigma_{R\theta} = 0 & \textbf{(9.33d)} \\ \sigma_{\theta z} = 0 & \textbf{(9.33e)} \\ \sigma_{Rz} = \mu \left(\dfrac{dw}{dR}\right) & \textbf{(9.33f)} \end{cases}$$

where μ and η are functions of II, which is given by

$$\text{II} = \frac{1}{2}\left(\frac{dw}{dR}\right)^2 \qquad \textbf{(9.34)}$$

For the equations of motion, we could first transform Eqs. (9.10a,b,c) into (R, θ, z) coordinates and then specialize to the case at hand. But this would be too roundabout, and without any side benefits. Actually, as you will show in one of the Problems, it is much easier to derive the equations of motion for the Poiseuille flow *ad hoc*. The resulting equations express only the force balance, since the fluid particles are not accelerating:

$$\begin{cases} 0 = \dfrac{1}{R}\dfrac{\partial}{\partial R}(R\sigma_{RR}) - \dfrac{\sigma_{\theta\theta}}{R} & \textbf{(9.35a)} \\ 0 = \dfrac{\partial \sigma_{zz}}{\partial z} + \dfrac{1}{R}\dfrac{\partial}{\partial R}(R\sigma_{Rz}) & \textbf{(9.35b)} \end{cases}$$

Substituting Eqs. (9.33a,b,c,f) into Eqs. (9.35a,b), we have

$$\begin{cases} \dfrac{\partial p}{\partial R} = \dfrac{1}{6}\dfrac{1}{R}\dfrac{d}{dR}(R\eta \text{II}) + \dfrac{\eta \text{II}}{3R} & \textbf{(9.36a)} \\ \dfrac{\partial p}{\partial z} = \dfrac{1}{R}\dfrac{d}{dR}\left(R\mu\dfrac{dw}{dR}\right) & \textbf{(9.36b)} \end{cases}$$

Differentiating Eq. (9.36a) with respect to z yields

$$\frac{\partial^2 p}{\partial R\, \partial z} = 0$$

Therefore

$$\frac{\partial p}{\partial z} = fct\langle z \rangle, \text{ only}$$

Then, the left-hand side of Eq. (9.36b) would be a function of z alone, while the

[7] As implied at the end of Section 9.1, there *is* in fact a general stress-rate-of-strain relationship, too cumbersome to quote here, which reduces to the present simple form when the particular flow situation is substituted. (To write down the general relation, some abbreviating devices, such as dyadics or tensors, are needed.)

right-hand side is a function of R alone. So, both sides must be constant; that is,

$$\frac{\partial p}{\partial z} = F = \frac{\Delta p}{L} \tag{9.37}$$

where $\Delta p = p_2 - p_1$ is measured across[8] $L = z_2 - z_1$ at a given R, say R_0; and,

$$\frac{1}{R}\frac{d}{dR}\left(R\mu\frac{dw}{dR}\right) = F \tag{9.38}$$

Integrating Eq. (9.38) once, we have

$$\mu\frac{dw}{dR} = \frac{1}{2}FR \tag{9.39}$$

To go further from here, we must first know μ as a function of $(dw/dR)^2$, or of $|dw/dR|$. This makes Eq. (9.39) nonlinear unless μ = constant.

For any μ-function, Eq. (9.37) always yields

$$p = Fz + f\langle R\rangle \tag{9.40}$$

where $f\langle R\rangle$ is to be determined through Eq. (9.36a):

$$\begin{aligned}\frac{df}{dR} &= \frac{1}{6}\frac{1}{R}\frac{d}{dR}(R\eta\text{II}) + \frac{\eta\text{II}}{3R}\\ &= \frac{1}{2}\frac{1}{R}\frac{d}{dR}(R\eta\text{II}) - \frac{1}{3}\frac{d}{dR}(\eta\text{II})\end{aligned}$$

that is,

$$f\langle R\rangle = \frac{1}{2}\int\frac{1}{R}\frac{d}{dR}(R\eta\text{II})\,dR - \frac{1}{3}\eta\text{II} \tag{9.41}$$

where η is a given function of $\text{II} = \frac{1}{2}(dw/dR)^2$. Furthermore, for any μ-function, the volume flow rate is

$$\begin{aligned}Q &= \int_0^{R_0} w(2\pi R)\,dR\\ &= \pi\int_0^{R_0} wd(R^2)\end{aligned}$$

which, through an integration by parts, yields

$$\begin{aligned}Q &= \pi\left\{(wR^2)\Big|_0^{R_0} - \int_0^{R_0}\left(\frac{dw}{dR}\right)R^2\,dR\right\}\\ &= \frac{\pi(-F)}{2}\int_0^{R_0}\frac{R^3}{\mu}\,dR\end{aligned} \tag{9.42}$$

where the fluid is assumed to satisfy the no-slip condition, and where Eq. (9.39) is used.

---

[8] The measurement of average pressure is actually very difficult, if not impossible. But luckily, from Eq. (9.33a), $\Delta p = -\Delta\sigma_{RR}$ at a fixed R; and the measurement of a normal stress can be easily carried out.

It is of extreme interest and importance here to note that, if $\mu =$ constant, Eq. (9.42) yields exactly the Hagen-Poiseuille efflux formula, Eq. (9.31). But $\mu =$ constant does not mean that the fluid is Newtonian! A Newtonian fluid must have *both* $\mu =$ constant and $\eta = 0$. This is just what we had in mind when we stated in footnote 4 of Chapter 3 that "other relations may give you the same formula for the volume flow rate." Thus, a Reiner-Rivlin fluid with constant μ is indistinguishable from a Newtonian fluid in the measurement of flow rate through a straight, circular pipe.

Furthermore, when $\mu =$ constant, Eq. (9.39) can be solved, subject to the boundary condition $w\langle R_0\rangle = 0$, to yield exactly Eq. (9.30). Therefore, a comparison of the velocity profiles across a cross section is of no help either. To decide whether a fluid with constant μ is Newtonian or not, we must go to Eq. (9.40), which reads (if η as well as μ is a constant)

$$p = Fz + \frac{5}{96}\frac{\eta F^2}{\mu^2}R^2 + \text{constant} \tag{9.43}$$

Equation (9.43) shows that $p = p\langle z, R\rangle$ unless $\eta = 0$. So if the fluid is non-Newtonian, there will be a change of pressure radially. As to how this radial change in the average pressure is to be detected in an experimental setup, we will leave the worry to the rheologists.[9]

9.4 THE ROTATING DISK

In our review of Chapters 1 through 8 with an eye on generalizations, we have come now to the steady stagnation-point flow (Section 4.2). The modification needed here in going from the plane to the axisymmetric case is so obvious and simple that we have put it into the Problems.[10] In its place, we will discuss the case where the plane $z = 0$, which represents physically a large (infinite) disk, is rotating about the z-axis with speed ω_0. The fluid above the disk (that is, $z > 0$), which would otherwise be at rest, is dragged into a *pseudoaxisymmetric* flow with $\partial/\partial\theta = 0$, but a $q_\theta\langle R, z\rangle$ out of the representative plane (Fig. 9.9). We will treat only the steady case (again with gravitation neglected). The key equations here are Eqs. (9.27) and (9.28a,b,c) with $\partial/\partial t$- and g-terms omitted. These equations are to be solved subject to the boundary conditions

$$z = 0: \quad q_R = 0,\ q_\theta = R\omega_0,\ w = 0 \tag{9.44a,b,c}$$

---

[9] It is customary in the literature to use the Poiseuille flow to explain a fantastic (to a Newtonian-fluid dynamicist) phenomenon—namely, the swelling (or shrinking) in diameter of a jet of non-Newtonian fluid leaving a straight, circular pipe. Similarly, the Couette flow (Section 8.2) is used to explain the climbing (and falling) of a free surface of a non-Newtonian fluid, in a manner amazingly different from that of a Newtonian fluid (for example, it *climbs* up an inner cylinder at rest when the outer cylinder rotates; it also climbs up a very thin inner cylinder rotating in a large outer cylinder at rest). But the explanation seems to stretch the validity of the Poiseuille or Couette flow too much.

[10] To be exact, only the axisymmetric flow should be called the stagnation-*point* flow; the plane case is actually the stagnation-*line* flow since the point in the representative plane is really a line perpendicular to the plane.

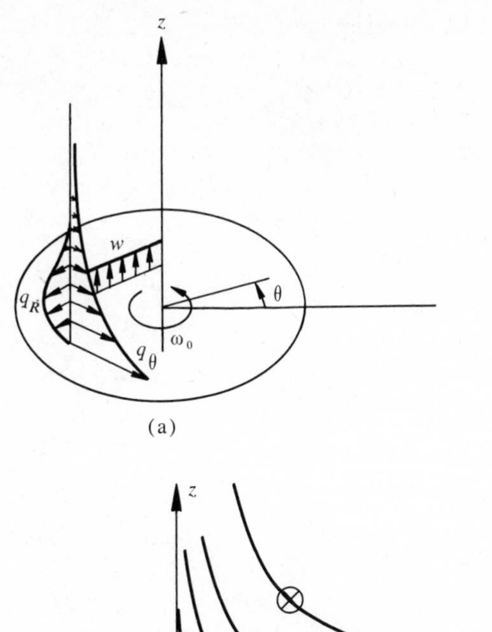

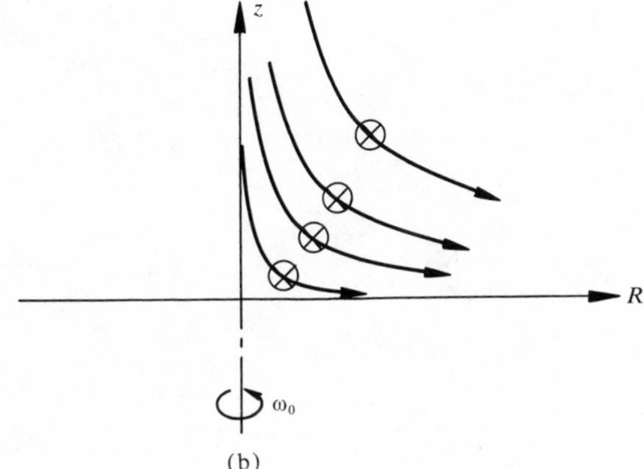

Figure 9.9 A rotating disk.

$$z = \infty: \quad q_R,\ q_\theta, \text{ and } w \text{ behave properly} \tag{9.45a,b,c}$$

Introducing the ansatz[11]

$$w = k\langle z\rangle \tag{9.46}$$

we have, from Eq. (9.27),

$$\frac{1}{R}\frac{\partial}{\partial R}(Rq_R) = -\frac{dk}{dz}$$

Therefore

$$q_R = -\frac{R}{2}\frac{dk}{dz} + \frac{c_1\langle z\rangle}{R}$$

---

[11] Boundary condition (9.44c) may seem to suggest an ansatz $w \equiv 0$; however, it does not work. $w = k\langle z\rangle$ is then the next best thing; it was first tried by Batchelor; see G. K. Batchelor, "Note on a class of solutions of the Navier-Stokes equations representing steady nonrotationally symmetric flow," *Quart. J. Mech. Appl. Math.*, **4**, 29–41, 1951. G. K. Batchelor is professor of applied mathematics at Cambridge University.

But q_R must remain finite at $R=0$ for any z; therefore

$$c_1\langle z\rangle \equiv 0$$

Then

$$q_R = R\left(-\frac{1}{2}\frac{dk}{dz}\right)$$
$$= Rf\langle z\rangle \tag{9.47}$$

where

$$f = -\frac{1}{2}\frac{dk}{dz} \tag{9.48}$$

Substituting into Eq. (9.28c), we get

$$\frac{1}{\rho}\frac{\partial p}{\partial z} = \nu\frac{d^2k}{dz^2} - k\left(\frac{dk}{dz}\right)$$

and thus

$$\frac{p}{\rho} = \nu\frac{dk}{dz} - \frac{k^2}{2} + P\langle R\rangle \tag{9.49}$$

Substituting into Eq. (9.28a), we have

$$\frac{1}{R}\frac{dP}{dR} - \frac{q_\theta^2}{R^2} = -f^2 - k\left(\frac{df}{dz}\right) + \nu\frac{d^2f}{dz^2}$$
$$= F\langle z\rangle \tag{9.50}$$

At $z=0$, $q_\theta = R\omega_0$; therefore

$$\frac{dP}{dR} - R\omega_0^2 = RF\langle 0\rangle$$
$$= Rc_2$$

where c_2 stands for $F\langle 0\rangle$. Therefore,

$$P\langle R\rangle = \tfrac{1}{2}(\omega_0^2 + c_2)R^2 + c_3 \tag{9.51}$$

Furthermore,

$$F\langle z\rangle = \frac{1}{R}\frac{dP}{dR} - \frac{q_\theta^2}{R^2}$$
$$= (\omega_0^2 + c_2) - \frac{q_\theta^2}{R^2}$$

Thus, q_θ^2/R^2 or q_θ/R is seen to be a function of z only; that is,

$$q_\theta = RG\langle z\rangle \tag{9.52}$$

Now, substituting back into Eq. (9.50), we obtain

$$\omega_0^2 + c_2 - (G\langle z\rangle)^2 = -f^2 - k\frac{df}{dz} + \nu\frac{d^2f}{dz^2} \tag{9.53}$$

The proper behavior of f and k at ∞ requires that

$$\lim_{z\to\infty} \frac{df}{dz}, \frac{d^2f}{dz^2}, \frac{dk}{dz} = 0$$

From Eq. (9.48), we have

$$\lim_{z\to\infty} f = 0$$

Thus, Eq. (9.53) evaluated at $z = \infty$ yields

$$\omega_0{}^2 + c_2 = (G\langle\infty\rangle)^2 = \omega_\infty{}^2$$

If $G\langle\infty\rangle \neq 0$, the fluid at $z = \infty$ must be rotating with an angular speed ω_∞; see Eq. (9.52). If the fluid at $z = \infty$ is not being rotated, $G\langle\infty\rangle = \omega_\infty = 0$. Then, $\omega_0{}^2 + c_2 = 0$. Equation (9.53) thus becomes

$$-(G\langle z\rangle)^2 = \nu\frac{d^2f}{dz^2} - k\frac{df}{dz} - f^2 \tag{9.53A}$$

Equations (9.49) and (9.51) together yield[12]

$$\frac{p}{\rho} = \nu\frac{dk}{dz} - \frac{k^2}{2} + c_3 \tag{9.54}$$

that is,

$$\frac{p}{\rho} = fct\langle z\rangle, \text{ only} \tag{9.54A}$$

We can now make a short summary by introducing the nondimensional quantities:

$$f' = \frac{f}{\omega_0} \tag{9.55a}$$

$$k' = \frac{k}{\sqrt{\nu\omega_0}} \tag{9.55b}$$

$$G' = \frac{G}{\omega_0} \tag{9.55c}$$

$$p' = \frac{p}{\rho\nu\omega_0} \tag{9.55d}$$

$$z' = \frac{z}{\sqrt{\nu/\omega_0}} \tag{9.55e}$$

The problem is then to solve the system of (nonlinear) ordinary differential

[12] Von Kármán, who first posed this important problem of rotating disk, introduced originally the four ansatz, Eqs. (9.46), (9.47), (9.52), and (9.54A), all at once with one stroke of genius. We, of course, prefer the step-by-step interpretation of Batchelor.

equations:

$$\left\{\begin{aligned}
&\frac{dk'}{dz'}+2f'=0 && \textbf{(9.56a)}\\
&\frac{d^2f'}{dz'^2}-k'\frac{df'}{dz'}-f'^2=-G'^2 && \textbf{(9.56b)}\\
&\frac{d^2G'}{dz'^2}-k'\frac{dG'}{dz'}-2f'G'=0 && \textbf{(9.56c)}\\
&p'=\frac{dk'}{dz'}-\frac{k'^2}{2}+c_3' && \textbf{(9.56d)}
\end{aligned}\right.$$

where $c_3' = c_3/\nu\omega_0$, subject to the boundary conditions

$$z'=0:\ f'=0,\ G'=1,\ k'=0 \qquad \textbf{(9.57a, b, c)}$$

and the conditions

$$\lim_{z'\to\infty}\ f', G', \frac{df'}{dz'}, \frac{dG'}{dz'}, \frac{d^2f'}{dz'^2}, \ldots = 0 \qquad \textbf{(9.58a, b, ...)}$$

Please pay special attention to the fact that there is no condition to be enforced on k' at infinity. On the one hand, only the first-order derivative of k' appears in Eqs. (9.56a, b, c, d); so one boundary condition on k' is enough. On the other hand, $k'\langle\infty\rangle$ actually comes out of the final solution of the problem; whether it vanishes or not is inherently determined by the problem.

That the ansatz (9.46) really works was shown by Cochran,[13] who actually solved the system of ordinary differential equations with the specified conditions. The result is shown in Figure 9.10. It is seen that $k' < 0$ (that is, $w < 0$). So the axial flow is actually *toward* the disk. Furthermore,

$$k'\langle\infty\rangle = -0.886 \neq 0$$

That is to say, the fluid at infinity is *not* really at rest, but is moving toward the disk with a velocity

$$w\langle\infty\rangle = -0.886\sqrt{\nu\omega_0} \qquad \textbf{(9.59)}$$

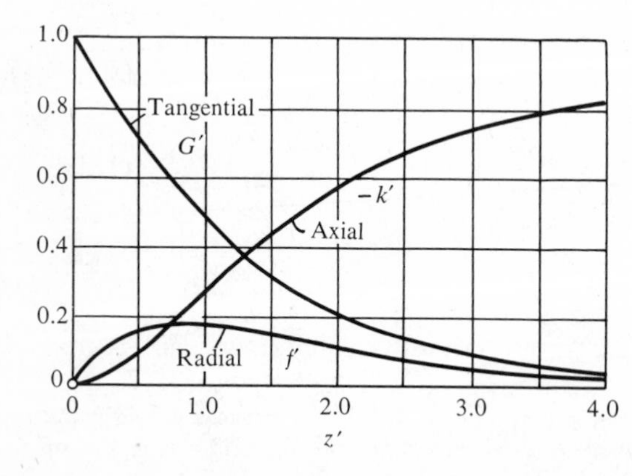

Figure 9.10 Solution of the rotating-disk problem.

[13] W. G. Cochran, "The flow due to a rotating disc," *Proc. Cambridge Phil. Soc.*, **30**, 365–375, 1934.

This axial flow at infinity is physically necessary since the rotating disk is throwing fluid out radially, and there must be an inflow to balance it.

9.5 BOUNDARY LAYER ON A YAWED FLAT WALL

As a three-dimensional generalization or modification of Chapter 6, we consider again Figure 6.6; but this time with a constant velocity component W in the z-direction (perpendicular to and coming out of the representative plane) superimposed on the approaching stream Υ. In other words (Fig. 9.11), the wedge-shaped body is yawed with respect to the total approaching stream, so the leading edge of the body shows a swept-back angle of β degrees. The resulting flow is obviously pseudoplane with $\partial/\partial z \equiv 0$ and $w = w\langle x, y\rangle$.

For a steady pseudoplane flow (neglecting gravity), the governing equations can be easily established from Eqs. (9.3) and (9.13a, b, c):

$$\frac{\partial u}{\partial x}+\frac{\partial v}{\partial y}=0 \tag{1.9}$$

$$\begin{cases} u\dfrac{\partial u}{\partial x}+v\dfrac{\partial u}{\partial y}=-\dfrac{1}{\rho}\dfrac{\partial p}{\partial x}+\nu\left(\dfrac{\partial^2 u}{\partial x^2}+\dfrac{\partial^2 u}{\partial y^2}\right) & \textbf{(9.60a)} \\[2ex] u\dfrac{\partial v}{\partial x}+v\dfrac{\partial v}{\partial y}=-\dfrac{1}{\rho}\dfrac{\partial p}{\partial y}+\nu\left(\dfrac{\partial^2 v}{\partial x^2}+\dfrac{\partial^2 v}{\partial y^2}\right) & \textbf{(9.60b)} \\[2ex] u\dfrac{\partial w}{\partial x}+v\dfrac{\partial w}{\partial y}=-\dfrac{1}{\rho}\dfrac{\partial p}{\partial z}+\nu\left(\dfrac{\partial^2 w}{\partial x^2}+\dfrac{\partial^2 w}{\partial y^2}\right) & \textbf{(9.60c)} \end{cases}$$

The same nondimensionalization (that is, using Υ) and arguments as in Section 6.2 yield again two limiting flows: (a) outer flow and (b) inner or boundary-layer flow.

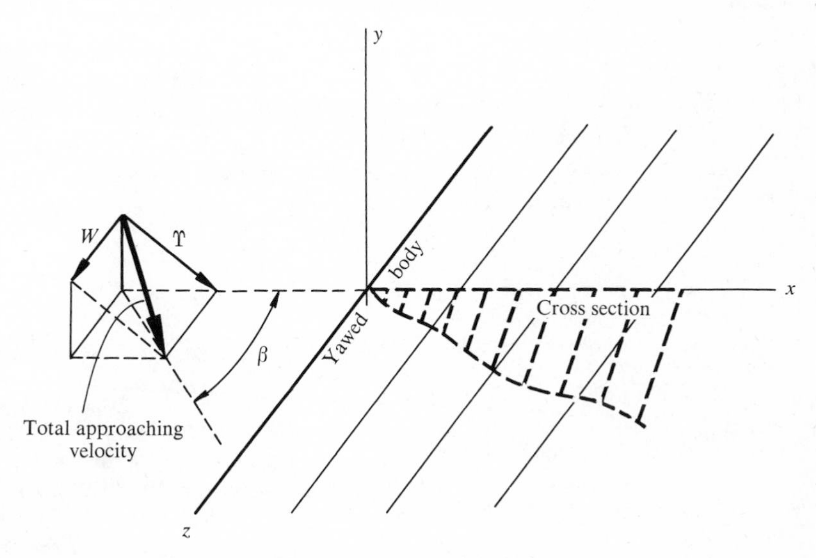

Fig. 9.11 A yawed body.

For the outer flow, Eq. (9.60c) becomes

$$u'^o\frac{\partial w'^o}{\partial x'}+v'^o\frac{\partial w'^o}{\partial y'}=-\frac{1}{\rho}\frac{\partial p'^o}{\partial z'}$$

which has the obvious solution

$$\begin{cases} w'^o=\dfrac{W}{\Upsilon}\ (=\text{constant}) \\ p'^0=fct\langle x',y'\rangle,\ \text{only} \end{cases}$$

while Eqs. (1.9) and (9.60a,b) with appropriate conditions become again the system quoted at the very beginning of Section 6.3, *independently* of w'^o. In other words, the effect of yaw in the outer region is only the superposition of a constant velocity component W out of the representative plane, while the situation in the representative plane stays exactly the same as without the yaw. Thus, all of Section 6.3 can be taken over without change. The fact that the flow in the representative plane is independent of that out of the plane is known as the *principle of independence.*[14]

For the boundary-layer flow, Eq. (9.60c) becomes

$$u'\frac{\partial w'}{\partial x'}+v''\frac{\partial w'}{\partial y''}=\frac{\partial^2 w'}{\partial y''^2} \tag{9.61}$$

subject to the no-slip condition on the wall, and the matching condition

$$w'\langle x',\infty\rangle=\frac{W}{\Upsilon}$$

while Eqs. (1.9) and (9.60a,b) with appropriate conditions become exactly the system quoted at the very beginning of Section 7.1. It is seen, then, that the boundary-layer flow in the representative plane is also independent of that out of the plane, and the principle of independence again holds. Therefore Chapter 7 can be employed verbatim. Once the situation in the representative plane is settled, one can proceed to solve[15] for w' with the aid of the known u' and v''. To this end, we may point out that Eq. (9.61) is relatively easy to handle since it is linear in w'. In the special case of a yawed (semiinfinite) flat plate, the boundary-layer (Blasius) flow in the representative plane satisfies the equation

$$u'\frac{\partial u'}{\partial x'}+v''\frac{\partial u'}{\partial y''}=\frac{\partial^2 u'}{\partial y''^2} \tag{6.28A}$$

[14] W. R. Sears, "Boundary layers in three-dimensional flow," *Appl. Mech. Rev.*, 7, 281–285, 1954. W. R. Sears is director of the Center of Applied Mathematics at Cornell University.

[15] The boundary-layer flow past a yawed body was first formulated by Prandtl in 1945; see L. Prandtl, "Über Reibungsschichten bei drei-dimensionalen Strömungen," *Betz 60th Birthday Anniversary Volume*, 134–141, 1945. But it was also attacked independently by W. R. Sears, V. V. Struminsky, and R. T. Jones during 1946–1948, due probably to the postwar difficulty in communication. Of these, W. R. Sears, "The boundary layer of yawed cylinders," *J. Aero. Sci.*, **15**, 49–52, 1948, is the most comprehensive.

subject to the conditions

$$u' = 0 \text{ on the plate}$$
$$u' = 1 \text{ at } y'' = \infty$$

Comparing this with Eq. (9.61) subject to the corresponding conditions, it is obvious that

$$w' = u'\left(\frac{W}{\Upsilon}\right) \tag{9.62}$$

satisfies Eq. (9.61) and all the conditions. This means two things: (1) Once u' is solved, w' is known at once. (2) w and u in the boundary layer bear the same ratio W/Υ as in the approaching stream (or, as in the outer flow); thus, the spatial boundary-layer flow past a yawed flat plate is in the same direction as the approaching stream, being deflected only vertically. These happy events do not occur for other yawed bodies, nor would they happen in other, more general three-dimensional flows.

9.6 AXISYMMETRIC JET[16]

In this section we will discuss a jet flow that is the axisymmetric counterpart of the one shown in Figure 7.13. With Section 7.5 as our guide, we can afford to be more brief here, omitting many explanations, which are essentially the same as those included in Section 7.5.

Figure 9.12 shows a narrow axisymmetric jet issuing from a small opening into the same fluid at rest. The arguments used in Section 7.5 to lead to the boundary-layer system can also be used here, with obvious modifications (mainly, changing x, y, u, v, into z, R, w, q_R, respectively; and extending the range of R to $-\infty < R < \infty$ in the matters of symmetry and antisymmetry about $R = 0$). The resulting boundary-layer system here is established easily by taking the inner

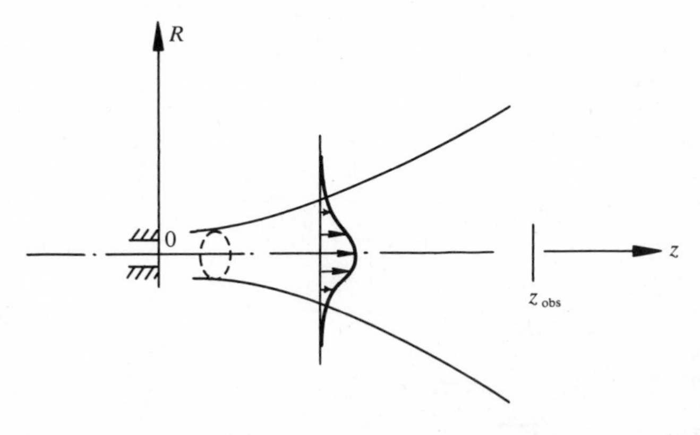

Figure 9.12 An axisymmetric jet.

[16] H. Schlichting, "Laminare Strahlausbreitung," *ZAMM*, **13**, 260–263, 1933.

limit of Eqs. (9.17A) and (9.18a,b):

$$\begin{cases} \dfrac{\partial(R''w')}{\partial z'}+\dfrac{\partial(R''q''_R)}{\partial R''}=0 & \text{(9.63)}\\[2ex] w'\dfrac{\partial w'}{\partial z'}+q''_R\dfrac{\partial w'}{\partial R''}=\dfrac{1}{R''}\dfrac{\partial}{\partial R''}\left(R''\dfrac{\partial w'}{\partial R''}\right) & \text{(9.64)}\\[2ex] z'>0,\ R''=0:\quad \dfrac{\partial w'}{\partial R''}=0 & \text{(9.65a)}\\[2ex] \qquad q''_R=0 & \text{(9.65b)}\\[1ex] R''=\infty:\quad w'=0 & \text{(9.66)} \end{cases}$$

In solving this system we again abandon the trivial solution w', $q''_R=0$. To search for similarity solutions, we will deviate from Section 7.5 a little by first introducing the stream function $\widehat{\psi}'$ such that

$$\begin{cases} w'=\dfrac{1}{R''}\dfrac{\partial\widehat{\psi}'}{\partial R''} & \text{(9.21a)}\\[2ex] q''_R=-\dfrac{1}{R''}\dfrac{\partial\widehat{\psi}'}{\partial z'} & \text{(9.21b)} \end{cases}$$

We then try the ansatz

$$\widehat{\psi}'=z'^{m}f\langle\eta\rangle \tag{9.67}$$

where

$$\eta=\frac{R''}{z'^{n}} \tag{9.68}$$

Equation (9.63) is now automatically satisfied. Without further ado, let us state that the ansatz turns out to be compatible with all the conditions. Furthermore, substituting Eq. (9.21a,b) into Eq. (9.64) shows that the ansatz will make it an ordinary differential equation while keeping $\int_0^\infty w''^2R''\,dR''$ constant, if

$$m=n=1$$

Thus

$$\widehat{\psi}'=z'f\left\langle\frac{R''}{z'}\right\rangle$$

and

$$\begin{cases} w'=\dfrac{df}{d\eta}\Big/\eta z' & \text{(9.69a)}\\[2ex] q''_R=\left(\dfrac{df}{d\eta}-\dfrac{f}{\eta}\right)\Big/z' & \text{(9.69b)} \end{cases}$$

The corresponding ordinary differential equation is

$$f\left(\frac{df}{d\eta}\right)\Big/\eta^2-\left(\frac{df}{d\eta}\right)^2\Big/\eta-f\left(\frac{d^2f}{d\eta^2}\right)\Big/\eta=\frac{d}{d\eta}\left[\frac{d^2f}{d\eta^2}-\left(\frac{df}{d\eta}\right)\Big/\eta\right] \tag{9.70}$$

The conditions to be satisfied by f must now be determined. First, from Eq. (9.69a), we must have

$$\lim_{\eta \to 0} \left(\frac{df}{d\eta} \Big/ \eta \right) = w'\langle 0, z' \rangle z' \tag{9.71}$$

which must remain finite for any given z'. This is possible only if

$$\eta = 0: \quad \frac{df}{d\eta} = 0 \tag{9.71A}$$

is satisfied. On the other hand, boundary condition (9.65a) demands that

$$\lim_{\eta \to 0} \left(\eta \frac{d^2 f}{d\eta^2} - \frac{df}{d\eta} \right) \Big/ \eta^2 z'^2 = 0$$

that is,

$$\lim_{\eta \to 0} \left(\eta \frac{d^2 f}{d\eta^2} - \frac{df}{d\eta} \right) = 0$$

which yields, in view of condition (9.71),

$$\lim_{\eta \to 0} \frac{d^2 f}{d\eta^2} = w'\langle 0, z' \rangle z'$$

Since $w'\langle 0, z' \rangle z'$, or $(df/d\eta)/\eta$ evaluated at $\eta = 0$, comes out of the final solution, boundary condition (9.65a) does not yield any condition on f. Its place is taken up by (9.71A). $d^2f/d\eta^2$ at $\eta = 0$ turns out to be a part of the final result, in contrast to the plane case; see boundary condition (7.39a). For the rest of the boundary conditions, we see that, with (9.71A) in force, (9.65b) demands that[17]

$$\eta = 0: \quad f = 0 \tag{9.72}$$

Condition (9.66) leads to $df/d\eta < \infty$ at $\eta = \infty$. Then, a *finite* $R''q_R''$ at $R'' = \infty$ (or, a finite mass-flux dragged into the jet) dictates that

$$\eta = \infty: \quad \frac{df}{d\eta} = 0 \tag{9.73}$$

Now, to solve Eq. (9.70), we notice that the left-hand side is just

$$-d\left[f\left(\frac{df}{d\eta}\right) \Big/ \eta \right] \Big/ d\eta$$

So it can be integrated once to yield

$$f\left(\frac{df}{d\eta}\right) = \frac{df}{d\eta} - \eta \frac{d^2 f}{d\eta^2} \tag{9.74}$$

---

[17] To be more precise, we should require

$$\eta = 0: \quad \frac{f}{\eta} = 0$$

that is, $f \sim o(\eta)$ as $\eta \to 0$.

where boundary conditions (9.71A) and (9.72) are used to argue that the constant of integration must vanish.[18] Multiplying Eq. (9.74) by η, we have the equivalent equation

$$\eta^2 \frac{d^2 f}{d\eta^2} - \eta \frac{df}{d\eta} = -\eta f\left(\frac{df}{d\eta}\right) \tag{9.75}$$

The left-hand side looks like terms in the famous Euler (homogeneous) equation,[19] and should suggest to you the equally famous Euler's transformation:

$$\eta = e^{\xi} \tag{9.76}$$

Then,

$$\eta \frac{df}{d\eta} = \frac{df}{d\xi}$$

$$\eta^2 \frac{d^2 f}{d\eta^2} = \frac{d^2 f}{d\xi^2} - \frac{df}{d\xi}$$

Substituting these into Eq. (9.75), we have

$$\frac{d^2 f}{d\xi^2} + (f-2)\frac{df}{d\xi} = 0 \tag{9.77}$$

Writing F for $df/d\xi$, we would get

$$\frac{d^2 f}{d\xi^2} = \frac{dF}{d\xi} = \frac{dF}{df}\left(\frac{df}{d\xi}\right) = F\left(\frac{dF}{df}\right)$$

Equation (9.77) can then be written as

$$F\frac{dF}{df} + F(f-2) = 0$$

which leads to either $F = 0$ or

$$\frac{dF}{df} = 2 - f \tag{9.78}$$

$F = 0$ is obviously the trivial solution of no flow whatsoever, and is therefore abandoned, whereas Eq. (9.78) is easily integrated:

$$F = 2f - \frac{f^2}{2} + C_1$$

or

$$\frac{df}{d\xi} = 2f - \frac{f^2}{2} + C_1$$

In terms of η this becomes

$$\eta\frac{df}{d\eta} = 2f - \frac{f^2}{2} + C_1$$

---

[18] The fact that $d^2f/d\eta^2$ stays finite at $\eta = 0$ for any finite z' is also implicitly used.

[19] See any elementary text on ordinary differential equations. (Some authors associate it with the name Bernoulli.) It is always difficult to name something after the originator if he is someone as versatile as Euler (or Bernoulli); there are many Euler's equations.

By boundary conditions (9.71A) and (9.72), the constant C_1 must vanish. Therefore

$$\frac{df}{f(f-4)} = -\frac{1}{2}\frac{d\eta}{\eta}$$

or

$$\frac{df}{f} - \frac{df}{f-4} = 2\frac{d\eta}{\eta}$$

This equation integrates easily and the result is

$$\frac{f}{(f-4)\eta^2} = \text{constant, } C_2$$

that is,

$$f = \frac{-4C_2\eta^2}{1-C_2\eta^2}$$

This is usually written with C_2 replaced[16] by $-\alpha^2/4$:

$$f = \frac{(\alpha\eta)^2}{1+\frac{1}{4}(\alpha\eta)^2} \qquad \textbf{(9.79)}$$

where the constant α is still to be determined. It is to be noted that Eq. (9.79) automatically satisfies boundary condition (9.73). The determination of α is similar to that in Section 7.5, and is left as one of the Problems.

The plane wall jet of Figure 7.15 also has an axisymmetric counterpart, which can be produced by rotating Fig. 7.15 about the y-axis, and which is known as a radial jet. For details, see Glauert's paper, cited in footnote 11 of Chapter 7.

9.7 SMALL-*Re* FLOW PAST A SPHERE

Our scope does not allow us to touch all bases in this chapter, but we certainly cannot leave it without presenting the axisymmetric counterpart of Section 8.6.

Figure 9.13 shows a sphere of radius r_0 fixed in a steady stream with

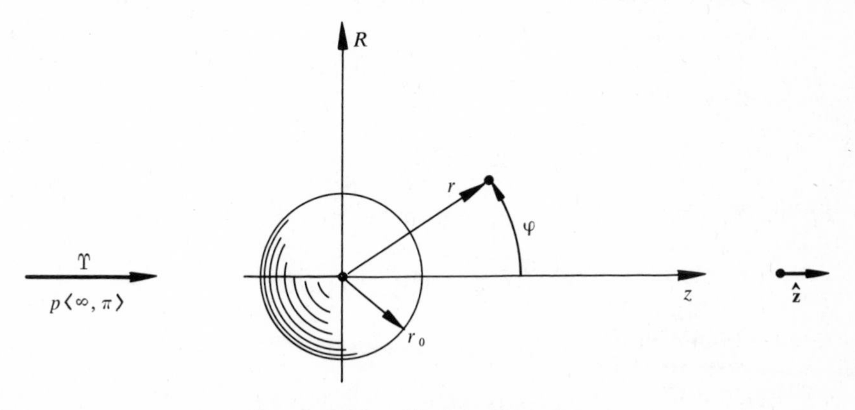

Figure 9.13 Flow past a sphere.

velocity Υ in the direction of $\varphi = 0$. With $Re = r_0\Upsilon/\nu$, we let $Re \to 0$. The discussion in Section 8.6 that led to Eq. (8.88) should be restudied with R replaced by r, θ by φ, x by z, y by R, Eqs. (8.24a,b) by Eqs. (9.25a,b), Eq. (8.16B) by Eq. (9.23A), and with other associated modifications.

The inner limit near $p'' = \infty$ is again the uniform approaching flow in the z-direction; that is,

$$\tilde{p}' = 0, w' = 1, q_R' = 0$$

or in the (r, φ) coordinates,

$$\tilde{p}' = 0, q_r' = \cos\varphi, q_\varphi' = -\sin\varphi \tag{9.80a,b,c}$$

The corresponding Stokes limit or flow turns out to be governed by the following set of equations:

$$\frac{\partial(r'^2 q_r'^o \sin\varphi)}{\partial r'} + \frac{\partial(r' q_\varphi'^o \sin\varphi)}{\partial\varphi} = 0 \tag{9.23B}$$

$$\frac{1}{r'^2}\frac{\partial}{\partial r'}\left(r'^2\frac{\partial q_r'^o}{\partial r'}\right) + \frac{1}{r'^2\sin\varphi}\frac{\partial}{\partial\varphi}\left(\sin\varphi\frac{\partial q_r'^o}{\partial\varphi}\right) - \frac{2q_r'^o}{r'^2} - \frac{2}{r'^2}\frac{\partial q_\varphi'^o}{\partial\varphi} - \frac{2q_\varphi'^o\cot\varphi}{r'^2} = \frac{\partial\tilde{p}''^o}{\partial r'} \tag{9.25aA}$$

$$\frac{1}{r'^2}\frac{\partial}{\partial r'}\left(r'^2\frac{\partial q_\varphi'^o}{\partial r'}\right) + \frac{1}{r'^2\sin\varphi}\frac{\partial}{\partial\varphi}\left(\sin\varphi\frac{\partial q_\varphi'^o}{\partial\varphi}\right) + \frac{2}{r'^2}\frac{\partial q_r'^o}{\partial\varphi} - \frac{q_\varphi'^o}{r'^2\sin^2\varphi} = \frac{1}{r'}\frac{\partial\tilde{p}''^o}{\partial\varphi} \tag{9.25bA}$$

subject to the boundary conditions

$$r' = 1: \quad q_r'^o, q_\varphi'^o = 0 \tag{9.81a,b}$$

and the matching condition

$$\lim_{\substack{Re\to 0\\(r''\text{ fixed})}} q_r'^o = \cos\varphi \tag{9.82}$$

(or, the similar one for q_φ). Equation (9.23B) is, of course, automatically satisfied when we introduce a stream function $\widetilde{\psi}'^o$ such that

$$q_r'^o = \frac{1}{r'^2\sin\varphi}\frac{\partial\widetilde{\psi}'^o}{\partial\varphi} \tag{9.24aA}$$

$$q_\varphi'^o = -\frac{1}{r'\sin\varphi}\frac{\partial\widetilde{\psi}'^o}{\partial r'} \tag{9.24bA}$$

Substituting these into Eqs. (9.25aA,bA), and eliminating $\tilde{p}''^o$, we have

$$\left[\frac{\partial^2}{\partial r'^2} + \frac{\sin\varphi}{r'^2}\frac{\partial}{\partial\varphi}\left(\frac{1}{\sin\varphi}\frac{\partial}{\partial\varphi}\right)\right]^2\widetilde{\psi}'^o = 0 \tag{9.83}$$

subject to the boundary conditions

$$r' = 1: \quad \frac{\partial\widetilde{\psi}'^o}{\partial\varphi}, \frac{\partial\widetilde{\psi}'^o}{\partial r'} = 0 \tag{9.81aA,bA}$$

and the matching condition

$$\lim_{\substack{Re\to 0\\(r''\text{ fixed})}} \left(\frac{1}{r'^2}\frac{\partial\widehat{\psi}'^o}{\partial\varphi}\right) = \cos\varphi\sin\varphi \qquad \textbf{(9.82A)}$$

which suggests the ansatz

$$\widehat{\psi}'^o = f\langle r'\rangle \sin^2\varphi \qquad \textbf{(9.84)}$$

Substituting this ansatz into Eq. (9.83) leads to a certain linear ordinary differential equation, subject to the boundary conditions

$$r' = 1:\quad f, \frac{df}{dr'} = 0 \qquad \textbf{(9.85a,b)}$$

The solution is easily seen to be

$$f\langle r'\rangle = \frac{A}{r'} + Br' + Cr'^2 + Dr'^4 \qquad \textbf{(9.86)}$$

where

$$\begin{cases} A = \frac{1}{2}C + 2D \\ B = -\frac{3}{2}C - \frac{5}{2}D \end{cases}$$

The matching condition then requires that

$$\lim_{\substack{Re\to 0\\(r''\text{ fixed})}} \frac{f\langle r'\rangle}{r'^2} = \frac{1}{2} \qquad \textbf{(9.87)}$$

Now

$$\lim_{\substack{Re\to 0\\(r''\text{ fixed})}} \frac{f\langle r'\rangle}{r'^2} = \lim_{\substack{Re\to 0\\(r''\text{ fixed})}} \left[\frac{A(Re)^3}{r''^3} + \frac{BRe}{r''} + C + \frac{Dr''^2}{(Re)^2}\right]$$

$$\sim \begin{cases} Dr'^2, & \text{if } D \neq 0 \\ C, & \text{if } D = 0 \end{cases}$$

Thus, a matching is possible only if

$$D = 0,\ C = \tfrac{1}{2}$$

That is to say, the matching principle yields

$$\widehat{\psi}'^o = \left(\frac{1}{4r'} - \frac{3r'}{4} + \frac{r'^2}{2}\right)\sin^2\varphi \qquad \textbf{(9.88)}$$

This is the famous Stokes solution; but our approach here follows in spirit the 1957 paper by Kaplun and Lagerstrom,[20] cited in footnote 24 of Chapter 8. The reason that Stokes succeeded in solving the flow past a sphere as $Re \to 0$, within the classical framework of Stokes linearization, is as follows:

From Eq. (9.88), it is easily seen that

$$\lim_{r'\to\infty} (q_r'^o, q_\varphi'^o) = (\cos\varphi, -\sin\varphi)$$

---

[20] See also Proudman and Pearson, cited in footnote 24 of Chapter 8.

Thus it so happens that the Stokes solution approaches the uniform oncoming flow as $r \to \infty$, which fact was actually imposed by Stokes as a condition to be satisfied. This happy coincidence accounted for Stokes' success; he was not so lucky with the flow past a cylinder.

The drag formula based on Eq. (9.88) can be established as

$$\text{drag} = \rho \Upsilon^2 r_0^2 \left(\frac{6\pi}{Re}\right) \tag{9.89}$$

This is known as Stokes drag formula[21] and was used by R. Millikan in his determination of the charge of the electron (oil-drop experiment).

The success of Stokes linearization in the case of the sphere should not lead you to doubt that there is trouble in the asymptotics of $Re \to 0$ with the seat of trouble located at infinity. The trouble does exist; only it is so weak for the flow past a sphere that it does not show up in the inner and outer limits. In an attempt to improve on the Stokes flow, the trouble will appear;[22] it can be cured only by proper matching with the corresponding improvement on the inner limit.

[21] Today this formula is refined to read

$$\text{drag} = \rho \Upsilon^2 r_0^2 \left(\frac{6\pi}{Re}\right)\left\{1 + \frac{3}{8} Re + \frac{9}{40} Re^2 \left[\ln Re + 0.5772 + \frac{5}{3} \ln 2 - \frac{323}{360}\right] + \frac{27}{80} Re^3 \ln Re + O(Re^3)\right\}$$

See W. Chester and D. R. Breach, "On the flow past a sphere at low Reynolds number," *J. Fluid Mech.*, **37**, 751–758, 1969.

[22] A. N. Whitehead, "Several approximations to viscous fluid motion," *Quart. J. Math.*, **23**, 143–152, 1889. Whitehead, however, gave a wrong explanation of the cause of the trouble.

CHAPTER 10

INTRODUCTION TO TURBULENT FLOWS

10.1 RANDOM NATURE OF TURBULENCE

Every elementary text of fluid mechanics quotes routinely the classical experiment in which Reynolds[1] demonstrated that the flow in a circular pipe is *laminar* for smaller values of $Re = \Upsilon R_0/\nu$ (where Υ is the average flow speed and R_0 is the pipe radius), but becomes *turbulent* when Re is large. We have no intention here of describing this justly famous experiment again. Instead we will go further and ask what we mean *precisely* when we refer to a flow as being *turbulent* or *laminar*.[2]

To answer this question, let us perform a "thought experiment." We will consider, just to fix ideas, the flat Couette flow of footnote 3, Chapter 8; that is, the flow between $y = 0$ and Y, without pumping and gravity, subject to the following boundary conditions on $\mathbf{q}\langle x, y, z; t\rangle$:

$$y = 0: \quad \mathbf{q} = 0 \tag{10.1a}$$

$$y = Y: \quad \mathbf{q} = U\hat{\mathbf{x}} \tag{10.1b}$$

---

[1] See his 1883 paper, cited in footnote 5 of Chapter 5.

[2] We will not ask, in this book, how and why a laminar flow becomes turbulent. The transition from a laminar to a turbulent flow is too special a topic to be embraced in our scope. The first step toward its explanation seems to be the instability of a laminar flow against very small disturbances (a linear study); the next step seems to be that against larger disturbances (a nonlinear study). If interested, you must enter the field of *hydrodynamic stability*.

Imagine now that we have an *enormous* number N of identical setups with the same Y and U, and using the same fluid. There are also N experimentalists who measure the velocity distribution $\mathbf{q}\langle x, y, z; t\rangle$ for each of these setups, where (x, y, z) are rectangular Cartesian coordinates attached in the same way to each of the setups. If all the measured velocity distributions turn out to be the same, we say that the flat Couette flow for that fluid, with the specified values of Y and U, is laminar. We are then able to tell, with certainty, the flow in any of the N setups. The flow is thus completely *determined* by the given data; that is, a *laminar flow is deterministic* in nature. On the other hand, if the measured velocity distributions differ among the N setups, we say that the flow is *turbulent.* We will then never be sure about the flow in a setup picked out at random; any of the N velocity distributions *may* occur there, but none will *definitely* occur. Thus, a *turbulent flow is stochastic or random* in nature, and can be described only in terms of its mean properties and fluctuations of properties. To be more precise, the N velocity distributions can be added up, and divided by N, to form a mean velocity, $\overline{\mathbf{q}}\langle x, y, z; t\rangle$. Then, in any one setup, we must have

$$\mathbf{q}\langle x, y, z; t\rangle = \overline{\mathbf{q}}\langle x, y, z; t\rangle + \mathbf{q}^+\langle x, y, z; t\rangle \tag{10.2}$$

where $\mathbf{q}^+$ is the deviation from the mean, or the *fluctuation*, of the velocity. If we pick out one setup at random, we can be certain about its mean velocity, but not its velocity fluctuation. The totality of the N fluctuations $\mathbf{q}^+$ will tell us how random, or how turbulent, the flow situation is. The *degree, level,* or *intensity* of turbulence can therefore be measured by some quantities that represent these N fluctuations properly. It is convenient here to count the three components of $\mathbf{q}^+$ (that is, u^+, v^+, and w^+), since they will give us details about the situation in the three perpendicular directions. But, the mean values $\overline{u^+}$, $\overline{v^+}$, and $\overline{w^+}$ certainly do not represent the intensity of turbulence since u^+ (or v^+ or w^+) can be positive for one setup and negative for another, and may cancel out one another when added up. To remedy this, one can square the fluctuations before taking the mean: $\overline{(u^+)^2}$, $\overline{(v^+)^2}$ and $\overline{(w^+)^2}$. Here the numbers being added up are all positive, and cannot cancel. As a matter of fact,

$$\overline{(u^+)^2} = \frac{1}{N}[(u_1{}^+)^2 + (u_2{}^+)^2 + \cdots + (u_N{}^+)^2]$$

vanishes *only* if $u_1{}^+, u_2{}^+, \ldots, u_N{}^+$ *all* vanish; and similarly for v^+ or w^+. It is therefore proper to represent the *levels of turbulence* in the x-, y-, and z-directions, respectively, by

$$\frac{\sqrt{\overline{(u^+)^2}}}{U}, \quad \frac{\sqrt{\overline{(v^+)^2}}}{U}, \quad \text{and} \quad \frac{\sqrt{\overline{(w^+)^2}}}{U}$$

A flow is then said to be turbulent if at least one of these levels is nonzero. If all the three levels vanish, the flow is laminar.[3]

[3] In practice, one must regard a flow as laminar when all the three levels are smaller than a certain small number—for example, 10^{-4}. A level beyond this tolerance indicates turbulence.

At this juncture, we must point out that a laminar, flat Couette flow is not necessarily the simple shear flow of Section 3.2, Special Case 1. It may be, for example, the simple shear flow with a certain tiny, periodic (in x) flow superimposed on it. As long as the same combined flow shows up in all N setups, it is laminar. We wish to emphasize this because a laminar flow, the name notwithstanding, is not necessarily in nice, simple laminae. As a matter of fact, many laminar flows are extremely complicated: full of vortices, and yet laminar.

It is also to be noticed that the levels of turbulence are calculated at given points and instants. In other words, they are themselves functions of x, y, z, and t. It is therefore possible to have a flow that is turbulent or laminar at a certain point, on a certain curve, in a certain region, and at a certain instant. For example, our flat Couette flow must be laminar at the two plates, $y = 0$ and Y, because of the boundary conditions (10.1a,b) that must be satisfied in all the N setups. It is also expected that the two layers adjacent to the plates would remain laminar even if the core is turbulent; the layers are then known as laminar sublayers. On the other hand, in a laminar flow, intermittent bursts of turbulence can be made to originate from certain points by artificial, or natural, random disturbances. We will then have a laminar flow with turbulent subregions.

To summarize, we will abandon the example of flat Couette flow, and consider an arbitrary flow problem which we call a system. We will then consider an enormous number of replicas of the system, known as an *ensemble* of systems, and proceed to determine $u\langle x, y, z; t\rangle$, $v\langle x, y, z; t\rangle$, and $w\langle x, y, z; t\rangle$ for all the systems in the ensemble. Their arithmetic mean (called the *ensemble mean*[4]) $\bar{u}$, and so on, describe the behavior of the *mean flow*. The deviations from the mean are measured by the quantities $\overline{(u^+)^2}, \ldots$, where $u^+ = u - \bar{u}, \ldots$. If at least one of the three quantities, $\overline{(u^+)^2}$, $\overline{(v^+)^2}$, and $\overline{(w^+)^2}$, is nonzero, the flow is turbulent; if all three vanish, it is laminar.

To study the details of the fluctuation is a very difficult task. The variations of the mean properties, however, are much easier to ascertain. For example, the velocity field in a turbulent flow is inherently time-dependent and three-dimensional; but the mean velocity may very well turn out to be steady and plane, since the time-dependence and three-dimensionality of the fluctuations may cancel out in the mean. Fortunately for us, many engineering problems are solvable on the basis of the mean flow alone. A knowledge of the detailed deviations from the mean is needed only (a) in special problems (such as meteorology, diffusion, and mixing) and (b) in complementing the study of the mean flow. This kind of knowledge can only be gained through the *statistical theory of turbulence*. In the past, the statistical theory has been successful in meeting the first need, but not as much so in meeting the second. Since we are not particularly interested in those special problems and also since we are not mathematically sophisticated enough, we will

[4] The concepts of ensemble and ensemble mean (or average) are both borrowed from statistical mechanics, a brainchild of Gibbs. Josiah Willard Gibbs (1839–1903) was the first *great* physicist the United States ever produced. Although he was professor of mathematical physics at Yale until his death, he was appreciated in the States only after he became famous in Europe. Interestingly enough, he started out with a Ph.D. dissertation on the strength of a gear tooth.

keep away from the statistical theory. As a consequence, in our study of the mean flow a gap will come to exist, which can only be closed up through some semi-empirical (or phenomenological) theories.

10.2 REYNOLDS EQUATIONS AND REYNOLDS STRESSES

In an ensemble of flow systems, the three-dimensional, time-dependent, Navier-Stokes equations, Eqs. (9.13a,b,c), and the equation of continuity, Eq. (9.3), govern the flow in each system. If we add them up and divide by the number of systems in the ensemble, we will get equations that govern the (ensemble) mean flow. To formalize this procedure of taking the mean, let us note the following four basic properties of the ensemble mean $\overline{(\ )}$:

Let F and G be two independent random quantities, and c be a deterministic quantity, then

$$\begin{cases} \overline{F+G} = \bar{F}+\bar{G} & \textbf{(10.3a)} \\ \overline{c \cdot G} = c \cdot \bar{G} & \textbf{(10.3b)} \\ \overline{\bar{F} \cdot G} = \bar{F} \cdot \bar{G} & \textbf{(10.3c)} \\ \overline{\dfrac{\partial F\langle \xi, \eta \rangle}{\partial \xi}} = \dfrac{\partial \bar{F}\langle \xi, \eta \rangle}{\partial \xi} & \textbf{(10.3d)} \end{cases}$$

The validity of these are obvious when we regard the ensemble mean as an arithmetic mean over an enormous number (N) of entries.[5] Actually, the second property is but a special case of the third when F becomes deterministic. The fourth property also follows directly from the first two, since

$$\begin{aligned} \overline{\frac{\partial F}{\partial \xi}} &= \overline{\lim_{\Delta\xi \to 0} \frac{1}{\Delta \xi}\{F\langle \xi+\Delta\xi, \eta\rangle - F\langle \xi, \eta \rangle\}} \\ &= \lim_{\Delta\xi \to 0} \frac{1}{\Delta \xi}\{\bar{F}\langle \xi+\Delta\xi, \eta\rangle - \bar{F}\langle \xi, \eta \rangle\} \\ &= \frac{\partial \bar{F}}{\partial \xi} \end{aligned}$$

There are many interesting consequences of these properties. For example, since

$$F = \bar{F} + F^{+}$$

by definition, we must have

$$\bar{F} = \bar{\bar{F}} + \overline{F^{+}}$$

[5] They are obvious, since we have regarded N as a large, but *finite*, number. This definition of the ensemble mean is not rigorously correct; but we will let it go at that.

But $G = 1$ in Eq. (10.3c) gives

$$\bar{\bar{F}} = \bar{F}$$

and therefore

$$\overline{F^+} = 0 \tag{10.4}$$

In the previous section, we remarked that $u^+, \ldots$ may very well come up with zero $\overline{u^+}, \ldots$; now, we see that it is indeed an inescapable fate for all fluctuations!

As another example, we see that

$$\begin{aligned}
\overline{F\left(\frac{\partial G}{\partial \xi}\right)} &= \overline{(\bar{F}+F^+)\frac{\partial G}{\partial \xi}} \\
&= \bar{F}\left(\frac{\partial \bar{G}}{\partial \xi}\right) + \overline{F^+\left(\frac{\partial \bar{G}}{\partial \xi}+\frac{\partial G^+}{\partial \xi}\right)} \\
&= \bar{F}\left(\frac{\partial \bar{G}}{\partial \xi}\right) + \overline{F^+}\left(\frac{\partial \bar{G}}{\partial \xi}\right) + \overline{F^+\left(\frac{\partial G^+}{\partial \xi}\right)} \\
&= \bar{F}\left(\frac{\partial \bar{G}}{\partial \xi}\right) + \overline{F^+\left(\frac{\partial G^+}{\partial \xi}\right)} \\
&= \bar{F}\left(\frac{\partial \bar{G}}{\partial \xi}\right) + \frac{\partial \overline{F^+G^+}}{\partial \xi} - \overline{G^+\left(\frac{\partial F^+}{\partial \xi}\right)}
\end{aligned} \tag{10.5}$$

With these properties in mind, we can now present the results[6] of taking the ensemble mean of Eqs. (9.3) and (9.13a,b,c):

$$\frac{\partial \bar{u}}{\partial x}+\frac{\partial \bar{v}}{\partial y}+\frac{\partial \bar{w}}{\partial z}=0 \tag{10.6}$$

$$\frac{\partial \bar{u}}{\partial t}+\bar{u}\frac{\partial \bar{u}}{\partial x}+\bar{v}\frac{\partial \bar{u}}{\partial y}+\bar{w}\frac{\partial \bar{u}}{\partial z} = -\frac{1}{\rho}\frac{\partial \bar{p}}{\partial x}+\nu\left(\frac{\partial^2 \bar{u}}{\partial x^2}+\frac{\partial^2 \bar{u}}{\partial y^2}+\frac{\partial^2 \bar{u}}{\partial z^2}\right) - \frac{\partial \overline{(u^+)^2}}{\partial x} - \frac{\partial \overline{u^+v^+}}{\partial y} - \frac{\partial \overline{u^+w^+}}{\partial z} \tag{10.7a}$$

$$\frac{\partial \bar{v}}{\partial t}+\bar{u}\frac{\partial \bar{v}}{\partial x}+\bar{v}\frac{\partial \bar{v}}{\partial y}+\bar{w}\frac{\partial \bar{v}}{\partial z} = -\frac{1}{\rho}\frac{\partial \bar{p}}{\partial y}+\nu\left(\frac{\partial^2 \bar{v}}{\partial x^2}+\frac{\partial^2 \bar{v}}{\partial y^2}+\frac{\partial^2 \bar{v}}{\partial z^2}\right) - \frac{\partial \overline{u^+v^+}}{\partial x} - \frac{\partial \overline{(v^+)^2}}{\partial y} - \frac{\partial \overline{v^+w^+}}{\partial z} \tag{10.7b}$$

$$\frac{\partial \bar{w}}{\partial t}+\bar{u}\frac{\partial \bar{w}}{\partial x}+\bar{v}\frac{\partial \bar{w}}{\partial y}+\bar{w}\frac{\partial \bar{w}}{\partial z} = -\frac{1}{\rho}\frac{\partial \bar{p}}{\partial z}+\nu\left(\frac{\partial^2 \bar{w}}{\partial x^2}+\frac{\partial^2 \bar{w}}{\partial y^2}+\frac{\partial^2 \bar{w}}{\partial z^2}\right) - \frac{\partial \overline{u^+w^+}}{\partial x} - \frac{\partial \overline{v^+w^+}}{\partial y} - \frac{\partial \overline{(w^+)^2}}{\partial z} \tag{10.7c}$$

The only explanation needed here is that Eqs. (10.5) and (10.6) have been used in obtaining Eqs. (10.7a,b,c).

[6] Gravity is again neglected.

It thus turns out that the mean flow satisfies the same kind of equation of continuity as the laminar flow. But it no longer satisfies the Navier-Stokes equations. Equations (10.7a,b,c) are called the Reynolds equations.[7] They differ from the N-S equations by nine additional terms associated with the fluctuations of **q**. Compared to Eqs. (9.10a,b,c), it is seen that the nine (actually six) terms

$$\begin{array}{ccc} -\rho\overline{(u^{+})^2}, & -\rho\overline{u^{+}v^{+}}, & -\rho\overline{u^{+}w^{+}}, \\ -\rho\overline{u^{+}v^{+}}, & -\rho\overline{(v^{+})^2}, & -\rho\overline{v^{+}w^{+}}, \\ -\rho\overline{u^{+}w^{+}}, & -\rho\overline{v^{+}w^{+}}, & -\rho\overline{(w^{+})^2}, \end{array}$$

take on the same role as the nine (six) stresses, σ_{xx}, σ_{xy}, and so forth. They are therefore called the Reynolds stresses,[7] and are denoted by σ_{xx}^{+}, σ_{xy}^{+}, and so forth:

$$\begin{cases} \sigma_{xx}^{+} = -\rho\overline{(u^{+})^2} & \textbf{(10.8a)} \\ \sigma_{xy}^{+} = -\rho\overline{u^{+}v^{+}} & \textbf{(10.8b)} \\ \sigma_{xz}^{+} = -\rho\overline{u^{+}w^{+}} & \textbf{(10.8c)} \\ \sigma_{yy}^{+} = -\rho\overline{(v^{+})^2} & \textbf{(10.8d)} \\ \sigma_{yz}^{+} = -\rho\overline{v^{+}w^{+}} & \textbf{(10.8e)} \\ \sigma_{zz}^{+} = -\rho\overline{(w^{+})^2} & \textbf{(10.8f)} \end{cases}$$

For a laminar flow, we have $\overline{(u^{+})^2}$, $\overline{(v^{+})^2}$, $\overline{(w^{+})^2} = 0$, which implies that $u^{+}, v^{+}, w^{+} = 0$. So all the Reynolds stresses disappear in a laminar flow; they must represent additional or apparent stresses due to the nonzero levels of turbulence in a turbulent flow.

As to the initial-boundary conditions that go with Eqs. (10.6) and (10.7a,b,c), there should not be any difficulty in getting them by taking the mean of the original conditions on u, v, and so on. The trouble with the whole thing is that, even with proper initial-boundary conditions, Eqs. (10.6) and (10.7a,b,c) do not determine completely the mean flow, since the relations between the Reynolds stresses and the mean flow properties are not known. This is analogous to the situation in the laminar flow before the stress-rate-of-strain relationship is given; there are simply too many unknowns. To proceed, there are only two routes: (a) To study the details of the fluctuation by the statistical theory, hoping to obtain the missing link here. This has not been entirely successful up to now. (b) To observe actual turbulent flows, hoping to induce laws of regularity from all kinds of data. This has turned out to be partially successful (only partially, since all the proposed laws have restrictions on their validity, and all lack the necessary, fully satisfactory rationale). We are thus left with no choice but to follow the second route, which we will regard as semiempirical.

---

[7] Reynolds first derived them in his 1895 paper, "On the dynamic theory of incompressible fluid and the determination of the criterion."

To simplify matters, we will treat only turbulent flows with steady mean properties (including the Reynolds stresses). The simplification goes much further than the dropping of the $(\partial/\partial t)$-terms. In a flow system like this, the same system at different instants can be regarded as the replicas in an ensemble since the same mean flow is shared. Thus an observation of the same system over an extended time interval is equivalent to the investigation of an ensemble, and the ensemble mean can be replaced by the time mean. (This is known as the ergodic hypothesis.) In other words,

$$\overline{(\ )} = \frac{1}{\Delta t} \int_{t_0}^{t_0+\Delta t} (\ )\, dt$$

where Δt is large. This is good news to the experimentalists since an ensemble of systems is very difficult, if not impossible, to realize. Once it is decided that a single system at different instants is to be probed instead, the instrumentation becomes quite simple; sometimes, even our bare eyes can be used in a qualitative manner (for example, to distinguish laminar from turbulent flows), if the flow is first made visible by some technique of visualization.[8] Actually, the experimentalists like this replacement so much that they keep on using it even when the mean flow is clearly unsteady. The justification for this common practice lies in their ability to obtain good mean values of the flow properties over an interval Δt that is very small compared to the time interval required for the mean flow to vary appreciably. It is certainly not recommended for rapidly changing mean flow with very slow (sluggish) fluctuations.

Our set scope does not allow us to dwell too long on the subject of turbulence. In the rest of this chapter, the mean behavior of two representative turbulent flows of practical importance will be explored in what we call a semi-empirical manner. The first flow to be treated is the turbulent counterpart of the plane Poiseuille flow (Section 3.2, Special Case 2); the other one is the turbulent boundary-layer flow past a flat plate. One distinguishing feature of our discussion will be the complete absence of any pseudophysical concept; you will *not* find any mention of "mixing length" and such. Only *dimensional analysis* is thrown in, nothing else. We hope that this will be more in keeping with the spirit of the rest of the book.

10.3 TURBULENT FLOW BETWEEN FIXED PARALLEL PLATES

We wish to solve here Eqs. (10.6) and (10.7a,b,c) (without the $\partial/\partial t$-terms, of course). For the boundary conditions, we take the ensemble mean of

---

[8] A jet of water from a household faucet is usually a steady flow system in the mean. If it appears to be the same over any length of time, it is laminar. If it appears to fluctuate from one instant to another, it is turbulent. (If the level of turbulence is low, instruments other than our eyes must be used to detect it.)

boundary conditions (10.1a,b) with $U = 0$:

$$y = 0: \quad \bar{u}, \bar{v}, \bar{w} = 0 \tag{10.9a,b,c}$$

$$y = Y: \quad \bar{u}, \bar{v}, \bar{w} = 0 \tag{10.9d,e,f}$$

Furthermore, since boundary conditions (10.1a,b) apply to all the replicas in the ensemble, $\mathbf{q}^+$ must be zero at $y = 0$ and Y; that is,

$$y = 0, Y: \quad u^+, v^+, w^+ = 0$$

which, in turn, implies the boundary conditions

$$y = 0, Y: \quad \overline{(u^+)^2}, \overline{u^+v^+}, \ldots = 0 \tag{10.10}$$

The fact that the same boundary conditions hold for all z-values suggests the ansatz

$$\frac{\partial \overline{(\ )}}{\partial z} \equiv 0 \tag{10.11}$$

Just as $\partial \overline{(\ )}/\partial t \equiv 0$ (steady mean flow) asserts that the turbulent flow is *statistically* the same from one instant to another, this ansatz conjectures that our flow here is statistically the same from one z-plane to another.

Next, the boundary conditions on $\bar{w}$ naturally suggest that

$$\bar{w} \equiv 0 \tag{10.12}$$

Ansatz (10.11) and (10.12) together surmise that the mean flow is plane. If these ansatz work, we would have

$$\frac{\partial \overline{u^+w^+}}{\partial x} + \frac{\partial \overline{v^+w^+}}{\partial y} = 0 \tag{10.13}$$

from Eq. (10.7c), and

$$\frac{\partial \bar{u}}{\partial x} + \frac{\partial \bar{v}}{\partial y} = 0 \tag{10.14}$$

from Eq. (10.6).

Finally, the fact that all the listed boundary conditions (of which the boundary condition for $\bar{p}$ is missing) hold in the same way for all x-values would suggest the additional ansatz:

$$\frac{\partial \overline{(\ )}}{\partial x} \left(\text{except } \frac{\partial \bar{p}}{\partial x}\right) \equiv 0 \tag{10.15}$$

(This is the ansatz of fully developed mean flow.) With this, Eq. (10.13) becomes

$$\frac{d\overline{v^+w^+}}{dy} = 0$$

that is,

$$\overline{v^+w^+} = \text{constant}$$
$$= 0$$

because of boundary condition (10.10), and Eq. (10.14) yields $\bar{v} = 0$ because of boundary condition (10.9b).

The remaining two equations, Eqs. (10.7a,b) then become

$$\begin{cases} \dfrac{d\overline{u^+v^+}}{dy} = -\dfrac{1}{\rho}\dfrac{\partial \bar{p}}{\partial x} + \nu \dfrac{d^2\bar{u}}{dy^2} & (10.16) \\ \dfrac{d\overline{(v^+)^2}}{dy} = -\dfrac{1}{\rho}\dfrac{\partial \bar{p}}{\partial y} & (10.17) \end{cases}$$

Whether our ansatz of plane, fully developed mean flow works or not now depends on whether we can solve these two equations (semiempirically) under the posed conditions.

Equation (10.17) can be immediately integrated:

$$\frac{\bar{p}}{\rho} + \overline{(v^+)^2} = G\langle x\rangle$$

Now, boundary condition (10.10) requires that $\overline{(v^+)^2} = 0$ at $y = 0$; therefore

$$G\langle x\rangle = \frac{\bar{p}\langle x, 0\rangle}{\rho}$$

where $\bar{p}\langle x, 0\rangle$ (the mean pressure along the lower plate) can be measured or prescribed. Therefore,

$$\frac{\bar{p}}{\rho} + \overline{(v^+)^2} = \frac{\bar{p}\langle x, 0\rangle}{\rho} \tag{10.18}$$

From Eq. (10.18), we have

$$\frac{1}{\rho}\frac{\partial \bar{p}}{\partial x} = \frac{1}{\rho}\frac{d\bar{p}\langle x, 0\rangle}{dx}$$

remembering that $\partial\overline{(v^+)^2}/\partial x = 0$. Substituting into Eq. (10.16), we have

$$-\frac{d\overline{u^+v^+}}{dy} + \nu\frac{d^2\bar{u}}{dy^2} = \frac{1}{\rho}\frac{d\bar{p}\langle x, 0\rangle}{dx}$$

$$= \text{a constant, } F$$

Therefore

$$\frac{d\bar{p}\langle x, 0\rangle}{dx} = \rho F$$

$$= \frac{\bar{p}\langle x_0 + L, 0\rangle - \bar{p}\langle x_0, 0\rangle}{L}$$

or

$$F = \frac{1}{\rho}\frac{\bar{p}\langle x_0 + L, 0\rangle - \bar{p}\langle x_0, 0\rangle}{L} \tag{10.19}$$

Thus there is no need to measure or prescribe $\bar{p}\langle x, 0\rangle$ along the lower plate; its difference across a distance L anywhere in the fully developed region would be enough. Also,

$$\frac{d\overline{u^+v^+}}{dy} = -F + \nu\frac{d^2\bar{u}}{dy^2} \tag{10.20}$$

The constant F is of course a measure of the performance of the pumping device. Equation (10.20) then yields

$$\overline{u^+v^+} = \nu \frac{d\bar{u}}{dy} - Fy + C_1 \tag{10.21}$$

For the special case of a laminar flow, $\bar{u} = u$; and $u^+, v^+ = 0$. Equation (10.21) can be integrated once more, subject to the boundary conditions (10.9a,d). The result is naturally just the plane Poiseuille flow of Section 3.2, Special Case 2. For the turbulent case, we use boundary condition (10.10) and get

$$\nu \left(\frac{d\bar{u}}{dy}\right)\bigg|_{y=0} = -C_1$$

$$\nu \left(\frac{d\bar{u}}{dy}\right)\bigg|_{y=Y} = FY - C_1$$

But $\bar{u}$ must be symmetric about $y = Y/2$,

$$\frac{d\bar{u}}{dy}\bigg|_{y=0} = -\frac{d\bar{u}}{dy}\bigg|_{y=Y}$$

This yields $C_1 = FY/2$, and Eq. (10.21) becomes

$$\overline{u^+v^+} = \nu \frac{d\bar{u}}{dy} - F\left(y - \frac{Y}{2}\right) \tag{10.21A}$$

Remembering that $\mu\ (d\bar{u}/dy)$ is the local *viscous* shear stress σ_{xy} in the mean flow while $-\rho\overline{u^+v^+}$ is the local Reynolds shear stress σ^+_{xy}, we can lump the two together to form the total shear stress

$$\tau = \mu \frac{d\bar{u}}{dy} - \rho\overline{u^+v^+} \tag{10.22}$$

Then, Eq. (10.21A) can be written as

$$\frac{\tau}{\rho} = F\left(y - \frac{Y}{2}\right) \tag{10.23}$$

Denoting τ at $y = 0$ by τ_0 (the wall shear stress), we can use $\tau_0 = -\rho FY/2$ as the parameter characterizing the problem.

To go on from here, let us introduce a few experimental facts:[9]

1. Since $\overline{u^+v^+}$ vanishes on the walls, there must be a thin laminar sublayer, in which $\overline{u^+v^+} \cong 0$, adjacent to each wall.
2. Outside the laminar sublayer, there is another thin (turbulent) layer in which $\bar{u}$ depends on the wall shear stress τ_0, the distance from the wall, the density and viscosity of the fluid; but not explicitly on the distance Y between the walls (or the distance from the other wall), as though the thin layer knew only of the wall close

[9] In the following, it goes without saying that the *Re* involved is large enough that the flow is unmistakably turbulent.

to it. In this book, we will refer to this layer as the "μ-layer" to emphasize the fact that it is still under the influence of viscosity, though no longer laminar.

3. In the rest of the region covering the large core of the channel, $(\bar{u}\langle y\rangle - \bar{u}\langle Y/2\rangle)$ depends on the wall shear stress, the distance from one wall, the distance between the two walls (or, equivalently, the distance from the other wall), and the density of the fluid, but not explicitly on the viscosity of the fluid. We will call this region the "μ-free region" to show that it is free from the *explicit* influence of viscosity (because the large Reynolds stress there simply overwhelms the viscous stress).
4. Close to the center of the channel, $\overline{u^+v^+} \propto d\bar{u}/dy$.

In the laminar sublayer near $y = 0$, Eq. (10.21 A) yields

$$\bar{u} = \frac{\tau_0 y}{\mu} \tag{10.24}$$

In the μ-layer outside of this sublayer, we have

$$\bar{u} = \bar{u}\langle \tau_0, y, \rho, \mu\rangle$$

A simple dimensional analysis shows that this must be of the form[10]

$$\frac{\bar{u}}{u^*} = \mathscr{F}\left\langle \frac{u^* y}{\nu} \right\rangle \tag{10.25}$$

where the characteristic speed

$$u^* = \sqrt{\frac{\tau_0}{\rho}} \tag{10.26}$$

In the μ-free region, we have

$$\bar{u}\langle Y/2\rangle - \bar{u}\langle y\rangle = fct\langle \tau_0, y, Y, \rho\rangle$$

which yields, after a simple dimensional analysis,

$$\frac{\bar{u}\langle Y/2\rangle - \bar{u}\langle y\rangle}{u^*} = g\langle y/Y\rangle \tag{10.27}$$

where g vanishes at $y = Y/2$.

Equation (10.25) is known as the *law of the wall*, and Eq. (10.27) as the *velocity defect law*. In practice, it is expected that portions of the μ-layer and μ-free region should overlap, so both laws are valid to within limits of experimental accuracy (or the two laws are practically undistinguishable) in the overlapping strip. We will assume that this is so. Then

$$\begin{aligned}\mathscr{F}\langle u^* y/\nu\rangle &= \mathscr{F}\langle (u^* Y/\nu) \cdot (y/Y)\rangle \\ &= \frac{\bar{u}\langle Y/2\rangle}{u^*} - g\langle y/Y\rangle\end{aligned}$$

---

[10] C. B. Millikan, "A critical discussion of turbulent flows in channels and circular tubes," *Proc. 5th Int. Congr. for Appl. Mech.*, 386–392, 1938.

in the overlapping strip. Now, the left-hand side is a function of u^*Y/ν and y/Y, whereas the first term on the right-hand side is definitely independent of y. Therefore it must follow that

$$\frac{\bar{u}\langle Y/2\rangle}{u^*} = \mathcal{G}\langle u^*Y/\nu\rangle$$

and thus

$$\mathcal{F}\langle Re^*y'\rangle = \mathcal{G}\langle Re^*\rangle - g\langle y'\rangle \tag{10.28}$$

where $Re^* = u^*Y/\nu$ and $y' = y/Y$. Then

$$\begin{cases} \dfrac{\partial\mathcal{F}}{\partial Re^*} = \dfrac{d\mathcal{F}}{d(Re^*y')}\cdot y' = \dfrac{d\mathcal{G}}{d(Re^*)} \\ \dfrac{\partial\mathcal{F}}{\partial y'} = \dfrac{d\mathcal{F}}{d(Re^*y')}\cdot Re^* = -\dfrac{dg}{dy'} \end{cases}$$

and therefore

$$Re^*\frac{d\mathcal{G}}{d(Re^*)} = -y'\frac{dg}{dy'} = \text{a constant}, A$$

since the left-hand side is a function of Re^* only and the right-hand side is one of y' only.

Thus[10]

$$\begin{cases} -g\langle y'\rangle = A\ln y' + B & (10.29) \\ \mathcal{G}\langle Re^*\rangle = A\ln Re^* + C & (10.30) \end{cases}$$

or

$$\bar{u}' = \frac{\bar{u}}{u^*} = A\ln(Re^*y') + A_1 \tag{10.31}$$

in the overlapping strip. Experiments then show that

$$A = 2.44 \tag{10.32}$$

Also, substituting into Eq. (10.27), we have $B = A\ln 2$; therefore $A_1 = B + C = A\ln 2 + C$.

Now, what about (a) the part of the μ-layer between the laminar sublayer and the overlapping strip and (b) the part of the μ-less region outside the overlapping strip?

For part (a), let us notice that

$$\bar{u} = \frac{\tau_0 y}{\mu} \tag{10.24}$$

is probably true not only in the extremely narrow laminar sublayer, but also in the adjacent portion of the turbulent μ-layer. We naturally take it to be valid throughout part (a) until it patches up with Eq. (10.31), for lack of something better. In other words, the straight-line velocity distribution from the wall outward will be patched up with the logarithmic distribution where they meet.

The patching[11] is to be effected at $y = y_1$, so Eqs. (10.24) and (10.31) yield the same $\bar{u}\langle y_1 \rangle$ (continuity of $\bar{u}$), and the viscous stress $\mu(d\bar{u}/dy)$ at $y = y_1$, calculated from Eq. (10.31), is much smaller than τ_0 (y_1 is way inside the turbulent layer). Thus

$$\begin{cases} \dfrac{\tau_0 y_1}{\mu} = u^* \left[A \ln \left(\dfrac{Re^* y_1}{Y} \right) + A_1 \right] & \quad (10.33) \\[2ex] \dfrac{\mu u^* A}{y_1} \ll \tau_0 & \quad (10.34) \end{cases}$$

The second requirement yields

$$\frac{u^* y_1}{\nu} = Re^* y_1' \gg A \tag{10.35}$$

If we take $5A$ as practically large enough,[11] we have

$$Re^* y_1' = 5A = 5 \times 2.44 = 12.2 \tag{10.36}$$

or

$$y_1' = \frac{y_1}{Y} = \frac{12.2}{Re^*} \tag{10.37}$$

as the place where the two velocity distributions meet. The first requirement then yields

$$A_1 = 6.1 \tag{10.38}$$

Thus

$$C = A_1 - A \ln 2 = 4.41 \tag{10.39}$$

Figure 10.1 shows a plot of the above semiempirical, composite velocity profile for $0 \leqslant Re^* y' < 40$. A typical measured profile is also plotted for comparison. The large deviation from the measurements around y_1 is not as bad as it looks, since it certainly does not influence the shear stress on the wall; and, as far

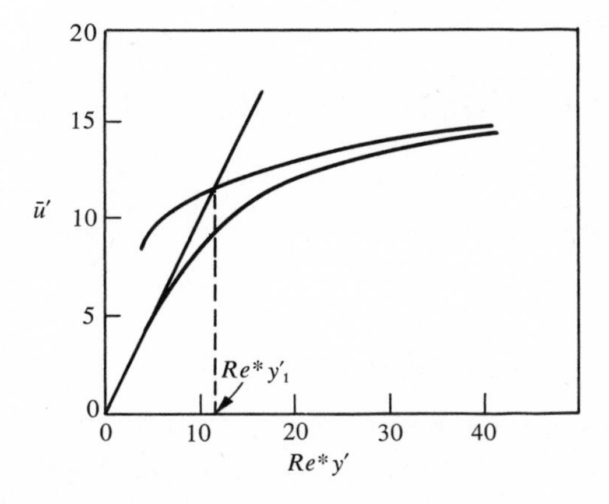

Figure 10.1 Comparison of semiempirical and measured velocity distributions.

[11] A. A. Townsend, *The Structure of Turbulent Shear Flow*, Cambridge University Press, New York, 1956.

as velocity profile is concerned, it occupies only a very narrow region in terms of y' $(10/Re^* < y' < 20/Re^*)$. Actually, on an ordinary plot of $\bar{u}'$ vs. y', this portion of the curve cannot even be seen; it shows up only when y' is magnified by Re^* as in Figure 10.1. As a matter of fact, the range of validity of Eq. (10.24) is as wide as we hoped, thanks to the structure of turbulence near the wall.

For part (b), the portion of the μ-less region close to the center of the channel, an extrapolation of Eq. (10.31) is not as satisfactory as one might hope; for one thing, it yields a nonzero $d\bar{u}/dy$ at $y = Y/2$, which means a sharp bend of the velocity profile at the center because of the symmetry. To improve on this affair we will employ the experimental fact (4) listed before; that is,

$$\overline{u^+v^+} \propto \frac{d\bar{u}}{dy}$$

or[12]

$$-\rho\,\overline{u^+v^+} = \mu_t\left(\frac{d\bar{u}}{dy}\right) \tag{10.40}$$

where μ_t is a constant. Furthermore, near the center, the turbulent stress should completely overwhelm the viscous stress. Thus Eq. (10.21A) becomes approximately

$$-\rho\overline{u^+v^+} = \mu_t\left(\frac{d\bar{u}}{dy}\right) = \rho F\left(y - \frac{Y}{2}\right) \tag{10.41}$$

Therefore

$$\begin{aligned}\bar{u} &= \frac{\rho F}{\mu_t}\left(\frac{y^2}{2} - \frac{Yy}{2}\right) + C_2 \\ &= \frac{\tau_0}{\mu_t}\left(y - \frac{y^2}{Y}\right) + \bar{u}\left\langle\frac{Y}{2}\right\rangle - \frac{\tau_0 Y}{4\mu_t}\end{aligned}$$

or

$$\bar{u}\left\langle\frac{Y}{2}\right\rangle - \bar{u} = \frac{\tau_0 Y}{\mu_t}\,\left(\tfrac{1}{4} - y' + y'^2\right) \tag{10.42}$$

Thus, the velocity profile is parabolic near the center, which fact is experimentally verified (see Fig. 10.2).

This velocity profile should match with Eq. (10.31), or its equivalence

$$\bar{u}\langle Y/2\rangle - \bar{u} = -\,u^* A \ln 2y' \tag{10.31A}$$

so that both $\bar{u}$ and $d\bar{u}/dy$ are continuous at $y = y_2$:

$$\begin{cases}\dfrac{\tau_0 Y}{\mu_t}\,\left(\tfrac{1}{4} - y_2' + y_2'^2\right) = -\,u^* A \ln 2y_2' \\[2ex] \dfrac{\tau_0 Y}{\mu_t}\,(1 - 2y_2') = \dfrac{u^* A}{y_2'}\end{cases}$$

---

[12] C. Ferrari, "Wall Turbulence," translated as NASA Republication No. 2-8-59W, 1959.

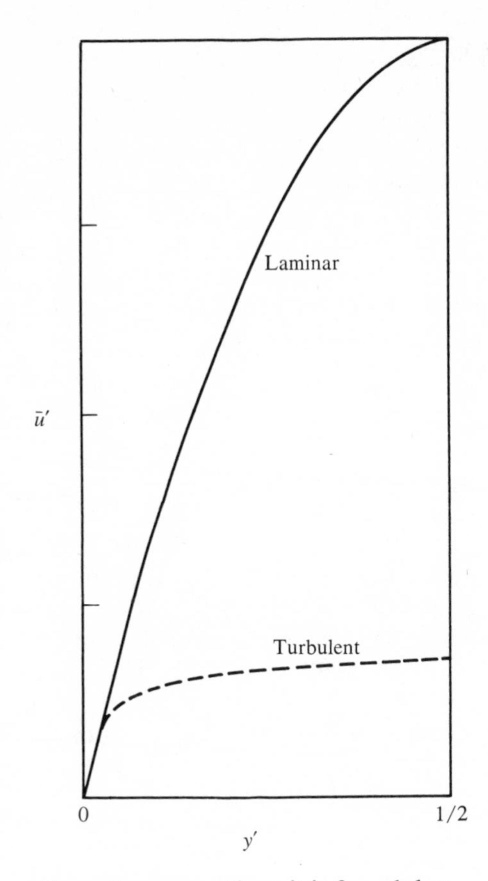

Figure 10.2 A typical comparison of turbulent and laminar flows in a channel, with the same wall shear stress. (The central portion of the turbulent profile is parabolic.)

From these equations it is found that

$$y_2' = 0.143$$

$$\frac{\tau_0 Y/\mu_t}{u^*} = \frac{A}{y_2'(1-2y_2')}$$

$$= 23.8$$

Then, we have, for $0.143 \leqslant y' \leqslant 1/2$,

$$\bar{u}\left\langle\frac{Y}{2}\right\rangle - \bar{u} = 23.82u^* \left(\tfrac{1}{4} - y' + y'^2\right) \tag{10.43}$$

This result may not look very reliable on two counts: (a) $y_2' = 0.143$ is certainly not close to the center of the channel; (b) in such a semiempirical setting, there is no reason why the slope $d\bar{u}/dy$ should be matched so closely over other matchable quantities. However, comparison with actual observations shows that it is in quite good agreement with reality. The turbulent structure in the core obviously allows an extrapolation of Eq. (10.43) into the not-so-central region.[13]

[13] Just as Eq. (10.24) is extrapolated out of the laminar sublayer.

To summarize, we have obtained

$$\begin{cases} \bar{u} = \dfrac{\tau_0 y}{\mu} & 0 \leqq y \leqq \dfrac{12.2Y}{Re^*} \\ \bar{u} = 2.44u^* \ln\left(\dfrac{Re^* y}{Y}\right) + 6.1u^* & \dfrac{12.2Y}{Re^*} \leqq y \leqq 0.143Y \\ \bar{u} = \bar{u}\langle Y/2\rangle - 23.82u^*\left[\dfrac{1}{4} - \dfrac{y}{Y} + \left(\dfrac{y}{Y}\right)^2\right] & 0.143Y \leqq y \leqq \dfrac{Y}{2} \end{cases}$$

where $u^* = \sqrt{\tau_0/\rho}$, $Re^* = \rho u^* Y/\mu$, and

$$\bar{u}\langle Y/2\rangle = 2.44u^* \ln Re^* + 4.41u^* \qquad \textbf{(10.30A)}$$

10.4 TURBULENT BOUNDARY-LAYER FLOW PAST A FLAT PLATE

It has already been implied in point 3 of Section 7.2 that the laminar Blasius flow will eventually become turbulent sufficiently far downstream from the leading edge. Again, we cannot say anything here about this transition from a laminar to a turbulent boundary-layer flow; we will only look at the region where the flow is already turbulent.

We will start with the steady, plane forms of the mean equation of continuity and the Reynolds equations:

$$\begin{cases} \dfrac{\partial \bar{u}}{\partial x} + \dfrac{\partial \bar{v}}{\partial y} = 0 & \textbf{(10.14)} \\ \bar{u}\dfrac{\partial \bar{u}}{\partial x} + \bar{v}\dfrac{\partial \bar{u}}{\partial y} = -\dfrac{1}{\rho}\dfrac{\partial \bar{p}}{\partial x} + \nu\left(\dfrac{\partial^2 \bar{u}}{\partial x^2} + \dfrac{\partial^2 \bar{u}}{\partial y^2}\right) - \dfrac{\partial \overline{(u^+)^2}}{\partial x} - \dfrac{\partial \overline{u^+ v^+}}{\partial y} & \textbf{(10.44a)} \\ \bar{u}\dfrac{\partial \bar{v}}{\partial x} + \bar{v}\dfrac{\partial \bar{v}}{\partial y} = -\dfrac{1}{\rho}\dfrac{\partial \bar{p}}{\partial y} + \nu\left(\dfrac{\partial^2 \bar{v}}{\partial x^2} + \dfrac{\partial^2 \bar{v}}{\partial y^2}\right) - \dfrac{\partial \overline{u^+ v^+}}{\partial x} - \dfrac{\partial \overline{(v^+)^2}}{\partial y} & \textbf{(10.44b)} \end{cases}$$

Without going into details, we may state that, just as for the laminar case, the mean flow splits up into an outer and an inner flow when $Re = \Upsilon x_{obs}/\nu$ is large, where $\Upsilon\hat{\mathbf{x}}$ is the given uniform flow far upstream. The outer flow is just $\mathbf{q} = \Upsilon\hat{\mathbf{x}}$. For the inner or boundary-layer flow, it is again necessary to magnify the region near $y = 0$ as well as the transverse velocity. The corresponding Prandtl problem in terms of *dimensional* quantities turns out to be[14]

$$\begin{cases} \dfrac{\partial \bar{u}}{\partial x} + \dfrac{\partial \bar{v}}{\partial y} = 0 & \textbf{(10.14)} \\ \bar{u}\dfrac{\partial \bar{u}}{\partial x} + \bar{v}\dfrac{\partial \bar{u}}{\partial y} + \dfrac{\partial \overline{u^+ v^+}}{\partial y} = \nu\dfrac{\partial^2 \bar{u}}{\partial y^2} & \textbf{(10.45)} \end{cases}$$

[14] $\bar{p}$ is constant throughout the boundary layer, as in the laminar case.

subject to the conditions

$$x > 0, y = 0: \quad \bar{u} = 0 \tag{10.46a}$$

$$\bar{v} = 0 \tag{10.46b}$$

$$\overline{u^+v^+} = 0 \tag{10.46c}$$

$$y = \infty: \quad \bar{u} = \Upsilon \tag{10.46d}$$

The derivation follows Section 6.2 closely. Only one remark is needed here: It has been assumed that

$$\frac{\overline{u^+v^+}}{\Upsilon^2} \sim \Box\left(\frac{1}{\sqrt{Re}}\right) \tag{10.47}$$

while

$$\frac{\overline{(u^+)^2}}{\Upsilon^2}, \frac{\overline{(v^+)^2}}{\Upsilon^2} \sim O\left(\frac{1}{\sqrt{Re}}\right)$$

(If not for the semiempiricism involved in the solution of the problem, we would have dictated that this be regarded as an ansatz rather than an assumption. As it stands, we might at least point out that, should $\overline{u^+v^+}/\Upsilon^2$ be any smaller, the level of turbulence would be so low as to make the boundary layer effectively laminar; should it be any larger, the principle of least degeneracy would be violated.)

The four experimental facts listed in Section 10.3 hold also in boundary-layer flow with proper modifications. To be more specific, we will introduce a boundary-layer thickness $\delta\langle x\rangle$ such that $\bar{u}$ at $y = \delta\langle x\rangle$ is, for all practical purposes, equal to Υ. (Let us say that $\bar{u} = 0.999\Upsilon$ is close enough to Υ; δ is then the 99.9 percent thickness referred to in Section 7.2.) In the four facts of Section 10.3, δ will take the place of $Y/2$, and the "edge" of the boundary will play the role of the center of the channel.

We again start from the laminar sublayer near $y = 0$:

$$\bar{u}\langle x, y\rangle = \frac{\tau_0\langle x\rangle y}{\mu} \tag{10.49}$$

Then, the same argument as in Section 10.3 would lead to[15]

$$\bar{u} = u^*\langle x\rangle \left\{A \ln\left(\frac{u^*\langle x\rangle y}{\nu}\right) + A_1\right\}$$

or

$$\Upsilon - \bar{u}\langle x, y\rangle = -u^*\langle x\rangle A \ln\left(\frac{y}{\delta\langle x\rangle}\right) \tag{10.51}$$

(that is, $B = 0$), and

$$\Upsilon = u^*\langle x\rangle A \ln\left(\frac{u^*\langle x\rangle \delta\langle x\rangle}{\nu}\right) + A_1 u^*\langle x\rangle \tag{10.52}$$

---

[15] As they stand now, A, A_1, and C may all depend on x.

(that is, $C = A_1$), where

$$u^*\langle x\rangle = \sqrt{\frac{\tau_0\langle x\rangle}{\rho}} \tag{10.53}$$

(Note that the argument is to be put forward for a given arbitrary value of x.) Experiments show that $A = 2.44$, just as for the channel flow.

The above result originally holds only in the overlapping zone of validity of the law of the wall and the velocity defect law. But, as in Section 10.3, it is to be patched up with Eq. (10.49) in an extrapolated manner. The patching is done at

$$\frac{y_1}{\delta\langle x\rangle} = \frac{12.2}{u^*\langle x\rangle\delta\langle x\rangle/\nu} \tag{10.54}$$

which yields also $A_1 = 6.1$. On the other hand, near $y = \delta\langle x\rangle$,

$$-\rho\overline{u^+v^+} = \mu_t\left(\frac{\partial\bar{u}}{\partial y}\right)$$

where μ_t may depend on x. Furthermore, the viscous stress $\mu(\partial\bar{u}/\partial y)$ should be negligible near $y = \delta\langle x\rangle$. So, Eq. (10.45) becomes[12]

$$\begin{aligned}\bar{u}\frac{\partial\bar{u}}{\partial x} + \bar{v}\frac{\partial\bar{u}}{\partial y} &= \frac{\mu_t}{\rho}\left(\frac{\partial^2\bar{u}}{\partial y^2}\right)\\ &= \nu_t\left(\frac{\partial^2\bar{u}}{\partial y^2}\right)\end{aligned} \tag{10.55}$$

near $y = \delta\langle x\rangle$.

Now Eq. (10.55) is to be solved together with Eq. (10.14), subject to the boundary condition that $\bar{u} = 0.999\Upsilon$ at $y = \delta\langle x\rangle$. Comparison with the (laminar) Blasius flow shows that $\bar{u}\langle x, y\rangle$ here must be the Blasius velocity profile[12,16] with ν replaced by ν_t and with its 99.9%-thickness properly adjusted to the given $\delta\langle x\rangle$. (How $\delta\langle x\rangle$ is given will be discussed presently.) But, because of this replacement and adjustment, the profile $\bar{u}\langle x, y\rangle$ will not end up zero[16] at $y = 0$. This profile is then to match with Eq. (10.50). The matching is in principle the same as in Section 10.3, but it is a little messy in its execution.[12,16] Because of numerical uncertainties in fitting experimental data, we will refrain from actually carrying out the matching, except to conjecture that the place where the matching takes place —that is, $y_2/\delta\langle x\rangle$ —will be quite close to, but a bit lower than, the corresponding $y_2/(Y/2)$ (= 0.286) in the channel flow.

To summarize, for each x, $\bar{u}\langle x, y\rangle$ varies formally according to the channel-flow profile with $Y/2$ replaced by $\delta\langle x\rangle$, except that the logarithmic portion finally matches with a Blasius profile.

---

[16] F. H. Clauser, "The turbulent boundary layer," in *Advances in Applied Mechanics*, Vol. IV, edited by H. L. Dryden and Th. Von Kármán (G. Kuerti, managing editor), Academic Press, New York, 1956.

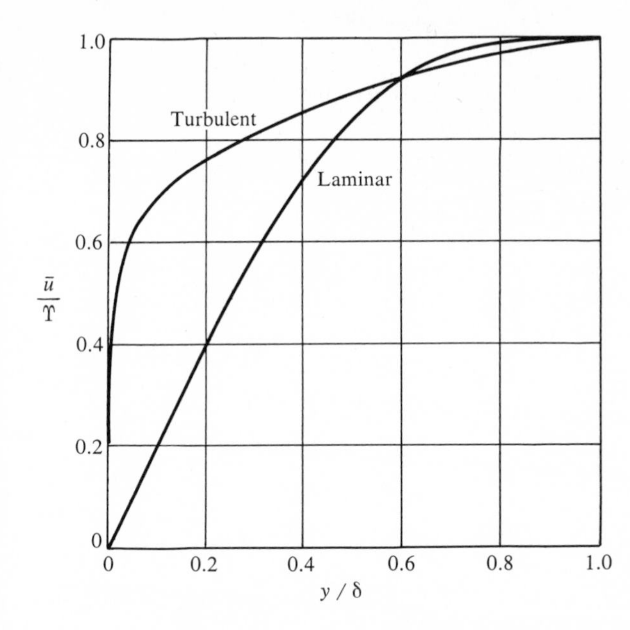

Figure 10.3 Comparison of turbulent and laminar boundary-layer flows past a flat plate with the same Υ.

Once $\bar{u}\langle x, y\rangle$ is known, $\bar{v}\langle x, y\rangle$ can be calculated through Eq. (10.14); but this is usually of little practical interest.

The above, of course, determines $\bar{u}\langle x, y\rangle$ only in terms of $\tau_0\langle x\rangle$ and $\delta\langle x\rangle$, which are still to be fixed: From Eq. (10.52),

$$\delta\langle x\rangle = \frac{\nu}{u^*\langle x\rangle}\exp\left(\frac{\Upsilon}{Au^*\langle x\rangle} - \frac{A_1}{A}\right) \tag{10.56}$$

On the other hand, similarly to Eq. (7.28), we have

$$\begin{aligned}\tau_0\langle x\rangle &= \mu\left.\frac{\partial \bar{u}}{\partial y}\right|_{y=0} \\ &= \rho\frac{d}{dx}\int_0^\infty \bar{u}(\Upsilon - \bar{u})\,dy\end{aligned} \tag{10.57}$$

Substituting into Eq. (10.57) the (composite) velocity profile just established, we will have another relationship between $\tau_0\langle x\rangle$ and $\delta\langle x\rangle$. This, together with Eq. (10.56), will eventually yield[12] both $\tau_0\langle x\rangle$ and $\delta\langle x\rangle$.

Figure 10.3 shows a comparison of a typical turbulent velocity profile and its laminar counterpart with the same upstream speed Υ. The turbulent profile is seen to be much flatter, with a much larger wall shear stress.

PART 3

FLOWS OF COMPRESSIBLE FLUIDS

CHAPTER 11

FROM VECTORS TO DYADICS

11.1 INTRODUCTION

In Part 3 the flow problem of a viscous fluid will be reformulated, with two general features added: (a) The fluid will be considered compressible and heat-conducting. (b) No specific coordinate system is erected (except in clarifying examples) until the very end, when we are about ready to solve specific problems. The first feature requires a basic knowledge of thermodynamics, which will be reviewed briefly in a later chapter. The second feature stems from a desire to reflect, in the formulation, the fact that a physical phenomenon occurs in the same manner no matter what fixed coordinate system man erects to describe it. In mathematical language, it is required that the formulation be *coordinate-invariant*. To achieve a coordinate-invariant formulation, one must use mathematical entities that have properties and operations defined in a manner that is independent of any coordinate system. *Vector* is one such entity. But, unfortunately, vector alone is not enough for the formulation of the flow of a viscous fluid. Another coordinate-invariant entity, together with some coordinate-invariant operations, must also be introduced—that is, *dyadic* and *dyadic analysis*.

In this chapter, vector analysis will be reviewed briefly in such a way as to lead naturally to the concept of *dyadic*. Dyadic analysis will then be studied extensively. Any reference to a coordinate system will be painstakingly avoided until all general discussions are finished.[1] The reader is of course assumed to have learned vector analysis elsewhere. But his previous knowledge may show an undue dependence on a coordinate system, and he may tend to think of vectors through their components.[2] This kind of "component- or coordinate-minded" vector analysis actually defeats the very purpose of the invention of the vector. It is then hoped that the following review will help him in getting things back to proper focus.

(1) Vector

Definition A vector is a quantity with both magnitude and direction, that *obeys the parallelogram law of addition.*

Remark There are many equivalent definitions available (for example, that a vector is a directed line segment, which *implies* the parallelogram law, or that a vector is a triplet of numbers with reference to a certain coordinate system, *and* the triplet transforms according to a certain rule against a coordinate transformation). But the definition that appears in so many texts—namely, that "a vector is a quantity with both magnitude and direction"—is certainly *not* one of them.

COUNTEREXAMPLE

A finite angular displacement is *not* a vector.[3]

EXAMPLE

An infinitesimal angular displacement, however, is a vector.[4] As a consequence of this, all angular velocities and accelerations are vectors.

---

[1] There are, of course, other alternatives to the dyadic, namely, the general and the Cartesian tensors. The general tensor is a bit too complicated for a beginner to grasp, especially when he is trying to learn fluid dynamics at the same time; furthermore, its power will not become apparent until nonaxisymmetric, three-dimensional, boundary-layer flows are encountered. It is therefore not very economical to employ general tensors in this book. Cartesian tensors, as the name implies, work only when rectangular Cartesian coordinate systems are used. To switch to cylindrical and spherical coordinates takes some doing. Because of the popularity of the Cartesian tensor, we will explain its symbolism in a later section; but no real use of it will be made in this book. It is decided to employ the dyadic because of (a) its close link to the vector, both in symbolism and in operation; (b) its generality over the Cartesian tensor; and (c) its relative ease in decomposing, at least with reference to rectangular (not necessarily Cartesian) coordinate systems.

[2] And, yet, *not* against a background of general coordinate transformations.

[3] See the beginning of Section 4–7 of Herbert Goldstein's *Classical Mechanics*, Addison-Wesley, Reading, Mass., 1950.

[4] It is difficult to prove this, since one must first establish the rule by which a finite angular displacement transforms against a coordinate transformation, and then let the displacement approach zero. If interested, read the rest of Section 4–7 of Goldstein, cited in footnote 3.

(2) Scalar

Definition A scalar is a quantity with only magnitude, *which is independent of the coordinate system used.*

Remark It is wrong to define a scalar as just a quantity with only magnitude. (This is, unfortunately, a common mistake in textbooks.)

COUNTEREXAMPLES

A component of a vector in the x-direction is *not* a scalar, since it changes as the x-axis changes. (It is *just* the x-component of the vector.) The direction of a vector (say, N 30° W) is not a scalar, since it is 30° only with respect to our N-W-S-E system.

EXAMPLES

Thirty degrees itself (for instance, one angle of a draftsman's triangle) is a scalar. The length of a vector is a scalar.

(3) Coordinate Invariance

The most important characteristic common to both the vector and the scalar is that their properties are independent of the coordinate system used. This basic characteristic will be called coordinate-invariance in this book.

Physical phenomena occur without bias toward or preference for any particular (fixed) coordinate system. In other words, physical laws are inherently coordinate-invariant. This is why scalars and vectors (and, later on, dyadics) must be introduced. In addition, we must also take care to introduce only those operations (on these quantities) that are also coordinate-invariant. In the following, all operations are coordinate-invariant except those to be introduced at the end of the chapter. By postponing thus the appearance of the noncoordinate-invariant operations (that is, those depending on the particular coordinate system used), it is hoped that unnecessary confusion is avoided.

(4) Dot Product

Definition The dot product of two vectors **A** and **B** is defined as the *scalar*

$$\mathbf{A} \cdot \mathbf{B} = AB \cos \measuredangle(A, B)$$

where $A = |\mathbf{A}|$, $B = |\mathbf{B}|$, and

$$0 \leq \measuredangle(A, B) \leq \pi$$

Remarks The range of $\measuredangle(A, B)$ is an important and integral part of the definition. The dot product is a commutative and distributive operation; that is,

$$\mathbf{B} \cdot \mathbf{A} = \mathbf{A} \cdot \mathbf{B}$$

$$\mathbf{A} \cdot (\mathbf{B} + \mathbf{C}) = \mathbf{A} \cdot \mathbf{B} + \mathbf{A} \cdot \mathbf{C}$$

If $\mathbf{A} \cdot \mathbf{B} = 0$, there are three possible consequences: $\mathbf{A} = 0$, $\mathbf{B} = 0$, and $\mathbf{A} \perp \mathbf{B}$. Note also that no reference to any coordinate system is implied in the definition.

(5) Cross Product

Definition The cross product of **A** and **B** is defined as the *vector*

$$\mathbf{A} \times \mathbf{B} = [AB \sin \measuredangle(A, B)]\,\hat{\mathbf{n}}$$

where $\hat{\mathbf{n}}$ is the unit vector normal to both **A** and **B**, and is so pointed that if we rotate **A** toward **B** through the angle

$$0 \leqslant \measuredangle(A, B) \leqslant \pi$$

a right-handed screw advances in the n-direction (Fig. 11.1).

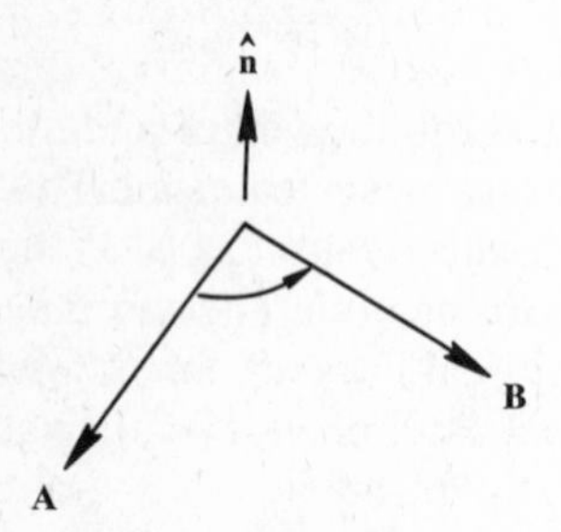

Figure 11.1 $\hat{\mathbf{n}}$ in the definition of cross product.

Remarks For brevity, we will say that the three vectors **A**, **B**, and $\hat{\mathbf{n}}$ (or the three directions A, B, and n) form a right-handed system. The cross product is an anticommutative, nonassociative, but distributive[5] operation; that is,

$$\mathbf{B} \times \mathbf{A} = -\mathbf{A} \times \mathbf{B}$$

$$\mathbf{A} \times (\mathbf{B} \times \mathbf{C}) \neq (\mathbf{A} \times \mathbf{B}) \times \mathbf{C}$$

$$\mathbf{A} \times (\mathbf{B} + \mathbf{C}) = \mathbf{A} \times \mathbf{B} + \mathbf{A} \times \mathbf{C}$$

If $\mathbf{A} \times \mathbf{B} = 0$, then either $\mathbf{A} = 0$ or $\mathbf{B} = 0$ or $\mathbf{A} \parallel \mathbf{B}$.

(6) Composite Products

The dot and cross products can be used repeatedly and compositely whenever allowable by the definitions. For example, $\mathbf{A} \cdot (\mathbf{B} \times \mathbf{C})$ yields a scalar;

[5] The distributiveness is not obvious and must be proved.

$\mathbf{A} \times (\mathbf{B} \times \mathbf{C})$, a vector; but $(\mathbf{A} \cdot \mathbf{B}) \times \mathbf{C}$ and $\mathbf{A} \cdot \mathbf{B} \cdot \mathbf{C}$ are meaningless. It is left as an exercise for the reader to prove that

$$\mathbf{A} \times (\mathbf{B} \times \mathbf{C}) = \mathbf{B}(\mathbf{A} \cdot \mathbf{C}) - \mathbf{C}(\mathbf{A} \cdot \mathbf{B}) \tag{11.1}$$

The above is the essence of vector algebra. We will take up vector calculus in a new section.

11.2 THE OPERATOR ∇

To fix ideas, let us consider a moving fluid in the physical space. At any instant t, there will be a definite velocity vector $\mathbf{q}$ at every space point represented by a position vector $\mathbf{r}$ from a chosen origin. That is to say, there will be a vectorial field

$$\mathbf{q} = \mathbf{q}\langle \mathbf{r} \rangle \text{ at } t$$

In this chapter the possible variation of $\mathbf{q}$ with t will be suppressed, so $\mathbf{q}$ is always instantaneous.

Consider a point P in the physical space and an arbitrary volume V surrounding it.[6] Let the boundary surface of V be S. In fluid mechanics (and also other branches of mathematical physics), only two-sided surfaces are needed.[7] Of these surfaces, one can call one side the positive side and the other, negative. But which side one calls positive is completely a matter of convention. In this book, whenever we deal with a surface that forms a part of the boundary of a volume V in which we are interested, we will call the side immediately adjacent to V the negative side and the side away from V the positive side (Fig. 11.2). If there is no adjacent volume of interest, we will just pick the positive side at random; it then does not make any difference, one way or the other, as long as we stick to our choice all the way through.

The *vectorial* area element $\hat{\mathbf{n}}\,\delta S$ at a point on the surface can now be defined as a vector having a magnitude equal to the area δS of the element, a direction normal to the surface S at that point, and a positive sense pointing from the negative to the positive side of S (Fig. 11.3). Thus, $\hat{\mathbf{n}}$ must be the local *unit normal vector* of S, pointing from the negative to the positive side of S (always pointing out of the volume of interest if S is the boundary, or a part of the boundary, of one).

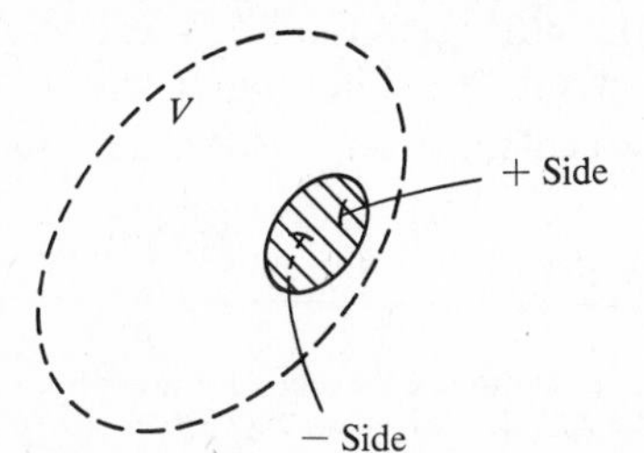

Figure 11.2 Two sides of a bounding surface.

[6] V and P are fixed in space.

[7] Thus excluding Möbius strip, Klein bottle, and the like.

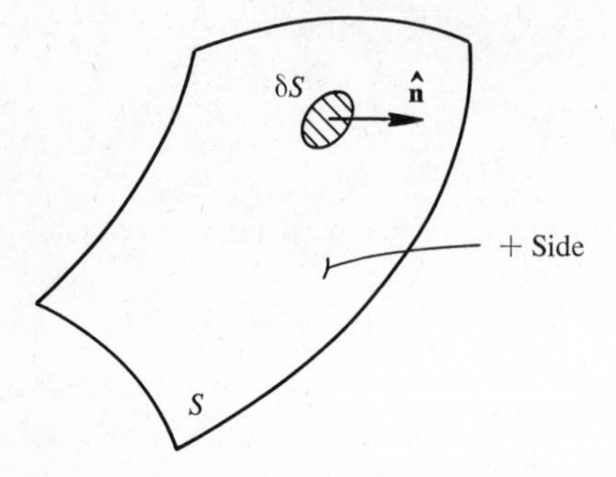

Figure 11.3 An area element.

Going back to the volume V surrounding the point P, and enclosed by S (Fig. 11.4), consider now the three surface integrals:[8]

$$\begin{cases} \oint_S \hat{\mathbf{n}} \cdot \mathbf{q}\, dS \\ \oint_S \hat{\mathbf{n}} \times \mathbf{q}\, dS \\ \oint_S \hat{\mathbf{n}}\phi\, dS \end{cases}$$

where ϕ is a scalar field[9]—that is, a function $\phi\langle\mathbf{r}\rangle$ of $\mathbf{r}$. Using a limiting process in which $V \to 0$ but with V always enclosing the point[10] P, three quantities are defined from these three integrals:[11]

$$\begin{cases} \text{div}\,\mathbf{q} = \nabla \cdot \mathbf{q} = \lim\limits_{\substack{V\to 0 \\ \text{Ⓟ}}} \dfrac{1}{V} \oint_S \hat{\mathbf{n}} \cdot \mathbf{q}\, dS \\ \text{curl}\,\mathbf{q} = \nabla \times \mathbf{q} = \lim\limits_{\substack{V\to 0 \\ \text{Ⓟ}}} \dfrac{1}{V} \oint_S \hat{\mathbf{n}} \times \mathbf{q}\, dS \\ \text{grad}\,\phi = \nabla\phi = \lim\limits_{\substack{V\to 0 \\ \text{Ⓟ}}} \dfrac{1}{V} \oint_S \hat{\mathbf{n}}\phi\, dS \end{cases}$$

All the three quantities are defined here locally at the point P. If the point P is allowed to vary, they will be functions of $\mathbf{r}$—that is, fields.

It is again to be noted that the above definitions make no reference whatsoever to any coordinate system; this is as it should be. The reader should not feel that to define something by a limiting process is at all strange. After all, the

[8] The circle around the integration sign denotes the fact that S here is a closed surface.

[9] For example, the temperature field.

[10] To be really careful, we also require that V should shrink to P without preference for any direction so that it would not first shrink into a disk or line in the process.

[11] The circle around P below the sign of limit emphasizes the fact that V must always surround the point P.

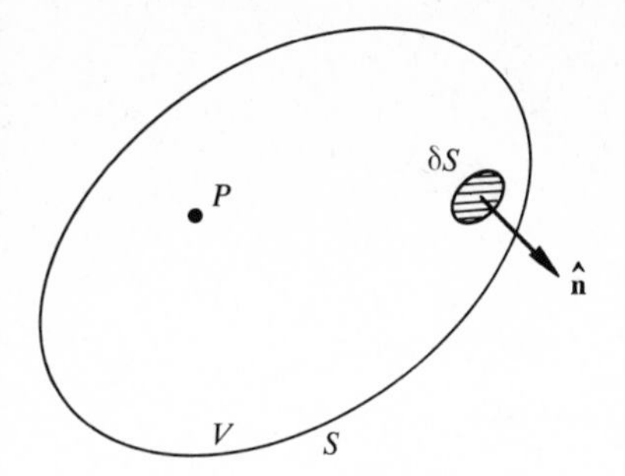

Figure 11.4 A volume surrounding point *P*.

ordinary derivative is defined that way. An even closer analogy is provided by the definition of the physical quantity, density at *P*:

$$\rho = \lim_{\substack{V\to 0 \\ (P)}} \frac{1}{V}(\text{total mass in } V)$$

With the velocity vector **q**, it is obvious that

$$\oint_S \hat{\mathbf{n}} \cdot \mathbf{q}\, dS$$

is just the time rate of net volume outflow of fluid across the closed surface S. Then div **q** is the local rate of volume outflow *per unit volume* of space. The converse implication is then that

$$\int_V \text{div}\, \mathbf{q}\, dV$$

must be the rate of volume flow of fluid out of V. We will elucidate this implication in a later section (on integral relations); at this point, we will say only that this physical interpretation accounts for the name "divergence" and the symbol "div."

The meaning of $\nabla \times \mathbf{q}$ will be taken up later. At this point, let us inquire only into the meaning of grad ϕ. This is made clear in the following manner.

In the **r**-space, $\phi\langle\mathbf{r}\rangle$ is a distribution of the ϕ-values from point to point. The locus of points where the values of ϕ (for example, temperature) are equal; that is,

$$\phi\langle\mathbf{r}\rangle = \text{a constant}$$

is an equi-ϕ surface. As an illustration, we have drawn three such surfaces in Figure 11.5. At every point on an equi-ϕ surface, one can erect a local coordinate line N normal to the surface. If one moves away from the surface along this line, ϕ will change; and, unless $\partial\phi/\partial N = 0$ at that point on $\phi = C$, ϕ will surely be found to increase as one moves away in one direction and decrease in the other direction. This latter fact provides a convention by which the positive direction of N can be picked. From now on, the $+N$ direction will be associated with the direction of increasing ϕ. With this convention, the unit normal vector $\hat{\mathbf{N}}$ *to an equi-ϕ surface*

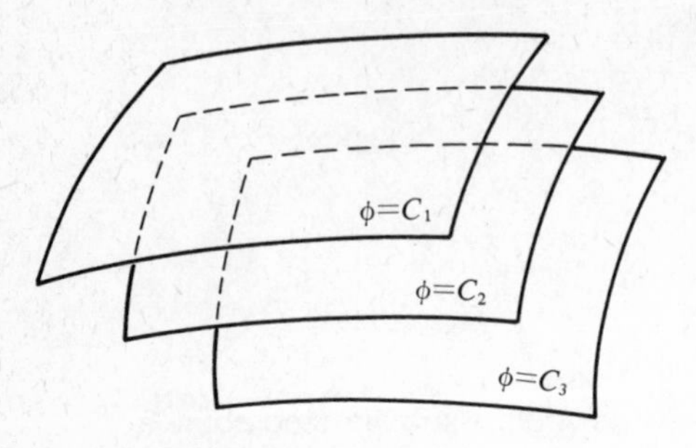

Figure 11.5 Equi-ϕ surfaces.

can be defined as the unit vector in the $+N$ direction. ($\hat{\mathbf{N}}$ may or may not be the same as $\hat{\mathbf{n}}$.) See Figure 11.6.

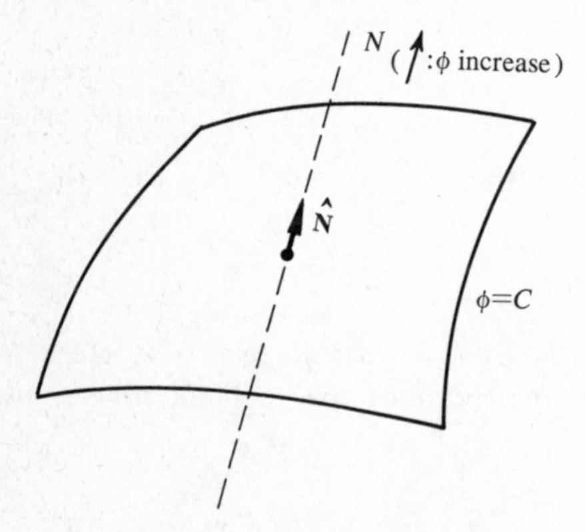

Figure 11.6 Positive-N direction.

Consider next a point P on an equi-ϕ surface $\phi = C$ (Fig. 11.7). Above and below $\phi = C$, there are two neighboring equi-ϕ surfaces as shown. In calculating grad ϕ at P according to the definition, we take the circular cylinder around P between these two neighboring surfaces as V. The corresponding S then consists of (a) the top δS with $\hat{\mathbf{n}}$ pointing upward, (b) the bottom δS with $\hat{\mathbf{n}}$ pointing down-

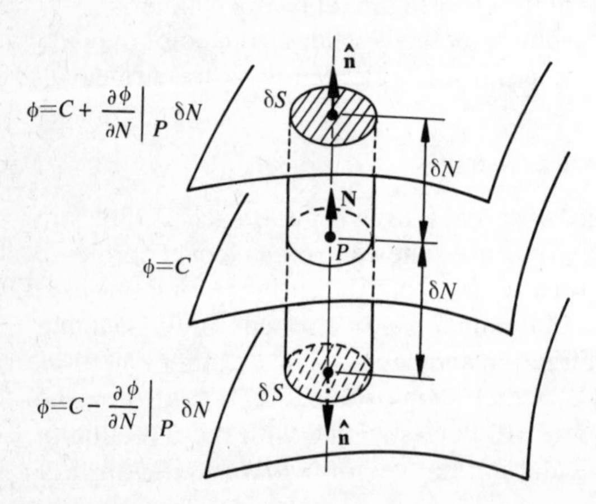

Figure 11.7 A cylindrical neighborhood.

ward, and (c) the lateral surface. The contributions to

$$\oint_S \phi \mathbf{n}\, dS$$

of (a) and (b) are obviously

$$\left(C + \frac{\partial \phi}{\partial N}\bigg|_P \delta N\right) \delta S\, \hat{\mathbf{N}}$$

and

$$\left(C - \frac{\partial \phi}{\partial N}\bigg|_P \delta N\right) \delta S\, (-\hat{\mathbf{N}})$$

The contribution of (c) is zero since, for every $\hat{\mathbf{n}}\, \delta S$, there is another diametrically opposed to it (Fig. 11.8). .

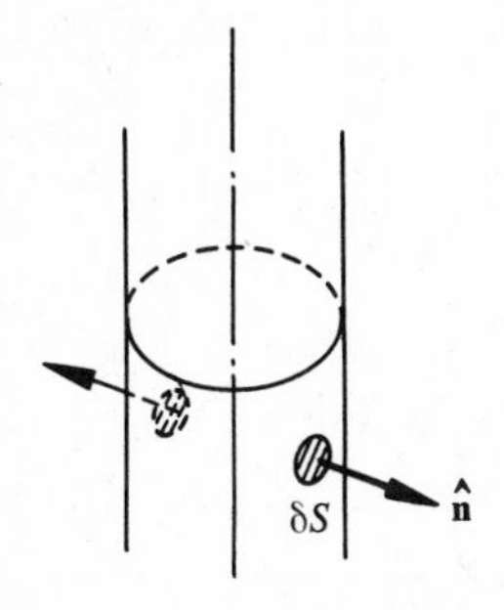

Figure 11.8 Cancellation of contributions from diametrically opposed elements.

So, for this V,

$$\oint_S \phi \hat{\mathbf{n}}\, dS = \left(2 \frac{\partial \phi}{\partial N}\bigg|_P \delta N\, \delta S\right) \hat{\mathbf{N}}$$

Now, let $V \to 0,\ P$, by making $\delta N,\ dS \to 0$. The result is

$$\begin{aligned}(\text{grad } \phi)|_P &= \lim_{\substack{V \to 0 \\ \textcircled{P}}} \left(2 \frac{\partial \phi}{\partial N}\bigg|_P \delta N\, \delta S\right) \hat{\mathbf{N}} \Big/ 2\delta N\, \delta S \\ &= \frac{\partial \phi}{\partial N}\bigg|_P \hat{\mathbf{N}}\end{aligned}$$

or, in general,

$$\text{grad } \phi = \boldsymbol{\nabla} \phi = \frac{\partial \phi}{\partial N} \hat{\mathbf{N}} \tag{11.2}$$

Conclusion At a point, grad ϕ is a vector in the positive normal direction ($+N$) of the local equi-ϕ surface, with a magnitude equal to the derivative of ϕ with respect to the local normal coordinate.

Furthermore, consider a point P again on $\phi = C$. Denoting the position of P by $\mathbf{r}$, the neighboring point Q on the neighboring equi-ϕ surface then has a

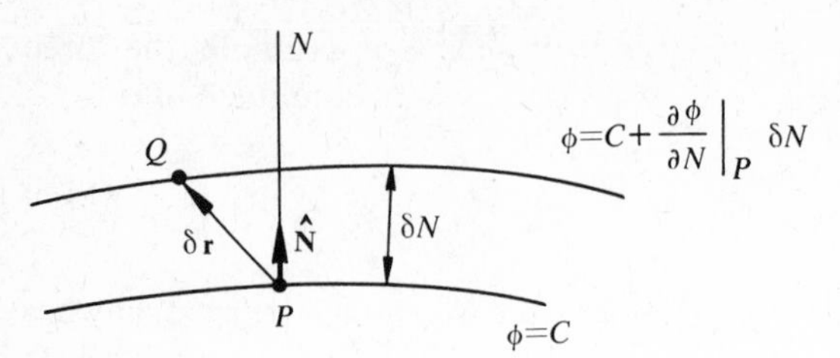

Figure 11.9 Two neighboring points.

position $\mathbf{r} + \delta\mathbf{r}$ (Fig. 11.9). Obviously,

$$\delta N = \hat{\mathbf{N}} \cdot \delta\mathbf{r}$$

and therefore

$$\begin{aligned}\delta\phi &= \frac{\partial\phi}{\partial N}\,\delta N \\ &= \frac{\partial\phi}{\partial N}\,\hat{\mathbf{N}} \cdot \delta\mathbf{r} \\ &= (\text{grad}\,\phi) \cdot \delta\mathbf{r}\end{aligned}$$

In the limit, we have

$$d\phi = (\text{grad}\,\phi) \cdot d\mathbf{r} = (\nabla\phi) \cdot d\mathbf{r} \tag{11.3}$$

One can also proceed in a slightly different fashion and define the *directional derivative* of ϕ in the direction $\overrightarrow{PQ}$ at P as

$$\begin{aligned}\left(\frac{\partial\phi}{\partial l}\right)_{P,\overrightarrow{PQ}} &= \lim_{Q\to P} \frac{\delta\phi}{\overline{PQ}} \\ &= \text{grad}\,\phi \cdot \frac{\delta\mathbf{r}}{\overline{PQ}} \\ &= \text{grad}\,\phi \cdot \frac{\overrightarrow{PQ}}{\overline{PQ}} \\ &= (\text{grad}\,\phi) \cdot \hat{\imath}\end{aligned}$$

where $\hat{\imath}$ is the unit vector in the direction of $\overrightarrow{PQ}$, and l is the distance measured in the direction of $\hat{\imath}$. Or, simply,

$$\frac{\partial\phi}{\partial l} = (\text{grad}\,\phi) \cdot \hat{\imath} = \hat{\imath} \cdot \text{grad}\,\phi \tag{11.4}$$

That is to say, $\hat{\imath} \cdot \text{grad}\,\phi$ is the *rate of change of* ϕ *in the direction of* $\hat{\imath}$. This then accounts for the name "gradient" or the symbol "grad."

The use of the symbols grad, div, and curl calls to mind some physical pictures which are helpful to the reader. But the alternative use of the operational symbol ∇ (pronounced "del") is actually even more suggestive for three reasons:

1. From Eqs. (11.2) and (11.3), it is seen that the operator ∇ works as if it were a vector. This fact is further reflected in the use of "$\nabla\cdot$" and "$\nabla\times$".

2. It enables one to write unifying formulas. For example, the three definitions of $\nabla\phi$, $\nabla \cdot \mathbf{q}$, and $\nabla \times \mathbf{q}$ can be easily combined into

$$\nabla \ast \mathscr{F}|_P = \lim_{\substack{V \to 0 \\ (P)}} \frac{1}{V} \oint_S \hat{\mathbf{n}} \ast \mathscr{F} \, dS \tag{11.5}$$

where $\ast$ stands for $\cdot$, $\times$, or nothing whatsoever, as the case may be; and $\mathscr{F}$ stands for a vector, or a scalar, again as the case may be.
3. From Eqs. (11.2) and (11.3), it is seen that ∇ is also a differential operator.[12] Three different symbols—div, curl, and grad—do not reflect this common fact clearly, yet one symbol ∇ does.

We may now summarize by stating that ∇ is a *vectorial differential operator*. The reader must pay special attention here to the word "operator." The symbol ∇ alone, all by itself, has no meaning attached to it. It takes on a meaning and comes to life only when *operating* on some quantity in a certain *defined* manner. The following groupings are, of course, meaningless (undefined): $\mathbf{q} \times \nabla$, $(\nabla \times \mathbf{q}) \cdot \nabla$, $\nabla \cdot \phi$. (The first two followed by appropriate quantities may, however, have meaning in a defined way.)

It is also necessary, from now on, to indicate where the differentiating power of ∇ ends. We introduce, in this respect, two conventions:

1. The differentiating power of ∇ extends only to its right; for example, $\phi_1 \nabla \phi_2$ means $\phi_1(\nabla\phi_2)$, *not* $(\phi_1\nabla)\phi_2$ with ∇ differentiating on ϕ_1, *nor* $(\nabla\phi_1)\phi_2$, *nor* $\nabla(\phi_1\phi_2)$.
2. Brackets, parentheses, and so forth, must be used to define the extent of the differentiating power of ∇; for example,

$(\nabla\phi_1)\phi_2$	differentiating on ϕ_1 only
$\nabla(\phi_1\phi_2)$	on $\phi_1\phi_2$
$\nabla(\mathbf{A} \cdot \mathbf{B})$	on $\mathbf{A} \cdot \mathbf{B}$
$(\nabla \times \mathbf{A}) \cdot \mathbf{B}$	on $\mathbf{A}$ only
$\nabla \times (\mathbf{A} \times \mathbf{B})$	on $\mathbf{A} \times \mathbf{B}$

The above discussion covers only the definition and basic characters of ∇. Its general properties will be discussed fully after the introduction of the concept of dyadics.

11.3 DYADICS AND DYADIC OPERATIONS

Let us look at the formal definition of ∇, Eq. (11.5), once more:

∇ is a mathematical entity *such that*, when operating on $\mathscr{F}$ to form

[12] This means that the limiting process involved in the definitions of div, curl, and grad is actually the same as that in the definition of differentiation.

$\nabla \times \mathscr{F}$, it yields

$$\nabla \times \mathscr{F} \stackrel{\text{def}}{=} \lim_{\substack{V \to 0 \\ (P)}} \frac{1}{V} \oint_S \hat{\mathbf{n}} \times \mathscr{F} \, dS$$

at point P. ($\stackrel{\text{def}}{=}$ denotes equality by definition.)

In other words, it is defined through its effect. The reader should not feel strange about this. Actually, many entities are defined this way. To mention one farfetched example, we can define a pump as a mechanical entity such that, when operating on a liquid, it raises the pressure of the liquid. We dwell a little bit on this point because we wish to define another operator or entity this way:

Definition[13] A dyadic—represented by two vectors put together (with nothing between) thus, **AB**—is an entity (or operator) such that

$$1. \quad (\mathbf{AB}) \cdot \mathbf{C} \stackrel{\text{def}}{=} \mathbf{A}(\mathbf{B} \cdot \mathbf{C}) \tag{11.6}$$

$$2. \quad \mathbf{C} \cdot (\mathbf{AB}) \stackrel{\text{def}}{=} (\mathbf{C} \cdot \mathbf{A})\mathbf{B} \tag{11.7}$$

Remarks (a) **AB** is known as the *dyadic product* of **A** and **B**. (b) The symbol **AB** comes to life only through the *man-made* fact that its dot product with a vector yields another vector; in Eq. (11.6), it yields a vector parallel to **A**, and in Eq. (11.7), one parallel to **B**. (c) It is also possible to look upon **AB** · ( ) or ( ) · **AB** as a function or machine; we feed in a vector and we get another vector.

From the definition, it follows that

$$\begin{aligned} (\mathbf{AB}) \cdot (a\mathbf{C} + b\mathbf{D}) &= \mathbf{A}[\mathbf{B} \cdot (a\mathbf{C} + b\mathbf{D})] \\ &= \mathbf{A}[(\mathbf{B} \cdot \mathbf{C})a + (\mathbf{B} \cdot \mathbf{D})b] \\ &= [\mathbf{A}(\mathbf{B} \cdot \mathbf{C})]a + [\mathbf{A}(\mathbf{B} \cdot \mathbf{D})]b \\ &= [(\mathbf{AB}) \cdot \mathbf{C}]a + [(\mathbf{AB}) \cdot \mathbf{D}]b \\ &= a(\mathbf{AB}) \cdot \mathbf{C} + b(\mathbf{AB}) \cdot \mathbf{D} \end{aligned} \tag{11.8}$$

and

$$(a\mathbf{C} + b\mathbf{D}) \cdot (\mathbf{AB}) = a\mathbf{C} \cdot (\mathbf{AB}) + b\mathbf{D} \cdot (\mathbf{AB}) \tag{11.9}$$

where a and b are scalars. Equations (11.8) and (11.9) are often referred to by saying that a dyadic is a *linear* operator.

Note also that, in the definition, there are two dot products distinguished from each other by pairs of parentheses. But once the definition is written down, there is no longer any need for such use of parentheses around **AB**. Thus $\mathbf{AB} \cdot \mathbf{C}$ means just $(\mathbf{AB}) \cdot \mathbf{C}$ or $\mathbf{A}(\mathbf{B} \cdot \mathbf{C})$; $\mathbf{C} \cdot \mathbf{AB}$ means just $\mathbf{C} \cdot (\mathbf{AB})$ or $(\mathbf{C} \cdot \mathbf{A})\mathbf{B}$;

[13] Gibbs developed his vector and dyadic analysis first in small pamphlets on *Elements of Vector Analysis* for the use of his students in 1881–1884. The fundamental ideas of his vector algebra came from Hamilton's theory of *quaternions* and Grassmann's "*extensive Grösse*"; his dyadic analysis apparently came out of his own thoughts.

there is no ambiguity whatsoever.[14] In short, Eqs. (11.6) and (11.7) will be written as

$$\mathbf{AB}\cdot\mathbf{C} \stackrel{\text{def}}{=} \mathbf{A}(\mathbf{B}\cdot\mathbf{C}) \tag{11.6A}$$

$$\mathbf{C}\cdot\mathbf{AB} \stackrel{\text{def}}{=} (\mathbf{C}\cdot\mathbf{A})\mathbf{B} \tag{11.7A}$$

from now on.

As a first application of the dyadic symbol, Eq. (11.1) can now be rewritten as

$$\mathbf{A}\times(\mathbf{B}\times\mathbf{C}) = \mathbf{BA}\cdot\mathbf{C}-\mathbf{CA}\cdot\mathbf{B} \tag{11.1A}$$

The above is then the basic concept behind dyadics. The author regrets that he must postpone physical examples of dyadics until later.

At this point, in order to make the concept more useful for the general development of fluid dynamics, we must proceed to define some dyadic operations.

(1) Distributiveness over a Sum of Dyadics[15]

$$(c_1\mathbf{A}_1\mathbf{B}_1 + c_2\mathbf{A}_2\mathbf{B}_2 + \cdots)\cdot\mathbf{C} \stackrel{\text{def}}{=} c_1\mathbf{A}_1\mathbf{B}_1\cdot\mathbf{C} + c_2\mathbf{A}_2\mathbf{B}_2\cdot\mathbf{C} + \cdots \tag{11.10}$$

$$\mathbf{C}\cdot(c_1\mathbf{A}_1\mathbf{B}_1 + c_2\mathbf{A}_2\mathbf{B}_2 + \cdots) \stackrel{\text{def}}{=} c_1\mathbf{C}\cdot\mathbf{A}_1\mathbf{B}_1 + c_2\mathbf{C}\cdot\mathbf{A}_2\mathbf{B}_2 + \cdots \tag{11.11}$$

In the spirit of this, Eq. (11.1) can be further written as

$$\begin{aligned}\mathbf{A}\times(\mathbf{B}\times\mathbf{C}) &= \mathbf{B}(\mathbf{A}\cdot\mathbf{C}) - \mathbf{C}(\mathbf{A}\cdot\mathbf{B})\\ &= \mathbf{B}(\mathbf{C}\cdot\mathbf{A}) - \mathbf{C}(\mathbf{B}\cdot\mathbf{A})\\ &= \mathbf{BC}\cdot\mathbf{A} - \mathbf{CB}\cdot\mathbf{A}\\ &= (\mathbf{BC}-\mathbf{CB})\cdot\mathbf{A}\end{aligned} \tag{11.1B}$$

(2) Linear Combinations

Because of definitions (11.10) and (11.11), the combination

$$c_1\mathbf{A}_1\mathbf{B}_1 + c_2\mathbf{A}_2\mathbf{B}_2 + \cdots + c_m\mathbf{A}_m\mathbf{B}_m$$

is also a dyadic in the sense[15] that its dot products with vectors are well-defined vectors. (This, in fact, generalizes the original definition of a dyadic.) We will refer to such a combination as a dyadic $\tilde{\boldsymbol{\tau}}$. (Note the symbolism: boldface *and* "tilde".) For the special case of only one term in the sum (and with $c_1 = 1$), $\tilde{\boldsymbol{\tau}} = \mathbf{A}_1\mathbf{B}_1$. The terms in parentheses on the right-hand side of Eq. (11.1B) thus form a single dyadic.

---

14 Dot product of a dyadic and a vector is in general not commutative; that is, $\mathbf{AB}\cdot\mathbf{C} \neq \mathbf{C}\cdot\mathbf{AB}$.

15 Many authors (following Gibbs) call $\mathbf{A}_1\mathbf{B}_1$, $\mathbf{A}_2\mathbf{B}_2$, ... *dyads*, and the sum a dyadic. This distinction serves no practical purpose at all, and will not be maintained in this book; they are all dyadics, summed or not.

(3) Equality

Two dyadics $\tilde{\tau}_1$ and $\tilde{\tau}_2$ are said to be equal (written $\tilde{\tau}_1 = \tilde{\tau}_2$) if and only if

$$\begin{cases} \tilde{\tau}_1 \cdot \mathbf{C} = \tilde{\tau}_2 \cdot \mathbf{C} & \quad (11.12) \\ \mathbf{C} \cdot \tilde{\tau}_1 = \mathbf{C} \cdot \tilde{\tau}_2 & \quad (11.13) \end{cases}$$

for an *arbitrary* vector $\mathbf{C}$.

From this definition, it is obvious that

$$\mathbf{BA} \neq \mathbf{AB} \tag{11.14}$$

that is, in general, the dyadic product is not commutative.

Furthermore, it can be shown that

$$\mathbf{AB} = \mathbf{DE}$$

if and only if

$$\mathbf{A} = k\mathbf{D} \quad \text{and} \quad \mathbf{B} = \frac{\mathbf{E}}{k}$$

where k is an arbitrary scalar.

The necessary condition (that is, the "only if" condition) for the equality of $(\mathbf{A}_1\mathbf{B}_1 + \mathbf{A}_2\mathbf{B}_2 + \cdots + \mathbf{A}_m\mathbf{B}_m)$ and $(\mathbf{D}_1\mathbf{E}_1 + \mathbf{D}_2\mathbf{E}_2 + \cdots + \mathbf{D}_l\mathbf{E}_l)$ is not so simple. The best way of stating it is to decompose first the vectors $\mathbf{A}_1, \mathbf{B}_1, \mathbf{D}_1, \ldots$ into components with respect to three orthogonal directions represented by the three unit vectors $\hat{\mathbf{X}}_1$, $\hat{\mathbf{X}}_2$, and $\hat{\mathbf{X}}_3$:

$$\begin{aligned}
\tilde{\tau} &= \mathbf{A}_1\mathbf{B}_1 + \mathbf{A}_2\mathbf{B}_2 + \cdots + \mathbf{A}_m\mathbf{B}_m \\
&= (A_{11}\hat{\mathbf{X}}_1 + A_{12}\hat{\mathbf{X}}_2 + A_{13}\hat{\mathbf{X}}_3)\,(B_{11}\hat{\mathbf{X}}_1 + B_{12}\hat{\mathbf{X}}_2 + B_{13}\hat{\mathbf{X}}_3) + \cdots \\
&= (A_{11}B_{11} + A_{21}B_{21} + \cdots)\hat{\mathbf{X}}_1\hat{\mathbf{X}}_1 \\
&\quad + (A_{11}B_{12} + A_{21}B_{22} + \cdots)\hat{\mathbf{X}}_1\hat{\mathbf{X}}_2 \\
&\quad + (\cdots)\hat{\mathbf{X}}_1\hat{\mathbf{X}}_3 \\
&\quad + (\cdots)\hat{\mathbf{X}}_2\hat{\mathbf{X}}_1 + (\cdots)\hat{\mathbf{X}}_2\hat{\mathbf{X}}_2 + (\cdots)\hat{\mathbf{X}}_2\hat{\mathbf{X}}_3 \\
&\quad + (\cdots)\hat{\mathbf{X}}_3\hat{\mathbf{X}}_1 + (\cdots)\hat{\mathbf{X}}_3\hat{\mathbf{X}}_2 + (\cdots)\hat{\mathbf{X}}_3\hat{\mathbf{X}}_3 \\
&= \tau_{11}\hat{\mathbf{X}}_1\hat{\mathbf{X}}_1 + \tau_{12}\hat{\mathbf{X}}_1\hat{\mathbf{X}}_2 + \tau_{13}\hat{\mathbf{X}}_1\hat{\mathbf{X}}_3 \\
&\quad + \tau_{21}\hat{\mathbf{X}}_2\hat{\mathbf{X}}_1 + \tau_{22}\hat{\mathbf{X}}_2\hat{\mathbf{X}}_2 + \tau_{23}\hat{\mathbf{X}}_2\hat{\mathbf{X}}_3 \\
&\quad + \tau_{31}\hat{\mathbf{X}}_3\hat{\mathbf{X}}_1 + \tau_{32}\hat{\mathbf{X}}_3\hat{\mathbf{X}}_2 + \tau_{33}\hat{\mathbf{X}}_3\hat{\mathbf{X}}_3
\end{aligned}$$

$$= \begin{array}{c|ccc} \curvearrowright & \hat{\mathbf{X}}_1 & \hat{\mathbf{X}}_2 & \hat{\mathbf{X}}_3 \\ \hline \hat{\mathbf{X}}_1 & \tau_{11} & \tau_{12} & \tau_{13} \\ \hat{\mathbf{X}}_2 & \tau_{21} & \tau_{22} & \tau_{23} \\ \hat{\mathbf{X}}_3 & \tau_{31} & \tau_{32} & \tau_{33} \end{array} \tag{11.15}$$

Conclusion A dyadic (in a three-dimensional space) can always be written as a linear combination of the nine basic dyadics $\hat{\mathbf{X}}_1\hat{\mathbf{X}}_1, \hat{\mathbf{X}}_1\hat{\mathbf{X}}_2, \ldots, \hat{\mathbf{X}}_3\hat{\mathbf{X}}_3$.

The nine coefficients involved, τ_{ij} $(i, j = 1, 2, 3)$, are called the nine *components* of $\tilde{\tau}$ with reference to the three orthogonal directions in $\mathbf{X}_1$, $\mathbf{X}_2$, and $\mathbf{X}_3$. In the form (11.15), we have fitted τ_{ij} into a matrix, abbreviated $\{\tau_{ij}\}$, for clarity.[16]

Now, in general, we can also write

$$\tilde{\sigma} = \sigma_{11}\hat{\mathbf{X}}_1\hat{\mathbf{X}}_1 + \cdots$$

Then, by the previous definition, $\tilde{\sigma}$ is said to be equal to $\tilde{\tau}$ if and only if $\tilde{\sigma} \cdot \mathbf{C} = \tilde{\tau} \cdot \mathbf{C}$ for any vector[17] $\mathbf{C}$. If we choose $\mathbf{C} = \hat{\mathbf{X}}_1 + 0 + 0$, we will have

$$\begin{aligned}\tilde{\tau} \cdot \mathbf{C} &= \tau_{11}\hat{\mathbf{X}}_1(\hat{\mathbf{X}}_1 \cdot \hat{\mathbf{X}}_1) + \tau_{12}\hat{\mathbf{X}}_1(\hat{\mathbf{X}}_2 \cdot \hat{\mathbf{X}}_1) + \cdots \\ &\quad + \tau_{21}\hat{\mathbf{X}}_2(\hat{\mathbf{X}}_1 \cdot \hat{\mathbf{X}}_1) + \cdots \\ &\quad + \tau_{31}\hat{\mathbf{X}}_3(\hat{\mathbf{X}}_1 \cdot \hat{\mathbf{X}}_1) + \cdots \\ &= \tau_{11}\hat{\mathbf{X}}_1 + \tau_{21}\hat{\mathbf{X}}_2 + \tau_{31}\hat{\mathbf{X}}_3\end{aligned}$$

since $\hat{\mathbf{X}}_1 \cdot \hat{\mathbf{X}}_1$, etc. $= 1$, and $\hat{\mathbf{X}}_1 \cdot \hat{\mathbf{X}}_2$, etc. $= 0$. Similarly,

$$\tilde{\sigma} \cdot \mathbf{C} = \sigma_{11}\hat{\mathbf{X}}_1 + \sigma_{21}\hat{\mathbf{X}}_2 + \sigma_{31}\hat{\mathbf{X}}_3$$

thus,

$$\tilde{\sigma} = \tilde{\tau} \quad \text{if and only if} \quad \sigma_{ij} = \tau_{ij} \tag{11.16}$$

for $i, j = 1, 2, 3$. That is, two dyadics are equal if and only if their components referred to $\hat{\mathbf{X}}_1$, $\hat{\mathbf{X}}_2$, and $\hat{\mathbf{X}}_3$ are equal.

We must emphasize here that the three directions $\hat{\mathbf{X}}_1$, $\hat{\mathbf{X}}_2$, and $\hat{\mathbf{X}}_3$ are chosen for convenience. The basic definitions of dyadics and their operations (including their equality) are still coordinate-invariant.

(4) Conjugate, Scalar, and Vector of a Dyadic

The conjugate of a dyadic

$$\tilde{\tau} = \mathbf{A}_1\mathbf{B}_1 + \mathbf{A}_2\mathbf{B}_2 + \cdots + \mathbf{A}_m\mathbf{B}_m$$

is defined as

$$\tilde{\tau}_c = \mathbf{B}_1\mathbf{A}_1 + \mathbf{B}_2\mathbf{A}_2 + \cdots + \mathbf{B}_m\mathbf{A}_m$$

It is then obvious that, if the components of $\tilde{\tau}$ referred to $\hat{\mathbf{X}}_1$, $\hat{\mathbf{X}}_2$, $\hat{\mathbf{X}}_3$ are $\{\tau_{ij}\}$, those of $\tilde{\tau}_c$ are $\{\tau_{ji}\}$, $i, j = 1, 2, 3$. It is also obvious that

$$(\tilde{\tau}_c)_c = \tilde{\tau}$$

and

$$\mathbf{C} \cdot \tilde{\tau} = \tilde{\tau}_c \cdot \mathbf{C}$$

for any $\mathbf{C}$.

If $\tilde{\tau} = \tilde{\tau}_c$, we say that $\tilde{\tau}$ is *symmetric*, and the corresponding matrix $\{\tau_{ij}\}$ is symmetric about its diagonal. If $\tilde{\tau} = -\tilde{\tau}_c$, we say that it is antisymmetric, and $\{\tau_{ij}\}$ must have zero elements on the diagonal.

---

[16] Note that a dyadic can be displayed in a matrix; but a matrix alone is *not* a dyadic. To say that "a dyadic (or a tensor of the second rank) is nothing but a matrix" is an outright fraud.

[17] The other requirement $\mathbf{C} \cdot \tilde{\sigma} = \mathbf{C} \cdot \tilde{\tau}$, leads to exactly the same conclusion.

The *scalar* of $\tilde{\tau}$ is defined as

$$\tilde{\tau}_s = \mathbf{A}_1 \cdot \mathbf{B}_1 + \mathbf{A}_2 \cdot \mathbf{B}_2 + \cdots + \mathbf{A}_m \cdot \mathbf{B}_m$$

Similarly, the *vector* of $\tilde{\tau}$ is

$$\tilde{\tau}_v = \mathbf{A}_1 \times \mathbf{B}_1 + \mathbf{A}_2 \times \mathbf{B}_2 + \cdots + \mathbf{A}_m \times \mathbf{B}_m$$

Obviously,

$$(\tilde{\tau}_c)_s = \tilde{\tau}_s$$

$$(\tilde{\tau}_c)_v = -\tilde{\tau}_v$$

(5) Dot Product and Double Dot Product

(a) The dot product of **AB** and **CD** is defined as a distributive operation:

$$\begin{aligned}(\mathbf{AB}) \cdot (\mathbf{CD}) &= \mathbf{A}(\mathbf{B} \cdot \mathbf{C})\mathbf{D} \\ &= (\mathbf{B} \cdot \mathbf{C})\mathbf{AD}\end{aligned}$$

which yields a dyadic. It is, in general, not commutative; that is,

$$(\mathbf{AB}) \cdot (\mathbf{CD}) \neq (\mathbf{CD}) \cdot (\mathbf{AB})$$

The *distributive* part of the definition refers to the *defined* fact that

$$\mathbf{AB} \cdot (\mathbf{CD} + \mathbf{EF}) = \mathbf{AB} \cdot \mathbf{CD} + \mathbf{AB} \cdot \mathbf{EF}$$

and that

$$(\mathbf{CD} + \mathbf{EF}) \cdot \mathbf{AB} = \mathbf{CD} \cdot \mathbf{AB} + \mathbf{EF} \cdot \mathbf{AB}$$

Note that (**AB**) · (**CD**) can be written as **AB** · **CD**; there is no ambiguity.

(b) The double dot product of **AB** and **CD** is defined as a distributive[18] operation:

$$\begin{aligned}(\mathbf{AB}) : (\mathbf{CD}) &= (\mathbf{B} \cdot \mathbf{C})(\mathbf{A} \cdot \mathbf{D}) \\ &\quad \text{or } \mathbf{B} \cdot (\mathbf{CA}) \cdot \mathbf{D} \\ &= (\mathbf{C} \cdot \mathbf{B})(\mathbf{D} \cdot \mathbf{A}) \\ &\quad \text{or } \mathbf{C} \cdot (\mathbf{BD}) \cdot \mathbf{A} \\ &= (\mathbf{D} \cdot \mathbf{A})(\mathbf{B} \cdot \mathbf{C}) \\ &\quad \text{or } \mathbf{D} \cdot (\mathbf{AB}) \cdot \mathbf{C} \\ &= (\mathbf{B} \cdot \mathbf{C})(\mathbf{D} \cdot \mathbf{A}) \\ &\quad \text{or } \mathbf{B} \cdot (\mathbf{CD}) \cdot \mathbf{A}\end{aligned}$$

which yields a scalar. In the "( )( )"-forms, the inner members of (**AB**):(**CD**)—that is, **B** and **C**—are in one pair of parentheses. This is easy to keep in mind; the "·( )·"-forms are then easily written out by regrouping.

---

[18] What does this adjective mean?

The operation of double dot product is commutative; that is,

$$(\mathbf{AB}):(\mathbf{CD}) = (\mathbf{CD}):(\mathbf{AB})$$

Caution In general,

$$(\mathbf{AB}):(\mathbf{CD}) \neq (\mathbf{A}\cdot\mathbf{C})(\mathbf{B}\cdot\mathbf{D})$$

(6) Admittance of ∇

The operator ∇ will now be admitted as a possible candidate for the first member of a dyadic product. All we have to do is to state exactly what we mean:

$$(\nabla\mathbf{A}) \stackrel{\text{def}}{=} \lim_{\substack{V\to 0\\ (P)}} \frac{1}{V}\oint_S \hat{\mathbf{n}}\mathbf{A}\,dS \tag{11.17}$$

where $\hat{\mathbf{n}}\mathbf{A}$ is a *bona fide* dyadic defined before.

This definition is in keeping with the vectorial character of ∇. (This character actually stems from the appearance of vector $\hat{\mathbf{n}}$ in the definitions involving ∇.) As a differential operator, ∇ in Eq. (11.17) governs only $\mathbf{A}$ as indicated by the parentheses.

Other dyadic operations involving ∇ are to be understood through

$$\lim_{\substack{V\to 0\\ (P)}} \frac{1}{V}\oint_S \hat{\mathbf{n}}(\ \ )\,dS$$

in a similar manner:

(a) $\mathbf{B}\cdot\nabla\mathbf{A}$

$$\begin{aligned}\mathbf{B}\cdot(\nabla\mathbf{A})|_P = \mathbf{B}\cdot\nabla\mathbf{A}|_P &= (\mathbf{B}\cdot\nabla_A)\mathbf{A}|_P\\ &= \lim_{\substack{V\to 0\\ (P)}} \frac{1}{V}\oint_S \mathbf{B}_P\cdot\hat{\mathbf{n}}\mathbf{A}\,dS\end{aligned}$$

where ∇_A is used to show that ∇ differentiates on $\mathbf{A}$ in spite of the parentheses (which are supposed to indicate the extent of the differentiating power of ∇), and $\mathbf{B}_P$ is $\mathbf{B}$ evaluated at P.

(b) $(\nabla\mathbf{A})\cdot\mathbf{B}$

$$(\nabla\mathbf{A})\cdot\mathbf{B}|_P = \nabla_A(\mathbf{A}\cdot\mathbf{B})|_P \stackrel{\text{def}}{=} \lim_{\substack{V\to 0\\ (P)}} \frac{1}{V}\oint_S \hat{\mathbf{n}}(\mathbf{A}\cdot\mathbf{B}_P)\,dS$$

where ∇_A is ∇ with its differentiating power restricted to $\mathbf{A}$ only.

(c) $\nabla\cdot(\mathbf{AB})$ or, more generally, $\nabla\cdot\tilde{\tau}$

$$\nabla\cdot\tilde{\tau} = \lim_{\substack{V\to 0\\ (P)}} \frac{1}{V}\oint_S \hat{\mathbf{n}}\cdot\tilde{\tau}\,dS$$

(d) $(\tilde{\boldsymbol{\tau}} \cdot \boldsymbol{\nabla}_q) \cdot \mathbf{q}$

$$(\tilde{\boldsymbol{\tau}} \cdot \boldsymbol{\nabla}_q) \cdot \mathbf{q} \overset{\text{def}}{=} \lim_{\substack{V \to 0 \\ Ⓟ}} \frac{1}{V} \oint_S (\tilde{\boldsymbol{\tau}}_p \cdot \hat{\mathbf{n}}) \cdot \mathbf{q}\, dS$$

Here, $\boldsymbol{\nabla}_q$ is used to show that $\boldsymbol{\nabla}$ differentiates on $\mathbf{q}$ in spite of the parentheses. $(\tilde{\boldsymbol{\tau}} \cdot \boldsymbol{\nabla} q) \cdot \mathbf{q}$ is definitely not equal to $\tilde{\boldsymbol{\tau}} \cdot (\boldsymbol{\nabla} \cdot \mathbf{q})$ which is meaningless.

We will now discuss fully the meaning of (a) and (b), which are frequently encountered.

(a) Choosing three *fixed* noncoplanar vectors $\mathbf{C}_1$, $\mathbf{C}_2$, and $\mathbf{C}_3$ in space, we can write

$$\mathbf{A} = A_1\mathbf{C}_1 + A_2\mathbf{C}_2 + A_3\mathbf{C}_3$$

Then,

$$\begin{aligned} \mathbf{B} \cdot \boldsymbol{\nabla}\mathbf{A}|_P &= \lim_{\substack{V \to 0 \\ Ⓟ}} \frac{1}{V} \oint_S \mathbf{B}_P \cdot \hat{\mathbf{n}}(A_1\mathbf{C}_1 + A_2\mathbf{C}_2 + A_3\mathbf{C}_3)\, dS \\ &= \mathbf{B}_P \cdot \left(\lim_{\substack{V \to 0 \\ Ⓟ}} \frac{1}{V} \oint_S \hat{\mathbf{n}} A_1\, dS \right) \mathbf{C}_1 + \cdots \\ &= \mathbf{B}_P \cdot (\text{grad}\, A_1)_P \mathbf{C}_1 + \cdots \end{aligned}$$

or

$$\begin{aligned} \mathbf{B} \cdot \boldsymbol{\nabla}\mathbf{A} &= (\mathbf{B} \cdot \text{grad}\, A_1)\mathbf{C}_1 + (\mathbf{B} \cdot \text{grad}\, A_2)\mathbf{C}_2 + (\mathbf{B} \cdot \text{grad}\, A_3)\mathbf{C}_3 \\ &= B\{(\hat{\mathbf{B}} \cdot \text{grad}\, A_1)\mathbf{C}_1 + \cdots\} \\ &= B\{(\text{rate of change of } A_1 \text{ in the } \mathbf{B}\text{-direction})\mathbf{C}_1 + \cdots\} \\ &= B\{(\text{rate of change of the } C_1\text{-component of } \mathbf{A} \text{ in the } \mathbf{B}\text{-direction})\mathbf{C}_1 + \cdots\} \end{aligned}$$

where Eq. (11.4) has been used.

Conclusion Therefore

$$\begin{aligned} \mathbf{B} \cdot \boldsymbol{\nabla}\mathbf{A} &= (\mathbf{B} \cdot \boldsymbol{\nabla}_A)\mathbf{A} \\ &= B(\text{rate of change of } \mathbf{A} \text{ in the direction of } \mathbf{B}) \end{aligned} \tag{11.18}$$

(b) $\boldsymbol{\nabla}_A(\mathbf{A} \cdot \mathbf{B})$

From Eq. (11.1),

$$\begin{aligned} \mathbf{B}_P \times (\hat{\mathbf{n}} \times \mathbf{A}) &= (\mathbf{B}_P \cdot \mathbf{A})\hat{\mathbf{n}} - (\mathbf{B}_P \cdot \hat{\mathbf{n}})\mathbf{A} \\ &= \hat{\mathbf{n}}(\mathbf{A} \cdot \mathbf{B}_P) - \mathbf{B}_P \cdot \hat{\mathbf{n}}\mathbf{A} \end{aligned}$$

Therefore

$$\begin{aligned} \boldsymbol{\nabla}_A(\mathbf{A} \cdot \mathbf{B})|_P &= \lim_{\substack{V \to 0 \\ Ⓟ}} \frac{1}{V} \oint_S \hat{\mathbf{n}}(\mathbf{A} \cdot \mathbf{B}_P)\, dS \\ &= \lim_{\substack{V \to 0 \\ Ⓟ}} \frac{1}{V} \oint_S \{\mathbf{B}_P \times (\hat{\mathbf{n}} \times \mathbf{A}) + \mathbf{B}_P \cdot \hat{\mathbf{n}}\mathbf{A}\}\, dS \end{aligned}$$

$$= \mathbf{B}_P \times \left(\lim_{\substack{V\to 0 \\ (P)}} \frac{1}{V} \oint_S \hat{\mathbf{n}} \times \mathbf{A}\, dS \right) + \mathbf{B}_P \cdot \left(\lim_{\substack{V\to 0 \\ (P)}} \frac{1}{V} \oint_S \hat{\mathbf{n}}\mathbf{A}\, dS \right)$$

$$= \mathbf{B}_P \times (\boldsymbol{\nabla} \times \mathbf{A}) + \mathbf{B}_P \cdot \boldsymbol{\nabla}\mathbf{A}$$

or

$$\boldsymbol{\nabla}_A(\mathbf{A} \cdot \mathbf{B}) = \mathbf{B} \times (\boldsymbol{\nabla} \times \mathbf{A}) + \mathbf{B} \cdot \boldsymbol{\nabla}\mathbf{A} \tag{11.19}$$

where $\mathbf{B} \cdot \boldsymbol{\nabla}\mathbf{A}$ is unambiguously defined in (a).

The reader must be warned in the matter of the conjugate of $\boldsymbol{\nabla}\mathbf{A}$:

$$(\boldsymbol{\nabla}\mathbf{A})_c \neq \mathbf{A}\boldsymbol{\nabla}$$

Actually, $\mathbf{A}\boldsymbol{\nabla}$ has no meaning since $\boldsymbol{\nabla}$ must, by definition, differentiate on something *to its right*; any expression that *ends with* $\boldsymbol{\nabla}$ is meaningless. To make up for this flaw, the author would like to introduce a private symbol for $(\boldsymbol{\nabla}\mathbf{A})_c$, namely,

$$(\boldsymbol{\nabla}\mathbf{A})_c = \mathbf{A}\overleftarrow{\boldsymbol{\nabla}} \tag{11.20}$$

where $\overleftarrow{\boldsymbol{\nabla}}$ denotes $\boldsymbol{\nabla}$ *differentiating* on $\mathbf{A}$ which is *to its left*.

EXAMPLES

$$\begin{aligned}
\mathbf{B} \cdot \boldsymbol{\nabla}\mathbf{A} &= (\mathbf{B} \cdot \boldsymbol{\nabla}_A)\mathbf{A} \\
&= (\boldsymbol{\nabla}\mathbf{A})_c \cdot \mathbf{B} \\
&= \mathbf{A}\overleftarrow{\boldsymbol{\nabla}} \cdot \mathbf{B} \\
&= \mathbf{A}(\overleftarrow{\boldsymbol{\nabla}}_A \cdot \mathbf{B}) \\
(\boldsymbol{\nabla}\mathbf{A}) \cdot \mathbf{B} &= \boldsymbol{\nabla}_A(\mathbf{A} \cdot \mathbf{B}) \\
&= \mathbf{B} \cdot (\mathbf{A}\overleftarrow{\boldsymbol{\nabla}}) \\
&= (\mathbf{B} \cdot \mathbf{A})\overleftarrow{\boldsymbol{\nabla}}_A
\end{aligned}$$

Obviously, no meaningful expression can begin with $\overleftarrow{\boldsymbol{\nabla}}$.

11.4 GENERAL PROPERTIES OF ∇

Definition (11.5) can now be generalized to allow $\mathscr{F}$ to be a scalar, or a vector, or a *dyadic*; and $\divideontimes$ to stand for a dot, a cross, or a "nothing," as the case may be. To be specific, we may list the following.

scalar $\mathscr{F}$: $\divideontimes$ means "nothing"

vector $\mathscr{F}$: $\divideontimes$ means "dot, cross, or nothing"

dyadic $\mathscr{F}$: $\divideontimes$ means "dot"

From this generalized definition, we can now gather in a convenient package the general properties of $\boldsymbol{\nabla}$.

(1) Vectorial Nature

∇ is a vectorial operator simply because $\hat{\mathbf{n}}$ in its definition is a vector. Symbolically, one might as well look upon ∇ as

$$\nabla(\ ) = \lim_{\substack{V\to 0\\ (P)}} \frac{1}{V} \oint_S \hat{\mathbf{n}}(\ )\, dS$$

Since $\hat{\mathbf{n}}$ is the only vector that appears in this symbolism, ∇ obeys all the laws that govern the vector $\hat{\mathbf{n}}$ *as long as* they are in keeping with the rest of its definition (that is. the surface integration and the limiting process). Thus, $\nabla\cdot(\nabla\times\mathbf{A}) = 0$ simply because $\hat{\mathbf{n}}\cdot(\hat{\mathbf{n}}\times\mathbf{A}) = 0$. But, $\nabla\cdot(\mathbf{A}\times\nabla)$ is not zero; it is not even defined, judging from the rest of the definition of ∇.

EXAMPLES

$$\nabla\cdot(\nabla\phi) = (\nabla_\phi\cdot\nabla_\phi)\phi = \nabla^2\phi$$

$$\mathbf{A}\cdot(\nabla\phi) = (\mathbf{A}\cdot\nabla_\phi)\phi$$

$$\mathbf{A}\cdot(\nabla\mathbf{B}) = (\mathbf{A}\cdot\nabla_B)\mathbf{B}$$

COUNTEREXAMPLES

$(\nabla\phi)\times(\nabla\psi) \neq 0$, although $(\hat{\mathbf{n}}\phi)\times(\hat{\mathbf{n}}\psi) = 0$. (This pitfall can be avoided if we realize that $\nabla\phi$ actually is $\nabla_\phi\phi$, $\nabla\psi$ is $\nabla_\psi\psi$; and one is not embedded in the other.)

$$\nabla^2\phi \neq \lim_{\substack{V\to 0\\ (P)}} \frac{1}{V} \oint_S (\hat{\mathbf{n}}\cdot\hat{\mathbf{n}})\phi\, dS$$

(This pitfall can be avoided if we realize that $\nabla^2\phi$ actually mean $\nabla\cdot(\nabla\phi)$ where the limiting process $\nabla\phi$ is embedded in the limiting process $\nabla\cdot(\ )$:

$$\nabla^2\phi = \lim_{\substack{V_2\to 0\\ (P)}} \frac{1}{V_2} \oint_{S_2} \hat{\mathbf{n}}_2\cdot\left\{\lim_{\substack{V_1\to 0\\ (P)}} \frac{1}{V_1} \oint_{S_1} \hat{\mathbf{n}}_1\phi\, dS_1\right\} dS_2$$

$$= \lim_{\substack{V_2\to 0\\ (P)}} \lim_{V_1\to 0} \frac{1}{V_2V_1} \oint_{S_2}\oint_{S_1} \hat{\mathbf{n}}_2\cdot\hat{\mathbf{n}}_1\phi\, dS_1\, dS_2$$

where subscripts 1 and 2 are used to show that two different, *independent* limiting processes are involved; $\mathbf{n}_2$ and $\mathbf{n}_1$ being totally unrelated. See Figure 11.10.)

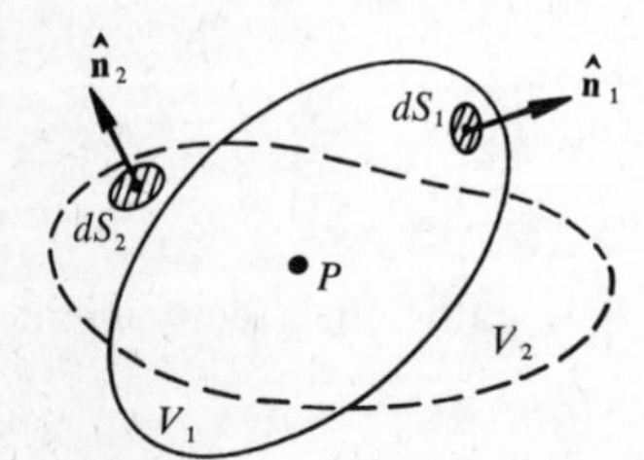

Figure 11.10 Two neighboring volumes.

(2) Differential Operator

Property (1) concentrates on the appearance of $\hat{\mathbf{n}}$ in the definition. But the rest of the definition dictates that ∇ be at the same time a limiting operator. Its operating power is by definition extended only to its right. We will see now that the limiting process involved is none other than the differentiation.

The first thing to be noticed is that ∇ possesses the property of linearity; that is,

$$\nabla \ast (a\mathscr{F} + b\mathscr{G}) = a\nabla \ast \mathscr{F} + b\nabla \ast \mathscr{G}$$

(where a and b are not involved in the process), as every limiting process must. Second, it can be shown that

$$\nabla \ast (\mathscr{F} \ast \mathscr{G}) = \nabla_{\mathscr{F}} \ast (\mathscr{F} \ast \mathscr{G}) + \nabla_{\mathscr{G}} \ast (\mathscr{F} \ast \mathscr{G}) \tag{11.21}$$

(The proof is left as a Problem.) A comparison of Eq. (11.21) with the corresponding expression for the differentiation operator, together with the gradient property of ∇, then gives ∇ the characteristics of a generalized differentiation operator.

With properties (1) and (2) established, all kinds of combinations involving ∇ can be carried out with ease. Care, however, must be constantly exercised to see that every step has meaning. It is especially important to remember that $\nabla_A(\mathbf{A}\cdot\mathbf{B}) \neq \mathbf{B}\cdot\nabla\mathbf{A}$!

EXAMPLE 1

$$\begin{aligned}\nabla(\mathbf{A}\cdot\mathbf{B}) &= \nabla_A(\mathbf{A}\cdot\mathbf{B}) + \nabla_B(\mathbf{A}\cdot\mathbf{B}) \\ &= \mathbf{B}\times(\nabla\times\mathbf{A}) + (\mathbf{B}\cdot\nabla_A)\mathbf{A} + \mathbf{A}\times(\nabla\times\mathbf{B}) + (\mathbf{A}\cdot\nabla_B)\mathbf{B}\end{aligned} \tag{11.22}$$

where Eq. (11.19) has been used.

EXAMPLE 2

$$\begin{aligned}\nabla\times(\mathbf{A}\times\mathbf{B}) &= \nabla_A\times(\mathbf{A}\times\mathbf{B}) + \nabla_B\times(\mathbf{A}\times\mathbf{B}) \\ &= (\nabla_A\cdot\mathbf{B})\mathbf{A} - (\nabla_A\cdot\mathbf{A})\mathbf{B} + (\nabla_B\cdot\mathbf{B})\mathbf{A} - (\nabla_B\cdot\mathbf{A})\mathbf{B}\end{aligned}$$

from Eq. (11.1). But $(\nabla_A\cdot\mathbf{B})\mathbf{A}$ and $(\nabla_B\cdot\mathbf{A})\mathbf{B}$ have no meaning unless one rearranges them into $(\mathbf{B}\cdot\nabla_A)\mathbf{A}$ and $(\mathbf{A}\cdot\nabla_B)\mathbf{B}$, respectively. Therefore

$$\nabla\times(\mathbf{A}\times\mathbf{B}) = (\mathbf{B}\cdot\nabla_A)\mathbf{A} - \mathbf{B}(\nabla\cdot\mathbf{A}) + \mathbf{A}(\nabla\cdot\mathbf{B}) - (\mathbf{A}\cdot\nabla_B)\mathbf{B} \tag{11.23}$$

Note that all terms in Eq. (11.23) are meaningful and unambiguous.

EXAMPLE 3

$$\begin{aligned}\nabla\cdot(\mathbf{A}\times\mathbf{B}) &= \nabla_A\cdot(\mathbf{A}\times\mathbf{B}) + \nabla_B\cdot(\mathbf{A}\times\mathbf{B}) \\ &= \nabla_A\times\mathbf{A}\cdot\mathbf{B} - \nabla_B\times\mathbf{B}\cdot\mathbf{A} \\ &= \mathbf{B}\cdot(\nabla\times\mathbf{A}) - \mathbf{A}\cdot(\nabla\times\mathbf{B})\end{aligned} \tag{11.24}$$

EXAMPLE 4

$$\begin{aligned}\nabla \cdot (a\mathbf{A}) &= \nabla_a \cdot (a\mathbf{A}) + \nabla_A \cdot (a\mathbf{A}) \\ &= (\nabla_a a) \cdot \mathbf{A} + a(\nabla_A \cdot \mathbf{A}) \\ &= \mathbf{A} \cdot (\nabla a) + a(\nabla \cdot \mathbf{A}) \\ &= \mathbf{A} \cdot (\text{grad } a) + a \text{ div } \mathbf{A} \qquad \textbf{(11.25)}\end{aligned}$$

EXAMPLE 5

$$\nabla \times (\nabla \times \mathbf{A}) = (\nabla_A \cdot \mathbf{A})\nabla_A - (\nabla_A \cdot \nabla_A)\mathbf{A}$$

from Eq. (11.1). But $(\nabla_A \cdot \mathbf{A})\nabla_A$ is meaningless, and has to be rearranged into $\nabla_A(\nabla_A \cdot \mathbf{A})$. Therefore

$$\begin{aligned}\nabla \times (\nabla \times \mathbf{A}) &= \nabla_A(\nabla_A \cdot \mathbf{A}) - (\nabla_A \cdot \nabla_A)\mathbf{A} \\ &= \nabla(\nabla \cdot \mathbf{A}) - \nabla^2\mathbf{A} \\ &= \text{grad (div } \mathbf{A}) - \nabla^2\mathbf{A} \qquad \textbf{(11.26)}\end{aligned}$$

where

$$\nabla^2\mathbf{A} = \nabla \cdot (\nabla\mathbf{A}) \qquad \textbf{(11.27)}$$

is sometimes referred to as the Laplacian of **A**. (Note that $\nabla^2\phi = \text{div (grad } \phi)$.) Because of Eq. (11.26), $\nabla^2\mathbf{A}$ can also be looked upon as just the shorthand for the quantity [grad (div **A**) – curl (curl **A**)]. Since this is often used, it is better to write it out:[19]

$$\nabla^2\mathbf{A} = \text{grad (div } \mathbf{A}) - \text{curl (curl } \mathbf{A}) \qquad \textbf{(11.28)}$$

EXAMPLE 6

$$\begin{aligned}\nabla \cdot (\mathbf{AB}) &= \nabla_A \cdot (\mathbf{AB}) + \nabla_B \cdot (\mathbf{AB}) \\ &= \mathbf{B}(\nabla \cdot \mathbf{A}) + (\mathbf{A} \cdot \nabla_B)\mathbf{B}\end{aligned}$$

Note that here the dot must always be located in between ∇ and **A**. In particular, $\nabla_B \cdot (\mathbf{AB})$ *cannot* be written as $\mathbf{A}(\nabla \cdot \mathbf{B})$.

EXAMPLE 7

$$\nabla \cdot (\mathbf{qq}) = \mathbf{q}(\nabla \cdot \mathbf{q}) + (\mathbf{q} \cdot \nabla_q)\mathbf{q} \qquad \textbf{(11.29)}$$

using the result of Example 6. Note that $\nabla \cdot (\mathbf{qq}) \neq 2\mathbf{q}(\nabla \cdot \mathbf{q})$!

EXAMPLE 8

$$\begin{aligned}\nabla \cdot (\mathbf{q} \cdot \tilde{\tau}) &= \nabla_q \cdot (\mathbf{q} \cdot \tilde{\tau}) + \nabla_\tau \cdot (\mathbf{q} \cdot \tilde{\tau}) \\ &= (\mathbf{q} \cdot \tilde{\tau}) \cdot \nabla_q + \nabla_\tau \cdot (\mathbf{q} \cdot \tilde{\tau}) \\ &= (\mathbf{q} \cdot \nabla) \cdot \tilde{\tau} + \mathbf{q} \cdot (\nabla \cdot \tilde{\tau}) \\ &= (\tilde{\tau} \cdot \nabla_q) \cdot \mathbf{q} + \mathbf{q} \cdot (\nabla \cdot \tilde{\tau}) \qquad \textbf{(11.30)}\end{aligned}$$

[19] Incidentally, div **q** and curl **q** are, respectively, the scalar and the vector of the dyadic $\nabla\mathbf{q}$.

where $(\tilde{\tau} \cdot \nabla_q) \cdot \mathbf{q}$ was defined in Section 11.3, subsection (6), part (d). One certainly cannot mistake $(\tilde{\tau} \cdot \nabla_q) \cdot \mathbf{q}$ to be $\tilde{\tau} \cdot (\nabla \cdot \mathbf{q})$ since this latter is not even defined.

(3) The Expression $(\hat{\mathbf{n}} \times \nabla) \ast \mathscr{F}$

This is actually a three-in-one expression:

$$\begin{cases} (\hat{\mathbf{n}} \times \nabla)\phi = \hat{\mathbf{n}} \times (\nabla\phi) \\ (\hat{\mathbf{n}} \times \nabla) \cdot \mathbf{q} = \hat{\mathbf{n}} \cdot (\nabla \times \mathbf{q}) \\ (\hat{\mathbf{n}} \times \nabla) \times \mathbf{q} \end{cases}$$

In the second expression, note that $\hat{\mathbf{n}} \times (\nabla \cdot \mathbf{q})$ is meaningless. In the third, note that it is certainly *not* the expression $\hat{\mathbf{n}} \times (\nabla \times \mathbf{q})$.

The other possibilities, $(\hat{\mathbf{n}} \times \nabla)\mathbf{q}$, $(\hat{\mathbf{n}} \times \nabla) \cdot \tilde{\tau}$, $(\mathbf{n} \times \nabla) \times \tilde{\tau}$, and so on, are not needed in this book and will not be discussed.

The vector $\hat{\mathbf{n}}$ here refers to the unit normal vector to an area element δS at the point P (Fig. 11.11). To avoid confusion, the definition of $\nabla \ast \mathscr{F}$ will be restated here as (Fig. 11.11)

$$\nabla \ast \mathscr{F} = \lim_{\substack{V_1 \to 0 \\ (P)}} \frac{1}{V_1} \oint_{S_1} \hat{\mathbf{n}}_1 \ast \mathscr{F} \, dS_1$$

(a) $$\hat{\mathbf{n}} \times (\nabla\phi) = \hat{\mathbf{n}} \times \left\{ \lim_{\substack{V_1 \to 0 \\ (P)}} \frac{1}{V_1} \oint_{S_1} \hat{\mathbf{n}}_1 \phi \, dS_1 \right\}$$

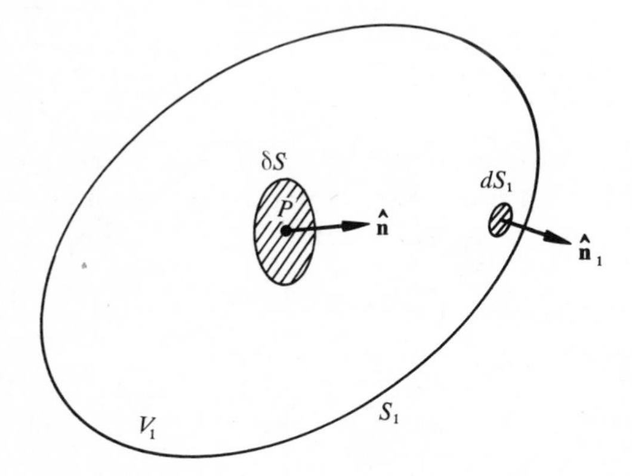

Figure 11.11 *dS_1 and δS.*

In the limit, V_1 can be taken as the cylinder with cross-sectional area δS and height $2\,\delta N$ (Fig. 11.12). Then,

$$\lim_{\substack{V_1 \to 0 \\ (P)}} \frac{1}{V_1} \oint_{S_1} \hat{\mathbf{n}}_1 \phi \, dS_1 = \frac{1}{2\,\delta n\,\delta S} \left\{ \int_{\text{top}} \hat{\mathbf{n}}_1 \phi \, dS_1 + \int_{\text{bottom}} \hat{\mathbf{n}}_1 \phi \, dS_1 + \int_{\text{side}} \hat{\mathbf{n}}_1 \phi \, dS_1 \right\}$$

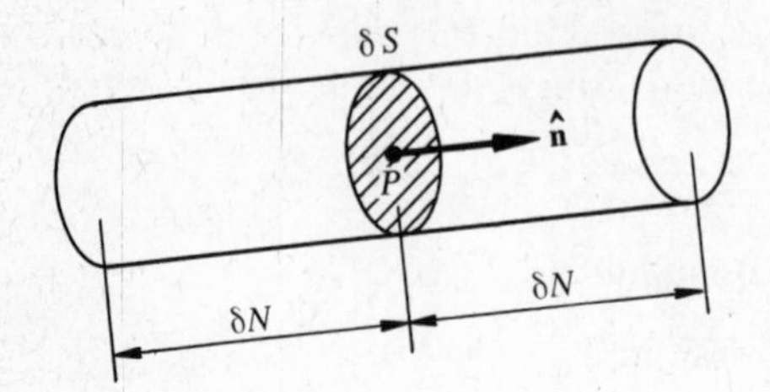

Figure 11.12 A cylindrical element.

The first two terms in the pair of braces are vectors parallel to $\hat{\mathbf{n}}$. Therefore,

$$\hat{\mathbf{n}}\times(\nabla\phi)=\hat{\mathbf{n}}\times\left(\frac{1}{2\,\delta n\,\delta S}\int_{\text{side}}\hat{\mathbf{n}}_1\phi\,dS_1\right)$$

$$=\hat{\mathbf{n}}\times\left(\frac{\not{2}\,\not{\delta n}}{\not{2}\,\not{\delta n}\,\delta S}\oint_{\delta C}\hat{\mathbf{n}}_1\phi\,dl\right)$$

neglecting higher-order terms, where δC is the bounding curve of δS, dl is the infinitesimal length along δC (whose positive sense remains to be chosen). $\hat{\mathbf{n}}_1$ on the side now becomes the normal unit vector of δC, positive pointing out of the area δS (Fig. 11.13). Since $\hat{\mathbf{n}}$ is fixed during the limit $V_1\to 0$, one has, in the eventual limit,

$$\hat{\mathbf{n}}\times(\nabla\phi)=\frac{1}{dS}\oint_{dC}\hat{\mathbf{n}}\times\hat{\mathbf{n}}_1\phi\,dl$$

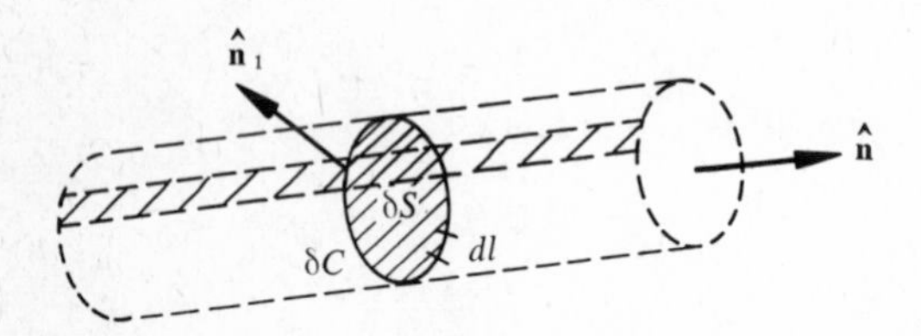

Figure 11.13 δC as bounding curve of δS.

Now, whether $\hat{\mathbf{n}}$ points upward or downward depends on which side of dS is regarded as positive. Each choice will yield a particular sense for the tangential vector $\hat{\mathbf{n}}\times\hat{\mathbf{n}}_1$ (Fig. 11.14). But, whatever the choice, the direction in which the line

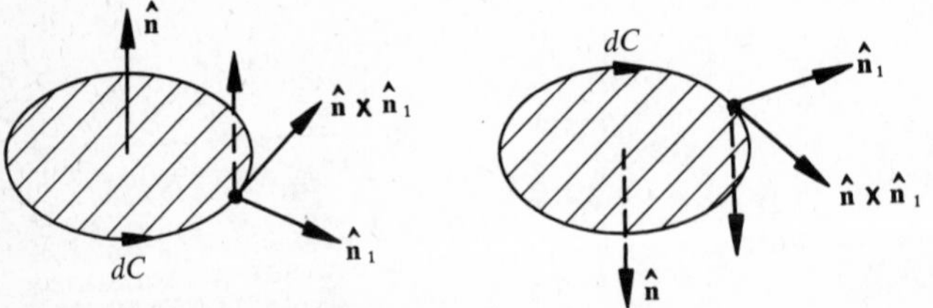

Figure 11.14 Orientation of dC.

integration is to proceed is always related to $\hat{\mathbf{n}}$ in the manner of a right-handed screw (Fig. 11.15). We therefore write

$$\hat{\mathbf{n}}\times(\nabla\phi)=\frac{1}{dS}\oint_{dC}(dl)\phi$$

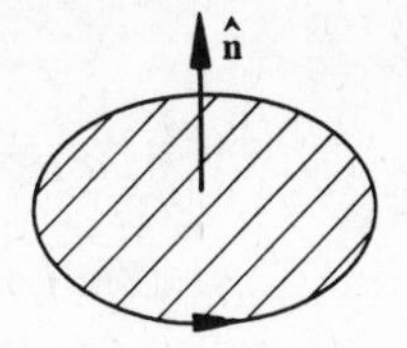

Figure 11.15 Right-handed orientation.

where the vectorial element of length dl around dC in the integration is *oriented* with relation to $\hat{\mathbf{n}}$ by the rule of the right-handed screw. (This will be referred to as the right-handed orientation.)

In keeping with the form in Eq. (11.5), the above result is better written as a limit:

$$\hat{\mathbf{n}} \times \nabla\phi = \lim_{\substack{S\to 0\\ (P)}} \frac{1}{S} \oint_C (dl)\phi \tag{11.31}$$

where C is the bounding curve of S, and the right-handed orientation is understood.

Similarly, one can prove that

$$\text{(b)}\quad \hat{\mathbf{n}} \cdot (\nabla \times \mathbf{q}) = \lim_{\substack{S\to 0\\ (P)}} \frac{1}{S} \oint_C (dl) \cdot \mathbf{q} \tag{11.32}$$

Incidentally, Eq. (11.32) provides a partial interpretation of curl $\mathbf{q}$; that is,

component of $\nabla \times \mathbf{q}$ in a certain direction = *local* $\oint (dl) \cdot \mathbf{q}$ per unit area perpendicular to that direction

where $\oint (dl) \cdot \mathbf{q}$ is known as the *circulation* around the local (infinitesimally small) circuit, *if* $\mathbf{q}$ is the velocity vector.

$$\text{(c)}\quad (\hat{\mathbf{n}} \times \nabla) \times \mathbf{q} = \lim_{\substack{S\to 0\\ (P)}} \frac{1}{S} \oint_C (dl) \times \mathbf{q} \tag{11.33}$$

The proof of this is contained in a Problem.

Conclusion

$$(\hat{\mathbf{n}} \times \nabla) \ast \mathscr{F} = \lim_{\substack{S\to 0\\ (P)}} \frac{1}{S} \oint_C (dl) \ast \mathscr{F} \tag{11.34}$$

with the right-handed orientation in force.

11.5 INTEGRAL RELATIONS

(1) Volume into Surface Integral

$$\int_V \nabla \ast \mathscr{F}\, dV = \oint_S \hat{\mathbf{n}} \ast \mathscr{F}\, dS \tag{11.35}$$

where S is the boundary of V.

Actually, this is a four-in-one relation:

$$\left\{\begin{array}{ll}\displaystyle\int_V \nabla\phi\, dV = \oint_S \hat{\mathbf{n}}\phi\, dS & \textbf{(11.35a)}\\ \displaystyle\int_V \nabla\cdot\mathbf{q}\, dV = \oint_S \hat{\mathbf{n}}\cdot\mathbf{q}\, dS & \textbf{(11.35b)}\\ \displaystyle\int_V \nabla\times\mathbf{q}\, dV = \oint_S \hat{\mathbf{n}}\times\mathbf{q}\, dS & \textbf{(11.35c)}\\ \displaystyle\int_V \nabla\cdot\tilde{\tau}\, dV = \oint_S \hat{\mathbf{n}}\cdot\tilde{\tau}\, dS & \textbf{(11.35d)}\end{array}\right.$$

where Eq. (11.35b) is the famous Gauss theorem.[20]

To show the *plausibility* of Eq. (11.35), one can divide the volume V into m small volume elements in an arbitrary manner. Let the kth element be denoted by $(\delta V)_k$, which has a boundary $(\delta S)_k$. Then, for vanishing $(\delta V)_k$, Eq. (11.5) yields

$$\int_{(\delta S)_k} \hat{\mathbf{n}} \divideontimes \mathscr{F}\, dS = (\nabla \divideontimes \mathscr{F})_k (\delta V)_k$$

where $(\nabla \divideontimes \mathscr{F})_k$ is evaluated at an arbitrary point inside $(\delta V)_k$. Summing over all volume elements and performing the limiting process $m \to \infty$ with $(\delta V)_k \to 0$, one obtains

$$\lim_{\substack{m\to\infty\\(\delta V)_k\to 0}} \sum_{k=1}^{m} \oint_{(\delta S)_k} \hat{\mathbf{n}} \divideontimes \mathscr{F}\, dS = \lim_{\substack{m\to\infty\\(\delta V)_k\to 0}} \sum_{k=1}^{m} (\nabla \divideontimes \mathscr{F})_k (\delta V)_k$$

The right-hand side is just the definition of the volume integral

$$\int_V \nabla \divideontimes \mathscr{F}\, dV$$

For the left-hand side we notice that, although $\hat{\mathbf{n}}$ on $(\delta S)_k$ points *out of* $(\delta V)_k$ and into one of its neighboring volume elements, say $(\delta V)_i$, $\hat{\mathbf{n}}$ for $(\delta V)_i$ on the interface will point *into* $(\delta V)_k$. Therefore, the contributions of adjacent elements to the left-hand side must cancel. This is the situation except for those elements in the neighborhood of the boundary S of V, since across S they are not adjacent to any

[20] Carl Friedrich Gauss (1777–1855) lives everywhere in mathematics; therefore he probably would not have minded sharing the credit of discovering this theorem with Lagrange, who implied it in 1762; G. Green, who arrived at it and Eqs. (11.35a,c) in 1828; and M. V. Ostrogradskii, who announced it in 1828 and published the (rather obscure) formulation in 1831. Gauss used the theorem in his researches on Newtonian attraction in 1813. His interest in the applications of mathematics was perpetuated by a long line of "Göttingen Mathematicians" after him: G. F. B. Riemann, who attacked the difficult problem of the motion of shock waves in his 1860 paper, "On the propagation of plane air-waves of finite amplitude" (in German); F. Klein, who, together with the physicist A. Sommerfeld, wrote the authoritative four-volume *Theory of Tops* (in German) during 1897–1910; D. Hilbert, who in 1912 applied his profound understanding of the theory of integral equations to the solution of the Boltzmann (integrodifferential) equation in the kinetic theory of gases; and finally, R. Courant, who, together with K. O. Friedrichs, taught the aerodynamicists the theory of supersonic flow through their celebrated 1948 text, *Supersonic Flow and Shock Waves*. The line was broken, however, when Courant migrated to the United States just before World War II.

other element of V. Obviously, then the only nonzero contributions to the left-hand side must come from S; that is, the left-hand side is equal to

$$\oint_S \hat{\mathbf{n}} \times \mathscr{F}\, dS$$

The plausibility of Eq. (11.35) is thus established.

Although any attempt at mathematical rigor is out of place in this book, the author must point out that even the above plausibility argument contains many hidden catches; if they are not carefully displayed, the reader will be eventually trapped by them. Any feeling on the part of the reader that this "proof" is easy and straightforward must be killed at once. In the following, three main catches will be mentioned first; and then, bits of detailed information will be offered in the form of cautions.

The first catch lies in the employment of $\nabla \times \mathscr{F}$. This presumes that the defining limit of $\nabla \times \mathscr{F}$ in Eq. (11.5) exists everywhere in V. (By existing we also imply that the limit is finite and single-valued.)

The second catch is in the statement that the contributions of adjacent elements to the left-hand side cancel. Now, this is true only when $\mathscr{F}$ is continuous on the interfaces of the elements. If $\mathscr{F}$ jumps in its value from one element to another across the interface, the contributions will not cancel.

The third catch lies in the existence of the defining limits of surface and volume integrals (again implying finiteness and single-valuedness). This then puts some restrictions on $\mathscr{F}$ and $\nabla \times \mathscr{F}$, and also on S and V. Some obvious restrictions here are that points in V and S must be situated within finite distances of one another (that is, no points at infinity[21]), that S should not form loops or knots, and that $\mathscr{F}$ must be bounded on S while $\nabla \times \mathscr{F}$ must be bounded in V.

Cautions

1. V must be a connected region—that is, one piece. Equation (11.35) must be applied piecewise to a multipiece region. However, V need *not* be simply connected; that is to say, V can assume the odd shapes of a spherical shell, a doughnut, a figure-8 doughnut, a sphere with a doughnut cavity, and deformations thereof.
2. S need *not* be connected. For example, the bounding surface of the volume in between two spheres is made of two disconnected pieces.
3. For the purposes of this book, it is safe to say that S need only be piecewise smooth—that is, smooth[22] except for a finite number of edges, corners, and points (Fig. 11.16). Equation (11.35) is either valid immediately or valid with insignificant modifications. (By insignificant modification we mean, for example, avoiding a point by going around it in its neighborhood and taking the limit as the point is approached.)

[21] With points at infinity, the integrals must be treated as improper.

[22] A surface is smooth if its normal varies continously.

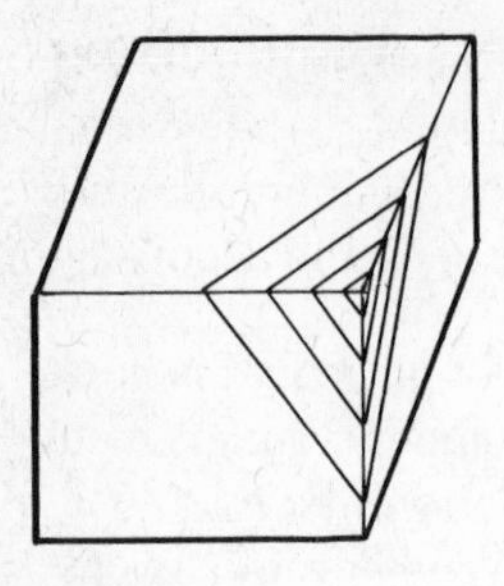

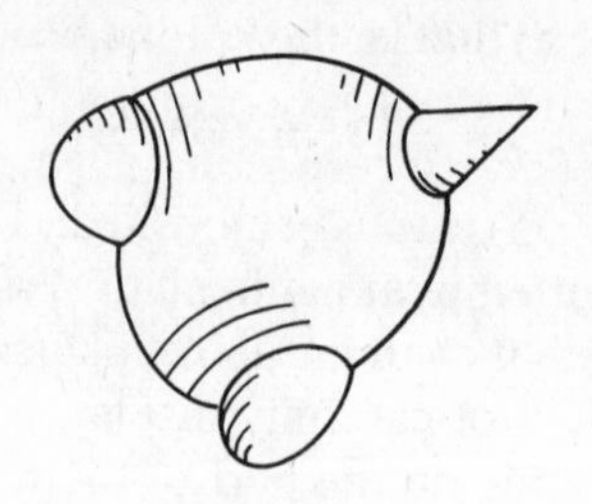

Figure 11.16 Piecewise smooth surfaces.

4. It is also safe to stick to cases where $\mathscr{F}$ and $\nabla \times \mathscr{F}$ are both single valued, finite, and continuous everywhere in V and on S.
5. Points, lines, and surfaces on which $\mathscr{F}$ or $\nabla \times \mathscr{F}$ are multivalued or unbounded are known as singular points, lines, and surfaces, respectively. In applying Eq. (11.35), one must take care to exclude such singularities from the volume V.
6. A surface across which $\mathscr{F}$ jumps is known as a surface of discontinuity (in $\mathscr{F}$). Equation (11.35) is, of course, *not* valid if the volume V contains such a surface of discontinuity. (However, it is possible to get around it; see one of the Problems.)

(2) Surface into Line Integral

$$\int_S (\hat{\mathbf{n}} \times \nabla) \times \mathscr{F}\, dS = \oint_C (dl) \times \mathscr{F} \tag{11.36}$$

where S extends the closed curve C, and right-handed orientation for $\hat{\mathbf{n}}$ and C is in force.

Here the surface S extends over C like a soap bubble. (S is said to be a *cap* of C.) The positive side of S is oriented in relation to the *general* direction along C according to the advancement of the right-handed screw (Fig. 11.17). To

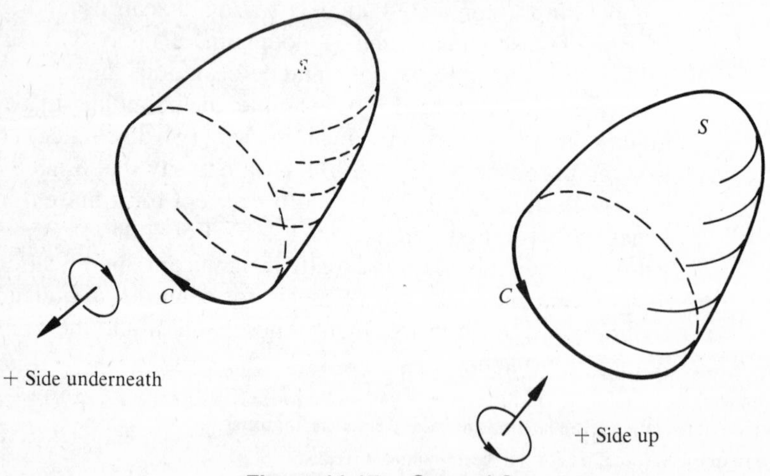

Figure 11.17 Caps of C.

understand this orientation a little bit more in detail, one can consider the curve C as a rubber band that can be rolled and shrunk toward any point on S while carrying with it the arrow (Fig. 11.18). The positive direction of the local unit normal vector $\hat{\mathbf{n}}$ of S is then oriented with respect to the shrunken rubber band according to the rule of the right-handed screw.

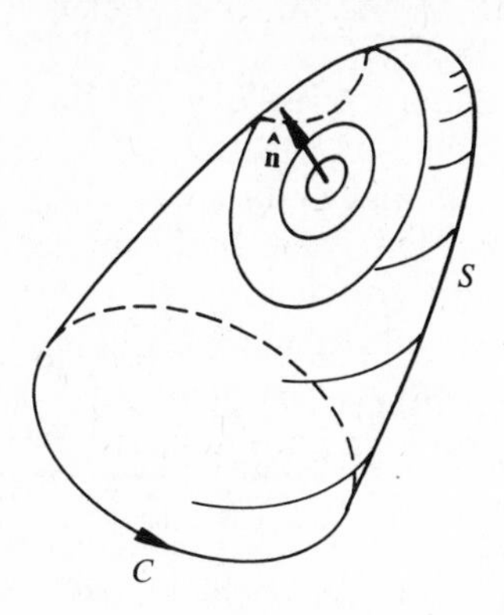

Figure 11.18 Orientation of C on its cap.

Equation (11.36) is actually a three-in-one relation:

$$\begin{cases} \int_S \hat{\mathbf{n}} \times (\nabla\phi)\, dS = \oint_C \phi\, dl & \text{(11.36a)} \\ \int_S \hat{\mathbf{n}} \cdot (\nabla \times \mathbf{q})\, dS = \oint_C \mathbf{q} \cdot dl & \text{(11.36b)} \\ \int_S (\hat{\mathbf{n}} \times \nabla) \times \mathbf{q}\, dS = \oint_C dl \times \mathbf{q} & \text{(11.36c)} \end{cases}$$

Equation (11.36b) is the famous Stokes theorem.[23] If $\mathbf{q}$ stands for the flow velocity, the right-hand side of Eq. (11.36b) is known as the circulation around the closed curve C. Thus, the Stokes theorem would provide fluid mechanics with a relation between the circulation and the curl of the velocity.

To show the plausibility of Eq. (11.36), one divides S into m small area elements in an arbitrary manner and apply Eq. (11.34). Then, proceeding in a similar way as that in subsection (1), the line integral along the interboundaries of adjacent area elements cancel out since it is integrated in one direction for one element and in the opposite direction for the adjacent element. This then leaves only the line integral over C for the right-hand side.

Discussions The closed curve C is a general space curve with a chosen *direction* (Fig. 11.19). dl is the line-element vector of C; that is, dl is tangent to C, and pointing in the *direction* of C. dl is, of course, just the line element (always positive). S extending C is an *open* surface.[24] But, it is quite arbitrary; any cap of C would do (Fig. 11.20).[25]

---

[23] It was really Kelvin who first established this theorem. He stated it fully in a postscript to a letter to Stokes, dated July 2, 1850. Stokes used it in 1854 as a Smith's prize examination question. In 1873 Maxwell referred to it as the Stokes theorem; and the name stuck.

[24] The sign of the surface integration therefore is not circled.

[25] However, see *Caution* below.

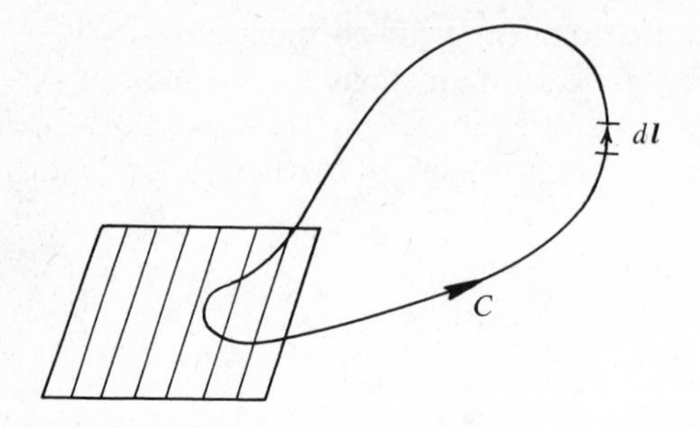

Figure 11.19 C as a space curve.

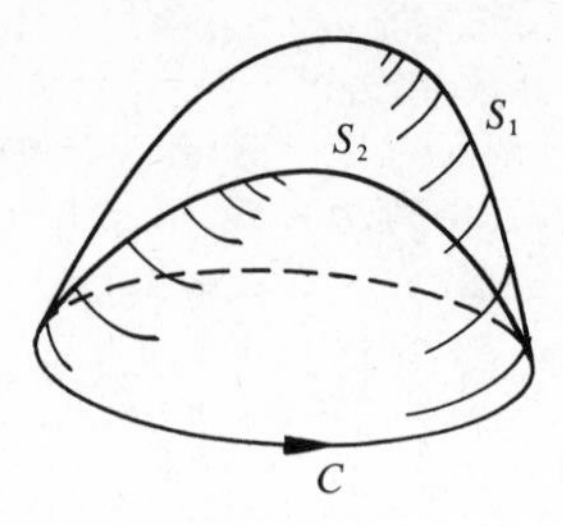

Figure 11.20 Two caps of C.

Discussions for subsection (1) also applies here with obvious modifications. For example, C must not have loops or the like; otherwise, it would be impossible to extend a surface, any surface, over C as a cap. (See Fig. 11.21.)

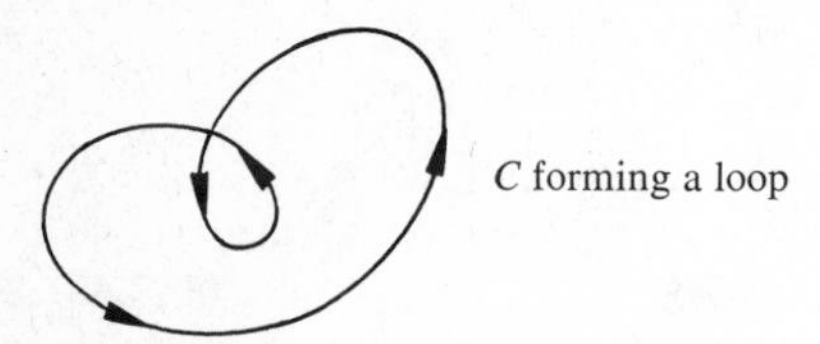

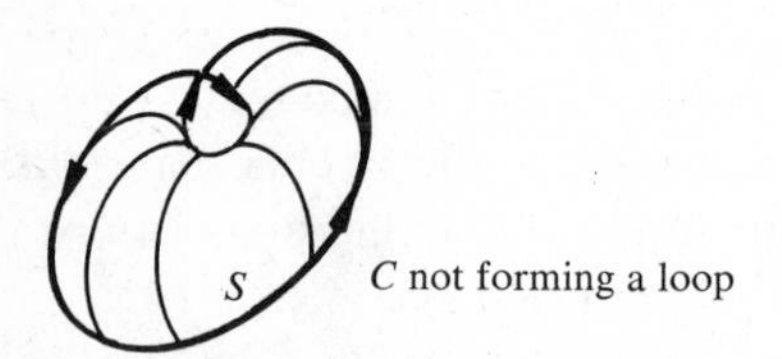

Figure 11.21 Looping and nonlooping curves.

Similarly, many of the previous cautions apply here also with appropriate modifications. For example, disconnected S must be treated separately, while disconnected C is admissible as long as the orientation is the same on every part of C. Figure 11.22 shows a counterexample where the two pieces of C are not consistently oriented. To see whether the different parts of C are consistently oriented, it is advised that the parts be connected up by cuts (Fig. 11.23). Then in

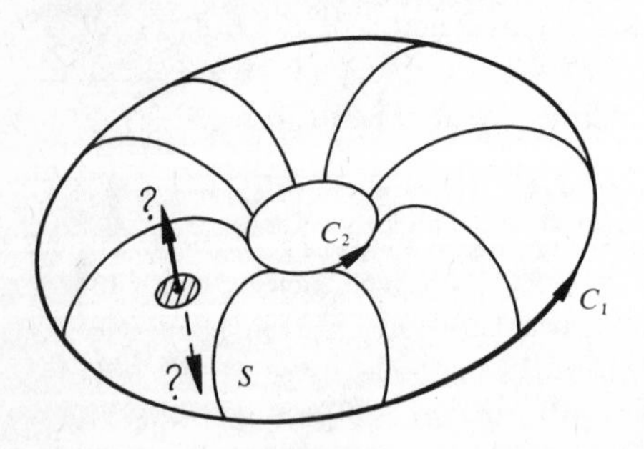

Figure 11.22 *Ill*-oriented curves.

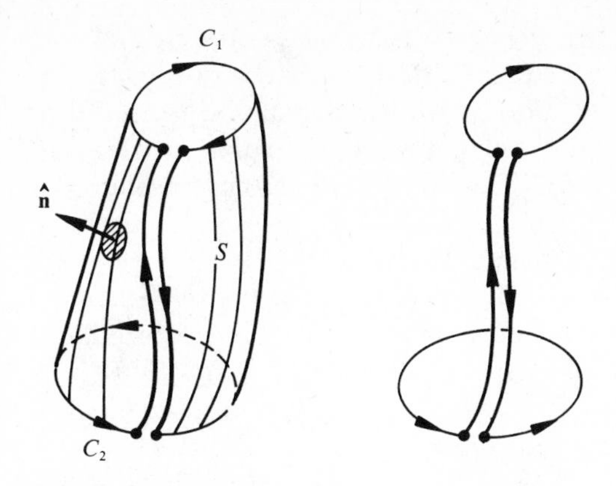

Figure 11.23 Proper orientation through cuts.

applying Eq. (11.36), the line integration along the cut, once this way and once that way, cancels itself, leaving only the line integrals along C_1 and C_2 with an unambiguous, consistent orientation.

Caution For the purposes of this book, it is safe to stick to piecewise smooth S and C. It is also advisable to stick to $\mathscr{F}$ and $\nabla \times \mathscr{F}$ that are single-valued, finite, and continuous *everywhere* on S and C. For a given C, this sometimes restricts the choice of S extending C.

COUNTEREXAMPLE

Consider a vector field $\mathbf{q}\langle\mathbf{r}\rangle$ that is zero for $x > 0$; on the plane $x = 0$, $\mathbf{q} = z\hat{\mathbf{y}}$; for $x < 0$, $\mathbf{q}$ varies in a certain fashion that is of no interest here. Now, take as C the square in the plane $x = 0$, with center at the origin and with sides parallel to the y- and z-axes. Then, clearly,

$$\oint_C \mathbf{q} \cdot dl \neq 0$$

But for any cap S of C, which lies in the half-space $x > 0$,

$$\int_S \hat{\mathbf{n}} \cdot (\nabla \times \mathbf{q})\, dS = 0$$

since $\mathbf{q} \equiv 0$ there. Thus, Stokes theorem does not hold here. The reason is simple: $\mathbf{q}$ is discontinuous on C when approached along the surface S. The theorem is valid, however, if S is restricted to lie entirely in the plane $x = 0$. (Whether it is also valid when S lies in the half-space $x < 0$ depends on the nature of variation of $\mathbf{q}$ there.)

11.6 VECTOR AND DYADIC RELATIONS IN GENERAL ORTHOGONAL COORDINATE SYSTEMS

Up to this point, all the definitions and relations are stated without reference to any coordinate system to show their coordinate-invariance. This is fine in *stating* a general problem. But, in *solving* the stated problem for individual situations, an appropriate coordinate system *must* be adopted in order to describe the specific geometry involved. The most frequently encountered coordinate systems are the orthogonal, curvilinear ones of which the rectangular Cartesian, the cylindrical, and the spherical are the three prominent members. In this section, all the previous coordinate-free relations will be rewritten for these orthogonal coordinate systems.

In general, an orthogonal coordinate system has three families of *coordinate surfaces*: ζ_1, ζ_2, ζ_3 = constants. The line of intersection of two surfaces, ζ_1 = constant and ζ_2 = constant, is then a *coordinate line* along which only ζ_3 changes. This is called a ζ_3-coordinate line. In Fig. 11.24, the ζ_1-, ζ_2-, and ζ_3-coordinate lines at a point P is shown together with a rectangular Cartesian coordinate system for comparison. Along the three coordinate lines at P, three unit vectors $\hat{\boldsymbol{\zeta}}_1$, $\hat{\boldsymbol{\zeta}}_2$, $\hat{\boldsymbol{\zeta}}_3$ are erected as shown. These three are mutually perpendicular since ζ_1, ζ_2, ζ_3 are orthogonal coordinates. In the figure, it is important to notice that $\hat{\boldsymbol{\zeta}}_1$, $\hat{\boldsymbol{\zeta}}_2$, $\hat{\boldsymbol{\zeta}}_3$ form a right-handed system.[26] (Actually, in this book, all coordinate systems are right-handed.)

Caution The directions of $\hat{\boldsymbol{\zeta}}_1$, $\hat{\boldsymbol{\zeta}}_2$, $\hat{\boldsymbol{\zeta}}_3$ change from point to point (but always remain mutually perpendicular); only $\hat{\mathbf{x}}$, $\hat{\mathbf{y}}$, $\hat{\mathbf{z}}$ referring to a rectangular Cartesian coordinate system stay fixed.

There is also a relation between $(\zeta_1, \zeta_2, \zeta_3)$ and (x, y, z):

$$\zeta_i = \zeta_i\langle x, y, z\rangle \qquad i = 1, 2, 3 \tag{11.37}$$

where $\zeta_i\langle x, y, z\rangle$ = constants, and $i = 1, 2, 3$, are exactly the three families of coordinate surfaces spoken of before. For example, in the cylindrical coordinate system, $\zeta_1 = R$, $\zeta_2 = \theta$, $\zeta_3 = z$; and the relation between $(\zeta_1, \zeta_2, \zeta_3)$ and (x, y, z) is

$$\begin{cases} \zeta_1 = \sqrt{x^2+y^2} & (\geqslant 0) \\ \zeta_2 = \tan^{-1}\dfrac{y}{x} & 0 \leqslant \zeta_2 < 2\pi \\ \zeta_3 = z \end{cases}$$

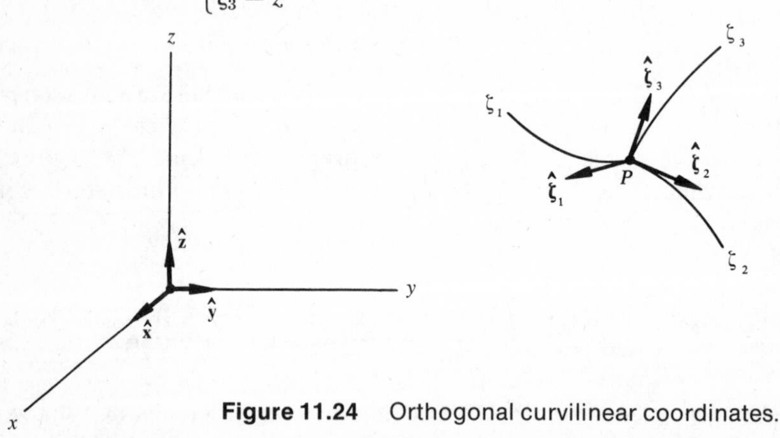

Figure 11.24 Orthogonal curvilinear coordinates.

[26] That is, rotating $\hat{\boldsymbol{\zeta}}_1$ toward $\hat{\boldsymbol{\zeta}}_2$ will advance a right-handed screw in the ζ_3-direction.

Relation (11.37) represents, of course, the transformation from the (x, y, z) system to the $(\zeta_1, \zeta_2, \zeta_3)$ system.

Now, considering the line element dl connecting P to a neighboring point, we have

$$(dl)^2 = (dx)^2 + (dy)^2 + (dz)^2$$

What, then, is the same expression in terms of $d\zeta_1$, $d\zeta_2$, and $d\zeta_3$?

From Eq. (11.37), one can solve for x, y, and z as functions of ζ_1, ζ_2, and ζ_3. Then,

$$\begin{cases} dx = \dfrac{\partial x}{\partial \zeta_1} d\zeta_1 + \dfrac{\partial x}{\partial \zeta_2} d\zeta_2 + \dfrac{\partial x}{\partial \zeta_3} d\zeta_3 \\ dy = \dfrac{\partial y}{\partial \zeta_1} d\zeta_1 + \dfrac{\partial y}{\partial \zeta_2} d\zeta_2 + \dfrac{\partial y}{\partial \zeta_3} d\zeta_3 \\ dz = \dfrac{\partial z}{\partial \zeta_1} d\zeta_1 + \dfrac{\partial z}{\partial \zeta_2} d\zeta_2 + \dfrac{\partial z}{\partial \zeta_3} d\zeta_3 \end{cases}$$

Therefore

$$\begin{aligned}(dl)^2 = &\left[\left(\frac{\partial x}{\partial \zeta_1}\right)^2 + \left(\frac{\partial y}{\partial \zeta_1}\right)^2 + \left(\frac{\partial z}{\partial \zeta_1}\right)^2\right](d\zeta_1)^2 \\ &+ \left[\left(\frac{\partial x}{\partial \zeta_2}\right)^2 + \left(\frac{\partial y}{\partial \zeta_2}\right)^2 + \left(\frac{\partial z}{\partial \zeta_2}\right)^2\right](d\zeta_2)^2 \\ &+ \left[\left(\frac{\partial x}{\partial \zeta_3}\right)^2 + \left(\frac{\partial y}{\partial \zeta_3}\right)^2 + \left(\frac{\partial z}{\partial \zeta_3}\right)^2\right](d\zeta_3)^2 \\ &+ [\cdots]\, d\zeta_1\, d\zeta_2 + [\cdots]\, d\zeta_2\, d\zeta_3 + [\cdots]\, d\zeta_1\, d\zeta_3 \end{aligned}$$

where the coefficients of $d\zeta_1\, d\zeta_2$, and so on, are not written out since they are of no interest in what follows.

On the other hand, from Fig. 11.25, we also have

$$(dl)^2 = (dl_1)^2 + (dl_2)^2 + (dl_3)^2 \qquad \textbf{(11.38)}$$

because of Pythagorean theorem. So, without further ado, we will conclude that the coefficients of $d\zeta_1\, d\zeta_2$, $d\zeta_2\, d\zeta_3$, and $d\zeta_1\, d\zeta_3$ must vanish (because of the mutual

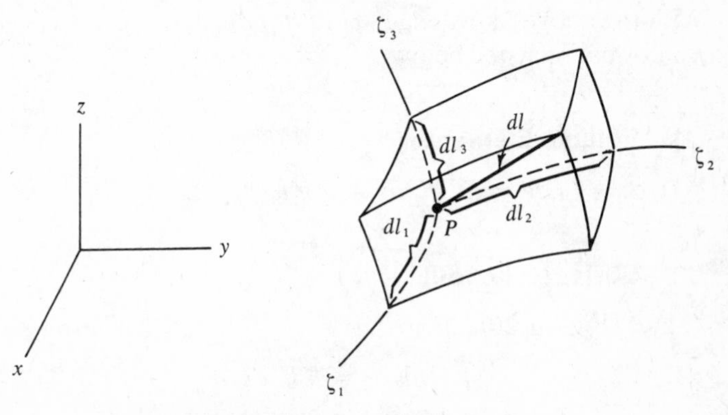

Figure 11.25 Line element.

orthogonality). Furthermore, since along a ζ_i-line only ζ_i changes, it must follow that

$$(dl_i)^2 = h_i^2(d\zeta_i)^2$$

where

$$h_i = \sqrt{\left(\frac{\partial x}{\partial \zeta_i}\right)^2 + \left(\frac{\partial y}{\partial \zeta_i}\right)^2 + \left(\frac{\partial z}{\partial \zeta_i}\right)^2} \qquad (> 0)$$

for $i = 1, 2, 3$.

Conclusion

$$\begin{aligned}(dl)^2 &= (dl_1)^2 + (dl_2)^2 + (dl_3)^2 \\ &= (h_1\, d\zeta_1)^2 + (h_2\, d\zeta_2)^2 + (h_3\, d\zeta_3)^2\end{aligned} \qquad \textbf{(11.39)}$$

For example, in the cylindrical coordinate system,

$$(dl)^2 = (dR)^2 + (R\, d\theta)^2 + (dz)^2$$

It is seen that, with $\zeta_1 = R$, $\zeta_2 = \theta$, and $\zeta_3 = z$,

$$\begin{aligned} h_1 &= 1 \qquad & dl_1 &= dR \\ h_2 &= R \qquad & dl_2 &= R\, d\theta \\ h_3 &= 1 \qquad & dl_3 &= dz \end{aligned}$$

Thus, h_i $(i = 1, 2, 3)$ are the factors one must use to multiply the differentials $d\zeta_i$ $(i = 1, 2, 3)$ in order to get the line elements (differential lengths) along the ζ_i-lines $(i = 1, 2, 3)$. They are called the scale factors for the three coordinates. For example, in the cylindrical coordinate system, $d\zeta_2 = d\theta$ is not a *length*; it is only a differential angle. The corresponding length is $R\, d\theta$; and R is thus the scale factor for the θ-coordinate.

Caution Remember that h_i $(i = 1, 2, 3)$ in general vary from point to point, and are thus functions of position (that is, of ζ_1, ζ_2, and ζ_3). For the special case of rectangular Cartesian coordinates, h_1, h_2, h_3 all assume the value unity, and stay constant throughout the entire space.

With the above knowledge about dl_1, dl_2, and dl_3, one can now proceed to gather the information listed below.

(1) Volume Element

$$dV = dl_1\, dl_2\, dl_3 = h_1 h_2 h_3\, d\zeta_1\, d\zeta_2\, d\zeta_3$$

(2) Surface Element

ζ_1-face (Fig. 11.26):

$$dS_1 = dl_2\, dl_3 = h_2 h_3\, d\zeta_2\, d\zeta_3$$

and so forth.

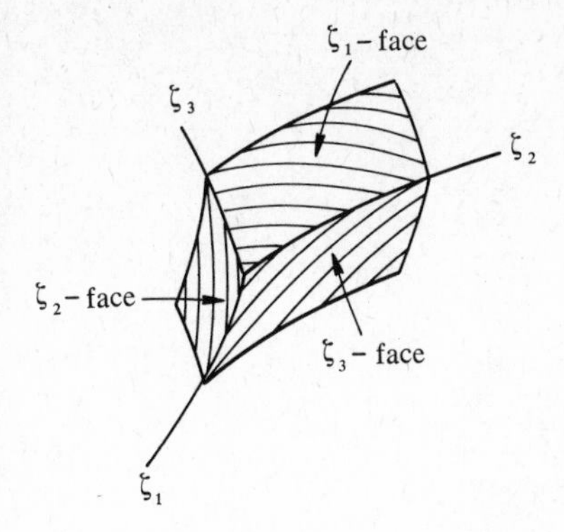

Figure 11.26 Coordinate surfaces.

(3) Dot Product

If $\mathbf{A} = A_1\hat{\zeta}_1 + A_2\hat{\zeta}_2 + A_3\hat{\zeta}_3$ and $\mathbf{B} = B_1\hat{\zeta}_1 + B_2\hat{\zeta}_2 + B_3\hat{\zeta}_3$ at a point P, then

$$\mathbf{A} \cdot \mathbf{B} = A_1B_1 + A_2B_2 + A_3B_3$$

(4) Cross Product

Similarly to (3),

$$\mathbf{A} \times \mathbf{B} = \begin{vmatrix} \hat{\zeta}_1 & \hat{\zeta}_2 & \hat{\zeta}_3 \\ A_1 & A_2 & A_3 \\ B_1 & B_2 & B_3 \end{vmatrix}$$

(5) Div q

Suppose that $\mathbf{q} = q_1\hat{\zeta}_1 + q_2\hat{\zeta}_2 + q_3\hat{\zeta}_3$ at a point P. Then

$$\operatorname{div} \mathbf{q} = \lim_{\substack{V \to 0 \\ \textcircled{P}}} \frac{1}{V} \oint_S \hat{\mathbf{n}} \cdot \mathbf{q} \, dS$$

by definition. As V, one may take the volume element bounded by the six coordinate surfaces: $\zeta_1 = \zeta_1|_P \pm \delta\zeta_1$, $\zeta_2 = \zeta_2|_P \pm \delta\zeta_2$, $\zeta_3 = \zeta_3|_P \pm \delta\zeta_3$, where $(\zeta_1|_P, \zeta_2|_P, \zeta_3|_P)$ are the coordinates of the point P. To find the contribution to the surface integral in the definition from the two sides $\zeta_1 = \zeta_1|_P \pm \delta\zeta_1$ we first calculate $\int \hat{\mathbf{n}} \cdot \mathbf{q} \, dS$ over the area on the ζ_1-surface passing through P, bounded by $\zeta_2|_P \pm \delta\zeta_2$ and $\zeta_3|_P \pm \delta\zeta_3$. The result is easily seen to be (Fig. 11.27)

$$(q_1h_2h_3)|_P \, (4 \, \delta\zeta_2 \, \delta\zeta_3)$$

for the positive side of which $\hat{\mathbf{n}} = \hat{\zeta}_1$, and

$$-(q_1h_2h_3)|_P \, (4 \, \delta\zeta_2 \, \delta\zeta_3)$$

for the negative side of which $\hat{\mathbf{n}} = -\hat{\zeta}_1$. The contribution from the face $\zeta_1 = \zeta_1|_P + \delta\zeta_1$ will then share the same factor $(4 \, \delta\zeta_2 \, \delta\zeta_3)$; but the factor $(q_1h_2h_3)|_P$ must now

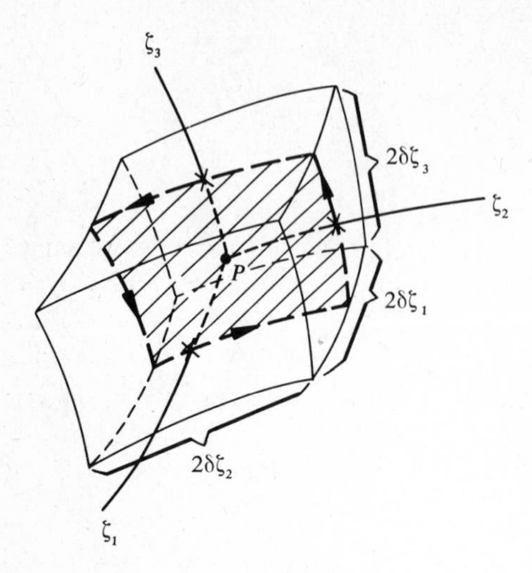

Figure 11.27 A small cell.

be replaced by

$$(q_1h_2h_3)\,|_P + \frac{\partial(q_1h_2h_3)}{\delta\zeta_1}\bigg|_P\,(\delta\zeta_1)$$

Similarly, the contribution from the face $\zeta_1|_P - \delta\zeta_1$ must be

$$\left[(-q_1h_2h_3)\,|_P - \frac{\partial(-q_1h_2h_3)}{\delta\zeta_1}\bigg|_P\,(\delta\zeta_1)\right](4\,\delta\zeta_1\,\delta\zeta_2)$$

Those from the other four faces are analogous. Adding all of them together for $\oint_S \hat{\mathbf{n}}\cdot\mathbf{q}\,dS$, one has the final result:

$$\text{div}\,\mathbf{q} = \frac{1}{h_1h_2h_3}\left[\frac{\partial(q_1h_2h_3)}{\partial\zeta_1} + \frac{\partial(q_2h_1h_3)}{\partial\zeta_2} + \frac{\partial(q_3h_1h_2)}{\partial\zeta_3}\right] \tag{11.40}$$

(6) Curl q

It would be very cumbersome to start here with the original definition of curl $\mathbf{q}$, since the surface integral now involves a vector $\hat{\mathbf{n}}\times\mathbf{q}$ instead of a scalar $\hat{\mathbf{n}}\cdot\mathbf{q}$ in (5). Then, the change in both the magnitude *and* the direction of this vector must be accounted for when we move from P to the six faces of V. Luckily, there is no need for us to perform such a messy job, since we can use instead the definition (11.32), where the left-hand side is seen to be just the n-component of curl $\mathbf{q}$; that is, $(\nabla\times\mathbf{q})_n$.

The integrand involved on the right-hand side is the scalar $(d\mathbf{l})\cdot\mathbf{q}$, which can be treated with ease.

For S, we will first choose the surface element on the positive side of the ζ_1-surface passing through P, bounded by $\zeta_2|_P \pm \delta\zeta_2$ and $\zeta_3|_P \pm \delta\zeta_3$ (Fig. 11.27). Here $\hat{\mathbf{n}} = \hat{\boldsymbol{\zeta}}_1$, and the boundary C carries the direction indicated in the figure. A similar argument as that in (5) will then lead to

$$(\nabla \times \mathbf{q})_1 = (\nabla \times \mathbf{q})_{\zeta_1} = \hat{\zeta}_1 \cdot (\nabla \times \mathbf{q})$$
$$= \frac{1}{h_2 h_3}\left[\frac{\partial (h_3 q_3)}{\partial \zeta_2} - \frac{\partial (h_2 q_2)}{\partial \zeta_3}\right]$$

The other two components of $\nabla \times \mathbf{q}$ can also be calculated in the same way with obvious modifications. The total result can be conveniently displayed as

$$\nabla \times \mathbf{q} = \frac{1}{h_1 h_2 h_3}\begin{vmatrix} h_1\hat{\zeta}_1 & h_2\hat{\zeta}_2 & h_3\hat{\zeta}_3 \\ \dfrac{\partial}{\partial \zeta_1} & \dfrac{\partial}{\partial \zeta_2} & \dfrac{\partial}{\partial \zeta_3} \\ h_1 q_1 & h_2 q_2 & h_3 q_3 \end{vmatrix} \tag{11.41}$$

(7) Grad ϕ

Here the original definition will also bring the same kind of trouble as in (6), since the integrand involved is the vector $\hat{\mathbf{n}}\phi$. To avoid the messy differentiation of a vectorial quantity, it is better to start with Eq. (11.3):

$$d\phi = (\text{grad}\ \phi) \cdot d\mathbf{r} = (\text{grad}\ \phi) \cdot dl$$
$$= (\text{grad}\ \phi)_1 (dl_1) + (\text{grad}\ \phi)_2 (dl_2) + (\text{grad}\ \phi)_3 (dl_3)$$
$$= (\text{grad}\ \phi)_1 (h_1\, d\zeta_1) + (\text{grad}\ \phi)_2 (h_2\, d\zeta_2) + (\text{grad}\ \phi)_3 (h_3\, d\zeta_3)$$

But, since $\phi = \phi\langle \zeta_1, \zeta_2, \zeta_3 \rangle$, we also have

$$d\phi = \frac{\partial \phi}{\partial \zeta_1}\, d\zeta_1 + \frac{\partial \phi}{\partial \zeta_2}\, d\zeta_2 + \frac{\partial \phi}{\partial \zeta_3}\, d\zeta_3$$

Therefore

$$\left[h_1\ (\text{grad}\ \phi)_1 - \frac{\partial \phi}{\partial \zeta_1}\right] d\zeta_1 + \left[h_2\ (\text{grad}\ \phi)_2 - \frac{\partial \phi}{\partial \zeta_2}\right] d\zeta_2 + \left[h_3\ (\text{grad}\ \phi)_3 - \frac{\partial \phi}{\partial \zeta_3}\right] d\zeta_3 = 0$$

for any *arbitrary* $(d\zeta_1, d\zeta_2, d\zeta_3)$. This is only true if

$$h_i\,(\text{grad}\ \phi)_i = \frac{\partial \phi}{\partial \zeta_i} \qquad i = 1, 2, 3$$

Therefore,

$$\text{grad}\ \phi = \left(\frac{1}{h_1}\frac{\partial \phi}{\partial \zeta_1}, \frac{1}{h_2}\frac{\partial \phi}{\partial \zeta_2}, \frac{1}{h_3}\frac{\partial \phi}{\partial \zeta_3}\right)$$
$$= \frac{\hat{\zeta}_1}{h_1}\frac{\partial \phi}{\partial \zeta_1} + \frac{\hat{\zeta}_2}{h_2}\frac{\partial \phi}{\partial \zeta_2} + \frac{\hat{\zeta}_3}{h_3}\frac{\partial \phi}{\partial \zeta_3} \tag{11.42}$$

A combined use of Eqs. (11.40) and (11.42) would yield

$$\nabla^2 \phi = \frac{1}{h_1 h_2 h_3}\left[\frac{\partial}{\partial \zeta_1}\left(\frac{h_2 h_3}{h_1}\frac{\partial \phi}{\partial \zeta_1}\right) + \frac{\partial}{\partial \zeta_2}\left(\frac{h_3 h_1}{h_2}\frac{\partial \phi}{\partial \zeta_2}\right) + \frac{\partial}{\partial \zeta_3}\left(\frac{h_1 h_2}{h_3}\frac{\partial \phi}{\partial \zeta_3}\right)\right] \tag{11.43}$$

(8) $\nabla^2\mathbf{q}$

First of all, do *not* make the mistake of using Eq. (11.43) here. Second, note that to express $\nabla^2\mathbf{q} = \nabla \cdot (\nabla\mathbf{q})$ in a general orthogonal coordinate system requires the difficult task of evaluating the surface integral involved in the definition of ∇. Although this is a possible way, it is best left alone. Instead, one can use Eq. (11.28):

$$\nabla^2\mathbf{q} = \text{grad}\,(\text{div}\,\mathbf{q}) - \nabla \times (\nabla \times \mathbf{q})$$

where the expressions on the right-hand side have already been established in the previous paragraphs.

(9) $\mathbf{q} \cdot \nabla\phi$

This is simple:

$$\mathbf{q} \cdot (\nabla\phi) = \frac{q_1}{h_1}\frac{\partial\phi}{\partial\zeta_1} + \frac{q_2}{h_2}\frac{\partial\phi}{\partial\zeta_2} + \frac{q_3}{h_3}\frac{\partial\phi}{\partial\zeta_3} \tag{11.44}$$

(10) $\mathbf{q} \cdot \nabla\mathbf{q}$

Use Eq. (11.22):

$$\mathbf{q} \cdot \nabla\mathbf{q} = \tfrac{1}{2}\nabla(q^2) - \mathbf{q} \times \text{curl}\,\mathbf{q}$$

(11) $\mathbf{A} \cdot \nabla\mathbf{B}$

Solve Eqs. (11.22) and (11.23) for $\mathbf{A} \cdot \nabla\mathbf{B}$, and add the results up.

Finally, before we close this section, let us here append three topics.

(a) The dyadic $\nabla\mathbf{r}$, where $\mathbf{r}$ is the position vector, has a very interesting property, namely

$$\mathbf{A} \cdot \nabla\mathbf{r} = (\nabla\mathbf{r}) \cdot \mathbf{A} = \mathbf{A}$$

for any $\mathbf{A}$. It is necessary to prove this property only with respect to a rectangular Cartesian coordinate system, since a *vectorial* equation must be true for *all* coordinate systems *if it is true for one*. Now, from the definition of ∇, it is easy to establish that

$$\nabla\mathbf{r} = \hat{\mathbf{x}}\hat{\mathbf{x}} + \hat{\mathbf{y}}\hat{\mathbf{y}} + \hat{\mathbf{z}}\hat{\mathbf{z}}$$

with respect to rectangular Cartesian coordinates (x, y, z). Also

$$\mathbf{A} = A_x\hat{\mathbf{x}} + A_y\hat{\mathbf{y}} + A_z\hat{\mathbf{z}}$$

Taking the dot products $\mathbf{A} \cdot \nabla\mathbf{r}$ and $(\nabla\mathbf{r}) \cdot \mathbf{A}$, and noting that $\hat{\mathbf{x}} \cdot \hat{\mathbf{x}} = 1$, $\hat{\mathbf{x}} \cdot \hat{\mathbf{y}} = 0$, and so forth, we have both equal to $\mathbf{A}$ with respect to (x, y, z).

Because of this interesting property, $\nabla\mathbf{r}$ is given a special name, the *idemfactor*; and a special symbol, $\tilde{\mathbf{I}}$. Thus

$$\tilde{\mathbf{I}} \cdot \mathbf{A} = \mathbf{A} \cdot \tilde{\mathbf{I}} = \mathbf{A} \tag{11.45}$$

for any **A**. The expression for $\tilde{\mathbf{I}}$ in the general orthogonal coordinates $(\zeta_1, \zeta_2, \zeta_3)$ is not so easy to obtain because of the change of directions of $\hat{\zeta}_1$, $\hat{\zeta}_2$, and $\hat{\zeta}_3$. But, in this book, all we need is the fact that there exists a dyadic idemfactor $\tilde{\mathbf{I}}$ with property (11.45).

(b) Although it is not adopted in this book, the reader must be aware of a convention called the Cartesian-tensorial notation, which is very often used in the literature. This convention is used *only with respect to rectangular Cartesian coordinate systems*. But, instead of (x, y, z), it uses (x_1, x_2, x_3), and instead of (q_x, q_y, q_z), it uses (q_1, q_2, q_3), and so on. Furthermore, (x_1, x_2, x_3) is abbreviated to just x_i, (q_1, q_2, q_3) to q_i, and so on, where the subscript i is understood to assume the numbers, 1, 2, 3, consecutively. Thus, for example,

$$\operatorname{div} \mathbf{q} = \frac{\partial q_1}{\partial x_1} + \frac{\partial q_2}{\partial x_2} + \frac{\partial q_3}{\partial x_3}$$

$$= \sum_{i=1}^{3} \frac{\partial q_i}{\partial x_i}$$

for rectangular Cartesian coordinates.

The shorthand is carried even further by omitting the above summation sign whenever it occurs

$$\operatorname{div} \mathbf{q} = \frac{\partial q_i}{\partial x_i}$$

(for rectangular Cartesian coordinates), where the summation is suggested through the recurrence of i in the term $\partial q_i/\partial x_i$, once at the foot of q and once at the foot of x. In other words, in this convention, whenever a subscript occurs more than once in one term, it is to be summed[27] over 1, 2, and 3.

(c) There is another way of obtaining expressions of various vectorial and dyadic operations with respect to general orthogonal coordinate systems. It starts with Eq. (11.42) which is seen to suggest that, in general, the operator ∇ is to be identified as

$$\nabla = \frac{\hat{\zeta}_1}{h_1}\frac{\partial}{\partial \zeta_1} + \frac{\hat{\zeta}_2}{h_2}\frac{\partial}{\partial \zeta_2} + \frac{\hat{\zeta}_3}{h_3}\frac{\partial}{\partial \zeta_3} \tag{11.46}$$

Now, this is perfectly in keeping with the operational nature of ∇, and is a true identification. But one must be careful when applying it to a vector or dyadic. For example, it is true that

$$\nabla \cdot \mathbf{q} = \left(\frac{\hat{\zeta}_1}{h_1}\frac{\partial}{\partial \zeta_1} + \frac{\hat{\zeta}_2}{h_2}\frac{\partial}{\partial \zeta_2} + \frac{\hat{\zeta}_3}{h_3}\frac{\partial}{\partial \zeta_3}\right) \cdot (q_1\hat{\zeta}_1 + q_2\hat{\zeta}_2 + q_3\hat{\zeta}_3)$$

---

27 This is known as Einstein's summation convention, since he first used it (in his relativistic papers). If one wishes to denote collectively the three quantities

$$\frac{\partial q_1}{\partial x_1}, \frac{\partial q_2}{\partial x_2}, \frac{\partial q_3}{\partial x_3}$$

one must write

$$\frac{\partial q_i}{\partial x_i} \textit{(no summation)}$$

Without the instruction "no summation," it would always mean div **q**.

One must, however, remember that ∇ governs not only q_1, q_2, q_3, but also $\hat{\zeta}_1, \hat{\zeta}_2, \hat{\zeta}_3$, which change in direction. So in order to be able to follow this operational route it is necessary first to inquire into the variations of $\hat{\zeta}_1$, $\hat{\zeta}_2$, and $\hat{\zeta}_3$.

To this end, one can apply Eq. (11.42) to a very special function $\phi\langle \zeta_1, \zeta_2, \zeta_3 \rangle$ – that is, to ζ_1 itself:

$$\nabla \zeta_1 = \frac{\hat{\zeta}_1}{h_1}$$

Furthermore, $\nabla \times (\nabla \zeta_1) = 0$. Therefore,

$$\begin{aligned} \nabla \times \left(\frac{\hat{\zeta}_1}{h_1}\right) &= \frac{1}{h_1} \nabla \times \hat{\zeta}_1 - \hat{\zeta}_1 \times \nabla \left(\frac{1}{h_1}\right) \\ &= 0 \end{aligned}$$

Thus,

$$\begin{aligned} \frac{1}{h_1}\left(\nabla \times \hat{\zeta}_1\right) &= \hat{\zeta}_1 \times \nabla \left(\frac{1}{h_1}\right) \\ &= \hat{\zeta}_1 \times \left[-\frac{\hat{\zeta}_1}{h_1^3} \frac{\partial h_1}{\partial \zeta_1} - \frac{\hat{\zeta}_2}{h_1^2 h_2} \frac{\partial h_1}{\partial \zeta_2} - \frac{\hat{\zeta}_3}{h_1^2 h_3} \frac{\partial h_1}{\partial \zeta_3} \right] \end{aligned}$$

where Eq. (11.42) is again used. Therefore

$$\nabla \times \hat{\zeta}_1 = \frac{\hat{\zeta}_2}{h_1 h_3} \frac{\partial h_1}{\partial \zeta_3} - \frac{\hat{\zeta}_3}{h_1 h_2} \frac{\partial h_1}{\partial \zeta_2} \tag{11.47}$$

Similarly, one can write down expressions for $\nabla \times \hat{\zeta}_2$ and $\nabla \times \hat{\zeta}_3$ by changing the subscripts in Eq. (11.47) cyclically from 1 to 2, 2 to 3, and 3 to 1.

Again, since $\hat{\zeta}_1 = \hat{\zeta}_2 \times \hat{\zeta}_3$, we have

$$\begin{aligned} \nabla \cdot \hat{\zeta}_1 &= \nabla \cdot (\hat{\zeta}_2 \times \hat{\zeta}_3) \\ &= \hat{\zeta}_3 \cdot (\nabla \times \hat{\zeta}_2) - \hat{\zeta}_2 \cdot (\nabla \times \hat{\zeta}_3) \\ &= \frac{1}{h_2 h_1} \frac{\partial h_2}{\partial \zeta_1} + \frac{1}{h_3 h_1} \frac{\partial h_3}{\partial \zeta_1} \\ &= \frac{1}{h_1 h_2 h_3} \frac{\partial (h_2 h_3)}{\partial \zeta_1} \end{aligned} \tag{11.48}$$

where the expressions for $\nabla \times \hat{\zeta}_2$, and so forth, have been dealt with in the preceding paragraph. Similarly, $\nabla \cdot \hat{\zeta}_2$ and $\nabla \cdot \hat{\zeta}_3$ can also be written down by cyclic changes of subscripts in Eq. (11.48). Furthermore,

$$d\mathbf{l} = h_1 (d\zeta_1) \hat{\zeta}_1 + h_2 (d\zeta_2) \hat{\zeta}_2 + h_3 (d\zeta_3) \hat{\zeta}_3$$

Therefore

$$\frac{\partial \mathbf{l}}{\partial \zeta_1} = h_1 \hat{\zeta}_1, \quad \frac{\partial \mathbf{l}}{\partial \zeta_2} = h_2 \hat{\zeta}_2, \quad \frac{\partial \mathbf{l}}{\partial \zeta_3} = h_3 \hat{\zeta}_3$$

and thus

$$\frac{\partial (h_1 \hat{\zeta}_1)}{\partial \zeta_2} = \frac{\partial (h_2 \hat{\zeta}_2)}{\partial \zeta_1}$$

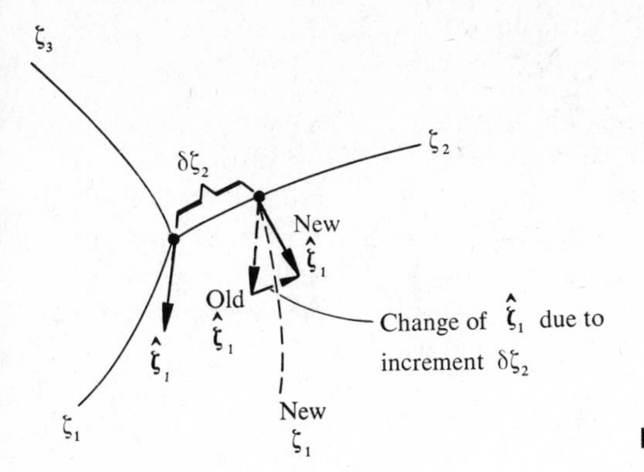

Figure 11.28 Variation of $\hat{\boldsymbol{\zeta}}_1$.

that is,

$$h_1 \frac{\partial \hat{\boldsymbol{\zeta}}_1}{\partial \zeta_2} - h_2 \frac{\partial \hat{\boldsymbol{\zeta}}_2}{\partial \zeta_1} = \hat{\boldsymbol{\zeta}}_2 \frac{\partial h_2}{\partial \zeta_1} - \hat{\boldsymbol{\zeta}}_1 \frac{\partial h_1}{\partial \zeta_2}$$

Referring to Fig. 11.28, it is seen that[28]

$$\frac{\partial \hat{\boldsymbol{\zeta}}_1}{\partial \zeta_2} \parallel \hat{\boldsymbol{\zeta}}_2$$

And, similarly,

$$\frac{\partial \hat{\boldsymbol{\zeta}}_2}{\partial \zeta_1} \parallel \hat{\boldsymbol{\zeta}}_1$$

Therefore

$$\begin{cases} h_1 \dfrac{\partial \hat{\boldsymbol{\zeta}}_1}{\partial \zeta_2} = \hat{\boldsymbol{\zeta}}_2 \dfrac{\partial h_2}{\partial \zeta_1} \\ h_2 \dfrac{\partial \hat{\boldsymbol{\zeta}}_2}{\partial \zeta_1} = \hat{\boldsymbol{\zeta}}_1 \dfrac{\partial h_1}{\partial \zeta_2} \end{cases}$$

that is,

$$\frac{\partial \hat{\boldsymbol{\zeta}}_1}{\partial \zeta_2} = \frac{\hat{\boldsymbol{\zeta}}_2}{h_1} \frac{\partial h_2}{\partial \zeta_1} \tag{11.49}$$

$$\frac{\partial \hat{\boldsymbol{\zeta}}_2}{\partial \zeta_1} = \frac{\hat{\boldsymbol{\zeta}}_1}{h_2} \frac{\partial h_1}{\partial \zeta_2} \tag{11.50}$$

Four more similar equations can be immediately written down by cyclic changes of subscripts.

Finally, starting with $\hat{\boldsymbol{\zeta}}_1 = \hat{\boldsymbol{\zeta}}_2 \times \hat{\boldsymbol{\zeta}}_3$, we have

$$\frac{\partial \hat{\boldsymbol{\zeta}}_1}{\partial \zeta_1} = \frac{\partial \hat{\boldsymbol{\zeta}}_2}{\partial \zeta_1} \times \hat{\boldsymbol{\zeta}}_3 + \hat{\boldsymbol{\zeta}}_2 \times \frac{\partial \hat{\boldsymbol{\zeta}}_3}{\partial \zeta_1} \tag{11.51}$$

$$= \frac{\hat{\boldsymbol{\zeta}}_1 \times \hat{\boldsymbol{\zeta}}_3}{h_2} \frac{\partial h_1}{\partial \zeta_2} + \frac{\hat{\boldsymbol{\zeta}}_2 \times \hat{\boldsymbol{\zeta}}_1}{h_3} \frac{\partial h_1}{\partial \zeta_3}$$

[28] This is so because orthogonal coordinate lines are *lines of curvature* (Dupin's theorem).

$$= -\frac{\hat{\boldsymbol{\zeta}}_2}{h_2}\frac{\partial h_1}{\partial \zeta_2} - \frac{\hat{\boldsymbol{\zeta}}_3}{h_3}\frac{\partial h_1}{\partial \zeta_3} \tag{11.52}$$

where Eq. (11.50) and the similar one for $\partial\hat{\boldsymbol{\zeta}}_3/\partial\zeta_1$ have been used. Out of Eq. (11.52), two more can be written down by cyclic changes of subscripts.

With the above formulas established, it is now possible to regain all the previous results, and to obtain new results, with reasonable effort. For example, the task of getting $\boldsymbol{\nabla}\cdot\mathbf{q}$, using Eq. (11.46), can now be continued; and the result would be exactly the same as Eq. (11.40). As another example, $\boldsymbol{\nabla}\mathbf{q}$ can be shown to contain nine terms, the first two of which are such that

$$\boldsymbol{\nabla}\mathbf{q} = \frac{\hat{\boldsymbol{\zeta}}_1\hat{\boldsymbol{\zeta}}_1}{h_1}\left(\frac{\partial q_1}{\partial \zeta_1} + \frac{q_2}{h_2}\frac{\partial h_1}{\partial \zeta_2} + \frac{q_3}{h_3}\frac{\partial h_1}{\partial \zeta_3}\right) + \frac{\hat{\boldsymbol{\zeta}}_1\hat{\boldsymbol{\zeta}}_2}{h_1 h_2}\left(h_2\frac{\partial q_2}{\partial \zeta_1} - q_1\frac{\partial h_1}{\partial \zeta_2}\right) + \cdots \tag{11.53}$$

CHAPTER 12

KINEMATICS OF FLUID FLOW

12.1 INTRODUCTION

In classical fluid dynamics a fluid is considered a *continuum*, with various properties that are well-defined functions of space $\mathbf{r}$ and time t, such as $\mathbf{q}\langle\mathbf{r}, t\rangle$, $\rho\langle\mathbf{r}, t\rangle$, and so on. In order to distinguish a fluid *occupying* the space from the *empty* space itself, we will consider fluid particles or infinitesimal fluid elements moving around in the empty space with individual velocities that change with time as they move. A fluid particle is simply an infinitesimal portion or sample of the fluid *as a continuum*, which possesses individuality. It is certainly *not* a fluid molecule.

The individuality of a fluid particle can be maintained—by coloring it with a dye, for example. This is actually done in many laboratory investigations. But for the sake of analysis it is better to label a particle according to some scheme.

In the so-called Lagrangian description,[1] the position $\mathbf{r}_0 = (a, b, c)$ of a fluid particle at an instant $t = t_0$ is used to label it. To fix ideas, let us say that a rectangular Cartesian coordinate system is used. Then, if $\mathbf{r} = (x, y, z)$ is the position of the same particle at time t, the flow geometry of the particle will be completely specified if the function $\mathbf{r} = \mathbf{r}\langle\mathbf{r}_0, t\rangle$ is known. Actually, $\mathbf{r} = \mathbf{r}\langle\mathbf{r}_0, t\rangle$ describes the particle trajectory or *pathline* of the particle (Fig. 12.1).

[1] The adjective "Lagrangian" is purely conventional; the scheme is actually attributable to Euler.

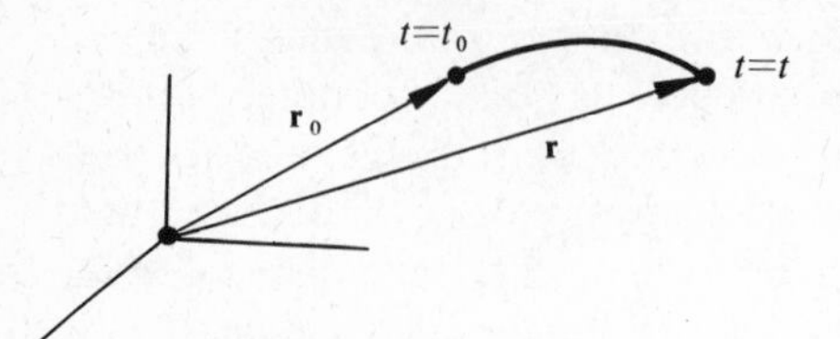

Figure 12.1 A pathline.

If, at $t = t_0$, there is a chunk of fluid $\mathcal{V}$, varying $\mathbf{r}_0$ in this region $\mathcal{V}$ will then cover all the particles in the chunk, and $\mathbf{r} = \mathbf{r}\langle \mathbf{r}_0, t\rangle = \mathbf{r}\langle a, b, c, t\rangle$ will give a bundle of pathlines (Fig. 12.2). The chunk $\mathcal{V}$ itself will move, distort, and change volume, but it will always consist of the *same* fluid particles. A volume, surface, or curve that always consists of the *same* fluid particles is called a *material volume*, *material surface*, or *material curve*. In this book, material volume, and so forth, will be denoted by script capital letters such as $\mathcal{V}$, $\mathcal{S}$, $\mathcal{C}$ to distinguish them from volume, and so forth, *fixed* in the empty space (denoted by V, S, C and the like).

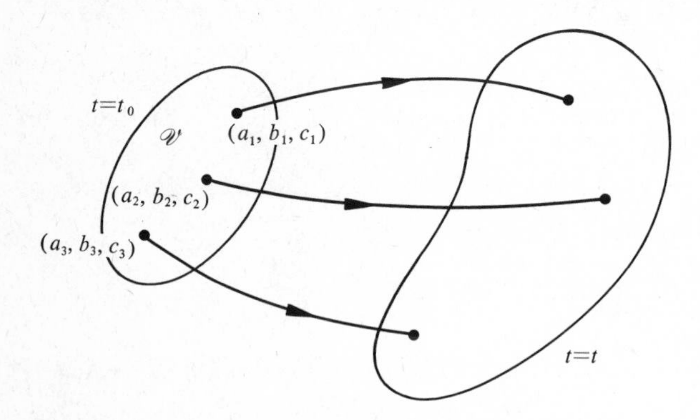

Figure 12.2 A bundle of pathlines.

In the Lagrangian description, the initial position $\mathbf{r}_0 = (a, b, c)$ and the time t are the independent variables. The later position $\mathbf{r} = (x, y, z)$ is a dependent, vectorial variable. Other fluid properties will also be functions of $\mathbf{r}_0$ (or a, b, c) and t. The most important properties here are the velocity and the acceleration:

$$\text{Velocity:} \quad \mathbf{q}\langle a, b, c, t\rangle = \frac{\partial \mathbf{r}\langle a, b, c, t\rangle}{\partial t}$$

$$\text{Acceleration:} \quad \frac{\partial \mathbf{q}\langle a, b, c, t\rangle}{\partial t} = \frac{\partial^2 \mathbf{r}\langle a, b, c, t\rangle}{\partial t^2}$$

This description may seem to be a natural way to set up fluid dynamics problems; but, in practice, it is not very convenient. For instance, a steady flow receives no special treatment and shows no simplification whatsoever in this description. Also, the expression for mass conservation turns out to be rather unwieldy. However, it does help in clarifying many aspects of fluid flow, especially in general discussions and physical interpretations.

In place of the Lagrangian description, we can also adopt the Eulerian description,[2] which is more powerful and more convenient, and hence is almost exclusively used in fluid dynamics. The Eulerian description abandons the tedious and often unnecessary task of following individual particles, and focuses attention on what happens at a fixed point in (the empty) space as different fluid particles go past the point. It considers the velocity $\mathbf{q}$, density ρ, and so on, of the fluid particles that pass that space point at different instants; that is, it gives

$$\mathbf{q} = \mathbf{q}\langle \mathbf{r}, t\rangle = \mathbf{q}\langle x, y, z, t\rangle$$

$$\rho = \rho\langle \mathbf{r}, t\rangle = \rho\langle x, y, z, t\rangle$$

and so on. Thus $\mathbf{r}$ is here an *independent* variable. It is important to notice that, in this description, individual particles are not labeled and are not distinguished from one another.

In the Euler description, a steady-$\mathbf{q}$ flow will have $\mathbf{q} = \mathbf{q}\langle \mathbf{r}\rangle$ independent of t. Similarly, a steady-ρ flow will have $\rho = \rho\langle \mathbf{r}\rangle$, and so on. A *steady flow* will then have *all* its flow properties independent of t, which amounts to a major simplification. (As mentioned before, such a flow does not have a special status in the Lagrangian description.)

In this book, Lagrangian description is only employed when it clarifies certain physical points, or when it paints a vivid physical picture. As the main tool, the Eulerian description is always used.

Before proceeding, our policies regarding surfaces of discontinuity and regarding singular points, lines, or surfaces must be clearly stated.

(1) Discontinuities

The possible existence of surfaces of discontinuity is *not* excluded.

In a real fluid flow it is true that all properties vary continuously. But there are narrow regions through which the properties change *very* rapidly. If the thickness of such a region is very small compared with the representative length of the field of interest, the region can be approximated by a surface of discontinuity across which fluid properties jump. This approximation will be good for all purposes unless one is particularly interested in what happens *inside* this narrow region. For example, Figure 12.3 shows a narrow region around $y = 0$ across which u changes rapidly from u_1 to u_2, being approximated by a surface of discontinuity at $y = 0$ across which u jumps from u_1 to u_2. A surface of discontinuity like this, across which the velocity component tangential to the surface jumps, is called a vortex sheet (the plane $y = 0$ in this example). The vortex-sheet approximation turns out to be indispensable in the airfoil theory. A similar approximation involving a surface of discontinuity, across which the velocity component normal to the surface jumps, turns out to be the cornerstone of the theory of supersonic flows. Such a surface of discontinuity is called a shock wave. It is in anticipation of these useful approximations that allowance is made for discontinuities right at the beginning.

---

[2] Since both descriptions are from Euler, the distinction by personal names here is certainly not appropriate. However, we will follow this conventional distinction for lack of a *popular* alternative.

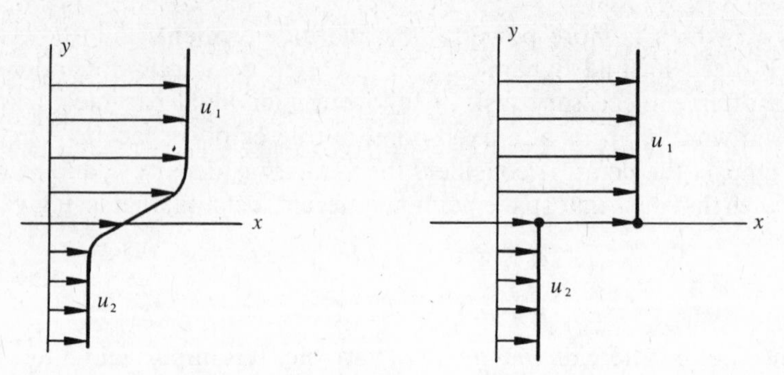

Figure 12.3 An approximation by a vortex sheet.

It is, however, important to realize three things: First, a surface of discontinuity may not always be a good approximation; it all depends on the purpose of the analysis. For instance, a vortex sheet would not do if viscous drag is to be calculated; a close look into the *structure* of the sheet is provided by the boundary-layer theory. Similarly, the shock-layer structure must be explored if the shock wave proves to be too rough an approximation. Second, a surface of discontinuity is not necessarily flat or stationary. For example, the shock wave accompanying a supersonic airplane is curved and moving. Third, a surface of discontinuity is not necessarily a material surface; that is, it is not necessarily composed always of the same fluid particles. Thus, the velocity $\mathbf{G}$ at which a point on a surface of discontinuity moves is not necessarily the same as the velocity $\mathbf{q}$ of the fluid particle that happens to be at that point. Only when $\mathbf{G} = \mathbf{q}$ *at every point* on the surface is the surface material.

(2) Singularities

The program of theoretical fluid dynamics must be divided into two parts: *formulation* of the general problem, and *solution* of specific cases of interest. It can then be stated that, *in the formulation, no singularities are allowed.*

But the solutions to the posed problem will often turn out to be singular (that is, multivalued, and/or infinite, and/or nondifferentiable) at certain points, along certain curves, or on certain surfaces. From a physical point of view, we can distinguish two cases as follows.

First, the singularities occur outside the region of interest. They, therefore, do not enter the physical picture; and one is not required to interpret them in physical terms.

Second, the singularities do occur inside the region of interest. Here, one can adopt three different attitudes depending on the situation.

(a) when the singularities have physical counterparts that occur or can be made to occur in reality, we will regard the singularities as close approximations to their corresponding physical counterparts. For instance, solutions may occur wherein fluid converges from all directions toward certain *sink* points with accompanying disappearance of mass at these points. We will not throw these solutions away since, in reality, it is conceivable that flows in the presence of suctions at the open ends of tiny pipes may be approximately described by just this kind of solution (Fig. 12.4). This sort of approximate realization can actually be done in a laboratory with due care and caution.

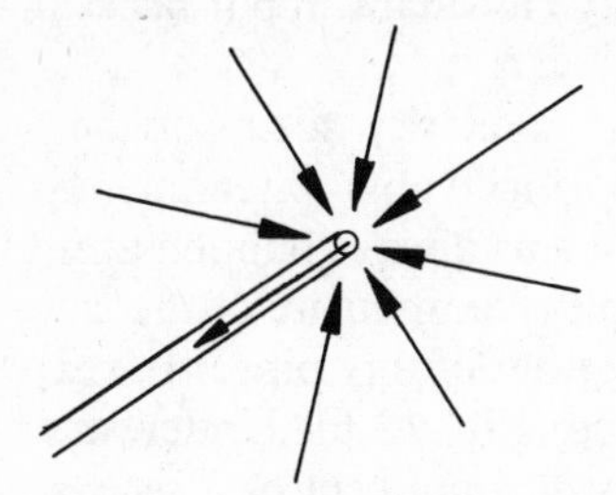

Figure 12.4 Suction through a tiny tube.

(b) The singularities may indicate certain happenings that are clearly impossible in the physical world or are definitely not in keeping with actual observations. For example, under certain conditions, solutions that predict infinite flow speeds at some points may appear. We know that this is unrealistic; and an actual observation shows a completely different story around these points. (Refer to the situation shown in Fig. 12.5.) When this happens, we again do not throw away the singularity-ridden solutions, since they may still agree with reality far

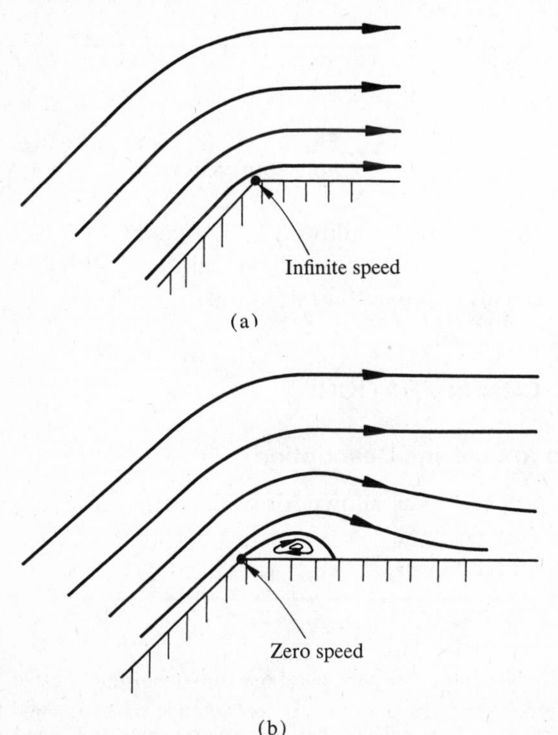

Figure 12.5 Flow past a corner, (a) potential flow and (b) real flow.

away from these singular points. If that is the case, we only have to be careful when approaching these points. As one possibility, a reexamination may show that the speed becomes unbounded at these points only because some dissipative mechanism has been ignored in posing or solving the problem. We could then accept the solutions as good approximations except near those points where a more careful investigation, with the neglected dissipative mechanism put back in, must be made.[3]

(c) Although the behavior indicated by a singularity seems to be a reasonable approximation to a certain physically conceivable phenomenon, nevertheless it is not observable because other mechanisms always happen at the same time to distort it beyond recognition. For example, the opposite of the sink (that is, the *source*) is not approximated by the blowing out of tiny pipes. In reality, such blowing always produces a jet, not a source (Fig. 12.6). Under such circumstances we may adopt the attitude that the physical concept of a source, though not directly observable, is too important conceptually to be brushed away. Solutions involving sources may not have any physical counterpart *in a laboratory*, they may nevertheless help us in understanding the general nature of a flow by showing us what might happen. This kind of conceptual understanding is even more important than the detailed investigation of some specific, real problems.

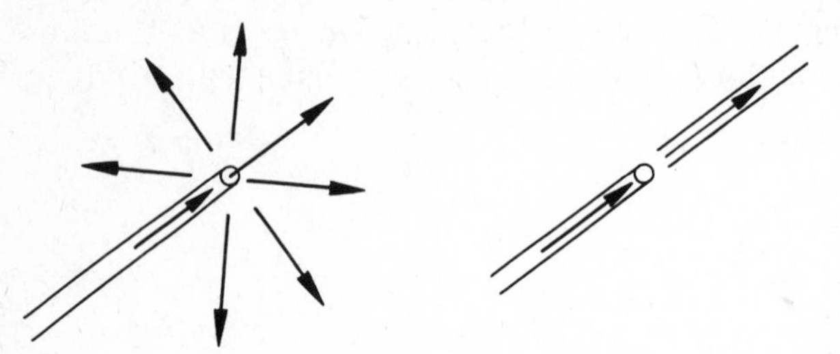

Figure 12.6 The unrealizable source flow.

Summary Discontinuities are always allowed for in a general exposition; singularities are excluded in the formulation, but will be embraced with due reservations when they are forced on us through certain solutions.

12.2 PRELIMINARY CONSIDERATIONS

(1) From Lagrangian to Eulerian Description

Suppose a certain fluid property $\mathscr{F}$ is known for each particle at every instant:

$$\begin{aligned}\mathscr{F} &= \mathscr{F}\langle \mathbf{r}_0, t\rangle \\ &= \mathscr{F}\langle a, b, c, t\rangle\end{aligned}$$

[3] To summarize, solutions with singularities that are either physically impossible or unsubstantiated may be looked upon as a piece of good material with some holes in it. These holes may indicate inadequate approximations introduced somewhere along the line. To fill them up, one must investigate more carefully, identify the trouble, and remedy it.

In the Lagrangian description, the positions of each particle at all times are known:

$$\mathbf{r} = \mathbf{r}\langle \mathbf{r}_0, t\rangle$$

or

$$\begin{cases} x = x\langle a, b, c, t\rangle \\ y = y\langle a, b, c, t\rangle \\ z = z\langle a, b, c, t\rangle \end{cases}$$

from which a, b, and c can be solved to yield

$$\begin{cases} a = a\langle x, y, z, t\rangle \\ b = b\langle x, y, z, t\rangle \\ c = c\langle x, y, z, t\rangle \end{cases}$$

or, $\mathbf{r}_0 = \mathbf{r}_0\langle \mathbf{r}, t\rangle$. Substituting this into $\mathscr{F}\langle a, b, c, t\rangle$, the corresponding Eulerian description of $\mathscr{F}$ is obtained:

$$\mathscr{F} = \mathscr{F}\langle a\langle x, y, z, t\rangle, b\langle x, y, z, t\rangle, c\langle x, y, z, t\rangle, t\rangle$$

which is a function of x, y, z, and t.

(2) From Eulerian to Lagrangian Description

Here, to start with, one has $\mathscr{F} = \mathscr{F}\langle x, y, z, t\rangle$, which includes

$$\mathbf{q} = \mathbf{q}\langle \mathbf{r}, t\rangle = \mathbf{q}\langle x, y, z, t\rangle$$

To go to the Lagrangian description, one must first obtain the pathlines:

$$\mathbf{r} = \mathbf{r}\langle \mathbf{r}_0, t\rangle = \mathbf{r}\langle a, b, c, t\rangle$$

If this were known at this stage, one would have

$$\mathbf{q} = \mathbf{q}\langle \mathbf{r}\langle \mathbf{r}_0, t\rangle, t\rangle$$

But, then, in the Lagrangian description,

$$\mathbf{q} = \frac{\partial \mathbf{r}\langle \mathbf{r}_0, t\rangle}{\partial t}$$

Therefore,

$$\frac{\partial \mathbf{r}\langle \mathbf{r}_0, t\rangle}{\partial t} = \mathbf{q}\langle \mathbf{r}\langle \mathbf{r}_0, t\rangle, t\rangle$$

that is,

$$\begin{cases} \dfrac{\partial x\langle a, b, c, t\rangle}{\partial t} = u\langle x\langle a, b, c, t\rangle, \ldots, t\rangle \\ \cdot \\ \cdot \\ \cdot \end{cases}$$

In this latter set (which we did not write out in full), a, b, and c occur only as parameters. As far as x, y, z, and t are concerned, it is just a set of ordinary differential equations:

$$\begin{cases} \dfrac{dx}{dt} = u\langle x, y, z, t\rangle & \textbf{(12.1a)} \\ \dfrac{dy}{dt} = v\langle x, y, z, t\rangle & \textbf{(12.1b)} \\ \dfrac{dz}{dt} = w\langle x, y, z, t\rangle & \textbf{(12.1c)} \end{cases}$$

the solution of which is

$$\begin{cases} x = x\langle A, B, C, t\rangle \\ y = y\langle A, B, C, t\rangle \\ z = z\langle A, B, C, t\rangle \end{cases}$$

where A, B, and C are the three constants of integration. For the particle that occupied $\mathbf{r}_0 = (a, b, c)$ at t_0, one can solve for A, B, and C from

$$\begin{cases} a = x\langle A, B, C, t_0\rangle \\ b = y\langle A, B, C, t_0\rangle \\ c = z\langle A, B, C, t_0\rangle \end{cases}$$

and substitute the result back to get

$$\begin{cases} x = x\langle a, b, c, t\rangle \\ y = y\langle a, b, c, t\rangle \\ z = z\langle a, b, c, t\rangle \end{cases}$$

with the known constant t_0 absorbed into the functional forms[4] $x\langle\ \rangle, y\langle\ \rangle, z\langle\ \rangle$. This set then describes the pathline passing through (a, b, c). Varying (a, b, c) will yield different pathlines.

Finally, $\mathscr{F} = \mathscr{F}\langle a, b, c, t\rangle$ can now be obtained from $\mathscr{F} = \mathscr{F}\langle x, y, z, t\rangle$ through $\mathscr{F} = \mathscr{F}\langle x\langle a, b, c, t\rangle, \ldots, t\rangle$.

EXAMPLE

A steady, plane flow is given in the Eulerian description by

$$\begin{cases} u = kx \\ v = -ky \\ w = 0 \end{cases}$$

Find its Lagrangian description.

---

[4] Note that the functional forms $x\langle A, B, C, t\rangle$ and $x\langle a, b, c, t\rangle$ are different. A mathematician would probably use two different symbols. But a physicist or engineer would like to display the fact that he is talking all the time about a single physical quantity—that is, the x-component of the particle position.

SOLUTION

$$\begin{cases} \dfrac{dx}{dt} = kx \\ \dfrac{dy}{dt} = -ky \\ \dfrac{dz}{dt} = 0 \end{cases}$$

Therefore

$$\begin{cases} x = Ae^{kt} \\ y = Be^{-kt} \\ z = C \end{cases}$$

If $x = a, y = b, z = c$, at $t = t_0$, we have

$$\begin{cases} a = Ae^{kt_0} \\ b = Be^{-kt_0} \\ c = C \end{cases}$$

Therefore

$$A = ae^{-kt_0}, \quad B = be^{kt_0}, \quad C = c$$

and

$$\begin{cases} x = ae^{k(t-t_0)} \\ y = be^{-k(t-t_0)} \\ z = c \end{cases}$$

are the pathlines if we vary the initial position (a, b, c).

(3) Streamlines

In the Eulerian description, a streamline at an instant t is defined as a curve with its line element $d\mathbf{r} \parallel \mathbf{q}$; that is,

$$\frac{dx}{u\langle x, y, z, t\rangle} = \frac{dy}{v\langle x, y, z, t\rangle} = \frac{dz}{w\langle x, y, z, t\rangle} \tag{12.2}$$

The solution of Eq. (12.2) is a family of space curves with three constants of integration. For a steady-**q** flow, we have

$$\frac{dx}{u\langle x, y, z\rangle} = \frac{dy}{v\langle x, y, z\rangle} = \frac{dz}{w\langle x, y, z\rangle} \tag{12.3}$$

the solution curves of which will stay the same as time changes.

Now, consider an arbitrary space curve C. Then, past every point of C there is a streamline. The surface formed by these streamlines is called a *stream surface*. If C is closed, the stream surface is termed a *stream tube*. The *strength* of a stream tube at a cross section is defined as the volume flow across that cross section; the strength in general varies from one cross section to another. A very thin stream tube can be looked upon as a filament of infinitesimal cross-sectional area, but still carrying a nonzero (and *not* infinitesimal) strength distribution; this is called a *stream filament*. The stream filament must be a line of singularity unless the strength happens to be zero, since the flow velocity on it must be unbounded in order to have a nonzero strength. A *streamline* is, then, just a stream filament with zero strength. A surface formed by stream filaments is called a *stream sheet*.

It is interesting to note that the equations for pathlines, Eqs. (12.1a,b,c), can also be written as

$$\frac{dx}{u\langle x, y, z, t\rangle} = \frac{dy}{v\langle x, y, z, t\rangle} = \frac{dz}{w\langle x, y, z, t\rangle} = dt$$

For a steady-**q** flow, this becomes

$$\left.\frac{dx}{u\langle x, y, z\rangle} = \frac{dy}{v\langle x, y, z\rangle} = \frac{dz}{w\langle x, y, z\rangle}\right| = dt$$

Here, t does not show up to the left of the vertical line. Then, the portion to the left of the vertical line can be solved in a decoupled manner. The result would, of course, be the same as the solution of Eqs. (12.1a,b,c). This shows that a streamline and a pathline starting from the same point must be identical in a steady-**q** flow. (The unused part of the pathline equation here only gives the time needed for a particle to travel a given distance along the pathline.) For an unsteady flow, such a decoupling is impossible, and pathlines are in general different from streamlines.

(4) Streaklines

If one keeps on supplying dye or smoke through a tiny opening at a *fixed* point in a flow field as often done in laboratories, one obtains a colored streak[5] called a *streakline*. Thus, a streakline at an instant t passing through a *fixed* point $\mathbf{r}_1$ is the locus of all fluid particles that passed $\mathbf{r}_1$ at some time between 0 and t (assuming that the dye was started at $t = 0$). Suppose a general particle passed $\mathbf{r}_1$ at t_1 where $0 \leq t_1 \leq t$. Then, the initial position of this particle at t_0 must be

$$\mathbf{r}_0 = \mathbf{r}_0\langle \mathbf{r}_1, t_1\rangle$$

which is the inversion of $\mathbf{r}_1 = \mathbf{r}_1\langle \mathbf{r}_0, t_1\rangle$. At time t this particle is at

$$\mathbf{r} = \mathbf{r}\langle \mathbf{r}_0, t\rangle = \mathbf{r}\langle \mathbf{r}_0\langle \mathbf{r}_1, t_1\rangle, t\rangle \tag{12.4}$$

where t_1 serves as a parameter between 0 and t. Equation (12.4) is, therefore, the defining equation of the streakline past $\mathbf{r}_1$ at time t. (See Fig. 12.7.)

For a steady-**q** flow, the streamline past $\mathbf{r}_1$ will stay the same for all times. Furthermore, this is also the pathline past $\mathbf{r}_1$. Then all particles that have passed $\mathbf{r}_1$ or will pass $\mathbf{r}_1$ must lie on this same line. So for such a flow the streakline is identical with the streamline and pathline past $\mathbf{r}_1$. This is why streaklines are often produced in laboratories to show pathlines or streamlines in *steady* flows.

[5] If the dye or smoke does not diffuse too fast.

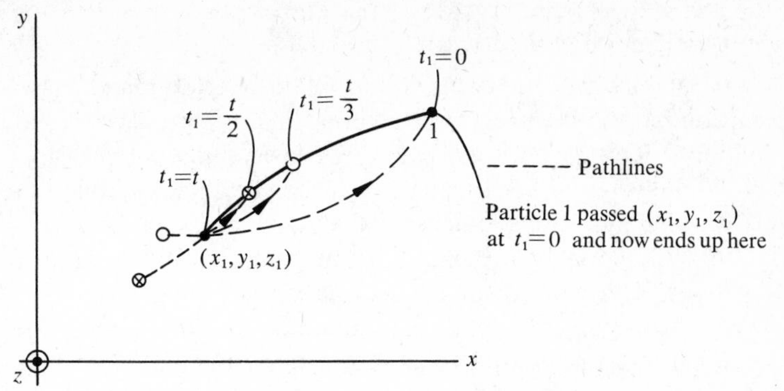

Figure 12.7 A streakline.

EXAMPLE Given

$$\begin{cases} u = \dfrac{x}{1+t} \\ v = y \\ w = 0 \end{cases}$$

Find the streaklines.

SOLUTION It is easy to show that the pathlines are

$$\begin{cases} x = a(1+t) \\ y = be^t \\ z = c \end{cases}$$

where t_0 has been taken as zero for convenience. Then, following the procedure outlined in arriving at Eq. (12.4), we first write

$$\mathbf{r}_1: \begin{cases} x_1 = a(1+t_1) \\ y_1 = be^{t_1} \\ z_1 = c \end{cases}$$

Inverting, we have

$$\begin{cases} a = \dfrac{x_1}{1+t_1} \\ b = y_1 e^{-t_1} \\ c = z_1 \end{cases}$$

Therefore

$$\begin{cases} x = x_1 \dfrac{1+t}{1+t_1} \\ y = y_1 e^{t-t_1} \\ z = z_1 \end{cases}$$

is the parametric form of the streakline at time t and passing through (x_1, y_1, z_1).

(5) Relative Motion

In dynamic investigations the problem is often made easier to handle when the motion is described relative to a coordinate system that is moving with a constant velocity. This is because Newton's law of motion are also valid with reference to such systems. So to calculate forces or accelerations in a fluid, one can use any moving system as long as it is not accelerating. Thus the force on an airplane is the same whether the plane is moving uniformly through still air or a uniform air stream is blowing against a stationary plane.

It is important, however, to remember that the velocity fields will be drastically different with respect to different moving (though not accelerating) systems.

12.3 DIFFERENTIATION FOLLOWING THE MOTION

In Section 12.1 we mentioned that, if there is a chunk of fluid at $t = t_0$, varying the initial position $\mathbf{r}_0$ in this region will cover all particles in the chunk. The fate of all the particles is then determined by

$$\mathbf{r} = \mathbf{r}\langle \mathbf{r}_0, t\rangle$$

for all $\mathbf{r}_0$ in the region. We have of course explained this equation as a bundle of pathlines. But there is another way of interpreting this equation, coached in the language of transformations. This new interpretation can clarify what was discussed before, and open up simple and direct ways of drawing further consequences.

From here on, let us call the coordinates represented by the position vector $\mathbf{r}$—for example, (x,y,z)—the *spatial coordinates*, and those represented by the initial position vector $\mathbf{r}_0$—for example, (a,b,c)—the *material coordinates* since they serve to distinguish different material points (or particles). It is important for the reader to realize that both coordinates *span the same space* (Fig. 12.8). To

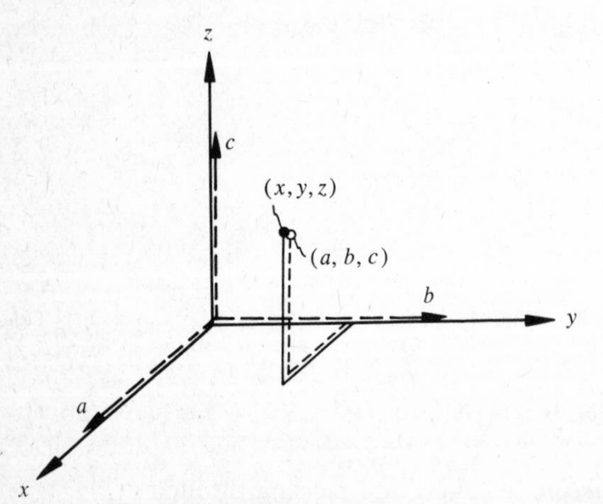

Figure 12.8 Spatial and material coordinates.

start with, there is the *empty* space. A point in it is described by $\mathbf{r}$ or (x,y,z). Now, at $t = t_0$, "throw" a fluid into this empty space. The position of each fluid particle at $t = t_0$ will be tagged onto the particle as its label. At $t = t$, the particle labeled with $\mathbf{r}_0$ or (a,b,c) is *transformed* to a new position $\mathbf{r}$ or (x,y,z) in the empty space; but, it is *still* identified as the material point with material coordinates $\mathbf{r}_0$ or (a,b,c).

The motion of all material points in the region of interest is described by

$$\mathbf{r} = \mathbf{r}\langle \mathbf{r}_0, t\rangle$$

Mathematically, this is nothing but a *transformation* of $\mathbf{r}_0$ into $\mathbf{r}$ with t as the parameter of the transformation. (To be exact, this is a transformation of points in a space to other points in the *same space*.)

If $\mathbf{r}_0$ is fixed, varying t will again give a pathline for a single material point. But, here, we are more interested in varying $\mathbf{r}_0$ while t is fixed consecutively at different instants. This would then transform a region originally occupied by the lump of fluid at $t = t_0$ to another spatial region at $t = t$ (Fig. 12.9).

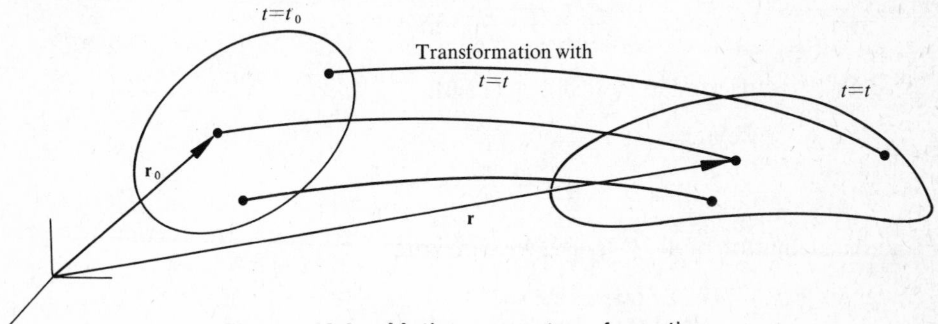

Figure 12.9 Motion as a transformation.

If we solve for $\mathbf{r}_0$ from $\mathbf{r} = \mathbf{r}\langle \mathbf{r}_0, t\rangle$, we will also have an *inverse transformation*, $\mathbf{r}_0 = \mathbf{r}_0\langle \mathbf{r}, t\rangle$. This inverse transformation serves to answer the question: Exactly which material point is now (at $t = t$) passing the spatial point $\mathbf{r}$?

Associated with this duality (that is, empty geometric space on the one hand and space filled with a fluid on the other), we see that the Lagrangian description is just the *material description:* $\mathscr{F} = \mathscr{F}\langle \mathbf{r}_0, t\rangle$; and the Eulerian, just the *spatial description:* $\mathscr{F} = \mathscr{F}\langle \mathbf{r}, t\rangle$. Furthermore,[6] to get to the Eulerian from the Lagrangian description, one simply performs the transformation of $\mathbf{r}_0$ into $\mathbf{r}$ at $t = t$; that is all.

Definitions

(a) Spatial *time* derivative

$$\frac{\partial \mathscr{F}}{\partial t} = \frac{\partial \mathscr{F}\langle \mathbf{r}, t\rangle}{\partial t} \tag{12.5}$$

[6] It would be helpful for the reader to remember the following: $\mathscr{F}\langle \mathbf{r}_0, t\rangle$ is $\mathscr{F}$ "felt" at t by that particle which was located at $\mathbf{r}_0$ at t_0; $\mathscr{F}\langle \mathbf{r}, t\rangle$ is $\mathscr{F}$ "felt" at t by the particle that happens to be at $\mathbf{r}$ at the moment.

(b) Material *time* derivative

$$\frac{D\mathscr{F}}{Dt}=\frac{\partial\mathscr{F}\langle\mathbf{r}_0,t\rangle}{\partial t} \tag{12.6}$$

Physically, (a) gives the time rate of change of $\mathscr{F}$ apparent to a viewer stationed at the fixed position $\mathbf{r}$; and (b) gives the time rate of change measured while following the motion of a particle. Thus, (b) is the one that is more interesting to a fluid dynamicist. But, in definition (b), $D\mathscr{F}/Dt$ is given as a function of $\mathbf{r}_0$ and t. As mentioned before, this is not very convenient; and, we have to put it in terms of $\mathbf{r}$ and t, starting from $\mathscr{F}\langle\mathbf{r},t\rangle$:

$$\begin{aligned}\frac{D\mathscr{F}}{Dt}&=\frac{\partial\mathscr{F}\langle\mathbf{r}\langle\mathbf{r}_0,t\rangle,t\rangle}{\partial t}\\&=\frac{\partial\mathscr{F}\langle\mathbf{r},t\rangle}{\partial t}+\frac{\partial\mathbf{r}\langle\mathbf{r}_0,t\rangle}{\partial t}\cdot\operatorname{grad}\mathscr{F}\langle\mathbf{r},t\rangle\\&=\frac{\partial\mathscr{F}\langle\mathbf{r},t\rangle}{\partial t}+\frac{D\mathbf{r}}{Dt}\cdot\operatorname{grad}\mathscr{F}\langle\mathbf{r},t\rangle\end{aligned}$$

where $D\mathbf{r}/Dt$ is inserted in view of Eq. (12.6).

(c) Velocity

$$\mathbf{q}=\frac{\partial\mathbf{r}\langle\mathbf{r}_0,t\rangle}{\partial t}=\frac{D\mathbf{r}}{Dt}$$

From this definition of the velocity, we can write

$$\frac{D\mathscr{F}}{Dt}=\frac{\partial\mathscr{F}\langle\mathbf{r},t\rangle}{\partial t}+\mathbf{q}\cdot\nabla\mathscr{F}\langle\mathbf{r},t\rangle \tag{12.7}$$

Note that D/Dt is composed of two parts, (1) $\partial/\partial t$ (that is, the time rate of change when the particle is regarded as fixed) and (2) $\mathbf{q}\cdot\nabla$ (that is, the rate of change at a fixed instant, in the direction of $\mathbf{q}$, due to the nonuniformity of $\mathscr{F}$ in space).

When applying Eq. (12.7) to a vectorial quantity, care must be exercised; one must remember that $\mathbf{q}\cdot\nabla\mathbf{A}$ is only a shorthand version of something rather complicated; see Problem 11.4.2, part (e); or Section 11.6, subsection (11). Most important of all, only for *rectangular Cartesian coordinates* is $\mathbf{q}\cdot\nabla\mathbf{A}$ equal to

$$\left(u\frac{\partial A_x}{\partial x}+v\frac{\partial A_x}{\partial y}+w\frac{\partial A_x}{\partial z}\right)\hat{\mathbf{x}}+\left(u\frac{\partial A_y}{\partial x}+v\frac{\partial A_y}{\partial y}+w\frac{\partial A_y}{\partial z}\right)\hat{\mathbf{y}}+\left(u\frac{\partial A_z}{\partial x}+v\frac{\partial A_z}{\partial y}+w\frac{\partial A_z}{\partial z}\right)\hat{\mathbf{z}}$$

(d) Acceleration

$$\begin{aligned}(\text{acceleration vector})=\frac{D\mathbf{q}}{Dt}&=\frac{\partial\mathbf{q}\langle\mathbf{r},t\rangle}{\partial t}+\mathbf{q}\langle\mathbf{r},t\rangle\cdot\nabla\mathbf{q}\langle\mathbf{r},t\rangle\\&=\frac{\partial\mathbf{q}}{\partial t}+\nabla(\tfrac{1}{2}q^2)-\mathbf{q}\times(\nabla\times\mathbf{q})\end{aligned} \tag{12.8}$$

See Section 11.6, subsection (10).

12.4 ROTATION AND DEFORMATION

In a given velocity field $\mathbf{q}\langle \mathbf{r}, t\rangle$, the spatial variation is of particular interest. Consider an arbitrary fluid particle P, which is located at $\mathbf{r}$ at the instant t. Then its velocity at that instant is $\mathbf{q}\langle \mathbf{r}, t\rangle$. At the same instant a fluid particle Q, $d\mathbf{r}$ away from P, must be moving with a velocity

$$\mathbf{q}\langle \mathbf{r}+d\mathbf{r}, t\rangle = \mathbf{q}\langle \mathbf{r}, t\rangle + (d\mathbf{r}) \cdot \nabla\mathbf{q}$$

See Eq. (11.18). Here, we have met with our very first physical quantity which is a dyadic – namely, $\nabla\mathbf{q}$.

The dyadic $\nabla\mathbf{q}$, sometimes called the derivative dyadic of velocity field, has altogether nine components with respect to three chosen directions. These nine components at $\mathbf{r}$ and t completely determine the relative motion of all the neighboring particles around P at t. A fluid particle in the neighborhood of the particle P will move relatively to P with a relative velocity that is equal to the dot product of $d\mathbf{r}$, the distance of the particle from P, and $\nabla\mathbf{q}$. (In other words, the dyadic $\nabla\mathbf{q}$ has the physical significance that, when dot-multiplied by the distance $d\mathbf{r}$, it yields the relative velocity $(d\mathbf{r}) \cdot \nabla\mathbf{q}$).

It is easy to split $\nabla\mathbf{q}$ into a sum of a symmetric part and an antisymmetric part:

$$\nabla\mathbf{q} = \tfrac{1}{2}(\nabla\mathbf{q} + \mathbf{q}\overleftarrow{\nabla}) + \tfrac{1}{2}(\nabla\mathbf{q} - \mathbf{q}\overleftarrow{\nabla})$$

where $\mathbf{q}\overleftarrow{\nabla} = (\nabla\mathbf{q})_c$. Thus, the relative velocity

$$(d\mathbf{r}) \cdot \nabla\mathbf{q} = \tfrac{1}{2}(d\mathbf{r}) \cdot (\nabla\mathbf{q} + \mathbf{q}\overleftarrow{\nabla}) + \tfrac{1}{2}(d\mathbf{r}) \cdot (\nabla\mathbf{q} - \mathbf{q}\overleftarrow{\nabla})$$

The second part may also be written as

$$\tfrac{1}{2}(\nabla\mathbf{q} - \mathbf{q}\overleftarrow{\nabla})_c \cdot d\mathbf{r} = \tfrac{1}{2}(\mathbf{q}\overleftarrow{\nabla} - \nabla\mathbf{q}) \cdot (d\mathbf{r})$$

which is seen to be just

$$-\tfrac{1}{2}(d\mathbf{r}) \times (\nabla \times \mathbf{q})$$

or

$$\tfrac{1}{2}(\nabla \times \mathbf{q}) \times (d\mathbf{r})$$

See Eq. (11.1B) in Section 11.3, subsection (1). As to the first part, we will denote it by a new symbol $\tilde{\boldsymbol{\epsilon}}$:

$$\tilde{\boldsymbol{\epsilon}} = \tfrac{1}{2}(\nabla\mathbf{q} + \mathbf{q}\overleftarrow{\nabla}) \tag{12.9}$$

Then

$$(d\mathbf{r}) \cdot \nabla\mathbf{q} = (d\mathbf{r}) \cdot \tilde{\boldsymbol{\epsilon}} + \tfrac{1}{2}(\nabla \times \mathbf{q}) \times (d\mathbf{r}) \tag{12.10}$$

(1) Rotation

Consider the motion of fluid particles situated on an infinitesimal sphere of radius dr around the particle P at a fixed instant. Establish at P a rectangular Cartesian coordinate system as well as a spherical coordinate system,

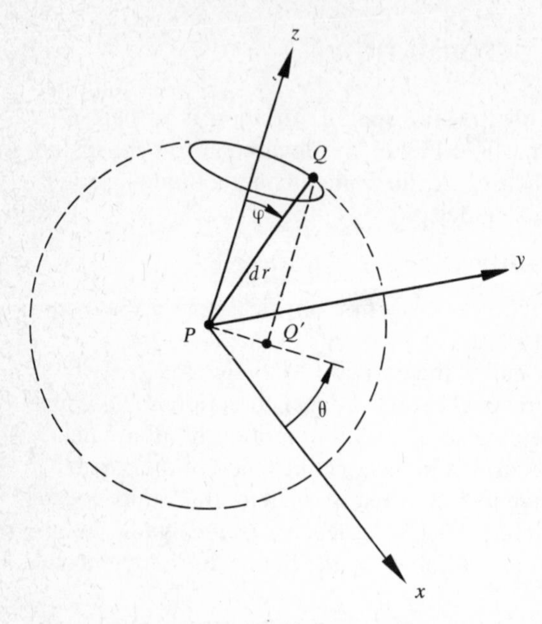

Figure 12.10 A spherical neighborhood.

oriented in an arbitrary fashion (Fig. 12.10). We then see that

$$\begin{cases} Q: \quad (dx, dy, dz) = (\cos\theta \sin\varphi\, dr, \sin\theta\sin\varphi\, dr, \cos\varphi\, dr) \\ \text{velocity of } Q \text{ relative to } P \end{cases}$$

$$= \left(\frac{\partial u}{\partial x}dx + \frac{\partial u}{\partial y}dy + \frac{\partial u}{\partial z}dz, \frac{\partial v}{\partial x}dx + \frac{\partial v}{\partial y}dy + \frac{\partial v}{\partial z}dz, \frac{\partial w}{\partial x}dx + \frac{\partial w}{\partial y}dy + \frac{\partial w}{\partial z}dz\right)$$

The infinitesimal material line $\overline{PQ}$ now rotates about the z-axis with a certain speed, which (from Fig. 12.11 and with $\overline{PQ'} = \sin\varphi\, dr$) is equal to

$$-\left(\frac{\partial u}{\partial x}dx + \frac{\partial u}{\partial y}dy + \frac{\partial u}{\partial z}dz\right)\frac{\sin\theta}{\overline{PQ'}} + \left(\frac{\partial v}{\partial x}dx + \frac{\partial v}{\partial y}dy + \frac{\partial v}{\partial z}dz\right)\frac{\cos\theta}{\overline{PQ'}}$$

$$= \left[\frac{\partial v}{\partial x}\cos^2\theta - \frac{\partial u}{\partial y}\sin^2\theta - \left(\frac{\partial u}{\partial x} - \frac{\partial v}{\partial y}\right)\sin\theta\cos\theta\right] + \left[\frac{\partial v}{\partial z}\cos\theta - \frac{\partial u}{\partial z}\sin\theta\right]\cot\varphi$$

similarly to Section 1.3, subsection (1).

The result is seen to be varying with θ and φ. However, from this, one can take a circle around the z-axis (Fig. 12.10), with a fixed φ, and integrate the above expression with respect to θ from 0 to 2π. The result divided by 2π would be an average over all the Q's on that circle. This average value turns out to be

$$\frac{1}{2}\left(\frac{\partial v}{\partial x} - \frac{\partial u}{\partial y}\right)$$

(In the calculation, note that all the partial derivatives are actually evaluated at **r**, and enter the integration around the circle only as constants.) Since this average rotational speed about the z-axis is independent of φ, there is no need to go from one circle to another.

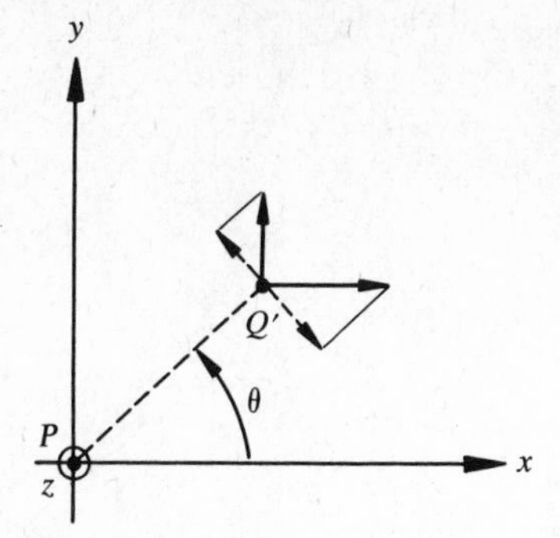

Figure 12.11 A projection onto the xy-plane.

Similar results can also be established for the components about the x- and y-axes.

Conclusion Average angular velocity of all material segments $\overline{PQ}$ within an infinitesimal sphere of center P, or briefly, the mean angular velocity of the fluid element around P is equal to

$$\frac{1}{2}\left(\frac{\partial w}{\partial y}-\frac{\partial v}{\partial z}\right)\hat{\mathbf{x}}+\frac{1}{2}\left(\frac{\partial u}{\partial z}-\frac{\partial w}{\partial x}\right)\hat{\mathbf{y}}+\frac{1}{2}\left(\frac{\partial v}{\partial x}-\frac{\partial u}{\partial y}\right)\hat{\mathbf{z}}$$

which we recognize as the component form of $\frac{1}{2}\nabla\times\mathbf{q}$ with respect to a rectangular Cartesian coordinate system. Now, since angular velocity is also a vector, it must be equal to $\frac{1}{2}\nabla\times\mathbf{q}$ in a coordinate-invariant manner (although the link was displayed against a background of rectangular Cartesian coordinates).

This conclusion can be used to kill two birds with one stone: (a) It gives a physical meaning to $\nabla\times\mathbf{q}$ or curl $\mathbf{q}$—namely, twice the mean angular velocity of the local fluid particle, (b) It provides an interpretation for the second term on the right-hand side of Eq. (12.10)—namely, the relative velocity due to the (mean) rotation of the fluid element around P.

Thus, if $\nabla\times\mathbf{q}=0$ at every point inside a region, the flow is irrotational in that region.

(2) Deformation

The nine components of the dyadic $\tilde{\epsilon}$ with respect to a rectangular Cartesian coordinate system can be easily displayed as

$$\begin{aligned}\tilde{\epsilon}&=\frac{1}{2}(\nabla\mathbf{q}+\mathbf{q}\nabla)\\&=\frac{\partial u}{\partial x}\hat{\mathbf{x}}\hat{\mathbf{x}}+\frac{1}{2}\left(\frac{\partial v}{\partial x}+\frac{\partial u}{\partial y}\right)\hat{\mathbf{x}}\hat{\mathbf{y}}+\frac{1}{2}\left(\frac{\partial w}{\partial x}+\frac{\partial u}{\partial z}\right)\hat{\mathbf{x}}\hat{\mathbf{z}}\\&\quad+\frac{1}{2}\left(\frac{\partial u}{\partial y}+\frac{\partial v}{\partial x}\right)\hat{\mathbf{y}}\hat{\mathbf{x}}+\frac{\partial v}{\partial y}\hat{\mathbf{y}}\hat{\mathbf{y}}+\frac{1}{2}\left(\frac{\partial w}{\partial y}+\frac{\partial v}{\partial z}\right)\hat{\mathbf{y}}\hat{\mathbf{z}}\\&\quad+\frac{1}{2}\left(\frac{\partial u}{\partial z}+\frac{\partial w}{\partial x}\right)\hat{\mathbf{z}}\hat{\mathbf{x}}+\frac{1}{2}\left(\frac{\partial v}{\partial z}+\frac{\partial w}{\partial y}\right)\hat{\mathbf{z}}\hat{\mathbf{y}}+\frac{\partial w}{\partial z}\hat{\mathbf{z}}\hat{\mathbf{z}}\end{aligned}\tag{12.11}$$

This symmetric dyadic can also be displayed in a matrix:

↗	$\hat{\mathbf{x}}$	$\hat{\mathbf{y}}$	$\hat{\mathbf{z}}$
$\hat{\mathbf{x}}$	ϵ_{xx}	ϵ_{xy}	ϵ_{xz}
$\hat{\mathbf{y}}$	ϵ_{xy}	ϵ_{yy}	ϵ_{yz}
$\hat{\mathbf{z}}$	ϵ_{xz}	ϵ_{yz}	ϵ_{zz}

where we have used the abbreviations

$$\epsilon_{xx}, \epsilon_{yy}, \epsilon_{zz} = \frac{\partial u}{\partial x}, \frac{\partial v}{\partial y}, \frac{\partial w}{\partial z}$$

$$\epsilon_{xy}, \epsilon_{yz}, \epsilon_{xz} = \frac{1}{2}\left(\frac{\partial v}{\partial x}+\frac{\partial u}{\partial y}\right), \frac{1}{2}\left(\frac{\partial w}{\partial y}+\frac{\partial v}{\partial z}\right), \frac{1}{2}\left(\frac{\partial u}{\partial z}+\frac{\partial w}{\partial x}\right)$$

To see the physical meaning of these six distinct components of the dyadic $\boldsymbol{\epsilon}$, let us take a small block of fluid with the point P as one of its corners (Fig. 12.12). The x-, y-, and z-axes are so oriented that they are parallel to the three edges of the block. It is only necessary to investigate the shaded face of the block, since the result can then be easily extended to the other faces. But this shaded face is exactly the same as that shown in Figure 1.28, so we can apply the entirety of Section 1.3, subsection (2), here, and extend it to cover also ϵ_{zz}, ϵ_{xz}, and ϵ_{yz}. The conclusion is then that the components of $\tilde{\epsilon}$, when referred to three perpendicular directions are simply the three rates of normal strain and three rates of shear strain of a fluid particle. Our dyadic quantity $\tilde{\epsilon}$ thus turns out to be just the rate of strain (dyadic),[7] which, when dot-multiplied by $d\mathbf{r}$, yields the relative velocity due to strain or deformation of the fluid element around P.

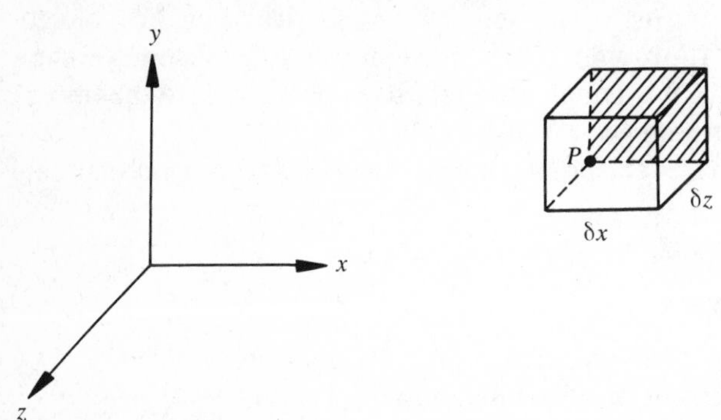

Figure 12.12 A small block of fluid.

Even if general orthogonal coordinates are used, the physical meaning of the components of $\tilde{\epsilon}$ would still be the same, since after all, any orthogonal coordinate system must look like a rectangular Cartesian one in the (infinitesimal)

[7] Also called the deformation dyadic. The reader is hereby warned that the symbolism and nomenclature vary from one author to another; he should always compare the definitions, not the symbols or names.

neighborhood of a point.[8] For example, ϵ_{RR} is still a normal rate of strain and $\epsilon_{R\theta}$ is still a shear rate of strain. The detailed expressions are, of course, different because velocity components with respect to general coordinates are more complicated. For example,

$$\epsilon_{RR} = \frac{\partial q_R}{\partial R}$$

$$\epsilon_{R\theta} = \frac{1}{2}\left(\frac{\partial q_\theta}{\partial R} - \frac{q_\theta}{R} + \frac{1}{R}\frac{\partial q_R}{\partial \theta}\right)$$

Further information about $\tilde{\boldsymbol{\epsilon}}$ can be deduced in a concise manner without referring to any coordinate system as follows.

Take the part of the relative velocity due to $\tilde{\boldsymbol{\epsilon}}$—that is, $(d\mathbf{r}) \cdot \tilde{\boldsymbol{\epsilon}}$. Dot-multiply it by $(d\mathbf{r})$ on the right and equate it to a constant:

$$(d\mathbf{r}) \cdot \tilde{\boldsymbol{\epsilon}} \cdot d\mathbf{r} = \text{constant} \tag{12.12}$$

For convenience of writing, let us denote $d\mathbf{r}$ by the symbol $\boldsymbol{\eta}$. Then,

$$\boldsymbol{\eta} \cdot \tilde{\boldsymbol{\epsilon}} \cdot \boldsymbol{\eta} = \text{constant} \tag{12.12A}$$

is a scalar equation in the three-dimensional space $\boldsymbol{\eta}$, and represents a surface in it.[9] What kind of surface?

1. It is a surface of the second degree; that is, it is represented by a quadratic algebraic equation.
2. It is a *centered* surface with P as its center; that is, if $\boldsymbol{\eta}$ is on the surface, so is $-\boldsymbol{\eta}$. The reason is that

$$(-\boldsymbol{\eta}) \cdot \boldsymbol{\epsilon} \cdot (-\boldsymbol{\eta}) = \boldsymbol{\eta} \cdot \tilde{\boldsymbol{\epsilon}} \cdot \boldsymbol{\eta}$$

Therefore, it must be either an ellipsoid or a hyperboloid.[10] As a consequence, the surface must have three mutually perpendicular axes of symmetry. Furthermore, these axes are normal to the surface where they meet it.

Since $\tilde{\boldsymbol{\epsilon}}$ is a symmetric dyadic, the surface has one more property of importance. Consider a neighboring point Q at $\boldsymbol{\eta}$. Passing through Q there must be a surface of the above kind. Next, consider an arbitrary point at $\boldsymbol{\eta} + d\boldsymbol{\eta}$ close to Q on the same surface. We have

$$(\boldsymbol{\eta} + d\boldsymbol{\eta}) \cdot \tilde{\boldsymbol{\epsilon}} \cdot (\boldsymbol{\eta} + d\boldsymbol{\eta}) = \text{constant}$$

Subtracting Eq. (12.12A) from this formula, and letting $d\boldsymbol{\eta} \to 0$, we have

$$\boldsymbol{\eta} \cdot \tilde{\boldsymbol{\epsilon}} \cdot d\boldsymbol{\eta} + (d\boldsymbol{\eta}) \cdot \tilde{\boldsymbol{\epsilon}} \cdot \boldsymbol{\eta} = 0$$

But $\tilde{\boldsymbol{\epsilon}}$ is symmetric; so

$$(d\boldsymbol{\eta}) \cdot \tilde{\boldsymbol{\epsilon}} \cdot d\boldsymbol{\eta} = \boldsymbol{\eta} \cdot \tilde{\boldsymbol{\epsilon}} \cdot d\boldsymbol{\eta}$$

---

[8] We say that all orthogonal coordinate systems are *locally* Cartesian.

[9] Incidentally, this is our first application of the double dot product.

[10] Including all possible degenerate cases. See also footnote 20 of Chapter 13.

Therefore

$$\boldsymbol{\eta} \cdot \tilde{\boldsymbol{\epsilon}} \cdot d\boldsymbol{\eta} = 0$$

that is, the vector $(\boldsymbol{\eta} \cdot \tilde{\boldsymbol{\epsilon}}) \perp d\boldsymbol{\eta}$. Now $d\boldsymbol{\eta}$ is arbitrary around Q, as long as it stays on the surface. Allowing $d\boldsymbol{\eta}$ to assume all possible directions would then generate a tangent plane to the surface at Q. So, $(\boldsymbol{\eta} \cdot \tilde{\boldsymbol{\epsilon}})$ must be normal to the surface itself. We thus arrive at the physical significance or the physical usefulness of the quadric, Eq. (12.12), as follows:

Any neighboring point Q would move relatively to P, because of $\tilde{\boldsymbol{\epsilon}}$, in a direction perpendicular to the (infinitesimal) quadric[11]

$$(d\mathbf{r}) \cdot \tilde{\boldsymbol{\epsilon}} \cdot d\mathbf{r} = \text{constant} \tag{12.12}$$

that passes through Q.

Furthermore, if Q happens to be on either one of the three axes of symmetry of the local quadric, it would move along this axis because of $\tilde{\boldsymbol{\epsilon}}$. If we repeat the argument of Section 1.3, subsection (2) with the three mutually perpendicular axes of symmetry, denoted by x_1, x_2, and x_3, as coordinate lines, we will find no shear rate of strain. Thus, referring locally to these three axes, the components of $\tilde{\boldsymbol{\epsilon}}$ are simply

↗	$\hat{\mathbf{x}}_1$	$\hat{\mathbf{x}}_2$	$\hat{\mathbf{x}}_3$
$\hat{\mathbf{x}}_1$	ϵ_{11}	0	0
$\hat{\mathbf{x}}_2$	0	ϵ_{22}	0
$\hat{\mathbf{x}}_3$	0	0	ϵ_{33}

These axes are called the (local) *principal axes* of the dyadic $\tilde{\boldsymbol{\epsilon}}$.

Summary At every point, and at every instant, there exist three mutually perpendicular principal axes, x_1, x_2, and x_3, of $\tilde{\boldsymbol{\epsilon}}$ such that

$$\tilde{\boldsymbol{\epsilon}} = \epsilon_{11}\hat{\mathbf{x}}_1\hat{\mathbf{x}}_1 + \epsilon_{22}\hat{\mathbf{x}}_2\hat{\mathbf{x}}_2 + \epsilon_{33}\hat{\mathbf{x}}_3\hat{\mathbf{x}}_3$$

If we refer to the x-, y-, and z-directions that are inclined to the principal axes, we must have

$$\tilde{\boldsymbol{\epsilon}} = \epsilon_{xx}\hat{\mathbf{x}}\hat{\mathbf{x}} + \epsilon_{xy}\hat{\mathbf{x}}\hat{\mathbf{y}} + \cdots$$

again. The two expressions are for the same physical quantity $\tilde{\boldsymbol{\epsilon}}$; their forms are different because the frames of reference are different.

To put it in another fashion: If we investigate the local rate of strain using any three perpendicular directions as reference, we would most likely see a very complicated picture of deformation, with both normal and shear components. But there are always three perpendicular principal axes, with reference to which we will not see any shear components at all. For example, if we look at a small sphere of fluid, and refer to the local principal axes of $\tilde{\boldsymbol{\epsilon}}$, we will see the sphere squashed and elongated in the directions of these axes into an ellipsoid (Fig. 12.13).

---

[11] Remembering that $(d\mathbf{r}) \cdot \tilde{\boldsymbol{\epsilon}}$ is the relative velocity due to $\tilde{\boldsymbol{\epsilon}}$.

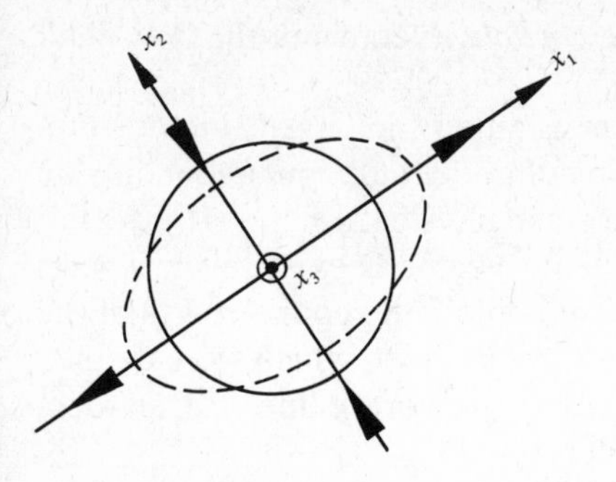

Figure 12.13 Deformation in the principal directions.

But, if we look at it referring to any other perpendicular axes, we will see it squashed, elongated, *and* sheared into an ellipsoid (Fig. 12.14).

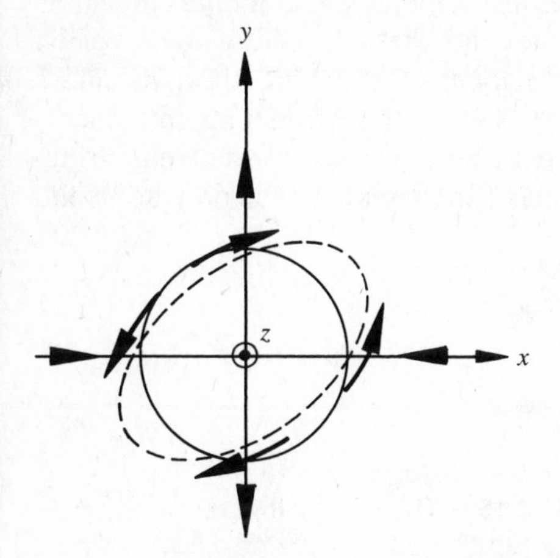

Figure 12.14 Deformation in nonprincipal directions.

Finally, we will close this section by relating *circulation* and *vorticity*.

Definition The *circulation* Γ around a closed spatial curve C, or a closed material curve $\mathscr{C}$, is defined as $\oint \mathbf{q} \cdot dl$ around C or $\mathscr{C}$.

Remarks The *closed curve* C or $\mathscr{C}$ is usually called the (spatial or material) *circuit*. To specify a circuit completely, one must also give the sense in which it is to be traversed in the line integration. In this section we will deal only with the instantaneous value of Γ, we can then stick to a spatial circuit C without any loss of generality, since a material circuit at *one particular* instant must be coincident with some spatial circuit C.[12]

Now, consider any *cap*, S, of the circuit C. If we set up positive directions of $\hat{\mathbf{n}}$ for S in such a fashion that it is *right-handedly* oriented with the given sense of C, we can apply Stokes theorem, Eq. (11.36b), and obtain

$$\begin{aligned}\Gamma = \oint_C \mathbf{q} \cdot dl &= \int_S (\nabla \times \mathbf{q}) \cdot \hat{\mathbf{n}}\, dS \\ &= \int_S \mathbf{\Omega} \cdot \hat{\mathbf{n}}\, dS \qquad \textbf{(12.13)}\end{aligned}$$

[12] At a later instant, $\mathscr{C}$ will move away from C which is fixed. But, at that instant, $\mathscr{C}$ is just C.

Definition $\boldsymbol{\Omega} = \text{curl } \mathbf{q}$ is called the *vorticity* (vector) of the flow field.

Equation (12.13) can be looked upon as either the physical interpretation of Γ in terms of the vorticity,[13] or $\boldsymbol{\Omega}$ in terms of the circulation; depending on which quantity is more difficult for the reader to grasp physically.

In a general rotational flow, the vorticity vector $\boldsymbol{\Omega}$ exists everywhere. One can then consider the vector field $\boldsymbol{\Omega}\langle \mathbf{r}, t\rangle$ and obtain the counterparts of the streamline, stream tube, stream filament, and the stream sheet, which were defined before for the **q**-field. These counterparts are called the vortex line, vortex tube, vortex filament, and the vortex sheet, in that order.

In one of the Problems, the reader will show that the strength of a vortex tube (or filament) is constant along the tube (or filament). It then follows that no vortex tube or filament can end inside the flow region; if it should disappear at a place, the strength would drop to zero beyond that place (no more vortex from there on)—a contradiction to the requirement of *constant* strength, unless the strength is zero to begin with (Fig. 12.15). Vortex tubes and filaments, therefore, must either be self-closed to form vortex ring-tubes and vortex rings (ring-filaments) or they must end on the boundaries of the region, which may be a solid or a free surface.

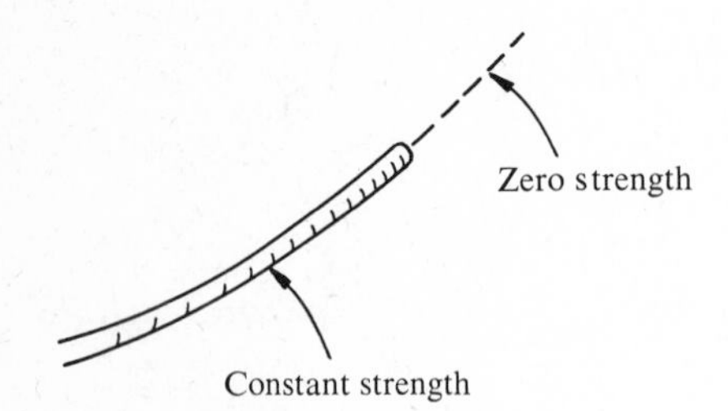

Figure 12.15 The impossible termination of a vortex tube or filament inside a flow region.

12.5 KINEMATICS OF A MOVING VOLUME INTEGRAL

Consider a field quantity $\mathscr{F}\langle \mathbf{r}, t\rangle$, and a volume Λ that changes its shape, position, and size in space as time goes on. To emphasize the change of Λ, we will denote Λ at an instant t by the symbol $\Lambda\langle t\rangle$. $\Lambda\langle t\rangle$, together with its bounding surface $\Sigma\langle t\rangle$, will then move about with local, instantaneous velocities **G** (Fig. 12.16).[14] Now, what is the time rate of change of the volume integral

$$\int_{\Lambda\langle t\rangle} \mathscr{F}\langle \mathbf{r}, t\rangle\, dV$$

as $\Lambda\langle t\rangle$ changes with t?

---

[13] That is, Γ is the outflow of $\boldsymbol{\Omega}$ through *any* cap S of C.

[14] If $\mathbf{G} = \mathbf{q}$ (the flow velocity) at every point on $\Sigma\langle t\rangle$, Λ and Σ become *material* volume and surface. When this is the case, symbols $\mathscr{V}$, $\mathscr{S}$, $\mathscr{V}\langle t\rangle$, and $\mathscr{S}\langle t\rangle$ will be used. Thus Λ, Σ include $\mathscr{V}$, $\mathscr{S}$ as a special case. We have in mind here, for example, the possibility of a volume bounded (at least partially) by a surface of discontinuity that propagates with its own velocity. To summarize, capital Greek letters denote motion that is independent of the fluid; capital script letters, motion with the fluid (material); and capital Roman letters, fixation.

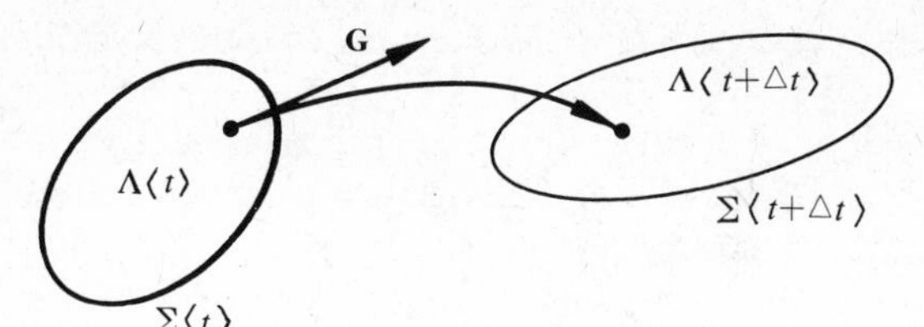

Figure 12.16 A changing volume.

THEOREM[15]

If V and S are the *spatial* volume and surface which coincide with $\Lambda\langle t\rangle$ and $\Sigma\langle t\rangle$, respectively, *at the instant t*, then

$$\frac{d}{dt}\int_{\Lambda\langle t\rangle} \mathscr{F}\langle \mathbf{r}, t\rangle\, dV = \frac{d}{dt}\int_V \mathscr{F}\, dV + \oint_S \mathscr{F}\mathbf{G}\cdot\hat{\mathbf{n}}\, dS \tag{12.14}$$

where $\mathbf{G}$ is the instantaneous velocities of advancement (or propagation) of points on $\Sigma\langle t\rangle$.

PROOF[16] For an increment of time from t to $t+\delta t$ (Fig. 12.17), two things change: (a) the integrand and (b) the domain over which the integration is to be taken. By definition,

$$\frac{d}{dt}\int_{\Lambda\langle t\rangle} \mathscr{F}\langle \mathbf{r}, t\rangle\, dV = \lim_{\delta t\to 0}\frac{1}{\delta t}\left[\int_{\Lambda\langle t+\delta t\rangle} \mathscr{F}\langle \mathbf{r}, t+\delta t\rangle\, dV - \int_{\Lambda\langle t\rangle} \mathscr{F}\langle \mathbf{r}, t\rangle\, dV\right]$$

$$= \lim_{\delta t\to 0}\frac{1}{\delta t}\left[\left\{\int_{\Lambda\langle t+\delta t\rangle} \mathscr{F}\langle \mathbf{r}, t+\delta t\rangle\, dV - \int_{\Lambda\langle t+\delta t\rangle} \mathscr{F}\langle \mathbf{r}, t\rangle\, dV\right\} + \left\{\int_{\Lambda\langle t+\delta t\rangle} \mathscr{F}\langle \mathbf{r}, t\rangle\, dV - \int_{\Lambda\langle t\rangle} \mathscr{F}\langle \mathbf{r}, t\rangle\, dV\right\}\right]$$

$$= \lim_{\delta t\to 0}\frac{1}{\delta t}[\{ ① \} + \{ ② \}]$$

The evaluation of the limit of $\{ ① \}/\delta t$ as $\delta t \to 0$ can be executed in three steps:

$$\lim_{\delta t\to 0}\frac{1}{\delta t}\{ ① \} = \lim_{\delta t\to 0}\frac{1}{\delta t}\left\{\int_{\Lambda\langle t\rangle} \mathscr{F}\langle \mathbf{r}, t+\delta t\rangle\, dV - \int_{\Lambda\langle t\rangle} \mathscr{F}\langle \mathbf{r}, t\rangle\, dV\right\}$$

$$= \lim_{\delta t\to 0}\frac{1}{\delta t}\left\{\int_V \mathscr{F}\langle \mathbf{r}, t+\delta t\rangle\, dV - \int_V \mathscr{F}\langle \mathbf{r}, t\rangle\, dV\right\}$$

$$= \frac{d}{dt}\int_V \mathscr{F}\langle \mathbf{r}, t\rangle\, dV$$

[15] This theorem was laid down first by Reynolds for a *material* volume as one of the basic postulates of his *Sub-Mechanics of the Universe,* cited in footnote 5 of Chapter 5.

[16] One proof using the Euler's expansion formula is particularly short; but the author prefers the following.

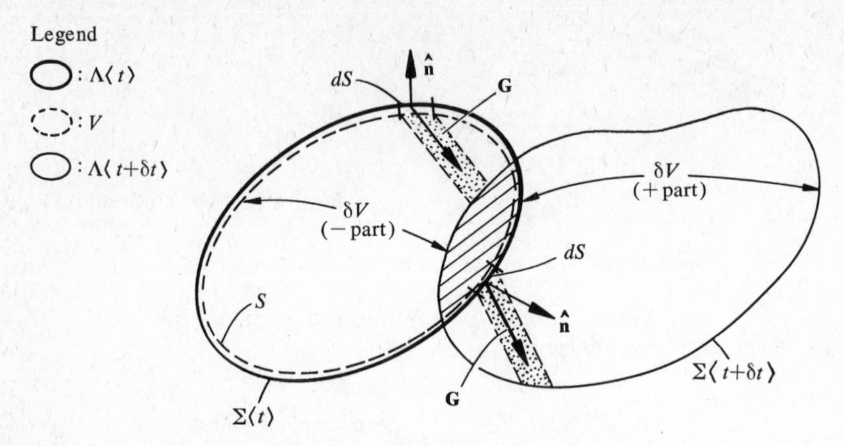

Figure 12.17 Changes in an interval δt.

Next, we see that

$$② = \int_{\delta V} \mathscr{F}\langle \mathbf{r}, t\rangle \, dV \text{ [to be calculated to } \square(\delta t)]$$

where δV is literally the spatial volume coincident with $\Lambda\langle t+\delta t\rangle$ *minus* that coincident with $\Lambda\langle t\rangle$. In Figure 12.17, it is the region that is not shaded. It is thus seen to be made up of two parts, the positive and the negative:

$$② = \int_{\substack{+\text{ portion}\\ \text{of } \delta V}} \mathscr{F}\, dV - \int_{\substack{-\text{ portion}\\ \text{of } \delta V}} \mathscr{F}\, dV$$

Referring again to Figure 12.17, we see that for the $+$portion of δV,

$$dV = (\mathbf{G}\,\delta t)\cdot \hat{\mathbf{n}}\, dS$$

and for the $-$portion of δV,

$$dV = -(\mathbf{G}\,\delta t)\cdot \hat{\mathbf{n}}\, dS$$

(since dV must always turn out a positive quantity). Therefore

$$② = (\delta t) \oint_S \mathscr{F}\langle \mathbf{r}, t\rangle \mathbf{G}\cdot \hat{\mathbf{n}}\, dS$$

and

$$\lim_{\delta t\to 0} \frac{1}{\delta t}\{ ② \} = \oint_S \mathscr{F}\langle \mathbf{r}, t\rangle \mathbf{G}\cdot \hat{\mathbf{n}}\, dS$$

This then completes the proof.

Behind this proof, there are a number of hidden implications. Cautions 1 through 3 of Section 11.5, subsection (1), also apply here. (For a multiply connected Λ, the proof is schematically the same except that one has to keep track of the geometric complications.) $\mathscr{F}$ and $\mathbf{G}$ must not be singular; otherwise, the integrals involved may fail to exist. But, there is *nothing against their being discontinuous* as long as

$$\int_V \mathscr{F}\, dV$$

is differentiable (with respect to time, of course).

If, in addition, $\partial\mathscr{F}/\partial t$ exists everywhere in V and the integral

$$\int_V \frac{\partial \mathscr{F}}{\partial t}\, dV$$

exists, Eq. (12.14) can also be written as

$$\frac{d}{dt}\int_{\Lambda\langle t\rangle} \mathscr{F}\langle \mathbf{r}, t\rangle\, dV = \int_V \frac{\partial \mathscr{F}}{\partial t}\, dV + \oint_S \mathscr{F}\mathbf{G}\cdot\hat{\mathbf{n}}\, dS \qquad \textbf{(12.15)}$$

If $\Lambda\langle t\rangle$ (and hence V) encloses a *moving* surface of discontinuity in $\mathscr{F}$, Eq. (12.15) certainly does not apply for the following reason: At a spatial point where the surface of discontinuity is passing through, $\mathscr{F}\langle t\rangle$ would jump. Equation (12.14) is, however, still valid.

If, in particular, Λ happens to be material (written as $\mathscr{V}$), **G** must be identical with **q**, and d/dt on the left-hand side should be changed to D/Dt according to convention; thus

$$\frac{D}{Dt}\int_{\mathscr{V}\langle t\rangle} \mathscr{F}\langle \mathbf{r}, t\rangle\, dV = \frac{d}{dt}\int_V \mathscr{F}\, dV + \oint_S \mathscr{F}\mathbf{q}\cdot\hat{\mathbf{n}}\, dS \qquad \textbf{(12.14A)}$$

or, if $\partial\mathscr{F}/\partial t$ exists everywhere in V (there being no moving surface of discontinuity),

$$\frac{D}{Dt}\int_{\mathscr{V}\langle t\rangle} \mathscr{F}\langle \mathbf{r}, t\rangle\, dV = \int_V \frac{\partial \mathscr{F}}{\partial t}\, dV + \oint_S \mathscr{F}\mathbf{q}\cdot\hat{\mathbf{n}}\, dS \qquad \textbf{(12.15A)}$$

where V is the spatial volume (with bounding surface S) that coincides with $\mathscr{V}\langle t\rangle$ at t. (This special case for a material volume is often referred to as the Reynolds transport theorem.)

12.6 CONSERVATION OF MASS

In the flow field of a fluid, consider an *arbitrary* material volume $\mathscr{V}\langle t\rangle$ with bounding surface $\mathscr{S}\langle t\rangle$. The law of mass conservation then *postulates* that, in the absence of mass creation or annihilation, the total mass of $\mathscr{V}$ is constant following its motion; that is,

$$\frac{D}{Dt}\int_{\mathscr{V}\langle t\rangle} \rho\langle \mathbf{r}, t\rangle\, dV = 0 \qquad \textbf{(12.16)}$$

Remarks

1. Equation (12.16) is definitely not true when applied to a spatial volume; that is,

$$\frac{d}{dt}\int_V \rho\langle \mathbf{r}, t\rangle\, dV \neq 0$$

2. The phrase "in the absence of mass creation or annihilation" may look overcautious. But, one must remember that, in the macroscopic description of a chemical reaction in a flowing fluid, one species of substance may actually disappear while another is created. Also, when singularities appear in the final solutions, our physical interpretation may have to include mass creation as an approximation of mass introduced from the outside through a very small opening.
3. Singularities are *not* allowed in the formulation here. (They may appear later in the final solutions as inherent in the mathematical structure of the posed problem.)
4. Discontinuities are, however, allowed in Eq. (12.16).
5. $\mathscr{V}$ is a finite volume of fluid, *not* a fluid particle. This kind of formulation is known as the integral formulation. It is *more general* than the particlewise formulation used in Parts 1 and 2, because it allows discontinuities in flow properties to appear. Furthermore, it is also more fundamental since, after all, one can only observe the behavior of finite volumes of fluid, not really fluid particles; then, when one wishes to postulate a general law, it is natural to take up a finite volume, not a particle.[17]

Using Eq. (12.14A), we can transform Eq. (12.16) into the *instantaneous* relation

$$\frac{d}{dt}\int_V \rho\, dV + \oint_S (\rho\mathbf{q}) \cdot \hat{\mathbf{n}}\, dS = 0 \tag{12.17}$$

where V (with boundary S) is the *spatial* volume that coincides with $\mathscr{V}$ at instant t. Actually, since $\mathscr{V}$ is arbitrary in Eq. (12.16), V can be interpreted as an arbitrary spatial volume; Eq. (12.17) is then true at any and all instants, for an arbitrary spatial volume. Here, we note that

$$\frac{d}{dt}\int_V \rho\, dV = -\oint_S (\rho\mathbf{q}) \cdot \hat{\mathbf{n}}\, dS \neq 0$$

In Eq. (12.17), it is obvious that

$$\oint_S \rho\mathbf{q} \cdot \hat{\mathbf{n}}\, dS$$

is just the mass outflow across the boundary of V. So, the equation only states that, at every instant, the rate of change of total mass inside a (fixed) spatial region is exactly balanced by the mass inflow (that is, the negative outflow) across the boundary of the region.

---

[17] This preference for the integral formulation was first advocated by Hilbert; and then by Kottler (F. Kottler, "Newton'sches Gesetz und Metrik," *Sitzgsber. Akad. Wiss. Wien (IIa)*, **131**, 1–14, 1922). David Hilbert (1862–1943), one of the Göttingen Mathematicians mentioned in footnote 20 of Chapter 11, posed 23 problems in 1900 as important tasks for the future mathematicians. The sixth of these problems calls for an *axiomatization* of physics.

In case that $\mathscr{V}$ (and hence V) is free of any moving surface of discontinuity, Eq. (12.15A) can be used to transform Eq. (12.16) into

$$\int_V \frac{\partial \rho}{\partial t}\, dV + \oint_S (\rho \mathbf{q}) \cdot \hat{\mathbf{n}}\, dS = 0 \tag{12.18}$$

In order to be able to apply the Gauss theorem, Eq. (11.35b), let us stipulate here that V is free of all surfaces of discontinuity, moving or stationary. Equation (12.18) then can be further transformed into

$$\int_V \left[\frac{\partial \rho}{\partial t} + \operatorname{div}(\rho \mathbf{q})\right] dV = 0 \tag{12.19}$$

Now, remembering that V is arbitrary, it must follow that[18]

$$\frac{\partial \rho}{\partial t} + \operatorname{div}(\rho \mathbf{q}) = 0 \tag{12.20}$$

at every point inside V, or at every point in the flow field, *as long as it is not a point of singularity and is not seated on a surface of discontinuity* at the moment.

Equation (12.20) is called the equation of continuity, where the word "continuity" refers to "continuity of material"—that is, mass conservation.

On a surface of discontinuity, Eq. (12.20) is *not* true. However, for a $\mathscr{V}$ that contains this surface, the original postulation, Eq. (12.16), is still valid, as is Eq. (12.17). (Surfaces of discontinuities will be studied in detail in Chapter 13.)

Two special consequences of Eq. (12.20) must be mentioned before we close this section.

(1) Steady-ρ Field

Here, $\rho = \rho\langle \mathbf{r} \rangle$, only. So, Eq. (12.20) becomes

$$\operatorname{div}(\rho \mathbf{q}) = 0 \tag{12.21}$$

In other words, the vector field $\rho\mathbf{q}$ (sometimes called the *mass velocity* or *mass-flux vector*) is divergence-free. Note that $\mathbf{q} = \mathbf{q}\langle \mathbf{r}, t \rangle$ in general; that is, a steady-ρ

---

[18] If

$$\int_V \mathscr{F}\, dV = 0 \text{ for an arbitrary } V$$

we can take any V surrounding an arbitrarily chosen point P and write

$$\frac{1}{V}\int_V \mathscr{F}\, dV = 0 \tag{A}$$

If $\mathscr{F}$ is single-valued, bounded, and continuous at point P, it would vary less and less as we take smaller and smaller V surrounding P. Then,

$$\lim_{\substack{V \to 0 \\ (P)}} \frac{1}{V}\int_V \mathscr{F}\, dV = \frac{1}{\not{V}}(\mathscr{F}_P \not{V})$$

From (A), $\mathscr{F}_P = 0$, or $\mathscr{F} = 0$ at every point where $\mathscr{F}$ is continuous and nonsingular.

field is not necessarily a steady-$\mathbf{q}$ field. Although this is true, a steady-ρ field is not of practical interest. We are actually more interested in a steady flow. Since a steady flow must also be a steady-ρ flow, Eq. (12.21) is all the more true for a steady flow (with $\mathbf{q} = \mathbf{q}\langle \mathbf{r} \rangle$, independent of t).

Conclusion $\rho\mathbf{q}$ is divergence-free in a steady flow.

(2) Isochoric Flow

A flow in which ρ stays constant on every pathline (or, what is the same thing, ρ of every particle stays constant) is called an isochoric flow.

Definition A flow in which $D\rho/Dt = 0$ is said to be isochoric.

One must notice that ρ may still assume different constant values on different pathlines. We mention this because the situation occurs in *nonhomogeneous* (or stratified) fluid dynamics.

A fluid may be inherently isochoric. For example, in the next chapter, we are going to define a class of fluids, called incompressible fluids, whose density is incapable of change no matter what. On the other hand, a compressible fluid such as air can conceivably flow in such a fashion that its density, though inherently changeable, is not changed.

Equation (12.20) can also be rewritten easily as

$$\frac{D\rho}{Dt} + \rho \operatorname{div} \mathbf{q} = 0 \tag{12.20A}$$

Then, for an isochoric flow,

$$\operatorname{div} \mathbf{q} = 0 \tag{12.22}$$

that is, the velocity field is divergence-free. Please note that $\mathbf{q} = \mathbf{q}\langle \mathbf{r}, t \rangle$ here; that is, whether the flow is steady or not does not enter into the argument at all.

Conclusion $\mathbf{q}$ is divergence-free in a (isochoric) flow of an incompressible fluid (or, less important, an isochoric flow of a compressible fluid), *steady or not*.

The mass velocity $\rho\mathbf{q}$ in Case (1) and the velocity $\mathbf{q}$ in Case (2) are divergence-free just like the vorticity curl $\mathbf{q}$. They, therefore, share many of the properties of the vorticity. If we denote the $\rho\mathbf{q}$-lines (that is, curves parallel to $\rho\mathbf{q}$) by the name *mass streamlines* (which are geometrically the same as the streamlines), we can state: The strength of a mass-stream tube (or filament) in Case (1), or a stream tube (or filament) in Case (2), is a constant along the tube (or filament); and they cannot end in the fluid region.

12.7 STREAM FUNCTIONS

Consider the following equation:[19]

$$\frac{\partial \mathscr{F}\langle \xi, \eta \rangle}{\partial \xi} + \frac{\partial \mathscr{G}\langle \xi, \eta \rangle}{\partial \eta} = 0 \tag{12.23}$$

The relations

$$\begin{cases} \mathscr{F} = \dfrac{\partial \psi}{\partial \eta} & \text{(12.24a)} \\ \mathscr{G} = -\dfrac{\partial \psi}{\partial \xi} & \text{(12.24b)} \end{cases}$$

then obviously satisfy Eq. (12.23). Thus we can replace two functions $\mathscr{F}$ and $\mathscr{G}$ by one, ψ. This single function will be called simply the ψ-function; and the curve $\psi\langle \xi, \eta \rangle =$ a constant, an equi-ψ-line. Giving different constant values to ψ, one will get a family of equi-ψ-lines on which

$$\begin{cases} \psi = \text{constant} \\ d\psi = 0 \\ \dfrac{d\xi}{\mathscr{F}} = \dfrac{d\eta}{\mathscr{G}} \end{cases}$$

Before going on, we must point out that ψ is inherently not unique, we could have used

$$\begin{cases} \mathscr{F} = \dfrac{\partial}{\partial \eta}(C_1\psi + C_2) \\ \mathscr{G} = -\dfrac{\partial}{\partial \xi}(C_1\psi + C_2) \end{cases}$$

instead of Eqs. (12.24a,b). Equation (12.23) will still be satisfied for any C_1 and C_2. As a matter of fact, many authors prefer the value $C_1 = -1$. In this book, $C_1 = 1$ will be adopted. The constant C_2 merely specifies the datum value of ψ and is not important.

Associated with this, we may also mention the fact that, while a given $\psi\langle \xi, \eta \rangle$ determines completely the geometric shape of the family of equi-ψ-lines, given the geometric shape of the family does not determine $\psi\langle \xi, \eta \rangle$ since any function $F\langle \psi\langle \xi, \eta \rangle \rangle =$ constants would give exactly the same lines.

Now, why do we discuss all these? Because the equation of continuity for a large class of problems assumes the form of Eq. (12.23). To be exact, let us enumerate them.[20]

---

[19] $\mathscr{F}$ and $\mathscr{G}$ can depend on variables in addition to ξ and η. Since only differentiations with respect to ξ and η appear, there is no need to write these additional variables out.

[20] The list is not exhaustive; for examples, pseudoplane and pseudoaxisymmetric flow are not included. The reader should have no difficulty in recognizing other cases when they occur.

Isochoric plane flow in rectangular Cartesian coordinates:

$$\frac{\partial u}{\partial x}+\frac{\partial v}{\partial y}=0 \tag{12.25}$$

Isochoric plane flow in polar coordinates:

$$\frac{\partial (Rq_R)}{\partial R}+\frac{\partial q_\theta}{\partial \theta}=0 \tag{12.26}$$

Isochoric axisymmetric flow in cylindrical coordinates:

$$\frac{\partial (Rq_R)}{\partial R}+\frac{\partial (Rw)}{\partial z}=0 \tag{12.27}$$

Isochoric axisymmetic flow in spherical coordinates:

$$\frac{\partial (r^2 \sin\varphi\, q_r)}{\partial r}+\frac{\partial (r \sin\varphi\, q_\varphi)}{\partial \varphi}=0 \tag{12.28}$$

The above, with "steady" replacing "isochoric":

Equations (12.25)–(12.28) with $\rho\mathbf{q}$ replacing $\mathbf{q}$ **(12.29)–(12.32)**

Unsteady unidirectional (not isochoric) flow:

$$\frac{\partial \rho}{\partial t}+\frac{\partial (\rho u)}{\partial x}=0 \tag{12.33}$$

Unsteady cylindrically radial (not isochoric) flow:

$$\frac{\partial(\rho R)}{\partial t}+\frac{\partial(\rho R q_R)}{\partial R}=0 \tag{12.34}$$

Unsteady spherically radial (not isochoric) flow:

$$\frac{\partial(\rho r^2)}{\partial t}+\frac{\partial(\rho r^2 q_r)}{\partial r}=0 \tag{12.35}$$

The corresponding ψ-function for plane flows is associated with the name d'Alembert;[21] that for axisymmetric flows, with Stokes.[22] On the other hand, ψ-functions for isochoric flows are called stream functions; for steady (but not isochoric) flows, mass stream functions. (For example, the ψ-function of an isochoric plane flow would be called the d'Alembert stream function.)

The ψ-functions for the unsteady (nonisochoric) flows are nameless. In the unidirectional case it is clear that

$$\psi = \text{constants:}\quad \frac{dx}{\rho u}=\frac{dt}{\rho} \qquad \text{or} \qquad \frac{dx}{dt}=u$$

---

[21] Appeared in his 1761 Remarks; cited in footnote 17 of Chapter 6.

[22] On the basis of his 1842 paper, "On the steady motion of incompressible fluids." Stokes' stream function was denoted by $\widehat{\psi}$ in Chapter 9.

that is, ψ = constants here are solution curves of $dx/dt = u$, which gives the position of particles at different times. (They are known as the *particle paths*, or *particle lines*, in gas dynamics.) Similar conclusions can also be drawn for the unsteady, cylindrically, or spherically radial flows. For all the other cases, it is obvious that constant (mass)-stream-function lines are streamlines.

The stream function for an isochoric flow may be time-dependent. This fact and the presence (or absence of) ρ are the only distinguishing factors between the stream and the mass-stream functions. We can, therefore, restrict ourselves to Eqs. (12.25)–(12.28) in the following.

(1) Plane Flows

Referring to a rectangular Cartesian coordinate system, we have here $\mathbf{q} = (u, v)$ as a *plane vector*, and $dl = (dx, dy)$ is the *plane line-element vector*. Clearly, then, $dl_1 = (dy, -dx)$ is dl rotated by $-90°$, and

$$d\psi = \mathbf{q} \cdot dl_1 \tag{12.36}$$

Equation (12.36) is true because it is a *scalar* equation,[23] and because it is true for rectangular Cartesian coordinates:

$$d\psi = u\,dy - v\,dx$$

From Eq. (12.36), we have

$$\begin{aligned}\psi_P - \psi_A &= \int_A^P \mathbf{q} \cdot dl_1 \\ &= \int_A^P \mathbf{q} \cdot \hat{l}_1\, dl \\ &= \int_A^P (q_{\perp dl})\, dl\end{aligned}$$

where $q_{\perp dl}$ denotes component of $\mathbf{q}$ perpendicular to dl (that is, in the direction of l_1) and the integration follows *any* curve from A to P in the field (Fig. 12.18).

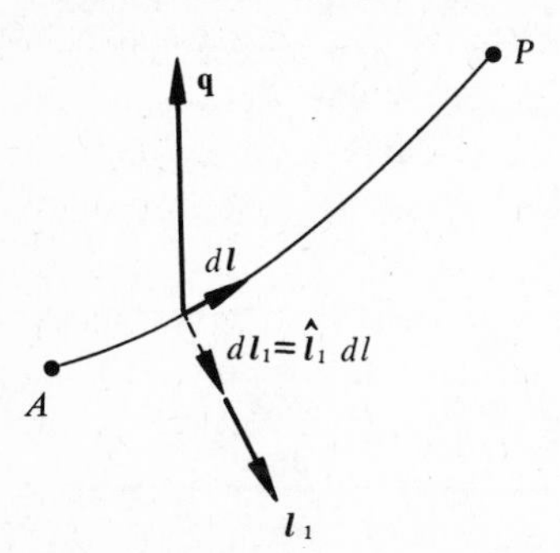

Figure 12.18 dl and dl_1.

[23] A scalar equation is coordinate-invariant, just as a vectorial equation. The reader should not mistake a *component* equation—for example, the x-component of Eq. (11.35a)—as a scalar equation.

Also,

$$\begin{aligned} d\psi &= (\text{grad } \psi) \cdot dl \\ &= [(\text{grad } \psi)_{\|dl}]\, dl \\ &= \mathbf{q} \cdot dl_1 \end{aligned}$$

Writing $dl_1 = l_1(dl)$, we have

$$(\text{grad } \psi)_{\|dl} = \mathbf{q} \cdot \hat{l}_1 = (q)_{\|dl_1}$$

Or, since dl is dl_1 rotated by $+90$ degrees, we have the following coordinate invariant *conclusion:*

To obtain the velocity component in a certain direction, differentiate ψ with respect to the direction 90 degrees ahead, counterclockwise.

(2) Axisymmetric Flows

The discussion here is left for the reader in one of the Problems.

CHAPTER 13

DYNAMICS AND ENERGETICS

13.1 EQUATION OF MOTION

Consider an arbitrary material volume $\mathscr{V}$ with boundary $\mathscr{S}$ in a flow field again. There are two kinds of forces acting on it: (a) body forces (for example, the gravitation) represented by their resultant *per unit mass*, $\mathbf{g}\langle\mathbf{r}, t\rangle$, and (b) surfaces forces represented by their resultant *per unit area* (called the force intensity)

$$\mathbf{f} = \mathbf{f}\langle\mathbf{r}, t; \hat{\mathbf{n}}\rangle \qquad \mathbf{r} \text{ on } \mathscr{S}$$

where $\hat{\mathbf{n}}$ is the unit normal of the infinitesimal area element dS of $\mathscr{S}$ on which $\mathbf{f}$ acts. (See the beginning part of Section 2.1.)

The total contact force on $\mathscr{V}$ is obviously

$$\oint_{\mathscr{S}} \mathbf{f}\langle\hat{\mathbf{n}}\rangle \, dS$$

where the dependence of $\mathbf{f}$ on $\hat{\mathbf{n}}$ is displayed for emphasis. The total body force is

$$\int_{\mathscr{V}} \rho \mathbf{g} \, dV$$

Referring to an unaccelerating frame of reference, we now postulate that

$$\frac{D}{Dt} \int_{\mathscr{V}\langle t\rangle} \rho \mathbf{q} \, dV = \int_{\mathscr{V}\langle t\rangle} \rho \mathbf{g} \, dV + \oint_{\mathscr{S}\langle t\rangle} \mathbf{f}\langle\hat{\mathbf{n}}\rangle \, dS \qquad \textbf{(13.1)}$$

for an arbitrary material volume $\mathscr{V}$ for any instant t.

Equation (13.1) is, of course, the balance of linear momentum, with the left-hand side representing the time rate of change of the linear momentum of $\mathscr{V}$; we have again presented it as a postulation.

If V and S are the spatial volume and surface that coincide with $\mathscr{V}$ and $\mathscr{S}$, respectively, at instant t, the right-hand side of Eq. (13.1) can be replaced by

$$\int_V \rho \mathbf{g}\, dV + \oint_S \mathbf{f}\langle \hat{\mathbf{n}} \rangle\, dS$$

and the left-hand side can be transformed into

$$\frac{d}{dt}\int_V \rho \mathbf{q}\, dV + \oint_S \rho \mathbf{q}\mathbf{q} \cdot \hat{\mathbf{n}}\, dS$$

by Eq. (12.14A). In this way we are confronted with another dyadic, $\rho \mathbf{q}\mathbf{q}$, whose physical significance we have to ascertain. As always, a dyadic quantity takes on meaning only when it is dot-multiplied by something vectorial. Here,

$$\begin{aligned} \rho \mathbf{q}\mathbf{q} \cdot (\hat{\mathbf{n}}\, dS) &= \rho \mathbf{q}(\mathbf{q} \cdot \hat{\mathbf{n}}\, dS) \\ &= \mathbf{q}(\rho \mathbf{q} \cdot \hat{\mathbf{n}}\, dS) \end{aligned}$$

is easily seen to be just (mass flow rate across dS from its negative to its positive side) times velocity. Then

$$\oint_S \rho \mathbf{q}\mathbf{q} \cdot \hat{\mathbf{n}}\, dS$$

is clearly the net rate of outflow of linear momentum across the closed surface S.

In the absence of any discontinuity, the left-hand side of Eq. (13.1) can also be written as

$$\int_V \frac{\partial(\rho \mathbf{q})}{\partial t}\, dV + \oint_S \rho \mathbf{q}\mathbf{q} \cdot \hat{\mathbf{n}}\, dS$$

by Eq. (12.15A), where the second term in turn can be transformed into the volume integral

$$\int_V \nabla \cdot (\rho \mathbf{q}\mathbf{q})\, dV = \int_V [\mathbf{q}\nabla \cdot (\rho \mathbf{q}) + \rho \mathbf{q} \cdot \nabla \mathbf{q}]\, dV$$

Then, if it is combined with the first term, we have

$$\frac{D}{Dt}\int_{\mathscr{V}(t)} \rho \mathbf{q}\, dV = \int_V \left\{\rho \frac{D\mathbf{q}}{Dt} + \mathbf{q}\left[\frac{\partial \rho}{\partial t} + \nabla \cdot (\rho \mathbf{q})\right]\right\} dV$$

Since we have already excluded discontinuities (and, of course, singularities) we might as well employ the equation of continuity, Eq. (12.20), to drop the terms inside the brackets; thus

$$\frac{D}{Dt}\int_{\mathscr{V}(t)} \rho \mathbf{q}\, dV = \int_V \rho \frac{D\mathbf{q}}{Dt}\, dV \tag{13.2}$$

in the absence of discontinuities.

Using Eq. (13.2), Eq. (13.1) becomes

$$\int_V \rho \frac{D\mathbf{q}}{Dt}\, dV = \int_V \rho \mathbf{g}\, dV + \oint_S \mathbf{f}\langle \hat{\mathbf{n}} \rangle\, dS \tag{13.3}$$

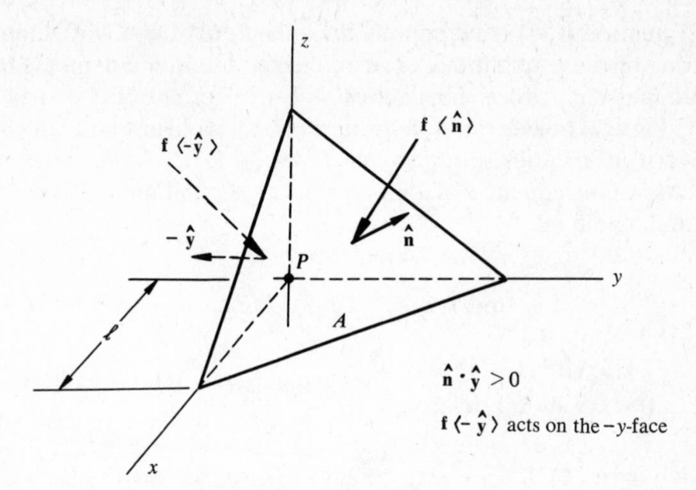

Figure 13.1 A small tetrahedron.

in the absence of discontinuities. Note that Eq. (13.3) is instantaneous, valid at any instant t; and V is thus an *arbitrary* spatial volume.

Now, consider V in the shape of a tetrahedron (Fig. 13.1) with edges parallel to three rectangular Cartesian coordinates. Although all the three components of $\hat{\mathbf{n}}$ are positive in the figure, this is not necessary. No matter which octant we are in, the areas (magnitudes only) of the side faces must be $A|\hat{\mathbf{n}}\cdot\hat{\mathbf{x}}|$, $A|\hat{\mathbf{n}}\cdot\hat{\mathbf{y}}|$, and $A|\hat{\mathbf{n}}\cdot\hat{\mathbf{z}}|$, which are all positive numbers; and Eq. (13.3) yields

$$\rho\left(\frac{D\mathbf{q}}{Dt}\right)\left(\frac{\ell A|\hat{\mathbf{n}}\cdot\hat{\mathbf{x}}|}{3}\right)=\rho\mathbf{g}\left(\frac{\ell A|\hat{\mathbf{n}}\cdot\hat{\mathbf{x}}|}{3}\right)+\mathbf{f}\langle\pm\hat{\mathbf{x}}\rangle\,(A|\hat{\mathbf{n}}\cdot\hat{\mathbf{x}}|)+\mathbf{f}\langle\pm\hat{\mathbf{y}}\rangle\,(A|\hat{\mathbf{n}}\cdot\hat{\mathbf{y}}|)$$
$$+\mathbf{f}\langle\pm\hat{\mathbf{z}}\rangle\,(A|\hat{\mathbf{n}}\cdot\hat{\mathbf{z}}|)+\mathbf{f}\langle\hat{\mathbf{n}}\rangle A$$

where $\mathbf{f}\langle\hat{\mathbf{x}}\rangle$ or $\mathbf{f}\langle-\hat{\mathbf{x}}\rangle$ is to be in force according to whether $\hat{\mathbf{n}}\cdot\hat{\mathbf{x}} < 0$ or > 0, and so on (Fig. 13.2).

Let the tetrahedron shrink toward the corner P by letting $\to 0$; we have (since $\mathbf{g}$ and $D\mathbf{q}/Dt$ are not singular)

$$\mathbf{f}\langle\hat{\mathbf{n}}\rangle=-\left[\mathbf{f}\langle\pm\hat{\mathbf{x}}\rangle\,(|\hat{\mathbf{n}}\cdot\hat{\mathbf{x}}|)+\mathbf{f}\langle\pm\hat{\mathbf{y}}\rangle\,(|\hat{\mathbf{n}}\cdot\hat{\mathbf{y}}|)+\mathbf{f}\langle\pm\hat{\mathbf{z}}\rangle\,(|\hat{\mathbf{n}}\cdot\hat{\mathbf{z}}|)\right] \qquad \mathbf{(13.4)}$$

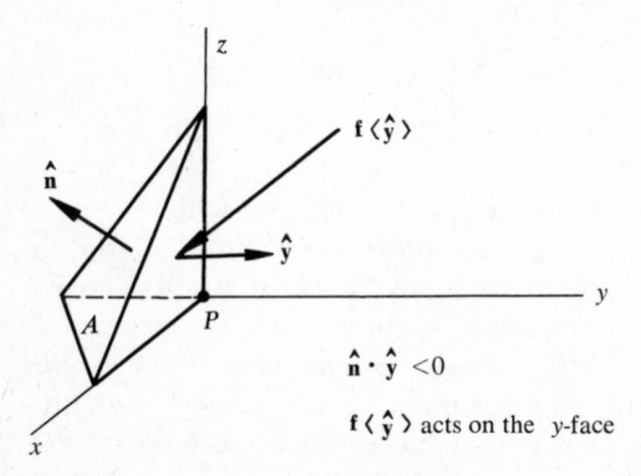

Figure 13.2 Tetrahedron in another octant.

Equation (13.4) is valid only when $\hat{\mathbf{n}} \cdot \hat{\mathbf{x}}, \hat{\mathbf{n}} \cdot \hat{\mathbf{y}}, \hat{\mathbf{n}} \cdot \hat{\mathbf{z}} \neq 0$. Geometrically, when any one of these quantities is zero, the tetrahedron collapses. However, if we *stipulate* that $\mathbf{f}\langle \mathbf{r}, t; \hat{\mathbf{n}}\rangle$ is continuous with respect not only to $\mathbf{r}$ but also to $\hat{\mathbf{n}}$, Eq. (13.4) must also hold for $\hat{\mathbf{n}} \cdot \hat{\mathbf{x}}, \hat{\mathbf{n}} \cdot \hat{\mathbf{y}}, \hat{\mathbf{n}} \cdot \hat{\mathbf{z}} = 0$ to avoid any sudden change in the behavior of $\mathbf{f}\langle \hat{\mathbf{n}}\rangle$ as $\hat{\mathbf{n}}$ varies.

As a consequence of this additional stipulation, we can put $\hat{\mathbf{n}} = \hat{\mathbf{x}}$ in Eq. (13.4) and obtain

$$\mathbf{f}\langle \hat{\mathbf{x}}\rangle = -\mathbf{f}\langle -\hat{\mathbf{x}}\rangle$$

Similarly, $\mathbf{f}\langle \hat{\mathbf{y}}\rangle = -\mathbf{f}\langle -\hat{\mathbf{y}}\rangle$ and $\mathbf{f}\langle \hat{\mathbf{z}}\rangle = -\mathbf{f}\langle -\hat{\mathbf{z}}\rangle$. Next, take the term $\mathbf{f}\langle \pm\hat{\mathbf{x}}\rangle\,(|\hat{\mathbf{n}} \cdot \hat{\mathbf{x}}|)$ in Eq. (13.4):

$$\left.\begin{array}{r}\mathbf{f}\langle \hat{\mathbf{x}}\rangle\,(|\hat{\mathbf{n}} \cdot \hat{\mathbf{x}}|),\ \hat{\mathbf{n}} \cdot \hat{\mathbf{x}} < 0 \\ \mathbf{f}\langle -\hat{\mathbf{x}}\rangle\,(|\hat{\mathbf{n}} \cdot \hat{\mathbf{x}}|),\ \hat{\mathbf{n}} \cdot \hat{\mathbf{x}} > 0\end{array}\right\} = \pm\mathbf{f}\langle \hat{\mathbf{x}}\rangle\,(|\hat{\mathbf{n}} \cdot \hat{\mathbf{x}}|) \qquad \hat{\mathbf{n}} \cdot \hat{\mathbf{x}} \lessgtr 0$$
$$= -\mathbf{f}\langle \hat{\mathbf{x}}\rangle\,(\hat{\mathbf{n}} \cdot \hat{\mathbf{x}}) \qquad \text{always}$$

since $|\hat{\mathbf{n}} \cdot \hat{\mathbf{x}}| = \mp(\hat{\mathbf{n}} \cdot \hat{\mathbf{x}})$ as $\hat{\mathbf{n}} \cdot \hat{\mathbf{x}} \lessgtr 0$. The other terms on the right-hand side of Eq. (13.4) can be similarly simplified. Therefore

$$\mathbf{f}\langle \hat{\mathbf{n}}\rangle = \mathbf{f}\langle \hat{\mathbf{x}}\rangle\,(\hat{\mathbf{n}} \cdot \hat{\mathbf{x}}) + \mathbf{f}\langle \hat{\mathbf{y}}\rangle\,(\hat{\mathbf{n}} \cdot \hat{\mathbf{y}}) + \mathbf{f}\langle \hat{\mathbf{z}}\rangle\,(\hat{\mathbf{n}} \cdot \hat{\mathbf{z}}) \tag{13.4A}$$

for any orientation of $\hat{\mathbf{n}}$. From Eq. (13.4A) it is easy to obtain Cauchy's lemma:

$$\mathbf{f}\langle -\hat{\mathbf{n}}\rangle = -\mathbf{f}\langle \hat{\mathbf{n}}\rangle \tag{13.5}$$

for any orientation of $\hat{\mathbf{n}}$.

Furthermore, the three vectors $\mathbf{f}\langle \hat{\mathbf{x}}\rangle$, $\mathbf{f}\langle \hat{\mathbf{y}}\rangle$, and $\mathbf{f}\langle \hat{\mathbf{z}}\rangle$ can also be written in their component forms:

$$\mathbf{f}\langle \hat{\mathbf{x}}\rangle = \sigma_{xx}\hat{\mathbf{x}} + \sigma_{xy}\hat{\mathbf{y}} + \sigma_{xz}\hat{\mathbf{z}}$$

and so on. Equation (13.4A) then becomes

$$\begin{aligned}\mathbf{f}\langle \hat{\mathbf{n}}\rangle &= (\sigma_{xx}\hat{\mathbf{x}} + \sigma_{xy}\hat{\mathbf{y}} + \sigma_{xz}\hat{\mathbf{z}})\,(\hat{\mathbf{n}} \cdot \hat{\mathbf{x}}) + \cdots \\ &= \hat{\mathbf{n}} \cdot (\sigma_{xx}\hat{\mathbf{x}}\hat{\mathbf{x}} + \sigma_{xy}\hat{\mathbf{x}}\hat{\mathbf{y}} + \sigma_{xz}\hat{\mathbf{x}}\hat{\mathbf{z}} + \sigma_{yx}\hat{\mathbf{y}}\hat{\mathbf{x}} + \sigma_{yy}\hat{\mathbf{y}}\hat{\mathbf{y}} + \sigma_{yz}\hat{\mathbf{y}}\hat{\mathbf{z}} + \sigma_{zx}\hat{\mathbf{z}}\hat{\mathbf{x}} + \sigma_{zy}\hat{\mathbf{z}}\hat{\mathbf{y}} + \sigma_{zz}\hat{\mathbf{z}}\hat{\mathbf{z}}) \\ &= \hat{\mathbf{n}} \cdot \tilde{\boldsymbol{\sigma}}\end{aligned} \tag{13.6}$$

where $\tilde{\boldsymbol{\sigma}}$ is a dyadic whose nine components *with respect to a rectangular* Cartesian coordinate system are

↗	$\hat{\mathbf{x}}$	$\hat{\mathbf{y}}$	$\hat{\mathbf{z}}$
$\hat{\mathbf{x}}$	σ_{xx}	σ_{xy}	σ_{xz}
$\hat{\mathbf{y}}$	σ_{yx}	σ_{yy}	σ_{yz}
$\hat{\mathbf{z}}$	σ_{zx}	σ_{zy}	σ_{zz}

Since Eq. (13.6) is a vectorial equation whose validity with respect to one particular coordinate system we have just demonstrated, it must be true *for all coordinate systems*. (This is tantamount to saying that if a vector $\mathbf{f}\langle \hat{\mathbf{n}}\rangle - \hat{\mathbf{n}} \cdot \tilde{\boldsymbol{\sigma}}$ is zero with respect to one coordinate system, it is zero for all coordinate systems.) We have, therefore, shown that there exists a dyadic $\tilde{\boldsymbol{\sigma}}$, called the *stress dyadic*, such that $\hat{\mathbf{n}} \cdot \tilde{\boldsymbol{\sigma}}$ yields the (resultant contact) force intensity $\mathbf{f}\langle \hat{\mathbf{n}}\rangle$ acting on an area element whose normal unit vector is $\hat{\mathbf{n}}$. The components of $\tilde{\boldsymbol{\sigma}}$ with respect to a certain coordinate system are called *stresses* with respect to that coordinate system. A stress happens to have the same physical dimension as the force intensity. But one is a component of a dyadic and the other a vector.

Summary Cauchy's fundamental theorem may be stated as follows: The stress situation in a flow field is completely determined by a dyadic field $\tilde{\boldsymbol{\sigma}}\langle \mathbf{r}, t\rangle$ such that

$$\mathbf{f}\langle \mathbf{r}, t; \hat{\mathbf{n}}\rangle = \hat{\mathbf{n}} \cdot \tilde{\boldsymbol{\sigma}}$$

At a point on a surface of discontinuity, Cauchy's fundamental theorem can still be seen to hold; but there may be two different stress dyadics and/or two different force intensities on the two sides of the surface. (The foregoing argument can be repeated, but keeping to one side of the surface at all times.)

Substituting Eq. (13.6) into Eq. (13.3) and transforming the surface integral into a volume integral, we have

$$\int_V \left(\rho \frac{D\mathbf{q}}{Dt} - \rho\mathbf{g} - \nabla \cdot \tilde{\boldsymbol{\sigma}}\right) dV = 0$$

Then, since V is arbitrary, we must have

$$\rho \frac{D\mathbf{q}}{Dt} = \rho\mathbf{g} + \nabla \cdot \tilde{\boldsymbol{\sigma}} \qquad \textbf{(13.7)}$$

at points not on surfaces of discontinuity (and not of singularity). Equation (13.7) is known as the equation of motion.

In the same spirit as the balance of linear momentum, we may also postulate the balance of angular momentum. Again, referring to an unaccelerating frame of reference, we postulate that, for an arbitrary $\mathscr{V}$,

$$\frac{D}{Dt}\int_{\mathscr{V}\langle t\rangle} \rho(\boldsymbol{\zeta} \times \mathbf{q})\, dV = \int_{\mathscr{V}\langle t\rangle} \rho(\boldsymbol{\zeta} \times \mathbf{g})\, dV + \oint_{\mathscr{S}\langle t\rangle} \boldsymbol{\zeta} \times \mathbf{f}\langle \hat{\mathbf{n}}\rangle\, dS \qquad \textbf{(13.8)}$$

where $\boldsymbol{\zeta}$ is the vectorial distance, from a fixed spatial point, of a *material point* in $\mathscr{V}$.

In the absence of discontinuities, we can write

$$\frac{D}{Dt}\int_{\mathscr{V}\langle t\rangle} \rho(\boldsymbol{\zeta} \times \mathbf{q})\, dV = \int_V \rho \frac{D(\boldsymbol{\zeta} \times \mathbf{q})}{Dt}\, dV$$

similarly to Eq. (13.2). Then,

$$\int_V \rho \frac{D(\boldsymbol{\zeta} \times \mathbf{q})}{Dt}\, dV = \int_V \rho(\boldsymbol{\zeta} \times \mathbf{g})\, dV + \oint_S \boldsymbol{\zeta} \times \mathbf{f}\langle \hat{\mathbf{n}}\rangle\, dS \qquad \textbf{(13.9)}$$

for any arbitrary V at any instant t. Applying Eq. (13.9) to an infinitesimal cubic volume element, one would get

$$\oint_S \boldsymbol{\zeta} \times \mathbf{f}\langle \hat{\mathbf{n}}\rangle\, dS = 0$$

around the six faces of the cube, similarly to the previous application of Eq. (13.3) to the infinitesimal tetrahedron. With respect to a rectangular Cartesian coordinate system, this conclusion demands that

$$\sigma_{xy} = \sigma_{yx}$$

$$\sigma_{xz} = \sigma_{zx}$$

$$\sigma_{yz} = \sigma_{zy}$$

(This is to be compared with the derivation of Eq. (2.10).) It follows then that $\tilde{\sigma}$ must be symmetric; that is,

$$\tilde{\boldsymbol{\sigma}}_c = \tilde{\boldsymbol{\sigma}} \qquad \textbf{(13.10)}$$

since Eq. (13.10) is a dyadic equation which has just been shown to be true with respect to one coordinate system. (This is tantamount to saying that, if the dyadic $\tilde{\boldsymbol{\sigma}}_c - \tilde{\boldsymbol{\sigma}}$ has zero components with respect to one coordinate system, it must have zero components in all coordinate systems.)

Because of the symmetry of $\tilde{\boldsymbol{\sigma}}$, it would share a number of properties with the rate-of-strain dyadic $\tilde{\boldsymbol{\epsilon}}$. For example, $\tilde{\boldsymbol{\sigma}}$ has really six independent components with respect to three mutually perpendicular directions; and it has three mutually perpendicular principal axes x_1, x_2, x_3 at every point such that

$$\tilde{\boldsymbol{\sigma}} = \sigma_{11}\hat{\mathbf{x}}_1\hat{\mathbf{x}}_1 + \sigma_{22}\hat{\mathbf{x}}_2\hat{\mathbf{x}}_2 + \sigma_{33}\hat{\mathbf{x}}_3\hat{\mathbf{x}}_3$$

without any shear stresses. Furthermore,

$$\begin{aligned} \mathbf{f}\langle\hat{\mathbf{n}}\rangle &= \hat{\mathbf{n}} \cdot \tilde{\boldsymbol{\sigma}} \\ &= \tilde{\boldsymbol{\sigma}}_c \cdot \hat{\mathbf{n}} \\ &= \tilde{\boldsymbol{\sigma}} \cdot \hat{\mathbf{n}} \end{aligned} \qquad \textbf{(13.6A)}$$

It is interesting to note that Eq. (13.9) cannot yield any more *local* information other than the symmetry of $\tilde{\boldsymbol{\sigma}}$. The local differential form of Eq. (13.9) must be just Eq. (13.7) cross-multiplied by $\boldsymbol{\zeta}$.[1] (Equation (13.9) itself definitely contains more information than Eq. (13.3) – namely, the symmetry of $\tilde{\boldsymbol{\sigma}}$.)

13.2 FIRST LAW OF THERMODYNAMICS

Dot-multiplying Eq. (13.7) by **q**, one gets the *scalar* equation

$$\rho \frac{D(q^2/2)}{Dt} = \rho\mathbf{q} \cdot \mathbf{g} + \mathbf{q} \cdot (\boldsymbol{\nabla} \cdot \tilde{\boldsymbol{\sigma}}) \qquad \textbf{(13.11)}$$

where every term has the dimension of energy per unit volume per unit time. This equation, although interesting and useful, contains no new information. To inject some new blood into Eq. (13.11), the first law of thermodynamics must be employed.

[1] This statement can be proved formally, but it needs some more complicated formulas in dyadic analysis. The author does not think it worthwhile to delve into this. See Serrin's article, cited in footnote 6.

A part of the content of the first law of thermodynamics states that every finite physical system in equilibrium possesses a quantity, called the internal energy, that depends only on the thermodynamic state of the system. We might also state here that, similarly, a part of the content of the second law of thermodynamics asserts the existence of a quantity called entropy. In a fluid flow, we have to extend the ideas of internal energy and entropy locally to every fluid particle. This is done in exactly the same manner as the concept of mass is extended to particles through the introduction of density (that is, local mass per unit volume). Thus, at every point in a flow field, there are internal energy per unit volume and entropy per unit volume. However, it is customary to use per-unit-mass values of internal energy and entropy (called the specific internal energy and specific entropy, denoted by e and s, respectively):

$$\left.\begin{matrix} e \\ s \end{matrix}\right\} \text{ per unit mass}$$

$$\left.\begin{matrix} \rho e \\ \rho s \end{matrix}\right\} \text{ per unit volume}$$

We will deal with entropy again in the next section. Here let us introduce the rest of the content of the first law by postulating that, for an arbitrary $\mathscr{V}$,

$$\begin{aligned} \frac{D}{Dt}\int_{\mathscr{V}(t)} \rho(\tfrac{1}{2}q^2+e)\,dV = & \oint_{\mathscr{S}(t)} \mathbf{f}\langle\hat{\mathbf{n}}\rangle\cdot\mathbf{q}\,dS \\ & +\int_{\mathscr{V}(t)} \rho\mathbf{g}\cdot\mathbf{q}\,dV - \oint_{\mathscr{S}(t)} \mathbf{H}\cdot\hat{\mathbf{n}}\,dS \\ & +\int_{\mathscr{V}(t)} Q\,dV \end{aligned} \tag{13.12}$$

where $\mathbf{H}$ is the heat flux vector (per unit area per unit time) and Q is the heat generation (per unit volume per unit time) due to chemical reactions and the like. The left-hand side of Eq. (13.12) is, of course, the rate of change of kinetic energy plus internal energy in $\mathscr{V}$.[2] The first term on the right-hand side is easily seen to be the work done (per unit time) on $\mathscr{S}$ by $\mathbf{f}$, the second term is the work done (per unit time) on $\mathscr{V}$ by $\mathbf{g}$, the third term (including the minus sign) is the total heat inflow (per unit time) into $\mathscr{V}$ across $\mathscr{S}$, and the last term is the total heat generated (per unit time) in $\mathscr{V}$.

Equation (13.12) is again true whether there is discontinuity or not. In the absence of discontinuities, it can also be written as

$$\begin{aligned} \int_V \rho\frac{D}{Dt}(\tfrac{1}{2}q^2+e)\,dV = & \oint_S (\hat{\mathbf{n}}\cdot\tilde{\boldsymbol{\sigma}})\cdot\mathbf{q}\,dS \\ & +\int_V \rho\mathbf{g}\cdot\mathbf{q}\,dV - \oint_S \mathbf{H}\cdot\hat{\mathbf{n}}\,dS + \int_V Q\,dV \end{aligned} \tag{13.13}$$

[2] The reader must be warned that the traditional "potential energy" has no place here or elsewhere. As a matter of fact, this term does not appear in this book at all outside of this footnote.

where $\hat{\mathbf{n}} \cdot \tilde{\boldsymbol{\sigma}}$ has replaced $\mathbf{f}\langle\hat{\mathbf{n}}\rangle$. We can, furthermore, transform the two surface integrals into volume integrals; thus

$$\int_V \left\{\rho \frac{D}{Dt}(\tfrac{1}{2}q^2 + e) - \operatorname{div}(\tilde{\boldsymbol{\sigma}} \cdot \mathbf{q}) - \rho \mathbf{g} \cdot \mathbf{q} + \operatorname{div} \mathbf{H} - Q\right\} dV = 0$$

Since V is arbitrary, this yields

$$\rho \frac{D}{Dt}(\tfrac{1}{2}q^2 + e) = \operatorname{div}(\tilde{\boldsymbol{\sigma}} \cdot \mathbf{q}) + \rho \mathbf{g} \cdot \mathbf{q} - \operatorname{div} \mathbf{H} + Q \tag{13.14}$$

at points not seated on any surface of discontinuity. Now

$$\begin{aligned} \operatorname{div}(\tilde{\boldsymbol{\sigma}} \cdot \mathbf{q}) &= \boldsymbol{\nabla} \cdot (\tilde{\boldsymbol{\sigma}} \cdot \mathbf{q}) \\ &= \mathbf{q} \cdot (\boldsymbol{\nabla} \cdot \tilde{\boldsymbol{\sigma}}) + \boldsymbol{\nabla}_q \cdot \tilde{\boldsymbol{\sigma}} \cdot \mathbf{q} \\ &= \mathbf{q} \cdot (\boldsymbol{\nabla} \cdot \tilde{\boldsymbol{\sigma}}) + \tilde{\boldsymbol{\sigma}} : \boldsymbol{\nabla}\mathbf{q} \end{aligned}$$

Using this relation and Eq. (13.11), we can reduce Eq. (13.14) to

$$\rho \frac{De}{Dt} = \tilde{\boldsymbol{\sigma}} : \boldsymbol{\nabla}\mathbf{q} - \operatorname{div} \mathbf{H} + Q$$

In addition, it has been established in one of the Problems of Chapter 11 that the double dot product of a symmetric and an antisymmetric dyadic vanishes. Therefore

$$\begin{aligned} \tilde{\boldsymbol{\sigma}} : \boldsymbol{\nabla}\mathbf{q} &= \tilde{\boldsymbol{\sigma}} : [\tilde{\boldsymbol{\epsilon}} + \tfrac{1}{2}(\boldsymbol{\nabla}\mathbf{q} - \mathbf{q}\boldsymbol{\nabla})] \\ &= \tilde{\boldsymbol{\sigma}} : \tilde{\boldsymbol{\epsilon}} \end{aligned}$$

and

$$\rho \frac{De}{Dt} = \tilde{\boldsymbol{\sigma}} : \tilde{\boldsymbol{\epsilon}} - \operatorname{div} \mathbf{H} + Q \tag{13.15}$$

Equation (13.15) is known as the *energy equation*. It is important to remember how the kinetic energy term and the **g**-term drop out by way of Eq. (13.11). The scope of Eq. (13.15) is very wide; it is *not* just for the case of no body force.

13.3 SECOND LAW OF THERMODYNAMICS

In the matter of the second law of thermodynamics, we will adopt Gibbs' approach[3] in the following manner.

---

[3] This is quite different from the usual practice in many texts. Although it is not expected that the reader will experience any difficulty in what follows, the author nevertheless takes this opportunity to recommend the first three chapters of H. B. Callen's excellent book, *Thermodynamics*, Wiley, New York, 1960. Gibbs laid down his approach to thermodynamics in three papers (two in 1873 and one in 1875), which appeared in the obscure *Transactions of the Connecticut Academy*. It was at once recognized by Maxwell (in the 1875 edition of his book, *Theory of Heat*) as of monumental importance. Hilbert also adopted this approach in his 1906–1907 lectures on *Mechanics of Continua* (in German), manuscript notes (by W. Marshall) of which are in the Purdue University Library. The definition of an ideal gas given later in this section can be generalized; see Callen's Sections D.1 and D.2.

We first isolate a portion of the fluid and investigate the equilibrium states of this portion, whose vertical dimension is assumed to be so small that the gravitational effect can be neglected. The result of such an investigation is contained in the existence of the following *fundamental equation*:

$$e = e\langle \rho^{-1}, s \rangle \quad \textbf{(13.16)}$$

where ρ^{-1} is the specific volume (that is, volume per unit mass). Note that e, and so on, do not vary from point to point in this portion since it is always in equilibrium.

Equation (13.16) actually contains all the thermodynamic information about the fluid in a nutshell. The derivative

$$\frac{\partial e\langle \rho^{-1}, s \rangle}{\partial s} = T \quad \textbf{(13.17)}$$

is called the (absolute) temperature of the fluid, and

$$-\frac{\partial e\langle \rho^{-1}, s \rangle}{\partial \rho^{-1}} = \Pi \quad \textbf{(13.18)}$$

is called the *thermodynamic pressure*. On the other hand, a fluid in equilibrium is also known to exhibit *mechanically* a stress dyadic

$$\tilde{\boldsymbol{\sigma}} = -\mathring{p}\tilde{\mathbf{I}} \quad \textbf{(13.19)}$$

where the scalar $\mathring{p}$ is known as the *hydrostatic pressure*, which can be measured by a manometer. It is a part of the structure of the Gibbs' approach to identify Π with $\mathring{p}$.

So much for the portion of fluid in equilibrium. To apply the above to a general flow field, which is certainly not in equilibrium, we *postulate* that at every point in the field (and at every instant), the *local* e, ρ^{-1}, and s are related by the same fundamental equation, Eq. (13.16), whether the particle accelerates or not. This is sometimes known as the *postulation of local equilibrium*, since it states that a fluid particle in a grossly nonequilibrium system behaves *thermodynamically the same* as the corresponding portion of (the same) fluid in equilibrium, which was taken out for investigation.

Equations (13.17) and (13.18) are now *locally* in effect. So, T and Π become the local temperature and the local thermodynamic pressure. However, as already emphasized in Section 2.3, the hydrostatic pressure $\mathring{p}$ does not exist at points of deformation (that is, where $\tilde{\boldsymbol{\epsilon}} \neq 0$); it is then senseless to talk about identification of Π with $\mathring{p}$ at these points. Without this link there can be only one meaning attached to Π—namely, a certain slope of the surface $e = e\langle \rho^{-1}, s \rangle$ in the thermodynamic space (e, ρ^{-1}, s). The thermodynamic pressure Π is thus only calculable (from the fundamental equation), not mechanically measurable, unless $\tilde{\boldsymbol{\epsilon}} = 0$ at the point.

As a very special case of Eq. (13.16), we will introduce a fictitious fluid called the incompressible fluid.

Definition A fluid is said to be incompressible if its intrinsic thermodynamic structure is such that the fundamental equation takes on the following singularly degenerate form:

$$\begin{cases} \rho^{-1} = \text{constant} \quad (\text{that is, } \rho = \text{constant}) & \textbf{(13.20a)} \\ e = e\langle s \rangle, \text{ only} & \textbf{(13.20b)} \end{cases}$$

For such a fluid, we see that the thermodynamic pressure Π defined by Eq. (13.18) *does not exist* because e is not dependent on ρ^{-1} at all. One might say that Π is not even defined for an incompressible fluid. The temperature is, of course, still there; being

$$T = \frac{de}{ds} \quad \textbf{(13.21)}$$

In a flow field of an incompressible fluid, Eq. (13.20a) should be interpreted to imply $D\rho/Dt = 0$, not $\partial\rho/\partial t = 0$ or $\nabla\rho = 0$. This then leads to Eq. (12.22) as the equation of continuity, as already anticipated in Section 12.6. Also, please note that Eqs. (13.20a,b) together define the incompressible fluid; Eq. (13.20a) alone would not do. In Figure 13.3, it is shown that $e = e\langle s \rangle$ represents a

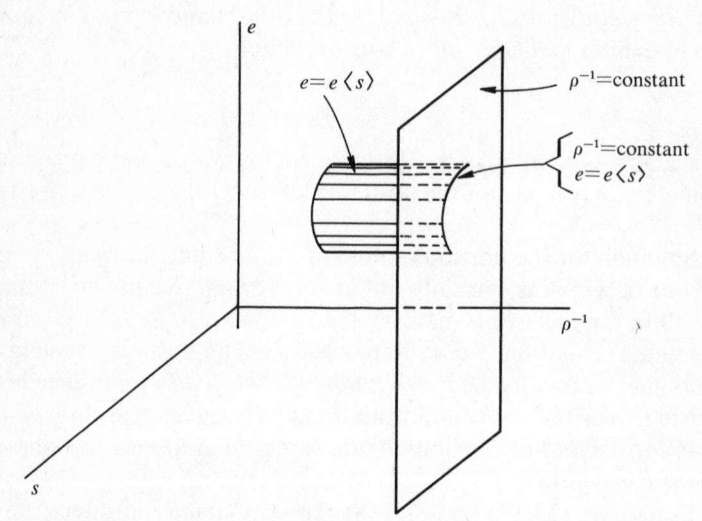

Figure 13.3 Thermodynamic surface of an incompressible fluid.

cylindrical surface, whereas $\rho^{-1} = \text{constant}$ maps out a plane. The geometric image of an incompressible fluid is then the curve of intersection of the cylindrical surface and the plane. In general the fundamental equation, Eq. (13.16), should be a surface in the thermodynamic space. For an incompressible fluid this surface degenerates into a curve. No real fluid (or solid, for that matter) exhibits a fundamental equation of this degenerate form; that is why we referred to incompressible fluids as fictitious. The importance of such fictitious fluids stems from the frequently encountered flow situation where the change of ρ is small even though the fluid involved is inherently compressible. In the thermodynamic space, the thermodynamic states of the fluid particles will show up inside a narrow strip of the surface $e = e\langle \rho^{-1}, s \rangle$, although inherently they *could* land anywhere else on the surface (Fig. 13.4). For such flow problems of a compressible fluid we could represent this narrow strip of the surface by a curve (indicated by dash-dots in Fig. 13.4),

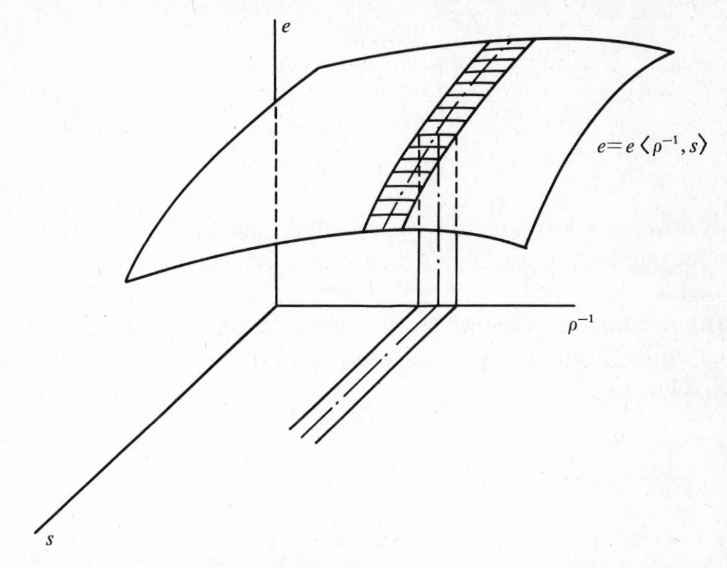

Figure 13.4 Approximation by an incompressible fluid.

and define a corresponding incompressible fluid as an approximation. Whether such an approximation is allowable for a given problem can be decided only by comparing the result of the approximation with either (a) actual observation, or (b) exact calculation without incurring this approximation. However, in the comparison one should not rely on the nominal change of ρ alone. For example, in an underwater explosion the change of ρ may be quite small; but treating water as incompressible here would miss the dynamic effect (that is, the explosion) completely. Another example of a different sort may also be quoted. In the propagation of sonic waves in liquid water, the changes of all properties are small. But in this problem it is exactly these small changes that are of interest. If we treat water as incompressible here, we will miss these small but important changes. Actually, in the explosion problem, pressure should be the thing to compare; and, in the sound propagation, the sonic speed should be compared.

Getting back to the second law, we have the following.

(1) Compressible Fluids

$$de = \frac{\partial e\langle \rho^{-1}, s \rangle}{\partial s} ds + \frac{\partial e\langle \rho^{-1}, s \rangle}{\partial \rho^{-1}} d\rho^{-1}$$
$$= T\,ds - \Pi d\rho^{-1} = T\,ds + \frac{\Pi}{\rho^2} d\rho \tag{13.22}$$

Therefore

$$\rho T \frac{Ds}{Dt} + \frac{\Pi}{\rho}\frac{D\rho}{Dt} = \tilde{\boldsymbol{\sigma}} : \tilde{\boldsymbol{\epsilon}} - \text{div}\,\mathbf{H} + Q \tag{13.23}$$

(2) Incompressible Fluids

$$\begin{cases} d\rho = 0 & (13.24\text{a}) \\ de = T\,ds & (13.24\text{b}) \end{cases}$$

Therefore

$$\rho T \frac{Ds}{Dt} = \tilde{\boldsymbol{\sigma}} : \tilde{\boldsymbol{\epsilon}} - \text{div}\,\mathbf{H} + Q \tag{13.25}$$

It now remains for us to write down the specific forms of the fundamental equation for the two kinds of fluids to be considered in this book.

(a) Incompressible or Nearly Incompressible Fluids The fundamental equation for these fluids has already been quoted. We will only add that, from Eq. (13.21),

$$T = T\langle s \rangle$$

or

$$s = s\langle T \rangle$$

Therefore

$$e = e\langle s \rangle = e\langle s\langle T \rangle \rangle$$

Then

$$de = \left(\frac{de}{dT}\right) dT = c_{in}\langle T \rangle\, dT \tag{13.26}$$

where $c_{in} = de/dT$ is known as the specific heat capacity of the *incompressible* fluid. The energy equation, Eq. (13.15), is now written as

$$c_{in}\langle T \rangle \frac{DT}{Dt} = \tilde{\boldsymbol{\sigma}} : \tilde{\boldsymbol{\epsilon}} - \text{div}\,\mathbf{H} + Q \tag{13.27}$$

Note that $c_{in}\langle T \rangle$ appears outside the differentiation.

(b) Ideal Gases The fundamental equation for these fluids assumes the form

$$e = \frac{A}{\gamma - 1}\rho^{\gamma-1} \exp\left[\frac{(\gamma-1)s}{\mathrm{I\!R}}\right] + B \tag{13.28}$$

where A, B, γ, $\mathrm{I\!R}$ are constants. It follows then that

$$\Pi = A \exp\left[\frac{(\gamma-1)s}{\mathrm{I\!R}}\right]\rho^{\gamma} \tag{13.29}$$

$$T = \frac{A}{\mathrm{I\!R}}\rho^{\gamma-1} \exp\left[\frac{(\gamma-1)s}{\mathrm{I\!R}}\right] \tag{13.30}$$

Therefore

$$\Pi = \rho \mathrm{I\!R} T \tag{13.31}$$

Equation (13.31) is usually referred to as the equation of state of the ideal gas and $\mathrm{I\!R}$ as its gas constant. Substituting Eq. (13.30) into Eq. (13.28), one gets

$$e = \frac{\mathrm{I\!R}T}{\gamma - 1} + B$$

and the specific enthalpy

$$h \stackrel{\text{def}}{=} e + \frac{\Pi}{\rho} = \frac{\gamma \mathrm{I\!R} T}{\gamma - 1} + B$$

both functions of temperature only. Also

$$c_v \stackrel{\text{def}}{=} \frac{\partial e\langle \rho, T\rangle}{\partial T} = \frac{de}{dT} = \frac{\mathrm{I\!R}}{\gamma - 1}$$

$$c_p \stackrel{\text{def}}{=} \frac{\partial h\langle \Pi, T\rangle}{\partial T} = \frac{dh}{dT} = \frac{\gamma \mathrm{I\!R}}{\gamma - 1}$$

where c_v and c_p are the specific heat capacities at constant *volume* and at constant (thermodynamic) *pressure*, respectively. Thus we see that γ is the ratio c_p/c_v, and B is the datum value of e, which is of no particular interest since e is always being differentiated in the energy equation. The constant A is also of no special interest, since it cancels out in the equation of state, Eq. (13.31).

The energy equation, Eq. (13.15), can also be written here as

$$\rho c_v \frac{DT}{Dt} = \tilde{\boldsymbol{\sigma}} : \tilde{\boldsymbol{\epsilon}} - \operatorname{div} \mathbf{H} + Q \tag{13.32}$$

As the rest of the content of the second law of thermodynamics, we postulate[4] that

$$\frac{D}{DT} \int_{\mathscr{V}\langle t\rangle} \rho s \, dV \geqslant -\oint_{\mathscr{S}\langle t\rangle} \frac{\mathbf{H} \cdot \hat{\mathbf{n}}}{T} \, dS + \int_{\mathscr{V}\langle t\rangle} \frac{Q}{T} \, dV \tag{13.33}$$

---

[4] If a material volume has no heat flux across its bounding surface, and has no heat generation within, it is said to be an *adiabatic* volume (for example, an insulated bag of nonreacting fluid). Then, the right-hand side of inequality (13.33) vanishes, and

$$\frac{D}{Dt} \int_{\substack{\mathscr{V}\langle t\rangle \\ \text{adiabatic}}} \rho s \, dV \geqslant 0$$

which is the usual statement of the principle of increase of entropy. The postulation of inequality (13.33) is thus more general than the usual statement of the second law or its corollaries. The case with $Q = 0$ can be traced back to Duhem's 1901 paper, "Researches on hydrodynamics" (in French). Pierre Duhem (1861–1916), French thermodynamicist, fluid dynamicist, elastician, was the founder of modern history of science. His writings are always careful, displaying the tacit assumptions usually glossed over by other authors.

for an arbitrary $\mathscr{V}$; or, if there are no discontinuities,

$$\int_V \rho \frac{Ds}{Dt}\, dV \geqslant -\oint_S \frac{\mathbf{H}\cdot\hat{\mathbf{n}}}{T}\, dS + \int_V \frac{Q}{T}\, dV \tag{13.34}$$

for an arbitrary V. At points not seated on surface of discontinuity, we have locally

$$\rho \frac{Ds}{Dt} \geqslant -\operatorname{div}\left(\frac{\mathbf{H}}{T}\right) + \frac{Q}{T} \tag{13.35}$$

Using Eq. (13.23) or Eq. (13.25), and noticing that the first term on the right-hand side of inequality (13.35) can also be written as

$$-\frac{1}{T}\operatorname{div}\mathbf{H} + \frac{\mathbf{H}\cdot\operatorname{grad} T}{T^2}$$

we see that

$$-\Phi + \frac{\mathbf{H}\cdot\operatorname{grad} T}{T} \leqslant 0 \tag{13.36}$$

where

$$\Phi = \tilde{\boldsymbol{\sigma}} : \tilde{\boldsymbol{\epsilon}} + \Pi \operatorname{div}\mathbf{q} \tag{13.37}$$

is called the *energy dissipation*. Note that, for an incompressible fluid, $\operatorname{div}\mathbf{q} = 0$ and Eq. (13.37) naturally reads $\Phi = \tilde{\boldsymbol{\sigma}} : \tilde{\boldsymbol{\epsilon}}$. The conclusion, inequality (13.36), plays no important role here other than serving as a criterion against which all experiments involving Φ and $\mathbf{H}$ must be tested.

13.4 CONSTITUTIVE EQUATIONS

The previous laws also dragged in more unknown quantities—that is, $\tilde{\boldsymbol{\sigma}}$ and $\mathbf{H}$. These quantities must be related to the other flow variables in order to complete the formulation. These relations are manifestations of the microscopic structure of the material being considered; and are called the constitutive equations. In this book the constitutive equations will be regarded as additional information supplied to us through guided (and indirect) experiments; but in the formal presentation they will appear as postulations (though motivated by some simple observations).

We will start with $\mathbf{H}$. Wherever there is space variation of T, the fluid will not be in thermal equilibrium, and there will be heat flow. If we consider a small area element dS with unit normal $\hat{\mathbf{n}}$, at a point P, the heat flow across dS in the n-direction is expected to be related to the rate of change of T along n—that is, $\partial T/\partial n$ or $(\operatorname{grad} T)\cdot\hat{\mathbf{n}}$. This consideration then leads to a relation between the local heat flux vector $\mathbf{H}$ (per unit area and time) and the local temperature gradient. It is an experimental fact that

$$H_n = \mathbf{H}\cdot\hat{\mathbf{n}} = -K\frac{\partial T}{\partial n} = -K(\operatorname{grad} T)\cdot\hat{\mathbf{n}} \qquad \text{for } any\ \hat{\mathbf{n}} \tag{13.38}$$

where K is a coefficient depending only on the local T and ρ. This coefficient is called the thermal conductivity of the fluid. The specific functional forms of $K\langle T, \rho\rangle$ for different fluids can only come from actual measurements or some microscopic theories.

Equation (13.38) can also be written as

$$(\mathbf{H} + K \operatorname{grad} T) \cdot \hat{\mathbf{n}} = 0 \qquad \text{for } any\ \hat{\mathbf{n}}$$

Therefore, it must follow that

$$\mathbf{H} = -K \operatorname{grad} T \tag{13.39}$$

Equation (13.39) is known as *Fourier's law*.[5] It states that $\mathbf{H}$ is parallel to grad T. For all real fluids, experiments show that $K \nleq 0$, so heat flows only in the direction of decreasing T.

The rest of this section will be devoted entirely to the relation between $\tilde{\boldsymbol{\sigma}}$ and $\tilde{\boldsymbol{\epsilon}}$. In this respect we shall restrict ourselves here to the Newtonian fluids only.

Definition[6] A *Newtonian* fluid is one that satisfies all the following requirements:

1. $\tilde{\boldsymbol{\sigma}}$ is a *linear*, continuous function of $\tilde{\boldsymbol{\epsilon}}$ and is independent of all the other *kinematic* quantities.
2. $\tilde{\boldsymbol{\sigma}}$ does not depend *explicitly* on the position $\mathbf{r}$. (It, of course, depends on $\mathbf{r}$ implicitly—that is, through $\tilde{\boldsymbol{\epsilon}}\langle \mathbf{r}, t\rangle$.)
3. It is *isotropic* (that is, shows no preference for any particular direction).
4. At points where $\tilde{\boldsymbol{\epsilon}} = 0$, $\tilde{\boldsymbol{\sigma}}$ must reduce to $-\mathring{p}\tilde{\mathbf{I}}$.

Requirements (1), (2), and (4) can be translated into

$$\begin{cases} \tilde{\boldsymbol{\sigma}} = \tilde{\mathbf{F}}\langle \tilde{\boldsymbol{\epsilon}}\rangle \\ \tilde{\mathbf{F}}\langle 0\rangle = -\mathring{p}\tilde{\mathbf{I}} \end{cases}$$

where $\tilde{\mathbf{F}}$ is a dyadic, continuous and *linear* function; and where $\mathring{p}$ is defined only at points of no deformation.

In this book[7] we will understand *isotropy* to mean that, in addition to

---

[5] Established in Fourier's 1822 treatise, "Analytical theory of heat" (in French). Jean-Baptiste-Joseph Fourier (1768–1830), friend of Napoleon Bonaparte, went to tell the Bourbons of Napoleon's escape from Elba, in the true spirit of a Dumas romance, but found himself the prisoner of his old friend instead. Although his great treatise deals, on the surface, with the theory of heat conduction, its greatest value lies in the introduction of the *Fourier series*. The theory of the Fourier series appeared inconceivable to the Academicians of Fourier's time, since it claims that a curve made up of broken lines and jumps can be represented by a series of certain *continuous* functions.

[6] J. Serrin, "Mathematical principles of classical fluid mechanics," *Encyclopedia of Physics*, Vol. VIII/1, edited by S. Flügge (co-edited by C. Truesdell), Springer-Verlag, Berlin, 1959. This definition came from Stokes; see his 1845 paper, cited in footnote 6 of Chapter 1.

[7] The following exposition is an adaptation of the author's "On the treatment of pressures in a moving Newtonian fluid," *Bulletin of Mechanical Engineering Education*, **9**, 307–311, 1970.

the fact that it has no preference of direction, the principal axes of $\tilde{\boldsymbol{\sigma}}$ coincide[8] with those of $\tilde{\boldsymbol{\epsilon}}$. In other words, purely normal stresses applied to an isotropic fluid particle induce only purely normal deformations. Then there exists at every point *one* rectangle Cartesian coordinate system (x_1, x_2, x_3) such that

$$\tilde{\boldsymbol{\sigma}} = \sigma_{11}\hat{\mathbf{x}}_1\hat{\mathbf{x}}_1 + \sigma_{22}\hat{\mathbf{x}}_2\hat{\mathbf{x}}_2 + \sigma_{33}\hat{\mathbf{x}}_3\hat{\mathbf{x}}_3$$

$$\tilde{\boldsymbol{\epsilon}} = \epsilon_{11}\hat{\mathbf{x}}_1\hat{\mathbf{x}}_1 + \epsilon_{22}\hat{\mathbf{x}}_2\hat{\mathbf{x}}_2 + \epsilon_{33}\hat{\mathbf{x}}_3\hat{\mathbf{x}}_3$$

So the third requirement boils down to saying that we can choose a coordinate system judiciously at any one point so that $\tilde{\boldsymbol{\sigma}}$ and $\tilde{\boldsymbol{\epsilon}}$ are each determined by three numbers.

Combined with the other requirements, a Newtonian fluid must have the following for an arbitrarily chosen particle:

$$\begin{cases} \sigma_{11} = F_{11}\langle \epsilon_{11}, \epsilon_{22}, \epsilon_{33} \rangle \\ \sigma_{22} = F_{22}\langle \epsilon_{11}, \epsilon_{22}, \epsilon_{33} \rangle \\ \sigma_{33} = F_{33}\langle \epsilon_{11}, \epsilon_{22}, \epsilon_{33} \rangle \end{cases}$$

where F_{11}, F_{22}, F_{33} are linear and continuous in $\epsilon_{11}, \epsilon_{22}, \epsilon_{33}$, and where

$$\begin{cases} F_{11}\langle 0,0,0 \rangle = -\mathring{p} \\ F_{22}\langle 0,0,0 \rangle = -\mathring{p} \\ F_{33}\langle 0,0,0 \rangle = -\mathring{p} \end{cases}$$

The reader must notice that one cannot take care of the last part of the demand by simply displaying $-\mathring{p}$ in

$$\begin{cases} \sigma_{11} = -\mathring{p} + G_{11}\langle \epsilon_{11}, \epsilon_{22}, \epsilon_{33} \rangle \\ \sigma_{22} = -\mathring{p} + G_{22}\langle \epsilon_{11}, \epsilon_{22}, \epsilon_{33} \rangle \\ \sigma_{33} = -\mathring{p} + G_{33}\langle \epsilon_{11}, \epsilon_{22}, \epsilon_{33} \rangle \end{cases}$$

by defining $G_{11} = F_{11} + \mathring{p}$, and so on, since $\mathring{p}$ is in general *not defined*, as explained before.

We are, however, at liberty to choose any *scalar* fluid property α that (a) always *exists*, (b) does not depend on any kinematic quantity other than ϵ_{11}, ϵ_{22}, and ϵ_{33}, (c) is a continuous, linear function of ϵ_{11}, ϵ_{22} and ϵ_{33}, and (d) reduces to $\mathring{p}$ when $\epsilon_{11}, \epsilon_{22}, \epsilon_{33} = 0$; and to display it, thus:

$$\begin{cases} \sigma_{11} = -\alpha + G_{11}\langle \epsilon_{11}, \epsilon_{22}, \epsilon_{33} \rangle \\ \sigma_{22} = -\alpha + G_{22}\langle \epsilon_{11}, \epsilon_{22}, \epsilon_{33} \rangle \\ \sigma_{33} = -\alpha + G_{33}\langle \epsilon_{11}, \epsilon_{22}, \epsilon_{33} \rangle \end{cases}$$

Although the choice of α may not be unique, this does not matter, since for differ-

[8] This is actually a consequence of "no preference of direction." But the proof requires some further knowledge of dyadic operations.

ent choices of α, $G_{11} = F_{11} + \alpha$, and so on, will take on different functional forms.

With this explanatory remark, we can now write down the *linear* continuous form that can meet the fourth requirement:

$$\sigma_{11} = -\alpha + a\epsilon_{11} + b\epsilon_{22} + c\epsilon_{33} \tag{13.40a}$$

This equation must hold not only for the (normal) stress in the x_1-direction but also for that in any principal direction, if only the right-handed cyclic relation among 1, 2, and 3 is preserved. In other words, Eq. (13.40a) must hold true if 1 is replaced by 2, 2 by 3, and 3 by 1:

$$\sigma_{22} = -\alpha + a\epsilon_{22} + b\epsilon_{33} + c\epsilon_{11} \tag{13.40b}$$

$$\sigma_{33} = -\alpha + a\epsilon_{33} + b\epsilon_{11} + c\epsilon_{22} \tag{13.40c}$$

where one must remember that α is a (coordinate-invariant) scalar. This is so because the $\tilde{\boldsymbol{\sigma}} \sim \tilde{\boldsymbol{\epsilon}}$ relation must have no preference of direction (isotropy). In the same token, isotropy also requires that Eq. (13.40a) remain valid when we interchange 1 and 2, while keeping 3 unchanged (that is, when right-handed labeling is changed into a left-handed one):

$$\sigma_{22} = -\alpha + a\epsilon_{22} + b\epsilon_{11} + c\epsilon_{33} \tag{13.40d}$$

Comparing Eqs. (13.40b,d), one obtains

$$b = c$$

So, finally, we have

$$\begin{cases} \sigma_{11} = -\alpha + (a-b)\epsilon_{11} + b(\epsilon_{11} + \epsilon_{22} + \epsilon_{33}) & (13.41a) \\ \sigma_{22} = -\alpha + (a-b)\epsilon_{22} + b(\epsilon_{11} + \epsilon_{22} + \epsilon_{33}) & (13.41b) \\ \sigma_{33} = -\alpha + (a-b)\epsilon_{33} + b(\epsilon_{11} + \epsilon_{22} + \epsilon_{33}) & (13.41c) \end{cases}$$

(a) Compressible Fluids For a compressible fluid, there always exists the thermodynamic pressure Π, which does not depend on kinematic quantities at all, except in the sense that, where $\epsilon_{11}, \epsilon_{22}, \epsilon_{33} = 0$, it is *suddenly* identifiable with $\mathring{p}$. We can therefore use Π for our α; then, the *corresponding* $\frac{1}{2}(a-b)$, denoted by the symbol μ, is called the viscosity, and b (denoted usually by the symbol λ) is called the *second viscosity*. Thus, Eqs. (13.41a,b,c) become

$$\begin{cases} \sigma_{11} = -\Pi + 2\mu\epsilon_{11} + \lambda(\epsilon_{11} + \epsilon_{22} + \epsilon_{33}) & (13.42a) \\ \sigma_{22} = -\Pi + 2\mu\epsilon_{22} + \lambda(\epsilon_{11} + \epsilon_{22} + \epsilon_{33}) & (13.42b) \\ \sigma_{33} = -\Pi + 2\mu\epsilon_{33} + \lambda(\epsilon_{11} + \epsilon_{22} + \epsilon_{33}) & (13.42c) \end{cases}$$

Of course, $\Pi + 3.17(\epsilon_{11} + \epsilon_{22} + \epsilon_{33})$, for example, can also be used for our α since it is also a scalar, linear in ϵ_{11}, ϵ_{22}, ϵ_{33}, and identical with $\mathring{p}$ where $\epsilon_{11}, \epsilon_{22}, \epsilon_{33} = 0$. However, if we do use $\Pi + 3.17(\epsilon_{11} + \epsilon_{22} + \epsilon_{33})$ as α, the corresponding $\frac{1}{2}(a-b)$ and b will have no right to be called the viscosity and second viscosity. Thus, using Π for α is but a *convention*.

μ and λ, just as K, are independent of the kinematic quantities; but may still depend on local thermodynamic quantities such as T and ρ. The functions $\mu\langle T, \rho\rangle$ and $\lambda(T, \rho)$ must be obtained from information sources other than fluid mechanics.

Finally, we claim that

$$\tilde{\sigma} = (-\Pi + \lambda\epsilon_s)\tilde{\mathbf{I}} + 2\mu\tilde{\epsilon} \tag{13.43}$$

at every point in the field, where

$$\epsilon_s = \operatorname{div} \mathbf{q} \tag{13.44}$$

is the scalar of the dyadic $\tilde{\epsilon}$. To substantiate this, we point out that Eq. (13.43) is dyadic in nature, and has just been shown to be true at every point with respect to the local principal axes.[9] (It may not be too superfluous to point out that the key step is to get λ and μ to appear in *all* of Eqs. (13.42a,b,c), and at the right places, too.)

Equation (13.43) is seen to involve two coefficients, μ and λ. As far as the macroscopic theory goes λ and μ are to be determined empirically, independently of each other.[10] Whether there is a connection between the two is of no concern to us. But, in practice, although μ can be measured quite easily, there is no direct and conclusive data on λ except that $\lambda \cong -2\mu/3$ for monatomic gases. We have, therefore, to assume *some* relation between λ and μ to tide us over until data on λ is available in the future. The one usually employed is the *Stokes hypothesis*:[11]

$$3\lambda + 2\mu = 0$$

that is,

$$\lambda = -\tfrac{2}{3}\mu \tag{13.45}$$

We must emphasize that, in principle, we can always leave λ and μ as they are. Whether Eq. (13.45) is valid or not[12] is only of practical interest, not of theoretical interest unless one approaches the problem from a microscopic point of view.

It is important to note that the *average pressure p* defined as the scalar

$$p = -\tfrac{1}{3}\sigma_s$$

can be calculated from Eq. (13.43):

$$\begin{aligned} p &= -\tfrac{1}{3}\{(-\Pi + \lambda\epsilon_s)I_s + 2\mu\epsilon_s\} \\ &= \Pi - \tfrac{1}{3}(3\lambda + 2\mu) \operatorname{div} \mathbf{q} \end{aligned} \tag{13.46}$$

Thus, the average pressure is *not* equal to the thermodynamic pressure, unless the Stokes hypothesis holds, or unless $\operatorname{div} \mathbf{q} = 0$ at that point.

The scalar of a dyadic is, by definition, *coordinate-invariant*. Yet, it

---

[9] This often-used trick should by now be very familiar to the reader.

[10] It is found that, for all real fluids, $\mu > 0$ and $3\lambda + 2\mu \geqslant 0$.

[11] See Stokes' 1845 paper, cited in footnote 6 of Chapter 1.

[12] Actually, it is already known to be *invalid* except for monatomic gases.

would be convenient to refer to a local rectangular Cartesian coordinate system (not necessarily the principal axes) in computing p, since then, from Eq. (13.6),

$$p = -\tfrac{1}{3}(\sigma_{xx} + \sigma_{yy} + \sigma_{zz}) \qquad \textbf{(13.47)}$$

Equation (13.47) would enable us to measure the average pressure by recording the local normal stresses in any three *mutually perpendicular directions* (see Section 2.3). (Note that σ_{xy}, and so on, are not in general zero; yet they do not enter the calculation of p at all.)

(b) Incompressible Fluids Here, we cannot use Π for our α since Π is not defined for an incompressible fluid. However, we remember that div $\mathbf{q} = 0$, always. So Eqs. (13.41a,b,c) become

$$\begin{cases} \sigma_{11} = -\alpha + (a-b)\epsilon_{11} & \textbf{(13.48a)} \\ \sigma_{22} = -\alpha + (a-b)\epsilon_{22} & \textbf{(13.48b)} \\ \sigma_{33} = -\alpha + (a-b)\epsilon_{33} & \textbf{(13.48c)} \end{cases}$$

involving only one coefficient $(a-b)$; the other one, b, is lost right from the beginning. Furthermore, α here is also completely independent of $\tilde{\boldsymbol{\epsilon}}$. This is proved as follows. α is (a) a scalar and (b) at most a linear function of ϵ_{11}, ϵ_{22}, ϵ_{33}; so at most, $\alpha = \mathcal{G}\langle \epsilon_{11} + \epsilon_{22} + \epsilon_{33}\rangle = \mathcal{G}\langle \epsilon_s \rangle$. Yet, $\epsilon_s =$ div $\mathbf{q}$ is always zero; therefore, α cannot depend on $\tilde{\boldsymbol{\epsilon}}$ in any way whatsoever.

From Eqs. (13.48a,b,c), we have also the following interesting result:

$$p = -\tfrac{1}{3}\sigma_s = \alpha$$

that is, for whatever choice of α, the average pressure will turn out to be just α, being independent of the coefficient $(a-b)$. To be more specific, if we have chosen two different quantities α_1 and α_2 for α, we *must* have

$$p = \alpha_1$$

and also

$$p = \alpha_2$$

Since $p \overset{\text{def}}{=} -\tfrac{1}{3}\sigma_s = -\tfrac{1}{3}(\sigma_{11} + \sigma_{22} + \sigma_{33})$ is unique at a point, we are forced to conclude that no matter how we choose α_1 and α_2 we will always wind up with the same

$$\alpha_1 = \alpha_2$$

In other words,[13] for an incompressible fluid there always exists a *unique* scalar property[14] $\mathring{\alpha}$, with the dimension of force/length^2 and independent of all kinematic quantities, such that

$$\tilde{\boldsymbol{\sigma}} = -\mathring{\alpha}\tilde{\mathbf{I}} + (a-b)\tilde{\boldsymbol{\epsilon}} \qquad \textbf{(13.49)}$$

[13] Thus, there is really only one candidate for α. So, the *choice* of α for the compressible fluids *now* degenerates into a *dictatorial election* with one candidate.

[14] $\mathring{\alpha}$ is used from here on to minimize the possibility of misunderstanding.

which is the coordinate-invariant form of Eqs. (13.48a,b,c). The scalar property $\mathring{\alpha}$, whose existence we have just proved, will be called the "incompressible" pressure in this book. The coefficient $\frac{1}{2}(a-b)$, denoted by μ, is known as the viscosity of the incompressible fluid.

Please notice very carefully that, in this interpretation, the conclusion $p=\mathring{\alpha}$ comes after the relation between $\boldsymbol{\sigma}$ and $\boldsymbol{\epsilon}$ is established in Eqs. (13.48a,b,c). Many authors immediately write

$$\begin{cases} \sigma_{11} = -p + (a-b)\epsilon_{11} \\ \quad = \frac{1}{3}\sigma_s + (a-b)\epsilon_{11} \\ \quad = \frac{1}{3}(\sigma_{11}+\sigma_{22}+\sigma_{33}) + (a-b)\epsilon_{11} \\ \text{and so forth} \end{cases}$$

in place of Eqs. (13.48a,b,c). This practice is not strictly in order, since the definition of a Newtonian fluid specifically called for an explicit relation $\tilde{\boldsymbol{\sigma}} = \tilde{\mathbf{F}}\langle\tilde{\boldsymbol{\epsilon}}\rangle$, not $\tilde{\boldsymbol{\sigma}} = \tilde{\mathbf{F}}\langle\tilde{\boldsymbol{\sigma}}, \tilde{\boldsymbol{\epsilon}}\rangle$.

Of course, once the fact that $p=\mathring{\alpha}$ *is* established, we can go ahead and use it as a vehicle to give the "incompressible" pressure an obvious physical meaning; namely, it is identical with the average pressure p. At places where $\tilde{\boldsymbol{\epsilon}}=0$, $\mathring{\alpha}$ (as well as p) reduces to the hydrostatic pressure $\mathring{p}$.

To summarize, for an incompressible fluid,

$$\begin{cases} \tilde{\boldsymbol{\sigma}} = -\mathring{\alpha}\tilde{\mathbf{I}} + 2\mu\tilde{\boldsymbol{\epsilon}} & \textbf{(13.50a)} \\ \mathring{\alpha} = p & \textbf{(13.50b)} \end{cases}$$

13.5 NAVIER-STOKES EQUATION

The normal component of the resultant (contact) force intensity $\mathbf{f}$ on a small surface element with normal unit vector $\hat{\mathbf{n}}$ is

$$\begin{aligned} \hat{\mathbf{n}}\cdot\mathbf{f}\langle\hat{\mathbf{n}}\rangle &= \hat{\mathbf{n}}\cdot(\tilde{\boldsymbol{\sigma}}\cdot\hat{\mathbf{n}}) \\ &= \hat{\mathbf{n}}\cdot(-\Pi+\lambda\epsilon_s)(\tilde{\mathbf{I}}\cdot\hat{\mathbf{n}}) + 2\mu\tilde{\boldsymbol{\epsilon}}:(\hat{\mathbf{n}}\hat{\mathbf{n}}) \\ &= -\Pi + \lambda\,\mathrm{div}\,\mathbf{q} + 2\mu\tilde{\boldsymbol{\epsilon}}:(\hat{\mathbf{n}}\hat{\mathbf{n}}) \\ &\qquad \text{(compressible)} \qquad\qquad \textbf{(13.51)} \\ &= -\mathring{\alpha} + 2\mu\tilde{\boldsymbol{\epsilon}}:(\hat{\mathbf{n}}\hat{\mathbf{n}}) \\ &= -p + 2\mu\tilde{\boldsymbol{\epsilon}}:(\hat{\mathbf{n}}\hat{\mathbf{n}}) \\ &\qquad \text{(incompressible)} \qquad\qquad \textbf{(13.52)} \end{aligned}$$

In general, this is equal to neither Π nor p, whether the fluid is compressible or not (and whether the Stokes hypothesis is valid or not). This point has two physical significances. (a) As mentioned before in Section 2.3, the so-called pressure gauge measures neither the average, nor the thermodynamic pressure in a general moving stream, unless the terms with λ and μ in Eqs. (13.51) and (13.52) are zero (for example, in a fluid at rest or in uniform flow) or negligibly small. (It would be much better practice, if at all possible, to measure p with three gauges that open to three perpendicular directions. This would give the "incompressible" pressure for incompressible fluids, and the thermodynamic pressure for compressible fluids *obeying the Stokes hypothesis*.) (b) A part of the force acting *normally* to a surface is due to viscosity. (See also Section 2.3.) We see, then, that the so-called "flow-work" term that appears in an energy balance, as presented in many texts on engineering thermodynamics, should *not* be $\Pi\rho^{-1}$ or $\mathring{\alpha}\rho^{-1}$ but must be ($\Pi+$ something due to viscosity) $\cdot\, \rho^{-1}$ or ($\mathring{\alpha}+$something due to viscosity) $\cdot\, \rho^{-1}$. The energy balance using $\Pi\rho^{-1}$ or $\mathring{\alpha}\rho^{-1}$ generally gives accurate enough results because the situation is usually such that the "something due to viscosity" is negligible.[15] (The same situation near the opening of an ordinary "pressure" gauge also saves the practice of pressure measurement.)

Now, substituting Eq. (13.43) into Eq. (13.7), we obtain

$$\rho\frac{D\mathbf{q}}{Dt} = \rho\mathbf{g} + \nabla\cdot[(-\Pi+\lambda\,\mathrm{div}\,\mathbf{q})\tilde{\mathbf{I}}] + \nabla\cdot(2\mu\tilde{\boldsymbol{\epsilon}})$$

which can be shown to lead to

$$\rho\frac{D\mathbf{q}}{Dt} = \rho\mathbf{g} - \nabla\Pi + \nabla(\lambda\,\mathrm{div}\,\mathbf{q}) + \nabla\cdot(2\mu\tilde{\boldsymbol{\epsilon}}) \tag{13.53}$$

where $2\tilde{\boldsymbol{\epsilon}} = (\nabla\mathbf{q}) + (\mathbf{q}\overleftarrow{\nabla})$. On the other hand, it is easy to show that

$$\nabla\cdot(2\mu\tilde{\boldsymbol{\epsilon}}) = \mu\nabla^2\mathbf{q} + \mu\nabla(\nabla\cdot\mathbf{q}) + [(\nabla\mathbf{q}) + (\mathbf{q}\overleftarrow{\nabla})]\cdot\nabla\mu \tag{13.54}$$

$$= -\mu\nabla\times\boldsymbol{\Omega} + 2\mu\nabla(\nabla\cdot\mathbf{q}) + [(\nabla\mathbf{q}) + (\mathbf{q}\overleftarrow{\nabla})]\cdot\nabla\mu \tag{13.54A}$$

where $\boldsymbol{\Omega} = \nabla\times\mathbf{q}$, and Eq. (11.26) has been used. Thus,

$$\rho\frac{D\mathbf{q}}{Dt} = \rho\mathbf{g} - \mathrm{grad}\,\Pi + (\lambda+\mu)\,\mathrm{grad}\,(\mathrm{div}\,\mathbf{q}) + \mu\nabla^2\mathbf{q} + (\mathrm{div}\,\mathbf{q})\,\mathrm{grad}\,\lambda + [(\nabla\mathbf{q}) + (\nabla\mathbf{q})_c]\cdot\mathrm{grad}\,\mu \tag{13.55}$$

or

$$\rho\frac{D\mathbf{q}}{Dt} = \rho\mathbf{g} - \mathrm{grad}\,\Pi + (\lambda+2\mu)\,\mathrm{grad}\,(\mathrm{div}\,\mathbf{q}) - \mu\nabla\times\boldsymbol{\Omega} + (\mathrm{div}\,\mathbf{q})\,\mathrm{grad}\,\lambda + [(\nabla\mathbf{q}) + (\nabla\mathbf{q})_c]\cdot\mathrm{grad}\,\mu \tag{13.55A}$$

Equation (13.55), or Eq. (13.55A), is called the Navier-Stokes equation. If the fluid satisfies the Stokes hypothesis, $\lambda = -2\mu/3$ should be substituted, and Π can be replaced by p. For an incompressible fluid, simply replace Π by p, and div $\mathbf{q}$ by zero. With Eq. (13.55), one must remember that ∇^2 operating on a vector is drastically different from ∇^2 operating on a scalar.

The energy equation, Eq. (13.32), with the aid of Eqs. (13.37) and (13.39), now reads

$$\rho c_v\frac{DT}{Dt} = -\Pi\,\mathrm{div}\,\mathbf{q} + \Phi + \mathrm{div}\,(K\,\mathrm{grad}\,T) + Q \tag{13.32A}$$

[15] See the translator's footnote on page 24 of K. Oswatitsch's *Gas Dynamics* (translated by G. Kuerti), Academic Press, New York, 1956.

where the energy dissipation

$$\begin{aligned}\Phi &= (\boldsymbol{\sigma}+\Pi\tilde{\mathbf{I}}) : \tilde{\boldsymbol{\epsilon}} \\ &= [\lambda(\text{div}\,\mathbf{q})\tilde{\mathbf{I}}+2\mu\tilde{\boldsymbol{\epsilon}}] : \tilde{\boldsymbol{\epsilon}} \\ &= \lambda(\text{div}\,\mathbf{q})^2+2\mu\tilde{\boldsymbol{\epsilon}}:\tilde{\boldsymbol{\epsilon}} \qquad \textbf{(13.37A)}\end{aligned}$$

Therefore

$$\rho c_v \frac{DT}{Dt} = -\Pi\,\text{div}\,\mathbf{q}+\lambda(\text{div}\,\mathbf{q})^2+2\mu\tilde{\boldsymbol{\epsilon}} : \tilde{\boldsymbol{\epsilon}}+\text{div}\,(K\,\text{grad}\,T)+Q \qquad \textbf{(13.32B)}$$

For incompressible fluids, simply replace c_v by c_{in}, and div **q** by zero.

For convenience, the equation of continuity, the Navier-Stokes equation, and the energy equation for ideal gases obeying the Stokes hypothesis have been written out in component forms with respect to orthogonal coordinate systems in the Appendix. There, they are first written for the general orthogonal system, and then for the rectangular Cartesian, the cylindrical, and the spherical systems. Again, for an incompressible fluid, simply replace c_v by c_{in}, and div **q** by zero.

To these *governing equations* we have to add the equation of state, Eq. (13.31); $K\langle\rho, T\rangle$, $\mu(\rho, T\rangle$ and $\lambda\langle\rho, T\rangle$ are given. We then have three scalar unknowns (ρ, p, and T) and one vectorial unknown (**q**). To solve for them, we have exactly three scalar equations (continuity, energy, and state) and one vectorial equation (Navier-Stokes). For an incompressible fluid, ρ is known; but then the equation of state would be missing. If the conditions to be satisfied in a specific situation are given, we can proceed to solve the governing equations and thereby solve that specific problem of fluid flow. In other words, we have now come to the end of our long *formulation*[16] of the problem of fluid flow. What remains to be done are: (a) to draw general conclusions from the governing equations and (b) to solve the governing equations for specific geometric configurations and under specific conditions. Unfortunately, our scope does not allow us to go into (a), except to point out that, for an incompressible fluid *with constant viscosity*, the energy equation is decoupled and can be solved *after* the Navier-Stokes equation (together with the equation of continuity) is solved.[17] As to (b), we will be able to present two examples in the very last part of this chapter (and of this book). But before that, we still wish to clear up two things: non-Newtonian fluids and surfaces of discontinuity.

13.6 NON-NEWTONIAN FLUIDS

In the definition of a Newtonian fluid (Section 13.4), if we omit the word "linear" and keep the rest of the definition intact, right away we have the definition of a Reiner-Rivlin fluid, which is the only type of non-Newtonian fluid treated in

[16] Or mathematical model-building.

[17] This actually created the "autonomous" discipline of heat convection. For compressible fluids, or incompressible fluids with temperature-dependent viscosity, heat transfer is tightly coupled with momentum and mass transfer; all things must be taken on at once.

this book (Sections 2.4, 8.1, 8.2, and 9.3). Again suppressing the word "linear" wherever it occurs in Section 13.4 following the said definition, we will be able to come all the way down to the three equations just ahead of Eq. (13.40a), namely

$$\begin{cases} \sigma_{11} = -\alpha + G_{11}\langle \epsilon_{11}, \epsilon_{22}, \epsilon_{33} \rangle \\ \sigma_{22} = -\alpha + G_{22}\langle \epsilon_{11}, \epsilon_{22}, \epsilon_{33} \rangle \\ \sigma_{33} = -\alpha + G_{33}\langle \epsilon_{11}, \epsilon_{22}, \epsilon_{33} \rangle \end{cases}$$

where α is any scalar fluid property that (a) always exists, (b) does not depend on any kinematic quantity other than ϵ_{11}, ϵ_{22}, and ϵ_{33}, (c) is a continuous function of ϵ_{11}, ϵ_{22}, and ϵ_{33}, and (d) reduces to $\mathring{p}$ when $\epsilon_{11}, \epsilon_{22}, \epsilon_{33} = 0$; and G_{11}, G_{22}, G_{33} are all continuous functions.

This is then our starting point from which we wish to build up the $\tilde{\boldsymbol{\sigma}} \sim \tilde{\boldsymbol{\epsilon}}$ relationship for a Reiner-Rivlin fluid. Since the functions involved are not necessarily linear, one cannot go to Eq. (13.40a) and proceed from there. Instead, we will claim that it is always possible to represent G_{11} in the form[18]

$$G_{11} = A\langle \epsilon_{11}, \epsilon_{22}, \epsilon_{33} \rangle + B\langle \epsilon_{11}, \epsilon_{22}, \epsilon_{33} \rangle \epsilon_{11} + C\langle \epsilon_{11}, \epsilon_{22}, \epsilon_{33} \rangle \epsilon_{11}^2 \qquad \textbf{(13.56a)}$$

where the functional values A, B, and C are invariant under any permutation[19] of ϵ_{11}, ϵ_{22}, and ϵ_{33} (for example, $A\langle \epsilon_{22}, \epsilon_{33}, \epsilon_{11} \rangle = A\langle \epsilon_{11}, \epsilon_{22}, \epsilon_{33} \rangle$).

Because of isotropy, this claim must also hold under all kinds of permutations of 1, 2, and 3. A cyclic change $1 \rightarrow 2 \rightarrow 3 \rightarrow 1$ transforms Eq. (13.56a) into

$$\begin{aligned} G_{22} &= A\langle \epsilon_{22}, \epsilon_{33}, \epsilon_{11} \rangle + B\langle \epsilon_{22}, \epsilon_{33}, \epsilon_{11} \rangle \epsilon_{22} + C\langle \epsilon_{22}, \epsilon_{33}, \epsilon_{11} \rangle \epsilon_{22}^2 \\ &= A\langle \epsilon_{11}, \epsilon_{22}, \epsilon_{33} \rangle + B\langle \epsilon_{11}, \epsilon_{22}, \epsilon_{33} \rangle \epsilon_{22} + C\langle \epsilon_{11}, \epsilon_{22}, \epsilon_{33} \rangle \epsilon_{22}^2 \qquad \textbf{(13.56b)} \end{aligned}$$

Similarly,

$$G_{33} = A\langle \epsilon_{11}, \epsilon_{22}, \epsilon_{33} \rangle + B\langle \epsilon_{11}, \epsilon_{22}, \epsilon_{33} \rangle \epsilon_{33} + C\langle \epsilon_{11}, \epsilon_{22}, \epsilon_{33} \rangle \epsilon_{33}^2 \qquad \textbf{(13.56c)}$$

This claim will have been substantiated if we can show how A, B, and C are to be found. But this is extremely simple, since Eqs. (13.56a,b,c) can be looked upon as three linear algebraic equations for three unknowns A, B, and C. Thus, by Cramer's rule,

$$A = \frac{1}{\Delta} \begin{vmatrix} G_{11} & \epsilon_{11} & \epsilon_{11}^2 \\ G_{22} & \epsilon_{22} & \epsilon_{22}^2 \\ G_{33} & \epsilon_{33} & \epsilon_{33}^2 \end{vmatrix}$$

$$B = \frac{1}{\Delta} \begin{vmatrix} 1 & G_{11} & \epsilon_{11}^2 \\ 1 & G_{22} & \epsilon_{22}^2 \\ 1 & G_{33} & \epsilon_{33}^2 \end{vmatrix}$$

$$C = \frac{1}{\Delta} \begin{vmatrix} 1 & \epsilon_{11} & G_{11} \\ 1 & \epsilon_{22} & G_{22} \\ 1 & \epsilon_{33} & G_{33} \end{vmatrix}$$

---

[18] See Serrin's article, cited in footnote 6. The name "Stokesian fluid" is used there.

[19] Not necessarily cyclic.

where

$$\Delta = \begin{vmatrix} 1 & \epsilon_{11} & \epsilon_{11}^2 \\ 1 & \epsilon_{22} & \epsilon_{22}^2 \\ 1 & \epsilon_{33} & \epsilon_{33}^2 \end{vmatrix} = (\epsilon_{11}-\epsilon_{22})(\epsilon_{22}-\epsilon_{33})(\epsilon_{33}-\epsilon_{11})$$

We see that A, B, C are determinate for any given G_{11}, G_{22}, G_{33}, if ϵ_{11}, ϵ_{22}, ϵ_{33} are all different. Since G_{11}, G_{22}, G_{33} are originally given functions of ϵ_{11}, ϵ_{22}, and ϵ_{33}, the above formulas give A, B, C as functions of ϵ_{11}, ϵ_{22}, and ϵ_{33}. Furthermore, since α is a scalar, G_{11} must be tied in with ϵ_{11} just as σ_{11}, and so on. That is when ϵ_{11} and ϵ_{22} are interchanged, so must be G_{11} and G_{22}, and so on. It is then obvious, from a basic property of determinants, that A, B, and C are invariant under any permutation of ϵ_{11}, ϵ_{22}, and ϵ_{33}, just as claimed.

In the event that two of $(\epsilon_{11}, \epsilon_{22}, \epsilon_{33})$ are identical, two corresponding G's must also be identical. Then, only two of Eqs. (13.56a,b,c) are independent. It is then obvious that $C = 0$ will enable A and B to be determined. Similarly, when $\epsilon_{11} = \epsilon_{22} = \epsilon_{33}$, $B = 0$ and $C = 0$ will enable A to be determined by the only one equation that remains independent. Thus, A, B, C are always calculable from any given set of G's; however, B and C may turn out to be zero. This is certainly within the general framework of the claim. In this way we have established the general validity of Eqs. (13.56a,b,c). It must then be true that

$$\boldsymbol{\sigma} = (-\alpha + A)\tilde{\mathbf{I}} + B\tilde{\boldsymbol{\epsilon}} + C\tilde{\boldsymbol{\epsilon}} \cdot \tilde{\boldsymbol{\epsilon}} \qquad \textbf{(13.57)}$$

since this dyadic equation has just been shown to hold for a special coordinate system (x_1, x_2, x_3).

At this point, we could have concluded the whole affair. The remaining business of determining the functions A, B, C as functions of ϵ_1, ϵ_2, ϵ_3 that are invariant under permutations of ϵ_1, ϵ_2, ϵ_3 is left to the experimentalists. We can, however, give them more help by making the following observation.

In Section 12.4, subsection (2), it was implied that a principal direction represented by the unit vector $\hat{\boldsymbol{\xi}}$ is such that the relative velocity due to $\tilde{\boldsymbol{\epsilon}}$ (that is, $\tilde{\boldsymbol{\epsilon}} \cdot \hat{\boldsymbol{\xi}}$) is parallel to $\hat{\boldsymbol{\xi}}$ (purely normal deformation):

$$\tilde{\boldsymbol{\epsilon}} \cdot \hat{\boldsymbol{\xi}} = \zeta \hat{\boldsymbol{\xi}} \qquad \textbf{(13.58)}$$

where ζ is, of course, just the (purely) normal rate of strain in this principal direction $\hat{\boldsymbol{\xi}}$. If, for convenience, we erect *locally* three arbitrary but orthogonal reference directions x, y, and z, Eq. (13.58) can be written as

$$\begin{cases} (\epsilon_{xx} - \zeta)\xi_x + \epsilon_{xy}\xi_y + \epsilon_{xz}\xi_z = 0 & \textbf{(13.59a)} \\ \epsilon_{xy}\xi_x + (\epsilon_{yy} - \zeta)\xi_y + \epsilon_{yz}\xi_z = 0 & \textbf{(13.59b)} \\ \epsilon_{xz}\xi_x + \epsilon_{yz}\xi_y + (\epsilon_{zz} - \zeta)\xi_z = 0 & \textbf{(13.59c)} \end{cases}$$

In order that these three equations yield nonzero $\hat{\boldsymbol{\xi}} = (\xi_x, \xi_y, \xi_z)$, the determinant of coefficients must vanish; or, after expanding the determinant, we must have

$$\zeta^3 - I_1\zeta^2 + I_2\zeta - I_3 = 0 \qquad \textbf{(13.60)}$$

where

$$I_1 = \epsilon_{xx} + \epsilon_{yy} + \epsilon_{zz} \tag{13.61a}$$

$$I_2 = \begin{vmatrix} \epsilon_{xx} & \epsilon_{xy} \\ \epsilon_{xy} & \epsilon_{yy} \end{vmatrix} + \begin{vmatrix} \epsilon_{yy} & \epsilon_{yz} \\ \epsilon_{yz} & \epsilon_{zz} \end{vmatrix} + \begin{vmatrix} \epsilon_{xx} & \epsilon_{xz} \\ \epsilon_{xz} & \epsilon_{zz} \end{vmatrix} \tag{13.61b}$$

$$I_3 = \begin{vmatrix} \epsilon_{xx} & \epsilon_{xy} & \epsilon_{xz} \\ \epsilon_{xy} & \epsilon_{yy} & \epsilon_{yz} \\ \epsilon_{xz} & \epsilon_{yz} & \epsilon_{zz} \end{vmatrix} \tag{13.61c}$$

with reference to *any* orthogonal (x, y, z). I_1, I_2, and I_3 are actually scalars, since the entire rate-of-strain situation, including Eq. (13.60), is coordinate-invariant. However, Eqs. (13.61a,b,c) are only for orthogonal coordinate directions. It is obvious that $I_1 = \text{div}\,\mathbf{q}$; the coordinate-free expressions for I_2 and I_3 are not so simple and obvious.

Thus, for each root of Eq. (13.60), there is a corresponding principal direction $\hat{\boldsymbol{\xi}}$. Conversely, each normal rate of strain in a principal direction is a root of Eq. (13.60); so ϵ_1, ϵ_2, ϵ_3 are just the three roots[20] of Eq. (13.60), and must be functions of the coefficients I_1, I_2, I_3. Therefore A, B, C in Eq. (13.57) are basically all functions of I_1, I_2, I_3. The experimentalists prefer I_1, I_2, I_3 over $\epsilon_1, \epsilon_2, \epsilon_3$.

From now on, let us denote A by κ, B by 2μ, and C by $-\eta$:

$$\tilde{\boldsymbol{\sigma}} = (-\alpha + \kappa\langle I_1, I_2, I_3\rangle)\tilde{\mathbf{I}} + 2\mu\langle I_1, I_2, I_3\rangle\tilde{\boldsymbol{\epsilon}} - \eta\langle I_1, I_2, I_3\rangle\tilde{\boldsymbol{\epsilon}}\cdot\tilde{\boldsymbol{\epsilon}} \tag{13.57A}$$

where μ is called the viscosity function; and η, the cross-viscosity function.

Equation (13.57A) contains, of course, the definition of Newtonian fluids as a special (linear) case, with $\eta = 0$, μ independent of I_1, I_2, I_3, and κ proportional to I_1.

(a) Compressible Fluids Π is used for α as before. The average pressure is then

$$p = \Pi - \kappa - \tfrac{2}{3}\mu\,\text{div}\,\mathbf{q} + \tfrac{1}{3}\eta\tilde{\boldsymbol{\epsilon}}:\tilde{\boldsymbol{\epsilon}}$$

where $\kappa\langle 0, 0, 0\rangle$ must vanish.

(b) Incompressible Fluids Again, there exists a unique "incompressible" pressure $\mathring{\alpha}$, which is used for α. Here, $I_1 = \text{div}\,\mathbf{q} = 0$; therefore, κ, μ, and η are all functions of I_2, I_3:

$$\tilde{\boldsymbol{\sigma}} = (-\mathring{\alpha} + \kappa\langle I_2, I_3\rangle)\tilde{\mathbf{I}} + 2\mu\langle I_2, I_3\rangle\tilde{\boldsymbol{\epsilon}} - \eta\langle I_2, I_3\rangle\tilde{\boldsymbol{\epsilon}}\cdot\tilde{\boldsymbol{\epsilon}} \tag{13.57B}$$

The average pressure

$$p = \mathring{\alpha} - \kappa + \tfrac{1}{3}\eta\tilde{\boldsymbol{\epsilon}}:\tilde{\boldsymbol{\epsilon}}$$

---

[20] Equation (13.60) can have only real roots. Should it have a complex root ζ, it must also have $\bar{\zeta}$ (the conjugate of ζ) as a root. Corresponding to ζ, there is the *complex* vector $\hat{\boldsymbol{\xi}} = (\xi_x, \xi_y, \xi_z)$; and corresponding to $\bar{\zeta}$, there is the conjugate $\bar{\hat{\boldsymbol{\xi}}} = (\bar{\xi}_x, \bar{\xi}_y, \bar{\xi}_z)$. From Eq. (13.58), we have $\tilde{\boldsymbol{\epsilon}}\cdot\hat{\boldsymbol{\xi}} = \zeta\hat{\boldsymbol{\xi}}$ and $\tilde{\boldsymbol{\epsilon}}\cdot\bar{\hat{\boldsymbol{\xi}}} = \bar{\zeta}\bar{\hat{\boldsymbol{\xi}}}$, from which it follows that $\bar{\hat{\boldsymbol{\xi}}}\cdot\tilde{\boldsymbol{\epsilon}}\cdot\hat{\boldsymbol{\xi}} = \hat{\boldsymbol{\xi}}\cdot\tilde{\boldsymbol{\epsilon}}\cdot\bar{\hat{\boldsymbol{\xi}}}$ (symmetry of $\tilde{\boldsymbol{\epsilon}}$) $= \zeta\bar{\hat{\boldsymbol{\xi}}}\cdot\hat{\boldsymbol{\xi}}$ (on the one hand) $= \bar{\zeta}\hat{\boldsymbol{\xi}}\cdot\bar{\hat{\boldsymbol{\xi}}}$ (on the other). Therefore, $\zeta = \bar{\zeta}$; that is, ζ is real.

unfortunately does not lend itself directly to a physical interpretation of $\mathring{\alpha}$, except where $\tilde{\boldsymbol{\epsilon}} = 0$. Again, $\kappa\langle 0, 0\rangle$ must vanish. In the literature, the combination $\mathring{\alpha} - \kappa$ is usually considered to be the ("incompressible") pressure. Although this is perfectly in order, the reader may prefer to see a display of the average pressure:

$$\tilde{\boldsymbol{\sigma}} = (-p + \tfrac{1}{3}\eta\tilde{\boldsymbol{\epsilon}} : \tilde{\boldsymbol{\epsilon}})\tilde{\mathbf{I}} + 2\mu\tilde{\boldsymbol{\epsilon}} - \eta\tilde{\boldsymbol{\epsilon}} \cdot \tilde{\boldsymbol{\epsilon}} \tag{13.57C}$$

This new form was used before in Sections 2.4, 8.1, 8.2, and 9.3. I_2 was denoted by $-\frac{1}{2}\text{II}$ and $\mathring{\alpha} - \kappa$ was denoted in Section 2.4 by β. The reader should show to himself that all the relations quoted in Sections 2.4, 8.1, 8.2, and 9.3 are deducible from Eq. (13.57C), if only he remembers to employ div $\mathbf{q} = 0$.

13.7 SURFACES OF DISCONTINUITY

In the previous sections a set of governing equations was established, valid wherever the flow variables are continuous. Once we introduce discontinuities we must go back to the original integral laws, Eqs. (12.16), (13.1), and (13.12), and inequality (13.33).

Consider now a material volume $\mathscr{V}\langle t\rangle$, which contains a surface of discontinuity $\Sigma\langle t\rangle$ (*not necessarily material*) moving with **G** (Fig. 13.5). Suppose that, across Σ, certain flow property $\mathscr{F}$ jumps.[21] We would like to know, besides Eq. (12.14A), what the following expression is equal to:

$$\frac{D}{Dt}\int_{\mathscr{V}\langle t\rangle} \mathscr{F}\, dV$$

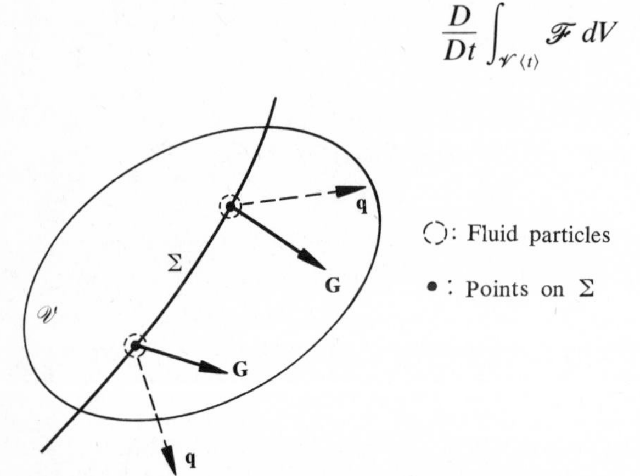

Figure 13.5 A surface of discontinuity.

Here, Eq. (12.15A) cannot be used. However, one can divide $\mathscr{V}$ into Λ_1 and Λ_2 and distinguish the 1-side from the 2-side of Σ (Fig. 13.6). Note that the capital Greek letter Λ here signifies the fact that Λ_1 and Λ_2 are not material (unless Σ is material). Then,

$$\frac{D}{Dt}\int_{\mathscr{V}\langle t\rangle} \mathscr{F}\, dV = \frac{d}{dt}\int_{\mathscr{V}\langle t\rangle} \mathscr{F}\, dV + \frac{d}{dt}\int_{\Lambda_2\langle t\rangle} \mathscr{F}\, dV$$

[21] The concept of surface of discontinuity with **q** as a possible candidate for $\mathscr{F}$ was due to Stokes (again!); see his 1848 paper, "On a difficulty in the theory of sound."

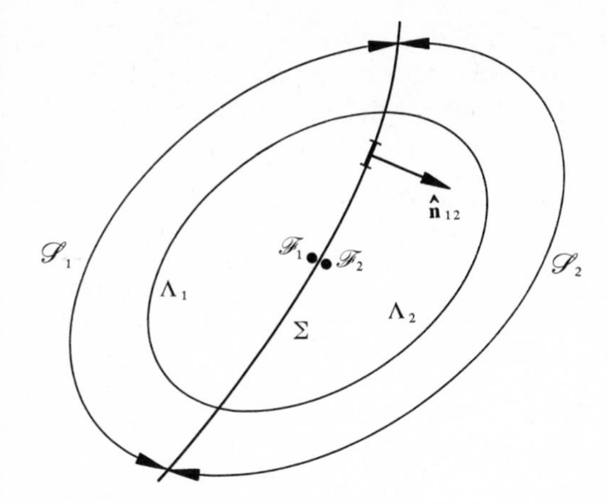

Figure 13.6 Λ_1 and Λ_2.

Next, applying Eq. (12.15) and introducing $\hat{\mathbf{n}}_{12}$ as the normal unit vector of Σ, positive from V_1 into V_2, we have

$$\frac{d}{dt}\int_{\Lambda_1\langle t\rangle} \mathscr{F}\, dV = \int_{V_1} \frac{\partial \mathscr{F}}{\partial t}\, dV + \int_{S_1} \mathscr{F}\mathbf{q}\cdot\hat{\mathbf{n}}\, dS + \int_{S_\Sigma} \mathscr{F}_1 \mathbf{G}\cdot\hat{\mathbf{n}}_{12}\, dS$$

$$\frac{d}{dt}\int_{\Lambda_2\langle t\rangle} \mathscr{F}\, dV = \int_{V_2} \frac{\partial \mathscr{F}}{\partial t}\, dV + \int_{S_2} \mathscr{F}\mathbf{q}\cdot\hat{\mathbf{n}}\, dS - \int_{S_\Sigma} \mathscr{F}_2 \mathbf{G}\cdot\hat{\mathbf{n}}_{12}\, dS$$

+)

Therefore

$$\frac{D}{Dt}\int_{\mathscr{V}\langle t\rangle} \mathscr{F}\, dV = \int_{V_{1,2}} \frac{\partial \mathscr{F}}{\partial t}\, dV + \oint_S \mathscr{F}\mathbf{q}\cdot\hat{\mathbf{n}}\, dS - \int_{S_\Sigma} [\![\mathscr{F}]\!]\mathbf{G}\cdot\hat{\mathbf{n}}_{12}\, dS \qquad \textbf{(13.62)}$$

where S_Σ is the *spatial* surface that coincides with $\Sigma\langle t\rangle$ at the instant t, and

$$[\![\mathscr{F}]\!] = \mathscr{F}_2 - \mathscr{F}_1 \qquad \textbf{(13.63)}$$

is just the jump in $\mathscr{F}$ across S_Σ from the 1-side to the 2-side.[22] It is seen that Eq. (13.62) reduces to Eq. (12.15A) for three special cases: (a) stationary surface of discontinuity (that is, $\mathbf{G} = 0$); (b) tangentially moving surface of discontinuity (that is, $\mathbf{G} \perp \hat{\mathbf{n}}_{12}$); and (c) no surface of discontinuity at all (that is, $[\![\mathscr{F}]\!] = 0$). For

[22] T. Y. Thomas, "The fundamental hydrodynamical equations and shock conditions for gases," *Math. Mag.*, **22**, 169–189, 1949. Tracy Y. Thomas is a professor at the Graduate School of Applied Mathematics, Indiana University.

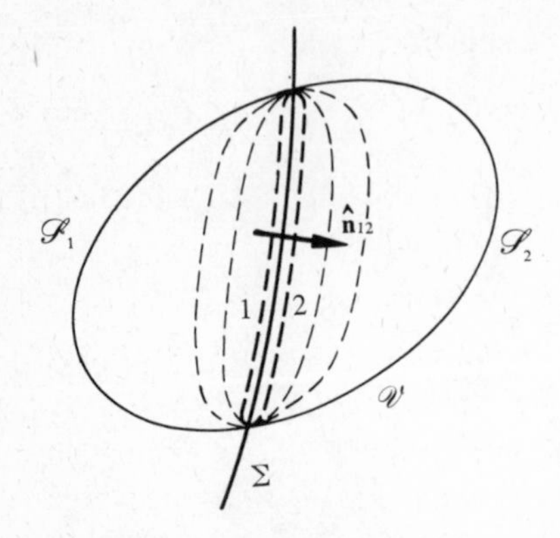

Figure 13.7 $\mathscr{S}_1$ and $\mathscr{S}_2$ shrinking toward Σ.

(a) and (b), the volume integral over V in Eq. (12.15A) is understood to be really that over $V_{1,2}$ (that is, $V_1 + V_2$ with Σ excluded).

Now, let us take thinner and thinner $\mathscr{V}\langle t\rangle$ surrounding $\Sigma\langle t\rangle$ (Fig. 13.7), or take the limit $\mathscr{V} \to 0$ in such a way that $\mathscr{S}_1$, $\mathscr{S}_2 \to$ 1-side, 2-side of Σ, respectively, with the area of Σ enclosed by $\mathscr{V}$ unchanged. If $\partial\mathscr{F}/\partial t$ is finite in $\mathscr{V}$ we will have

$$\begin{aligned}\lim\left\{\frac{D}{Dt}\int_{\mathscr{V}\langle t\rangle}\mathscr{F}\,dV\right\} &= \int_{S_\Sigma}(-\mathscr{F}_1\mathbf{q}_1\cdot\hat{\mathbf{n}}_{12}+\mathscr{F}_2\mathbf{q}_2\cdot\hat{\mathbf{n}}_{12})\,dS \\ &\quad -\int_{S_\Sigma}[\![\mathscr{F}]\!]\mathbf{G}\cdot\hat{\mathbf{n}}_{12}\,dS \\ &= \int_{S_\Sigma}[\![\mathscr{F}(\mathbf{q}\cdot\hat{\mathbf{n}}_{12}-\mathbf{G}\cdot\hat{\mathbf{n}}_{12})]\!]\,dS \\ &= \int_{S_\Sigma}[\![\mathscr{F}\mathbf{U}\cdot\hat{\mathbf{n}}_{12}]\!]\,dS \end{aligned} \qquad \textbf{(13.64)}$$

where $\mathbf{U} = \mathbf{q} - \mathbf{G}$ is the flow velocity relative to the motion of Σ. Note that $[\![\mathbf{G}]\!] = 0$, but $[\![\mathbf{U}]\!] = [\![\mathbf{q}]\!]$ in general is not zero.

Applying Eq. (13.64), with $\mathscr{F}$ identified with ρ, to Eq. (12.16), we get

$$\int_{S_\Sigma}[\![\rho\mathbf{U}\cdot\hat{\mathbf{n}}_{12}]\!]\,dS = 0$$

Since the original $\mathscr{V}$ is arbitrary as long as it encloses a portion of Σ, S_Σ is an arbitrary area on the surface of discontinuity at the instant t. Then, following similar arguments as in footnote 18 of Chapter 12, one must conclude that

$$[\![\rho\mathbf{U}\cdot\hat{\mathbf{n}}_{12}]\!] = 0 \qquad \textbf{(13.65)}$$

across the surface of discontinuity, wherever the *jump* $[\![\rho\mathbf{U}\cdot\hat{\mathbf{n}}_{12}]\!]$ is continuous.[23]

[23] This provides the starting point for an investigation of a *line of discontinuity on a surface of discontinuity*, a physically unlikely situation.

Similarly, Eqs. (13.1) and (13.12) can be shown to yield

$$[\![\rho(\mathbf{U}\cdot\hat{\mathbf{n}}_{12})\mathbf{q}-\hat{\mathbf{n}}_{12}\cdot\tilde{\boldsymbol{\sigma}}]\!]=0 \tag{13.66}$$

$$\left[\!\!\left[\rho(\mathbf{U}\cdot\hat{\mathbf{n}}_{12})\left(\frac{q^2}{2}+e\right)-\hat{\mathbf{n}}_{12}\cdot\tilde{\boldsymbol{\sigma}}\cdot\mathbf{q}+\mathbf{H}\cdot\hat{\mathbf{n}}_{12}\right]\!\!\right]=0 \tag{13.67}$$

And, the inequality (13.33) yields

$$\left[\!\!\left[\rho(\mathbf{U}\cdot\hat{\mathbf{n}}_{12})s+\mathbf{H}\cdot\frac{\hat{\mathbf{n}}_{12}}{T}\right]\!\!\right]\geqslant 0 \tag{13.68}$$

The above relations furnish the foundation of the shock theory in supersonic flows of compressible fluids. The inequality (13.68) is very remarkable on two accounts: (a) Although its counterpart in a continuous flow, inequality (13.36), yields nothing of prime importance, it turns out to be a key issue in the shock theory. (b) In the equalities (13.65)–(13.67), it does not matter whether we use $\hat{\mathbf{n}}_{12}$, or $\hat{\mathbf{n}}_{21}$ (that is, pointing from the 2- to the 1-side); in other words, the sense of the normal unit vector is irrelevant. But not so for the inequality; it is important to remember the sense of $\hat{\mathbf{n}}_{12}$ here and display the inequality thus:

$$\rho_2(\mathbf{U}_2\cdot\hat{\mathbf{n}}_{12})s_2+\mathbf{H}_2\cdot\frac{\hat{\mathbf{n}}_{12}}{T_2}\geqslant\rho_1(\mathbf{U}_1\cdot\hat{\mathbf{n}}_{12})s_1+\mathbf{H}_1\cdot\frac{\hat{\mathbf{n}}_{12}}{T_1} \tag{13.68A}$$

13.8 THE FLAT COUETTE FLOW OF COMPRESSIBLE FLUIDS—A SIMPLE EXAMPLE

The formulation of the general problem of fluid flow is now complete, even including possible surfaces of discontinuity. But before the subject can be satisfactorily brought to a close, it is deemed necessary to present some examples with simple geometry. We will treat again, in this section, the flat Couette flow, but this time for a compressible fluid (or, more specifically, for an ideal gas). The mathematics involved is extremely simple; yet, it is amazing how much light such a simple example can shed on the general problem.

Consider again two parallel plates of infinite extent at $y=0$, Y (Fig. 13.8), with a viscous compressible fluid between. For simplicity, the fluid is assumed to be an ideal gas, satisfying Stokes hypothesis. The lower plate is fixed, whereas the upper moves in the x-direction with a constant speed U. The start of the motion is long forgotten in the past so that the flow is now *steady*. The lower and the upper plates are kept at constant temperatures T_w and T_∞, respectively. It is also assumed that the fluid at $y=Y$ is being maintained at a constant (average) pressure p_∞.[24] (Only one condition for p is needed since only its

---

[24] One practical way of enforcing this approximately is to drill a number of small holes through the upper plate and connect them to a constant-pressure chamber; the distribution of the holes should be uniform, and the total area of the holes should be small so that the plate is still effectively impermeable. There, of course, should not be any pumping device.

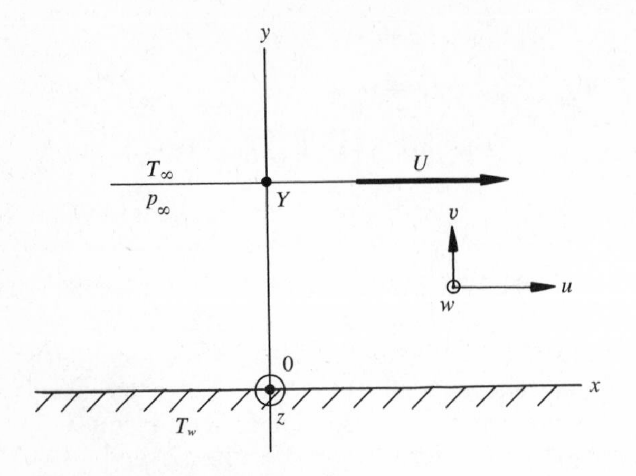

Figure 13.8 Flat Couette flow.

first derivatives appear in the governing equations.) The subscript ∞ is used because the problem can also be used qualitatively to shed light on a uniform flow past a flat wall with U, T_∞, and p_∞ at infinity. It is expected that at large Re (based on the distance along the wall), the flow will remain essentially uniform except for a thin boundary layer of "thickness" δ. The flat Couette flow will then qualify approximately for the boundary-layer flow if Y is replaced by δ.

Now, neglecting gravitation, we introduce the ansatz of pseudoplane and fully developed flow. There will then be only one independent variable, namely y. Choosing Y, U, p_∞, T_∞, $\rho_\infty = p_\infty/\mathbb{R}T_\infty$, $\mu_\infty = \mu\langle p_\infty, T_\infty\rangle$, $c_{p\infty} = c_p\langle T_\infty\rangle$, and $K_\infty = K\langle p_\infty, T_\infty\rangle$ as the characteristic quantities, one can nondimensionalize the governing equations. Let us start with the equation of continuity:

$$\frac{d\rho' v'}{dy'} = 0$$

This easily leads to

$$v' = 0$$

since $\rho' \neq 0$, and $v'\langle 0\rangle = 0$.

Next, the Navier-Stokes equation:

For the x-component,

$$0 = \mu' \frac{d^2u'}{dy'^2} + \left(\frac{du'}{dy'}\right)\frac{d\mu'}{dy'}$$

For the y-component,

$$0 = \frac{dp'}{dy'}$$

For the z-component,

$$0 = \mu' \frac{d^2w'}{dy'^2} + \left(\frac{dw'}{dy'}\right)\frac{d\mu'}{dy'}$$

The third of these yields

$$\mu' \frac{dw'}{dy'} = C_1$$

Therefore

$$w' = C_1 \int_0^{y'} \frac{dy'}{\mu'} + C_2$$

But $w' = 0$ at $y' = 0, 1$. Therefore

$$w' \equiv 0$$

This shows that the flow is really plane, not just pseudoplane. (The author refrained from jumping to the ansatz of plane flow right at the beginning for the sole purpose of making this point.)

The second of the previous triplet of equations gives simply

$$p' = 1$$

since $p'\langle 1 \rangle = 1$. The remaining equation of the triplet leads easily to

$$u' = \int_0^{y'} \frac{dy'}{\mu'} \Big/ \int_0^{1} \frac{dy'}{\mu'}$$

since $u'\langle 0 \rangle = 0$, and $u'\langle 1 \rangle = 1$. It must be emphasized that, here, u' is not yet solved; μ' as a function of y' is still to be found.

Finally, the energy equation reads (see Problem 13.5.2)

$$0 = \frac{1}{Pr}\frac{d}{dy'}\left(\mu' c_p' \frac{dT'}{dy'}\right) + (\gamma - 1) M^2 \mu' \left(\frac{du'}{dy'}\right)^2$$

where $Pr = \mu_\infty c_{p\infty}/K_\infty$ is the Prandtl number, and $M = U/\sqrt{\gamma \mathrm{I\!R}\, T_\infty}$ is the Mach number.[25] It is also assumed here that

$$\frac{\mu c_p}{K} = \text{constant}$$

that is,

$$\frac{\mu c_p}{K} = \frac{\mu_\infty c_{p\infty}}{K_\infty}$$

or

$$K' = \mu' c_p'$$

At this point, it is interesting to note that the usually important $Re = \rho_\infty UY/\mu_\infty$ plays no role whatsoever in this problem. The governing parameter is M; Pr is simply a given number depending on the gas used. (We follow here the common practice of allowing specific heat capacities of an ideal gas to vary with T; but the

[25] $\sqrt{\gamma \mathrm{I\!R} T_\infty}$ is, of course, the sonic speed of the ideal gas at T_∞. M is thus equal to the ratio of the flow speed to the sonic speed at the same point. The flow is then said to be supersonic, subsonic, or sonic as M is greater than, smaller than, or equal to 1. This Mach number must not be confused with the *local* Mach number $\underline{M}$ which is defined as $M = q/c$ where c is the sonic speed ($= \sqrt{\gamma \mathrm{I\!R} T}$ for ideal gases) at the point under consideration. For example, $\underline{M}$ for the lower plate is always zero. Ernst Mach (1838–1916), Austrian philosopher-physicist, did some experiments on projectile motion with his son Ludwig during 1878–1889. The motion is supersonic and his famous number showed up in the calculations. Although the importance of Mach number in the dynamics of compressible fluids can never be overstated, Mach's most prominent contribution is to the philosophy of science.

characteristic value $c_{p\infty}$ is still regarded to be equal to $\gamma \mathbb{R}/(\gamma-1)$ by properly choosing γ. Please also note that specific heat capacities, even varying ones, appear outside DT/Dt, not inside, in the energy equation. This latter fact is irrelevant in the present problem, but must be pointed out once and for all.)

It is easy to see that the energy equation yields

$$\frac{1}{Pr} c_p' \frac{dT'}{dy'} + (\gamma-1) M^2 \cdot \frac{1}{2} \frac{d(u'^2)}{dy'} = \frac{B_1}{\mu}$$

Finally, we also have the dimensionless equation of state (with $p' = 1$):

$$\rho' = \frac{1}{T'}$$

The relations

$$\mu' = \mu' \langle \rho', T' \rangle$$
$$c_p' = c_p' \langle T' \rangle,$$

are given functions for the gas considered.

For convenience one can introduce the specific enthalpy defined as

$$h = \int_0^T c_p \, dT + h_0$$

where h_0 is the datum value.[26] Using $h^* = c_{p\infty} T_\infty$ as the characteristic quantity, we have

$$h' = \int_0^{T'} c_p' \, dT'$$

In terms of h' instead of T', the energy equation now yields

$$\frac{h'}{Pr} + \frac{(\gamma-1) M^2}{2} (u')^2 = B_1 \int_0^{y'} \frac{dy'}{\mu'} + B_2$$

The boundary conditions for T can also be written in terms of h as follows:

$$y' = 0: \quad h' = h_w' \left(= \int_0^{T_w'} c_p' \, dT'\right)$$

$$y' = 1: \quad h' = h_\infty' \left(= \int_0^1 c_p' \, dT'\right)$$

With these boundary conditions, we have

$$h' = h_w' - \frac{(\gamma-1) M^2}{2} Pr \, (u')^2 + \left\{ (h_\infty' - h_w') + \frac{(\gamma-1) M^2 Pr}{2} \right\} \left[\int_0^{y'} \frac{dy'}{\mu'} \Big/ \int_0^1 \frac{dy'}{\mu'} \right]$$

$$= h_w' - \frac{(\gamma-1) M^2}{2} Pr (u')^2 + \left\{ (h_\infty' - h_w') + \frac{\gamma-1}{2} M^2 \, Pr \right\} u'$$

This equation expresses h' in terms of u'. Then $\mu' \langle \rho', T' \rangle$ can be put in the form of fct$\langle u' \rangle$ through the use of $\rho' = 1/T'$ and the above $h' \sim u'$ relation.

Next, the x-component of the Navier-Stokes equation can be solved

---

[26] In the following, h_0 will be omitted since it is always negligibly small.

for y' with u' regarded as the independent variable:

$$y' = \int_0^{u'} \mu' \, du' \Big/ \int_0^1 \mu' \, du'$$

where $\mu' = \mu'\langle u' \rangle$. Therefore, for any given viscosity law, we can obtain, at least numerically,

$$y' = y'\langle u' \rangle$$

This, when inverted, yields

$$u' = u'\langle y' \rangle$$

and the $h' \sim u'$ relation becomes

$$h' = h'\langle y' \rangle$$

The problem is *now* completely solved.[27]

It is interesting to investigate the heat flux across the fixed wall into the fluid:

$$\begin{aligned} H'_w &= \frac{H_w}{K_\infty T_\infty / Y} \\ &= -\left(K' \frac{dT'}{dy'}\right)\bigg|_{y'=0} \\ &= -\frac{1}{Pr} \tau'_w \left[\frac{\gamma - 1}{2} M^2 Pr + (h'_\infty - h'_w)\right] \end{aligned}$$

where

$$\begin{aligned} \tau'_w &= \frac{\tau_w}{\mu_\infty U / Y} \\ &= \int_0^1 \mu'\langle u' \rangle \, du' \end{aligned}$$

is the dimensionless shear stress on the fixed wall. From this result, one can calculate the temperature of the lower wall that gives zero heat transfer there; or, conversely, the temperature of the lower wall when it is *insulated.* The answer, coached in the latter form, is called the *recovery temperature*, and denoted by T_r. The corresponding recovery enthalpy is obtained by setting H'_w to zero:

$$\begin{aligned} h'_r &= \frac{h_r}{c_{p\infty} T_\infty} \\ &= h'_\infty + \frac{\gamma - 1}{2} M^2 \, Pr \end{aligned}$$

T_r is then calculable implicitly from

$$h'_r = \int_0^{T'_r} c'_p \, dT'$$

---

[27] See P. A. Lagerstrom, "Laminar flow theory," *Theory of Laminar Flows*, edited by F. K. Moore, Princeton University Press, Princeton, New Jersey, 1964.

If c_p = constant, T'_r can be directly obtained:

$$T'_r = 1 + \frac{\gamma - 1}{2} M^2 \, Pr$$

Incidentally, the ratio

$$\frac{T_r - T_\infty}{\left(1 + \frac{\gamma - 1}{2} M^2\right) T_\infty - T_\infty}$$

is called the *recovery factor* of the lower wall, and is very important in thermometry. (In the present problem, the recovery factor is just Pr.) It is impossible to measure directly the temperature of a moving stream (of ideal gas). In the *ideal* situation, one can only hope to measure the quantity $[1 + (\gamma - 1)M^2/2]T_\infty$. In practice, one can only measure T_r by an insulated thermometer. Then, if we know the recovery factor of the thermometer, we can calculate $[1 + (\gamma - 1)M^2/2]T_\infty$, and eventually T_∞ itself.

As $M \to 0$ (that is, U is very small compared to the sonic speed at the upper plate), the above discussion can be simplified somewhat, since the energy dissipation term in the energy equation drops out (there is *negligible energy dissipation*):

$$h' = h'_w + (h'_\infty - h'_w)\, u'$$

$$h'_r = h'_\infty$$

$$H'_w = -\frac{1}{Pr} \tau'_w (h'_\infty - h'_w)$$

All the other equations stay the same. If μ, c_p = constants, it is seen that

$$\begin{cases} u' = y' \\ T' = T'_w + (1 - T'_w)\, y' \\ (p' = 1, \rho' = 1/T') \end{cases}$$

For comparison, we will also list here the velocity and temperature distributions in a flat Couette flow of *incompressible fluids*:

$$\begin{cases} u' = y' \\ T' = T'_w + \left[(1 - T'_w) + \frac{Pr}{2}\left(\frac{U^2}{c_{in} T_\infty}\right)\right] y' - \frac{Pr}{2}\left(\frac{U^2}{c_{in} T_\infty}\right) y'^2 \\ (p' = 1, \rho' = 1) \end{cases}$$

where μ, c_{in} = constants. These distributions are valid for any $U^2/c_{in}T_\infty$. If $U^2/c_{in}T_\infty \ll 1$ (negligible energy dissipation), u', T', p' become the same as those for the compressible case with $M \ll 1$. We then say that there is a direct correlation of velocity, temperature, and pressure between a compressible fluid with small M and an incompressible one with small energy dissipation.

Note that the two cases are only *correlated*; no complete *analogy* is being implied. The common catch statement that "at low Mach numbers, a flow of a compressible fluid can be approximated by a corresponding one of an incompressible fluid" is particularly not true here, since, after all,

$$\rho' = \frac{1}{T'_w + (1 - T'_w)y'}$$

in one case and $\rho' = 1$ in the other. (A complete analogy occurs here only when $T'_w = 1$.) The fundamental difference lies really in the fact that $p = \rho \, \mathrm{I\!R} \, T$ for an ideal gas; yet, for an incompressible fluid, there is no relation whatsoever among p, ρ, and T.

(In general, it is safe to use the catch statement quoted above *only* in the outer regions of boundary-layer flows, where the viscous stresses and heat fluxes are negligible. However, partial and formal compressible-incompressible correlations, direct or indirect, do occur frequently in viscous flows, especially inside boundary layers.)

To get back to our main line of thought, the result for $M \to 0$ can also be improved to cover larger values of M. It will be established in one of the Problems that

$$u' = u'_{(0)} + u'_{(1)}M^2 + u'_{(2)}M^4 + O(M^6)$$

$$T' = T'_{(0)} + T'_{(1)}M^2 + T'_{(2)}M^4 + O(M^6)$$

Of these,

$$u'_{(0)} = \lim_{M \to 0} u' \quad \text{and} \quad T'_{(0)} = \lim_{M \to 0} T'$$

If we use these formulas for $M = 0.3$, the 1-term sum will have an error of $O(9\%)$; 2-term sum, $O(0.8\%)$; and 3-term sum, $O(0.0066\%)$.

On the other end of the spectrum, there is also the limit as $M \to \infty$ (or $1/M \to 0$). As an example, consider the case where $\mu \propto T$, $c_p =$ constant, with the lower wall insulated (that is, $h'_w = h'_r$ or $H'_w = 0$). It can be shown that, as $M \to \infty$,

$$\begin{cases} u' - \dfrac{u'^3}{3} = \frac{2}{3}y' + O(1/M^2) \\[2ex] h' - h'_r = -\dfrac{\gamma - 1}{2} M^2 \, Pr(u')^2 \end{cases}$$

from the exact solution. The second equation here is still exact. To take advantage of the fact that M is large, we first notice that $h' = h/c_{p\infty}T_\infty$ is no longer of $O(1)$, but much larger. This indicates that $c_{p\infty}T_\infty$ is no longer the proper characteristic quantity to use in ordering (Section 5.2). As a matter of fact, the energy dissipation term

$$\mu \left(\frac{du}{dy}\right)^2 \sim O\left(\frac{\mu_\infty U^2}{Y^2}\right)$$

now dominates the energy equation. The principle of least degeneracy (Section 5.5) then demands that this be balanced by the term

$$\frac{d}{dy}\left(K \frac{dT}{dy}\right) = \frac{1}{Pr} \frac{d}{dy}\left(\mu \frac{dh}{dy}\right)$$

$$\sim O\left(\frac{1}{Pr} \mu_\infty \frac{h^*}{Y^2}\right)$$

in the energy equation. With $Pr \sim O(1)$ (true for gases), we see that we must choose

$$h^* = U^2$$

in order to have the dimensionless $h \sim O(1)$ as $M \to \infty$. Let us write

$$h'' = \frac{h}{U^2}$$

Then

$$\begin{aligned} h'' - h''_r &= (h' - h'_r)\left(\frac{c_{p\infty} T_\infty}{U^2}\right) \\ &= -\frac{\gamma - 1}{2} M^2 \, Pr \left(\frac{c_{p\infty} T_\infty}{U^2}\right)(u')^2 \\ &= -\tfrac{1}{2} Pr \, (u')^2 \end{aligned}$$

and

$$\lim_{M\to\infty} (h'' - h''_r) = -\tfrac{1}{2} Pr [\lim_{M\to\infty} u']^2$$

where

$$[\lim_{M\to\infty} u'] - \tfrac{1}{3}[\lim_{M\to\infty} u']^3 = \tfrac{2}{3} y'$$

The flow at extremely large M is known as the hypersonic flow; the limit as $M \to \infty$ is then the hypersonic limit. The above introductory investigation of the hypersonic flat Couette flow is only of qualitative value, since many of the things that happen as $M \to \infty$, such as dissociation, combination, and so on, are not included in the formulation. Nevertheless, it does point out some particular trouble associated with large M. The importance of proper ordering shows up once more.

13.9 RAYLEIGH PROBLEM AT LARGE TIMES FOR COMPRESSIBLE FLUIDS—A BOUNDARY-LAYER FLOW

As our second example with simple geometry, and also as our concluding topic of the entire book, we will consider again the famous Rayleigh problem (Section 4.1), but with an ideal gas above the plate $y = 0$ (Fig. 4.1). At $t < 0$, the gas is at rest, and with a uniform temperature T_∞ and p_∞. At $t = 0$, the plate is suddenly jerked into motion with a constant speed U in the x-direction, and suddenly exposed to a temperature T_w; these are maintained for all $t > 0$. (At $y = \infty$, the initial rest state will prevail for any finite time; this is the reason behind the employment of subscript ∞.) For simplicity, the gas is assumed to satisfy Stokes hypothesis; to have constant c_p, constant $\mu c_p / K$, and $\mu \propto T$.

Similarly to the beginning of Section 4.1, we will introduce the ansatz of plane flow; but instead of $v \equiv 0$, we will try $\partial(\ )/\partial x \equiv 0$, which governs all flow properties including p. (In contrast to Section 4.1, $v \equiv 0$ and $\partial(\ )/\partial x \equiv 0$

are *not* equivalent here.) We favor the ansatz $\partial(\ )/\partial x \equiv 0$ because it reduces the independent variables to y and t, and is thus more likely to succeed.

For the nondimensionalization, the following characteristic quantities are chosen:

$$\begin{aligned} u^* &= v^* = U \\ p^* &= p_\infty \\ T^* &= T_\infty \\ \rho^* &= \rho_\infty (= p_\infty / \mathbb{R} T_\infty) \\ \mu^* &= \mu_\infty (= \mu\langle T_\infty \rangle) \\ K^* &= K_\infty (= K\langle T_\infty \rangle) \\ y^* &= U t^* \\ t^* &= t_{\text{obs}} \end{aligned}$$

where t_{obs} is the instant after $t = 0$, around which our interest is focused. We then have

$$Re = \frac{\rho_\infty U^2 t^*}{\mu_\infty}$$

$$Pr = \frac{\mu_\infty c_p}{K_\infty} \left(= \frac{\mu c_p}{K} \right)$$

$$M_\times = \frac{U}{\sqrt{\gamma \mathbb{R} T_\infty}}$$

where the Mach number is appended with a subscript cross to emphasize the fact that U and T_∞ are not taken at the same place in the flow.

The governing equations from the Appendix (and Problem 13.5.2) are then seen to reduce to the following:

$$\frac{\partial \rho'}{\partial t'} + \frac{\partial \rho' v'}{\partial y'} = 0 \tag{13.69}$$

$$\rho' \left(\frac{\partial u'}{\partial t'} + v' \frac{\partial u'}{\partial y'} \right) = \frac{1}{Re} \frac{\partial}{\partial y'} \left(\mu' \frac{\partial u'}{\partial y'} \right) \tag{13.70a}$$

$$\rho' \left(\frac{\partial v'}{\partial t'} + v' \frac{\partial v'}{\partial y'} \right) = -\frac{1}{\gamma} \frac{1}{M_\times^2} \frac{\partial p'}{\partial y'} + \frac{1}{Re} \frac{4}{3} \frac{\partial}{\partial y'} \left(\mu' \frac{\partial v'}{\partial y'} \right) \tag{13.70b}$$

$$\rho' \left(\frac{\partial T'}{\partial t'} + v' \frac{\partial T'}{\partial y'} \right) - \frac{\gamma - 1}{\gamma} \left(\frac{\partial p'}{\partial t'} + v' \frac{\partial p'}{\partial y'} \right) = \frac{1}{Pr\, Re} \frac{\partial}{\partial y'} \left(\mu' \frac{\partial T'}{\partial y'} \right) + \frac{(\gamma - 1) M_\times^2}{Re} \mu' \left[\left(\frac{\partial u'}{\partial y'} \right)^2 + \frac{4}{3} \left(\frac{\partial v'}{\partial y'} \right)^2 \right] \tag{13.71}$$

To these must be added the equation of state

$$p' = \rho' T' \tag{13.72}$$

and the linear variation of μ with T

$$\mu' = T' \tag{13.73}$$

($K' = \mu'$ has already been used in Eq. (13.71).) The conditions to be satisfied are

$$\begin{aligned} t' = 0&: \quad u', v' = 0; p', T' = 1 \\ y' = 0&: \quad u' = 1, v' = 0; T' = T'_w \\ y' \to \infty&: \quad u', v' \to 0; p', T' \to 1 \end{aligned}$$

At this point, one is tempted to try the additional ansatz $v' \equiv 0$; however, it fails (see Problem 13.9.2). We are then confronted with the very difficult problem of solving a system of four nonlinear partial differential equations. This system, unlike the incompressible case, cannot be solved exactly. As an approximation[28] we will consider only its asymptotic solution as $Re \to \infty$. (Physically, large Re is interpreted here as large time of observation, $t^* \gg \mu_\infty/\rho_\infty U^2$.) To this end, let us assume that $M_\times^2 \sim O(1)$ compared to Re; as to Pr, it is roughly 0.70 for all gases.

In taking the limit $Re \to \infty$, it is right away observed that the highest-order derivatives in the system are multiplied by $1/Re$ and will drop out. This is an unmistakable sign of a boundary layer appearing near the plate. (See Chapter 6.) Three special features must be pointed out here.

1. The boundary layer appears at large times over the entire plate. (See Section 6.2, where the boundary layer appears downstream from the leading edge.)
2. Temperature also changes rapidly across the layer.[29] (Heat transfer across a boundary layer was not studied at all in Parts 1 and 2.)
3. The outer flow is not only inviscid, but also nonconducting (that is, zero viscous stresses as well as zero heat fluxes).

Now, similarly to Section 6.2, the outer limit is the solution of the following system:

$$\begin{cases} \dfrac{\partial \rho'^o}{\partial t'} + \dfrac{\partial \rho'^o v'^o}{\partial y'} = 0 \\[2ex] \dfrac{\partial u'^o}{\partial t'} + v'^o \dfrac{\partial u'^o}{\partial y'} = 0 \\[2ex] \dfrac{\partial v'^o}{\partial t'} + v'^o \dfrac{\partial v'^o}{\partial y'} = -\dfrac{1}{\gamma}\dfrac{1}{M_\times^2}\dfrac{\partial p'^o}{\partial y'} \\[2ex] \rho'^o\left(\dfrac{\partial T'^o}{\partial t'} + v'^o \dfrac{\partial T'^o}{\partial y'}\right) - \dfrac{\gamma - 1}{\gamma}\left(\dfrac{\partial p'^o}{\partial t'} + v'^o \dfrac{\partial p'^o}{\partial y'}\right) = 0 \\[2ex] (p'^o = \rho'^o T'^o) \\[1ex] t' = 0: \quad u'^o, v'^o = 0; p'^o, T'^o = 1 \\[1ex] y' \to \infty: \quad u'^o, v'^o \to 0; p'^o, T'^o \to 1 \end{cases}$$

Without further ado, we will point out that the rest state

$$\begin{cases} u'^o, v'^o = 0 \\ p'^o, T'^o = 1 \end{cases}$$

satisfies all the governing equations and conditions. The outer "flow" is then just the gas at rest with temperature T_∞ and pressure p_∞; it is an inviscid and nonconducting "flow" without any response to the sudden motion and change of temperature of the plate.

This outer flow, being valid outside the immediate neighborhood of the plate, does not satisfy the boundary conditions for u and T at $y=0$. It actually leaves a discontinuity in velocity and temperature, which can be remedied by the inner (boundary layer) flow near the plate. The inner limit of the system is obtained, just as in Section 6.2, by introducing

$$\begin{cases} y'' = \sqrt{Re}\, y' \\ v'' = \sqrt{Re}\, v' \end{cases}$$

before taking $Re \to \infty$. The result is

$$\frac{\partial \rho'}{\partial t'} + \frac{\partial \rho' v''}{\partial y''} = 0 \tag{13.74}$$

$$\rho'\left(\frac{\partial u'}{\partial t'} + v''\frac{\partial u'}{\partial y''}\right) = \frac{\partial}{\partial y''}\left(\mu'\frac{\partial u'}{\partial y''}\right) \tag{13.75a}$$

$$0 = \frac{\partial p'}{\partial y''} \tag{13.75b}$$

$$\rho'\left(\frac{\partial T'}{\partial t'} + v''\frac{\partial T'}{\partial y''}\right) - \frac{\gamma-1}{\gamma}\left(\frac{\partial p'}{\partial t'} + v''\frac{\partial p'}{\partial y''}\right) = \frac{1}{Pr}\frac{\partial}{\partial y''}\left(\mu'\frac{\partial T'}{\partial y''}\right) + (\gamma-1)M_\infty^2\mu'\left(\frac{\partial u'}{\partial y''}\right)^2 \tag{13.76}$$

with

$$p' = \rho' T',\ \mu' = T'$$
$$t' = 0:\ u', v'' = 0;\ p', T' = 1$$
$$y'' = 0:\ u' = 1, v'' = 0;\ T' = T'_w$$

where the matching condition (6.26) must also be satisfied:

$$p'|_{y''=\infty} = p'^o_w = 1 \tag{13.77a}$$

$$T'|_{y''=\infty} = T'^o_w = 1 \tag{13.77b}$$

$$u'|_{y''=\infty} = u'^o_w = 0 \tag{13.77c}$$

$$0 = v'^o_w \tag{13.77d}$$

---

[28] M. D. Van Dyke, "Impulsive motion of an infinite plate in a viscous compressible fluid", *ZAMP*, **3**, 343–353, 1952. See also C. R. Illingworth, "Unsteady laminar flow of gas near an infinite flat plate," *Proc. Cambridge Phil. Soc.*, **46**, 603–613, 1950.

[29] This is known as the thermal boundary layer. It is here undistinguishable from the momentum boundary layer because $Pr \sim O(1)$. For large or small values of Pr (for example, for oils or liquid metals) the two layers may be of very different "thicknesses," and must then be treated separately.

Condition (13.77d) is arrived at in exactly the same way as condition (6.30), and is actually a boundary condition for the outer problem, which has already been (anticipated and) satisfied. Note also that there is no condition on v'' at $y'' = \infty$.

Equation (13.75b) yields

$$p' = p'\langle t' \rangle, \text{ only}$$

Condition (13.77a) then gives

$$p' \equiv 1 \; (\equiv p'^{o})$$

That is, for large times, the pressure stays constant throughout in the Rayleigh problem (or, the outside constant pressure is impressed on the boundary layer). The inner problem now takes on the simpler form:

$$\begin{cases} \dfrac{\partial \rho'}{\partial t'} + \dfrac{\partial \rho' v''}{\partial y''} = 0 & \textbf{(13.74)} \\[2ex] \rho'\left(\dfrac{\partial u'}{\partial t'} + v'' \dfrac{\partial u'}{\partial y''}\right) = \dfrac{\partial}{\partial y''}\left(T' \dfrac{\partial u'}{\partial y''}\right) & \textbf{(13.75aA)} \\[2ex] \rho'\left(\dfrac{\partial T'}{\partial t'} + v'' \dfrac{\partial T'}{\partial y''}\right) = \dfrac{1}{Pr}\dfrac{\partial}{\partial y''}\left(T' \dfrac{\partial T'}{\partial y''}\right) + (\gamma - 1) M_{\times}{}^{2} T' \left(\dfrac{\partial u'}{\partial y''}\right)^{2} & \textbf{(13.77A)} \\[2ex] (\rho' T' = 1) & \\ t' = 0\colon\; u', v'' = 0;\; T' = 1 & \\ y'' = 0\colon\; u' = 1, v'' = 0;\; T' = T'_{w} & \\ y'' = \infty\colon\; u' = 0;\; T' = 1 & \end{cases}$$

To solve this system, it is convenient to introduce the following change of variables (Dorodnitsyn transformation[30]):

$$\begin{cases} \zeta = \displaystyle\int_{0}^{y''} \rho' \, dy'' & \textbf{(13.78a)} \\[2ex] \xi = t' & \textbf{(13.78b)} \end{cases}$$

[30] A. A. Dorodnitsyn, "The boundary layer in a compressible gas" (in Russian), *Prikl. Mat. Mekh.*, 6, No. 6, 449, 1942. It is really analogous to the Mises transformation introduced by von Mises for incompressible boundary layer flows (R. von Mises; "Bermerkungen zur Hydrodynamik," *ZAMM*, 7, 425, 1927; see Problem 7.3.2). The original Dorodnitsyn (or modified Mises) transformation was used in attacking the compressible counterpart of Blasius flow. The first use on the present Rayleigh problem was due to Illingworth, cited in footnote 28. Richard von Mises (1883–1953) was Gordon McKay Professor of Aerodynamics at Harvard University when he died. He taught on three continents and in four languages. His contributions encompass an astonishing number of disciplines; some examples: airfoil theory (Mises' airfoils), plasticity (Mises' yield condition), probability (Mises' definition of probability), philosophy (positivism), and German poetry (studies on Rilke). He devoted special care to fundamentals, and advised "unceasing logical criticism" – a devotion apparently not well shared, and an advice not well heeded!

It can then be shown (Problem 13.9.4) that

$$\left\{\begin{aligned} \left(\frac{\partial y''}{\partial \zeta}\right)_\xi &= \frac{1}{\rho'} && \text{(13.79a)} \\ \left(\frac{\partial y''}{\partial \xi}\right)_\zeta &= v'' && \text{(13.79b)} \end{aligned}\right.$$

and

$$\left\{\begin{aligned} \left(\frac{\partial}{\partial y''}\right)_{t'} &= \rho'\left(\frac{\partial}{\partial \zeta}\right)_\xi && \text{(13.80a)} \\ \left(\frac{\partial}{\partial t'}\right)_{y''} + v''\left(\frac{\partial}{\partial y''}\right)_{t'} &= \left(\frac{\partial}{\partial \xi}\right)_\zeta && \text{(13.80b)} \end{aligned}\right.$$

Using Eqs. (13.80a,b), we can rewrite Eqs. (13.75aA) and (13.77A) as

$$\left\{\begin{aligned} \frac{\partial u'}{\partial \xi} &= \frac{\partial^2 u'}{\partial \zeta^2} && \text{(13.75aB)} \\ \frac{\partial T'}{\partial \xi} &= \frac{1}{Pr}\frac{\partial^2 T'}{\partial \zeta^2} + (\gamma-1)M_\times{}^2\left(\frac{\partial u'}{\partial \zeta}\right)^2 && \text{(13.77B)} \end{aligned}\right.$$

where $\rho' T' = 1$ has been used. The conditions to be satisfied are now

$$\xi = 0: \quad u' = 0; \; T' = 1$$

$$\zeta = 0: \quad u' = 1; \; T' = T'_w$$

$$\zeta = \infty: \quad u' = 0; \; T' = 1$$

Luckily, Eq. (13.75aB) and the associated conditions are mathematically the same as Eq. (4.5) and conditions in Section 4.1. So the similarity solution, Eq. (4.25), also applies here:

$$u' = 1 - \text{erf}\left(\frac{\zeta}{2\sqrt{\xi}}\right) \tag{13.81}$$

Warning This is *not* the solution for velocity yet. We still have to express ζ as $\zeta\langle y'', t'\rangle$.

To solve for the temperature, we notice that

$$\frac{\partial^2}{\partial \zeta^2}\left(\frac{1}{2}u'^2\right) = \left(\frac{\partial u'}{\partial \zeta}\right)^2 + u'\frac{\partial^2 u'}{\partial \zeta^2}$$

$$= \left(\frac{\partial u'}{\partial \zeta}\right)^2 + u'\frac{\partial u'}{\partial \xi}$$

or

$$\left(\frac{\partial u'}{\partial \zeta}\right)^2 = -\left(\frac{\partial}{\partial \xi} - \frac{\partial^2}{\partial \zeta^2}\right)\left(\frac{1}{2}u'^2\right)$$

So, if we approximate the true value of Pr (around 0.70) by rounding [31] it up to 1, we can write Eq. (13.77B) in the greatly simplified form:

$$\left(\frac{\partial}{\partial \xi}-\frac{\partial^2}{\partial \zeta^2}\right)\left(T'+\frac{\gamma-1}{2}M_\times{}^2u'^2\right)=0 \tag{13.82}$$

Furthermore, we have already the fact that

$$\left(\frac{\partial}{\partial \xi}-\frac{\partial^2}{\partial \zeta^2}\right)u'=0 \tag{13.75aB}$$

So, $T'+\frac{1}{2}(\gamma-1)M_\times{}^2u'^2=a+bu'$ must be a solution of the energy equation (13.82). Thus[32]

$$T'=a+bu'-\tfrac{1}{2}(\gamma-1)M_\times{}^2u'^2 \tag{13.83}$$

supplies a $T'\sim u'$ relation. (See also the $h'\sim u'$ relation of Section 13.8.) The constants a and b are easily determined from the boundary conditions for T' at $y''=0,\infty$:

$$T'=1+\left(T'_w-1+\frac{\gamma-1}{2}M_\times{}^2\right)u'-\frac{\gamma-1}{2}M_\times{}^2u'^2 \tag{13.84}$$

Although our solution is not complete yet, we can already determine the recovery temperature T_r as defined in Section 13.8. Remembering that the wall $y''=0$ is represented by $\zeta=0$, it is easy to show that the heat flux at the wall is zero if T'_w is equal to

$$T'_r=1+\tfrac{1}{2}(\gamma-1)M_\times{}^2 \tag{13.85}$$

(The recovery factor turns out to be 1 here.) On the other hand, when the plate is insulated, all the results throughout this section will still hold if T'_w is replaced by T'_r.

Let us continue our journey. In order to find $\zeta=\zeta\langle y'',t'\rangle$, we integrate Eq. (13.79a):

$$y''=\int_0^\zeta \frac{d\zeta}{\rho'}=\int_0^\zeta T'\,d\zeta \tag{13.86}$$

where the condition $y''\langle 0,\zeta\rangle=0$ has been used. Equation (13.86), when carried out, is of the form $y''=y''\langle\zeta,\xi\rangle$ which can be inverted to yield $\zeta=\zeta\langle y'',\xi\rangle$ or $\zeta=\zeta\langle y'',t'\rangle$. (In practice, there is no need to invert; one can calculate, for any chosen value of ζ and of $\xi=t'$, the values of u' and T', together with the *corresponding* value of y''. This *cross plotting* will easily give the distributions $u'\langle y'',t'\rangle$

---

[31] Although the error in Pr is thus about 40 percent, it would be wrong to think that the same kind of error is induced in the flow field. Actually, the error in using the boundary layer theory is $O(1/\sqrt{Pr\,Re})$, in which $Pr=0.7$ or 1 makes no significant difference whatsoever.

[32] Equation (13.84) also occurs in the compressible counterpart of Blasius flow. It was first discovered by A. Busemann in that context (see his "Gasdynamik" in *Handbuch der Experimentalphysik*, ed. by W. Wien and F. Harms, Vol. IV/1, Akademische Verlag, Leipzig, 1931), and then extended by L. Crocco ("Sulla transmissione del calore da una lamina piana a un fluido scorente ad alta velocitá," *L'Aerotecnica*, **12**, 1932; translated as NACA TM No. 690, 1932). L. Crocco is professor of jet propulsion at Princeton University.

and $T'\langle y'', t'\rangle$.) The explicit form of Eq. (13.86) will be developed in a Problem. The important thing to notice here is that the solutions for u' and T' are complete only after Eq. (13.86) is obtained.

Finally, Eq. (13.79b) can be used to calculate the transverse velocity:

$$v'' = \left(\frac{\partial y''}{\partial \xi}\right)_\zeta = v''\langle \zeta, \xi\rangle \tag{13.87}$$

The explicit form of Eq. (13.87) is again to be developed in a Problem. Attention must be called to the fact that

$$\lim_{y''\to\infty} v'' \propto \frac{1}{\sqrt{t'}} \neq 0$$

In other words, there is a nonzero residue v'' at the "edge" of the boundary layer just as in Section 7.2. In physical terms, this residue transverse velocity is of course very small, since

$$v' = 0 + \frac{v''}{\sqrt{Re}}$$

The error is only of $\square(1/\sqrt{Re})$. Also similarly to Section 7.2, the "edge" of the boundary layer turns out to be moving outward with its instantaneous position $\propto \sqrt{t'}$. This then corresponds to an outward speed $\propto 1/\sqrt{t'}$. It is thus seen that the nonvanishing v'' at the "edge" is due to the thickening of the layer. If one wishes to improve the accuracy beyond $\square(1/\sqrt{Re})$, one must first correct the outer flow so that it sees a "flat plate" pushing out with a speed $\propto 1/\sqrt{t}$ and of $\square(1/\sqrt{Re})$. (As a result, a sonic wave will be generated and will move into the gas at rest in the outer region; see Van Dyke.[28]) The nonzero value of $v''\langle\infty, t'\rangle$ is thus seen to furnish a starting point for refinements; if it *were* zero, the rationale behind the boundary-layer theory would become shaky.

But all these, of course, have nothing to do with the boundary layer theory proper, except to indicate an error of $\square(1/\sqrt{Re})$. (The reader should note the striking resemblance to the Blasius flow, described in Section 7.2.)

A READING LIST FOR FURTHER STUDIES

I GENERAL

1. Serrin, J., "Mathematical principles of classical fluid mechanics," *Encyclopedia of Physics*, Vol. VIII/1, edited by S. Flügge, (co-edited by C. Truesdell), Springer-Verlag, Berlin, 1959.
2. Milne-Thomson, L. M., *Theoretical Hydrodynamics*, 5th ed., Macmillan, New York, 1968
3. Friedrichs, K. O., and R. von Mises, *Fluid Dynamics* (Notes), Brown University, Providence, R. I., 1942; also, Springer-Verlag, Berlin, 1971.
4. Landau, L. D., and E. M. Lifshitz: *Fluid Mechanics*, translated by J. B. Sykes and W. H. Reid, Pergamon Press, New York, 1959.
5. Goldstein, S., *Lectures on Fluid Mechanics*, Interscience Publishers, New York, 1960.
6. Sommerfeld, A., *Mechanics of Deformable Bodies*, translated by G. Kuerti, Academic Press, New York, 1950.

II BOUNDARY-LAYER FLOWS

7. Lagerstrom, P. A., "Laminar flow theory," *Theory of Laminar Flows*, edited by F. K. Moore, Princeton University Press, Princeton, N.J., 1964.

III OUTER FLOWS

8. Courant, R., and K. O. Friedrichs, *Supersonic Flow and Shock Waves*, Interscience Publishers, New York, 1948.

9. von Mises, R., *Mathematical Theory of Compressible Fluid Flow*, Academic Press, New York, 1958.

Also References 1–6.

IV TWO COMMON REFERENCES

10. Goldstein, S. (ed.), *Modern Developments in Fluid Dynamics*, Oxford, University Press, London, 1938.

11. Schlichting, H., *Boundary-Layer Theory*, 6th ed., translated by J. Kestin, McGraw-Hill, New York, 1968.

A LIST OF PRE-1900 WORKS CITED

1687 Newton, I., Philosophiae Naturalis Principia Mathematica. London.

1738 Bernoulli, D., Hydrodynamica sive de Viribus et Motibus Fluidorum Commentarii. Argentorati.

1743 Bernoulli, J., Hydraulica nunc primum detecta ac demonstrata directe ex fundamentis pure mechanicis, Anno 1732, Opera **4**, 387–493.

1752 D'Alembert, J. L., Essai d'une Nouvelle Théorie de la Résistance des Fluides. Paris.
Euler, L., see 1761 Euler.

1755 Euler, L., see 1757 Euler.

1757 Euler, L., Principes généraux du mouvement des fluides. Mém. Acad. Sci. Berlin [**11**] (1755), 274–315 = Opera omnia (2) **12**, 54–91.

1761 D'Alembert, J. L., Remarques sur les lois du mouvement des fluides. Opusc. **1**, 137–168.
Euler, L., Principia motus fluidorum (1752–1755). Novi Comm. Acad. Sci. Petrop. **6** (1756–1757), 271–311 = Opera omnia (2) **12**, 133–168.

1769 Euler, L., see 1770 Euler.

1770 Euler, L., Sectio secunda de principiis motus fluidorum. Novi Comm. Acad. Sci. Petrop. **14** (1769), 270–386 = Opera omnia (2) **13**, 73–153. By a printer's error the volume is dated 1759.

1781 Lagrange, J. L., see 1783 Lagrange.

1783 Lagrange, J. L., Mémoire sur la théorie du mouvement des fluides. Nouv. mém. Acad. Sci. Berlin (1781), 151–198 = Oeuvres **4**, 695–748.

1788 Lagrange, J. L., Méchanique Analitique. Paris. Oeuvres **11**, **12**, are the 5th ed.

1822 Fourier, J., Théorie Analytique de la Chaleur. Paris = Oeuvres **1**.
Navier, C.-L.-M.-H., see 1827 Navier.

1823 Cauchy, A.-L., Recherches sur l'équilibre et le mouvement intérieur des corps solides ou fluides, élastiques ou non élastiques. Bull. Soc. Philomath. 9–13 = Oeuvres (2) **2**, 300–304.

1827 Cauchy, A.-L., Sur la condensation et la dilatation des corps solides. Ex. de math. **2**, 60–69 = Oeuvres (2) **7**, 82–83.

Navier, C.-L.-M.-H., Mémoire sur les lois du mouvement des fluides (1822). Mém. Acad. Sci. Inst. France (2) **6**, 389–440.

1841 Cauchy, A.-L., Mémoire sur les dilatations, les condensations et les rotations produits par un changement de forme dans un système de points matériels. Ex d'an. phys. math. **2**, 302–330 = Oeuvres (2) **12**, 343–377.
Poiseuille, J. L., Recherches expérimentales sur le mouvement des liquides dans les tubes de très petits diamètres. C. R. Acad. Sci., Paris **12**, 112.

1842 Stokes, G. G., On the steady motion of incompressible fluids. Trans. Cambridge Phil. Soc. **7** (1839–1842), 439–453 = Papers **1**, 1–16.

1845 Stokes, G. G., On the theories of the internal friction of fluids in motion, and of the equilibrium and motion of elastic solids. Trans. Cambridge Phil. Soc. **8** (1844–1849), 287–319 = Papers **1**, 75–129.

1848 Stokes, G. G., On a difficulty in the theory of sound: Phil. Mag (3) **33**, 349–356. A drastically condensed version, with an added note, appears in Papers **2**, 51–55.

1851 Stokes, G. G., On the effect of the internal friction of fluids on the motion of pendulums (1850). Trans. Cambridge Phil. Soc. 9^2, 8–106 = Papers **3**, 1–141.

1858 Helmholtz, H., Über Integrale der hydrodynamischen Gleichungen, welche den Wirbelbewegungen entsprechen. J. reine angew. Math. **55**, 25–55 = Wiss. Abh. **1**, 101–134. Transl., P. G. Tait, On integrals of the hydrodynamical equations, which express vortex-motion. Phil. Mag. (4) **33**, (1867) 485–512.

1860 Riemann, B., Ueber die Fortpflanzung ebener Luftwellen von endlicher Schwingungsweite. Gött. Abh., math. Cl. 8 (1858–1859), 43–65 = Werke, 145–164.

1868 Helmholtz, H., Sur le mouvement le plus général d'un fluide. Réponse à une communication précédente de M. J. Bertrand. C. R. Acad. Sci., Paris **67**, 221–225 = Wiss. Abh. **1**, 135–139.

Helmholtz, H., Sur le mouvement des fluides. Deuxième réponse à M. J. Bertrand. C. R. Acad. Sci., Paris **67**, 754–757 = Wiss. Abh. **1**, 140–144.

Helmholtz, H., Réponse à la note de M. J. Bertrand, du 19 Octobre. C. R. Acad. Sci., Paris **67**, 1034–1035 = Wiss. Abh. **1**, 145.

1869 Thomson, W. (Lord Kelvin), On vortex motion. Trans. Roy. Soc. Edinb. **25**, 217–260 = Papers **4**, 13–66.

1872 Froude, W., Experiments on the surface-friction experienced by a plane moving through water. Report to the 42nd Meeting of the British Association for the Advancement of Science.

1873 Gibbs, J. W., Graphical methods in the thermodynamics of fluids. Trans. Connecticut Acad. **2**, 309–342 = Works **1**, 1–32.

Gibbs, J. W., A method of geometrical representation of the thermodynamic properties of substances by means of surfaces. Trans. Connecticut Acad. **2**, 382–404 = Works **1**, 33–54.

1875 Gibbs, J. W., On the equilibrium of heterogeneous substances. Trans. Connecticut Acad. **3** (1875–1878), 108–248, 343–524 = Works **1**, 55–353.

1878 Strouhal, V. Über eine besondere Art der Tonerregung. Ann. Phys. und Chemie. New series, **5**, 216–251.

1881 Gibbs, J. W., Elements of Vector Analysis, Part I. New Haven = Works **2**, 17–36.

1883 Reynolds, O., An experimental investigation of the circumstances which determine whether the motion of water shall be direct or sinuous and of the law of resistance in parallel channels. Phil. Trans. Roy. Soc. **174**, 935–982 = Papers **2**, 51–105.

1884 Gibbs, J. W., Elements of Vector Analysis, Part II. New Haven = Works **2**, 50–90.

1889 Whitehead, A. N., Second approximations to viscous fluid motion. Quart. J. Math. **23**, 143–152.

1890 Couette, M., Études sur le frottement des liquides. Ann. Chim. Phys. (6) **21**, 433–510.

1895 Reynolds, O., On the dynamic theory of incompressible viscous fluids and the determination of the criterion. Phil. Trans. Roy. Soc. **186A**, 123–164 = Papers **2**, 535–577.

APPENDIX

GOVERNING EQUATIONS

The governing equations for ideal gases obeying the Stokes hypothesis are listed below. They are valid at points *not* seated on surfaces of discontinuity.

For incompressible fluids, simply replace c_v by c_{in}, div $\mathbf{q}$ by zero, and *equation of continuity* by div $\mathbf{q} = 0$.

Even with temperature-dependent heat capacities, the equations are valid just as they stand.)

I GENERAL ORTHOGONAL COORDINATES

$$\frac{\partial\rho}{\partial t}+\frac{1}{h_1h_2h_3}\left[\frac{\partial(h_2h_3\rho q_1)}{\partial\zeta_1}+\frac{\partial(h_3h_1\rho q_2)}{\partial\zeta_2}+\frac{\partial(h_1h_2\rho q_3)}{\partial\zeta_3}\right]=0$$

$$\rho\left[\frac{\partial q_i}{\partial t}+\sum_{m=1}^{3}\frac{q_m}{h_m}\left(\frac{\partial q_i}{\partial\zeta_m}+\frac{q_i}{h_i}\frac{\partial h_i}{\partial\zeta_m}-\frac{q_m}{h_i}\frac{\partial h_m}{\partial\zeta_i}\right)\right]$$

$$=\rho g_i-\frac{1}{h_i}\frac{\partial p}{\partial\zeta_i}+\frac{4\mu}{3}\frac{1}{h_i}\frac{\partial}{\partial\zeta_i}(\text{div }\mathbf{q})-\frac{\mu}{h_jh_k}\left\{\frac{\partial}{\partial\zeta_j}\left[\frac{h_k}{h_ih_j}\left(\frac{\partial h_jq_j}{\partial\zeta_i}-\frac{\partial h_iq_i}{\partial\zeta_j}\right)\right]\right.$$

$$\left.-\frac{\partial}{\partial\zeta_k}\left[\frac{h_j}{h_ih_k}\left(\frac{\partial h_iq_i}{\partial\zeta_k}-\frac{\partial h_kq_k}{\partial\zeta_i}\right)\right]\right\}+2\sum_{m=1}^{3}\epsilon_{im}\left(\frac{1}{h_m}\frac{\partial\mu}{\partial\zeta_m}\right)-\tfrac{2}{3}(\text{div }\mathbf{q})\left(\frac{1}{h_i}\frac{\partial\mu}{\partial\zeta_i}\right)$$

($i = 1, 2, 3$; ijk are cyclic in 1, 2, 3; for example, $j = 3$ and $k = 1$ if $i = 2$)

$$\rho c_v\left[\frac{\partial T}{\partial t} + \sum_{m=1}^{3} q_m\left(\frac{1}{h_m}\frac{\partial T}{\partial \zeta_m}\right)\right] + p \operatorname{div} \mathbf{q}$$

$$= \frac{1}{h_1h_2h_3}\left\{\frac{\partial}{\partial \zeta_1}\left[\frac{h_2h_3}{h_1}\left(K\frac{\partial T}{\partial \zeta_1}\right)\right] + \frac{\partial}{\partial \zeta_2}\left[\frac{h_3h_1}{h_2}\left(K\frac{\partial T}{\partial \zeta_2}\right)\right]\right.$$

$$\left. + \frac{\partial}{\partial \zeta_3}\left[\frac{h_1h_2}{h_3}\left(K\frac{\partial T}{\partial \zeta_3}\right)\right]\right\} - \frac{2\mu}{3}(\operatorname{div} \mathbf{q})^2 + 2\mu \sum_{\ell, m=1}^{3} \epsilon_{\ell m}^2 + Q$$

$$\operatorname{div} \mathbf{q} = \frac{1}{h_1h_2h_3}\left[\frac{\partial(h_2h_3q_1)}{\partial \zeta_1} + \frac{\partial(h_3h_1q_2)}{\partial \zeta_2} + \frac{\partial(h_1h_2q_3)}{\partial \zeta_3}\right]$$

II RECTANGULAR CARTESIAN COORDINATES[1]

$$\frac{\partial \rho}{\partial t} + \frac{\partial(\rho u)}{\partial x} + \frac{\partial(\rho v)}{\partial y} + \frac{\partial(\rho w)}{\partial z} = 0$$

$$\rho\left(\frac{\partial u}{\partial t} + u\frac{\partial u}{\partial x} + v\frac{\partial u}{\partial y} + w\frac{\partial u}{\partial z}\right)$$

$$= \rho g_x - \frac{\partial p}{\partial x} + \frac{\mu}{3}\frac{\partial}{\partial x}(\operatorname{div} \mathbf{q}) + \mu \nabla^2 u$$

$$-\tfrac{2}{3}(\operatorname{div} \mathbf{q})\frac{\partial \mu}{\partial x} + 2\epsilon_{xx}\left(\frac{\partial \mu}{\partial x}\right) + 2\epsilon_{xy}\left(\frac{\partial \mu}{\partial y}\right) + 2\epsilon_{xz}\left(\frac{\partial \mu}{\partial z}\right)$$

$$\rho\left(\frac{\partial v}{\partial t} + u\frac{\partial v}{\partial x} + v\frac{\partial v}{\partial y} + w\frac{\partial v}{\partial z}\right)$$

$$= \rho g_y - \frac{\partial p}{\partial y} + \frac{\mu}{3}\frac{\partial}{\partial y}(\operatorname{div} \mathbf{q}) + \mu \nabla^2 v$$

$$-\tfrac{2}{3}(\operatorname{div} \mathbf{q})\frac{\partial \mu}{\partial y} + 2\epsilon_{xy}\left(\frac{\partial \mu}{\partial x}\right) + 2\epsilon_{yy}\left(\frac{\partial \mu}{\partial y}\right) + 2\epsilon_{yz}\left(\frac{\partial \mu}{\partial z}\right)$$

$$\rho\left(\frac{\partial w}{\partial t} + u\frac{\partial w}{\partial x} + v\frac{\partial w}{\partial y} + w\frac{\partial w}{\partial z}\right)$$

$$= \rho g_z - \frac{\partial p}{\partial z} + \frac{\mu}{3}\frac{\partial}{\partial z}(\operatorname{div} \mathbf{q}) + \mu \nabla^2 w$$

$$-\frac{2}{3}(\operatorname{div} \mathbf{q})\frac{\partial \mu}{\partial z} + 2\epsilon_{xz}\left(\frac{\partial \mu}{\partial x}\right) + 2\epsilon_{yz}\left(\frac{\partial \mu}{\partial y}\right) + 2\epsilon_{zz}\left(\frac{\partial \mu}{\partial z}\right)$$

[1] The following forms differ, on the surface, from those under **I**. These alternative (but equivalent) forms are used to display the familiar operator ∇^2 *operating on a component of* a vector.

$$\rho c_v\left(\frac{\partial T}{\partial t}+u\frac{\partial T}{\partial x}+v\frac{\partial T}{\partial y}+w\frac{\partial T}{\partial z}\right)+p\,\mathrm{div}\,\mathbf{q}$$

$$=\frac{\partial}{\partial x}\left(K\frac{\partial T}{\partial x}\right)+\frac{\partial}{\partial y}\left(K\frac{\partial T}{\partial y}\right)+\frac{\partial}{\partial z}\left(K\frac{\partial T}{\partial z}\right)-\frac{2\mu}{3}(\mathrm{div}\,\mathbf{q})^2+2\sum_{\ell,m=x,y,z}\epsilon^2_{\ell m}+Q$$

$$\mathrm{div}\,\mathbf{q}=\frac{\partial u}{\partial x}+\frac{\partial v}{\partial y}+\frac{\partial w}{\partial z}$$

$$\nabla^2=\frac{\partial^2}{\partial x^2}+\frac{\partial^2}{\partial y^2}+\frac{\partial^2}{\partial z^2}$$

For ϵ_{xx} and so forth, see Eqs. (9.7a–f).

III CYLINDRICAL COORDINATES[1]

$$\frac{\partial\rho}{\partial t}+q_R\left(\frac{\partial\rho}{\partial R}\right)+\frac{q_\theta}{R}\left(\frac{\partial\rho}{\partial\theta}\right)+w\frac{\partial\rho}{\partial z}+\rho\,\mathrm{div}\,\mathbf{q}=0$$

$$\rho\left[\frac{\partial q_R}{\partial t}+q_R\left(\frac{\partial q_R}{\partial R}\right)+\frac{q_\theta}{R}\left(\frac{\partial q_R}{\partial\theta}\right)+w\frac{\partial q_R}{\partial z}-\frac{q_\theta^2}{R}\right]$$

$$=\rho g_R-\frac{\partial p}{\partial R}+\frac{\mu}{3}\frac{\partial}{\partial R}(\mathrm{div}\,\mathbf{q})+\mu\left(\nabla^2 q_R-\frac{q_R}{R^2}-\frac{2}{R^2}\frac{\partial q_\theta}{\partial\theta}\right)-\frac{2}{3}(\mathrm{div}\,\mathbf{q})\frac{\partial\mu}{\partial R}$$

$$+2\epsilon_{RR}\left(\frac{\partial\mu}{\partial R}\right)+2\epsilon_{R\theta}\left(\frac{1}{R}\frac{\partial\mu}{\partial\theta}\right)+2\epsilon_{Rz}\left(\frac{\partial\mu}{\partial z}\right)$$

$$\rho\left[\frac{\partial q_\theta}{\partial t}+q_R\left(\frac{\partial q_\theta}{\partial R}\right)+\frac{q_\theta}{R}\left(\frac{\partial q_\theta}{\partial\theta}\right)+w\frac{\partial q_\theta}{\partial z}+\frac{q_R q_\theta}{R}\right]$$

$$=\rho g_\theta-\frac{1}{R}\frac{\partial p}{\partial\theta}+\frac{\mu}{3}\frac{1}{R}\frac{\partial}{\partial\theta}(\mathrm{div}\,\mathbf{q})+\mu\left(\nabla^2 q_\theta+\frac{2}{R^2}\frac{\partial q_R}{\partial\theta}-\frac{q_\theta}{R^2}\right)-\frac{2}{3}(\mathrm{div}\,\mathbf{q})\frac{1}{R}\frac{\partial\mu}{\partial\theta}$$

$$+2\epsilon_{R\theta}\left(\frac{\partial\mu}{\partial R}\right)+2\epsilon_{\theta\theta}\left(\frac{1}{R}\frac{\partial\mu}{\partial\theta}\right)+2\epsilon_{\theta z}\left(\frac{\partial\mu}{\partial z}\right)$$

$$\rho\left[\frac{\partial w}{\partial t}+q_R\left(\frac{\partial w}{\partial R}\right)+\frac{q_\theta}{R}\left(\frac{\partial w}{\partial\theta}\right)+w\frac{\partial w}{\partial z}\right]$$

$$=\rho g_z-\frac{\partial p}{\partial z}+\frac{\mu}{3}\frac{\partial}{\partial z}(\mathrm{div}\,\mathbf{q})+\mu\nabla^2 w-\frac{2}{3}(\mathrm{div}\,\mathbf{q})\frac{\partial\mu}{\partial z}$$

$$+2\epsilon_{Rz}\left(\frac{\partial\mu}{\partial R}\right)+2\epsilon_{\theta z}\left(\frac{1}{R}\frac{\partial\mu}{\partial\theta}\right)+2\epsilon_{zz}\left(\frac{\partial\mu}{\partial z}\right)$$

$$\rho c_v\left[\frac{\partial T}{\partial t}+q_R\left(\frac{\partial T}{\partial R}\right)+\frac{q_\theta}{R}\left(\frac{\partial T}{\partial \theta}\right)+w\frac{\partial T}{\partial z}\right]+p\operatorname{div}\mathbf{q}$$

$$=\frac{\partial}{\partial R}\left(K\frac{\partial T}{\partial R}\right)+\frac{K}{R}\frac{\partial T}{\partial R}+\frac{1}{R^2}\frac{\partial}{\partial \theta}\left(K\frac{\partial T}{\partial \theta}\right)+\frac{\partial}{\partial z}\left(K\frac{\partial T}{\partial z}\right)-\frac{2\mu}{3}(\operatorname{div}\mathbf{q})^2$$

$$+2\mu\sum_{\ell,m=R,\theta,z}\epsilon_{\ell m}^2+Q$$

$$\operatorname{div}\mathbf{q}=\frac{\partial q_R}{\partial R}+\frac{q_R}{R}+\frac{1}{R}\frac{\partial q_\theta}{\partial \theta}+\frac{\partial w}{\partial z}$$

$$=\frac{1}{R}\left[\frac{\partial(Rq_R)}{\partial R}+\frac{\partial q_\theta}{\partial \theta}\right]+\frac{\partial w}{\partial z}$$

$$\nabla^2=\frac{\partial^2}{\partial R^2}+\frac{1}{R}\frac{\partial}{\partial R}+\frac{1}{R^2}\frac{\partial^2}{\partial \theta^2}+\frac{\partial^2}{\partial z^2}$$

$$=\frac{1}{R}\frac{\partial}{\partial R}\left(R\frac{\partial}{\partial R}\right)+\frac{1}{R^2}\frac{\partial^2}{\partial \theta^2}+\frac{\partial^2}{\partial z^2}$$

For ϵ_{RR} and so forth, see *answer* to Problem 12.4.2.

IV SPHERICAL COORDINATES[1]

$$\frac{\partial \rho}{\partial t}+q_r\left(\frac{\partial \rho}{\partial r}\right)+\frac{q_\varphi}{r}\left(\frac{\partial \rho}{\partial \varphi}\right)+\frac{q_\theta}{r\sin\varphi}\left(\frac{\partial \rho}{\partial \theta}\right)+\rho\operatorname{div}\mathbf{q}=0$$

$$\rho\left[\frac{\partial q_r}{\partial t}+q_r\left(\frac{\partial q_r}{\partial r}\right)+\frac{q_\varphi}{r}\left(\frac{\partial q_r}{\partial \varphi}\right)+\frac{q_\theta}{r\sin\varphi}\left(\frac{\partial q_r}{\partial \theta}\right)-\frac{q_\varphi^2+q_\theta^2}{r}\right]$$

$$=\rho g_r-\frac{\partial p}{\partial r}+\frac{\mu}{3}\frac{\partial}{\partial r}(\operatorname{div}\mathbf{q})+\mu\left(\nabla^2 q_r-\frac{2q_r}{r^2}-\frac{2}{r^2}\frac{\partial q_\varphi}{\partial \varphi}-\frac{2q_\varphi\cot\varphi}{r^2}-\frac{2}{r^2\sin\varphi}\frac{\partial q_\theta}{\partial \theta}\right)$$

$$-\tfrac{2}{3}(\operatorname{div}\mathbf{q})\frac{\partial \mu}{\partial r}+2\epsilon_{rr}\left(\frac{\partial \mu}{\partial r}\right)+2\epsilon_{r\varphi}\left(\frac{1}{r}\frac{\partial \mu}{\partial \varphi}\right)+2\epsilon_{r\theta}\left(\frac{1}{r\sin\varphi}\frac{\partial \mu}{\partial \theta}\right)$$

$$\rho\left[\frac{\partial q_\varphi}{\partial t}+q_r\left(\frac{\partial q_\varphi}{\partial r}\right)+\frac{q_\varphi}{r}\left(\frac{\partial q_\varphi}{\partial \varphi}\right)+\frac{q_\theta}{r\sin\varphi}\frac{\partial q_\varphi}{\partial \theta}+\frac{q_r q_\varphi-q_\theta^2\cot\varphi}{r}\right]$$

$$=\rho g_\varphi-\frac{1}{r}\frac{\partial p}{\partial \varphi}+\frac{\mu}{3}\frac{1}{r}\frac{\partial}{\partial \varphi}(\operatorname{div}\mathbf{q})+\mu\left(\nabla^2 q_\varphi+\frac{2}{r^2}\frac{\partial q_r}{\partial \varphi}-\frac{q_\varphi}{r^2\sin^2\varphi}-\frac{2\cot\varphi}{r^2\sin\varphi}\frac{\partial q_\theta}{\partial \theta}\right)$$

$$-\tfrac{2}{3}(\operatorname{div}\mathbf{q})\frac{1}{r}\frac{\partial \mu}{\partial \varphi}+2\epsilon_{r\varphi}\left(\frac{\partial \mu}{\partial r}\right)+2\epsilon_{\varphi\varphi}\left(\frac{1}{r}\frac{\partial \mu}{\partial \varphi}\right)+2\epsilon_{\varphi\theta}\left(\frac{1}{r\sin\varphi}\frac{\partial \mu}{\partial \theta}\right)$$

$$\rho\left[\frac{\partial q_\theta}{\partial t}+q_r\left(\frac{\partial q_\theta}{\partial r}\right)+\frac{q_\varphi}{r}\left(\frac{\partial q_\theta}{\partial \varphi}\right)+\frac{q_\theta}{r\sin\varphi}\left(\frac{\partial q_\theta}{\partial \theta}\right)+\frac{q_r q_\theta+q_\varphi q_\theta\cot\varphi}{r}\right]$$

$$= \rho g_\theta - \frac{1}{r \sin\varphi}\frac{\partial p}{\partial \theta} + \frac{\mu}{3}\frac{1}{r\sin\varphi}\frac{\partial}{\partial\theta}(\text{div}\,\mathbf{q})$$

$$+\mu\left(\nabla^2 q_\theta - \frac{q_\theta}{r^2\sin^2\varphi} + \frac{2}{r^2\sin\varphi}\frac{\partial q_r}{\partial\theta} + \frac{2\cot\varphi}{r^2\sin\varphi}\frac{\partial q_\varphi}{\partial\theta}\right)$$

$$-\tfrac{2}{3}(\text{div}\,\mathbf{q})\frac{1}{r\sin\varphi}\frac{\partial\mu}{\partial\theta} + 2\epsilon_{r\theta}\left(\frac{\partial\mu}{\partial r}\right) + 2\epsilon_{\varphi\theta}\left(\frac{1}{r}\frac{\partial\mu}{\partial\varphi}\right) + 2\epsilon_{\theta\theta}\left(\frac{1}{r\sin\varphi}\frac{\partial\mu}{\partial\theta}\right)$$

$$\rho c_v\left[\frac{\partial T}{\partial t} + q_r\left(\frac{\partial T}{\partial r}\right) + \frac{q_\varphi}{r}\left(\frac{\partial T}{\partial\varphi}\right) + \frac{q_\theta}{r\sin\varphi}\left(\frac{\partial T}{\partial\theta}\right)\right] + p\,\text{div}\,\mathbf{q}$$

$$= \frac{\partial}{\partial r}\left(K\frac{\partial T}{\partial r}\right) + \frac{2K}{r}\frac{\partial T}{\partial r} + \frac{1}{r^2}\frac{\partial}{\partial\varphi}\left(K\frac{\partial T}{\partial\varphi}\right) + \frac{K\cot\varphi}{r^2}\frac{\partial T}{\partial\varphi} + \frac{1}{r^2\sin^2\varphi}\frac{\partial}{\partial\theta}\left(K\frac{\partial T}{\partial\theta}\right)$$

$$-\frac{2\mu}{3}(\text{div}\,\mathbf{q})^2 + 2\mu\sum_{\ell,m=r,\varphi,\theta}\epsilon_{\ell m}^2 + Q$$

$$\text{div}\,\mathbf{q} = \frac{\partial q_r}{\partial r} + \frac{2q_r}{r} + \frac{1}{r}\frac{\partial q_\varphi}{\partial\varphi} + \frac{q_\varphi}{r}\cot\varphi + \frac{1}{r\sin\varphi}\frac{\partial q_\theta}{\partial\theta}$$

$$= \frac{1}{r^2}\frac{\partial}{\partial r}(r^2 q_r) + \frac{1}{r\sin\varphi}\frac{\partial}{\partial\varphi}(q_\varphi\sin\varphi) + \frac{1}{r\sin\varphi}\frac{\partial q_\theta}{\partial\theta}$$

$$\nabla^2 = \frac{\partial^2}{\partial r^2} + \frac{2}{r}\frac{\partial}{\partial r} + \frac{1}{r^2}\frac{\partial^2}{\partial\varphi^2} + \frac{\cot\varphi}{r^2}\frac{\partial}{\partial\varphi} + \frac{1}{r^2\sin^2\varphi}\frac{\partial^2}{\partial\theta^2}$$

$$= \frac{1}{r^2}\frac{\partial}{\partial r}\left[r^2\left(\frac{\partial}{\partial r}\right)\right] + \frac{1}{r^2\sin\varphi}\frac{\partial}{\partial\varphi}\left(\sin\varphi\frac{\partial}{\partial\varphi}\right) + \frac{1}{r^2\sin^2\varphi}\frac{\partial^2}{\partial\theta^2}$$

For e_{rr} and so forth, see *answer* to Problem 12.4.2.

PROBLEMS

(These problems are of varying degrees of difficulty. The instructor should select the ones that are within the reach of his class. The first two digits of a problem number refer to the corresponding section. The later problems of each section are *not* necessarily more difficult.)

1.1.1 Which of the following is a plane flow? (**a**) Flow in a circular pipe. (**b**) Flow induced by a circular disk rotating about an axis perpendicular to it. (**c**) Flow past a long, straight, but swept-back wing.

1.2.1 If we say that

$$\frac{\partial(\ )}{\partial x}, \frac{\partial(\ )}{\partial y} = 0$$

what do we mean? Can ( ) of a particle change as it moves with the above in force?

1.2.2 Why cannot water be treated as incompressible in (**a**) sound propagation, and (**b**) underwater explosion problems?

1.2.3 For the plane flow of an incompressible fluid, the velocity component in the x-direction is

$$u = ax^2 + by$$

Find the velocity component v in the y-direction. To evaluate the arbitrary function that may appear, assume (a) $v=0$ at $y=0$, (b) $v=ct$ at $y=0$.

1.2.4 Find the stream function ψ of the simple shear flow with

$$\begin{cases} u = \dfrac{Uy}{Y} \\ v = 0 \end{cases}$$

Since the constant of integration is immaterial, being a decision on which streamline is called $\psi=0$, let us take it to be zero. Find the streamlines $\psi=0$, $UY/8$, $UY/4$, $3UY/8$, $UY/2$, and sketch them in the xy-plane. Show that volume flows between the adjacent streamlines you just plotted are the same.

1.2.5 The x-component of the velocity in a plane flow of an incompressible fluid is

$$u = e^{-x}\cosh y + 1$$

(**a**) Find the y-component of the velocity v, assuming that $v=0$ at $y=0$. (**b**) Find the stream function (absorbing any constant of integration) from the velocity components and sketch the streamlines $\psi=0$ and 1.

1.2.6 Given $u=2kx$, $v=2ky$, find the streamlines passing through the point $(1, 0)$. Is this also an equi-ψ line? Is such a flow physically possible? (Plane flows of incompressible fluids are considered, of course.)

1.2.7 Show that the equation of continuity for the plane flows of a compressible fluid is

$$\frac{\partial\rho}{\partial t} + \frac{\partial(\rho u)}{\partial x} + \frac{\partial(\rho v)}{\partial y} = 0$$

(The law of mass conservation states now that the net rate of mass flow into the infinitesimal rectangular volume must equal the rate of increase of mass inside the volume.) What is the form of the equation for a steady flow?

1.2.8 Show that the general equation of continuity of Problem 1.2.7 can also be written as

$$\frac{D\rho}{Dt} + \rho\left(\frac{\partial u}{\partial x} + \frac{\partial v}{\partial y}\right) = 0$$

Now, specialize it for the case of incompressible fluids. Is this specialized result true only for steady flows?

1.3.1 Verify Eq. (1.24).

1.3.2 Show that the flow

$$\begin{cases} u = u\langle x, y\rangle \\ v = 0 \end{cases}$$

is, in general, rotational. In what special cases is the flow irrotational?

1.3.3 By actually averaging the angular speed of *all* line segments, without using the general formula, Eq. (1.24), show that $\Omega = -C$ for the simple shear flow.

1.3.4 Show that the average angular speed of *two* perpendicular line segments is the same as that of *all* segments. (Therefore, two instantaneously perpendicular segments at every point are all that is needed to determine Ω.)

1.3.5 Given a velocity field

$$\begin{cases} u = cx + 2\omega_0 y + u_0 \\ v = cy + v_0 \end{cases}$$

where c, ω_0, u_0, and v_0 are constants, determine ϵ_{xx}, ϵ_{yy}, ϵ_{xy}, and Ω.

1.3.6 Show that there are no shear rate-of-strain values in the wheel flow.

2.1.1 Derive Eq. (2.3) with $\hat{\mathbf{n}}$ pointing into the fourth quadrant.

2.3.1 Show that $\sigma_{nn} = \sigma_{xx} = \sigma_{yy}$ at a point where the fluid is in rigid-body motion (and hence, having no shear stresses) by an argument similar to that used in deriving Eq. (2.3). That is, go back to the first principles; do not apply Eq. (2.5).

2.3.2 Derive Eqs. (2.15a,b).

2.4.1 Show that the simple shear flow of a Reiner-Rivlin fluid, with μ and η independent of temperature, must have constant (average) pressure. (*Hint*: Substitute into equations of motion.)

3.1.1 To describe the flow in between two parallel plates completely, we must also add *two* initial conditions to boundary conditions (3.1a,b,c,d):

$$\begin{cases} u\langle x, y; 0\rangle = F_3\langle x, y\rangle & \textbf{(3.1e)} \\ v\langle x, y; 0\rangle = F_4\langle x, y\rangle & \textbf{(3.1f)} \end{cases}$$

The initial condition (3.11) is a special case. (The other initial condition $v\langle x, y; 0\rangle = 0$ is implied there.) Now, suppose the flow at $t = 0$ is not so obliging, but is such that the *general* forms (3.1e, f) hold. Would the ansatz $v \equiv 0$ still work?

3.1.2 Derive Eqs. (3.3b) and (3.4) by introducing the ansatz $\partial\mathbf{q}/\partial x = 0$ instead of $v \equiv 0$.

3.1.3 Derive Eqs. (3.13a,b), using the ansatz $v = v_0\langle t\rangle$.

3.1.4 Derive Eqs. (3.13a,b), using the ansatz $\partial\mathbf{q}/\partial x = 0$.

3.1.5 The ansatz used in deriving Eqs. (3.13a,b) narrows down the possible forms of the initial conditions (3.1e,f). Write down these possible forms. What does this mean physically?

3.2.1 Show that, for the plane Poiseuille flow, **(a)** $\bar{u} = 2u_{\max}/3$, **(b)** $dp/dx - \rho g_x = -12\mu(\bar{u}/Y^2)$, and **(c)** the friction coefficient C_f (= shear stress on the wall/$\frac{1}{2}\rho\bar{u}^2$) = $12/Re$, where $\bar{u}$ is the mean value of u based on the volume flow rate, and $Re = \bar{u}Y/\nu$.

3.2.2 A viscosity pump is made up of a stationary, cylindrical casing and a drum that rotates inside the casing at a constant speed ω (radians/sec); the drum and casing are concentric. Liquid with density ρ and viscosity μ enters a section A, flows through the annulus in between the drum and casing, and leaves at another section B. The pressure at B is higher than that at A by Δp. The circumferential length of the annulus from A to B is L. The gap of the annulus h is very small compared to the diameter of the drum, so the flow is approximately that between two flat plates. Assuming laminar flow, find Δp, the torque applied, the power input, the power output (in raising the pressure), and the efficiency of the pump, all in terms of the flow rate (per unit depth).

3.2.3 Derive Eq. (3.18), and show that P is dimensionless.

3.2.4 A stepped bearing is shown in Figure P1. The lower plate moves with respect to the bearing shoe and drags oil through the clearance space. The flow is plane and laminar, and the pressure is assumed to increase linearly from the entrance to the step and then to decrease linearly from the step to the exit.

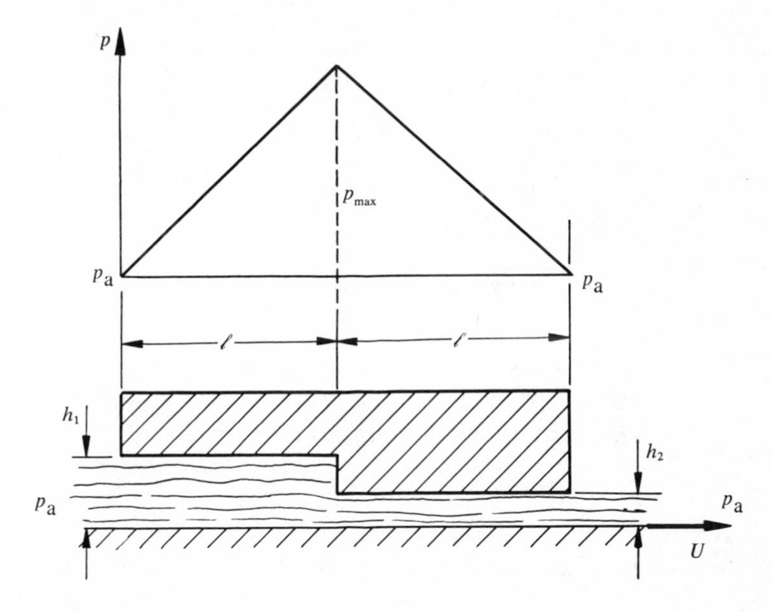

Figure P1

(**a**) Regarding this as two simple shear flows put back to back, find the flow rates for the left and the right halves in terms of p_{max}, U, ℓ, h_1, h_2, and μ.
(**b**) Equating the flow rates (why?), find p_{max}.
(**c**) Find the load the bearing can carry (per unit depth).
(**d**) What is the load capacity if $h_2 = \frac{1}{2}h_1$?

3.2.5 Repeat (b) and (c) of Problem 3.2.4 for $h_2 \to 0$, and with the length of the right half of the shoe shrunk to zero. (What kind of bearing do you have now?) Calculate the drag force on the moving surface exerted by the oil film, and the ratio of drag to load.

3.2.6 For Eq. (3.18), determine

(**a**) the volume rate of flow
(**b**) the average velocity (based on flow rate)
(**c**) the maximum (or minimum) velocity
(**d**) the location of u_{max} or u_{min}
(**e**) the shearing stress.

3.2.7 Consider the steady, laminar flow of two liquid layers between two parallel plates, as shown in Figure P2. The upper plate moves with velocity U, and the lower one is stationary. There is no pumping, and the gravitational force is negligible.

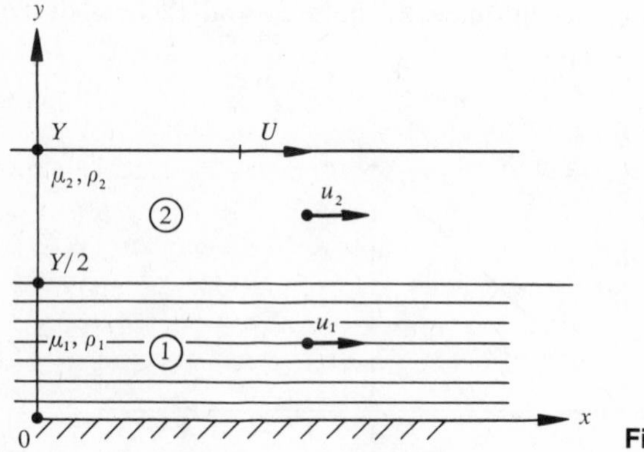

Figure P2

(**a**) State the conditions to be satisfied at the interface.
(**b**) Obtain $u_1\langle y\rangle$ and $u_2\langle y\rangle$ and sketch the results, for $\mu_1 > \mu_2$.
(**c**) Calculate the shear stress at the lower wall.
(**d**) What is the shear stress at an arbitrary point?
(**e**) What is the pressure distribution?

3.3.1 An endless belt passes upward through a chemical bath with speed V and picks up a film of liquid of thickness h, density ρ, and viscosity μ. Gravity tends to drain the liquid down, but the movement of the belt keeps it from completely running off. Assume that the flow is laminar, that the atmosphere exerts no shear on the film; and use the ansatz of fully developed flow.

(**a**) Solve for the velocity distribution across the film.
(**b**) Determine the rate at which the fluid is being dragged up by the belt.

3.4.1 Obtain Eq. (3.32) by solving Eq. (3.30a) under boundary conditions (3.15a,b).

3.4.2 Solve Eq. (3.30a) with boundary condition (3.15a) and $u\langle Y\rangle = 0$. (Define here $Re = \rho u^* Y/\mu$, where $u^* = -Y^2(F-\rho g_x)/2\mu$.)

3.4.3 Plot u/U vs. y/Y for Eq. (3.32) with $v_0 = -U$, $C = 0.2$, and $Re = 0, 1, 10, 100, 1000$.

3.4.4 Consider again the system of equations (2.15a,b) and (1.9). Introducing the stream function ψ, let us substitute Eqs. (1.10a,b) into Eqs. (2.15a, b). Now, differentiate the resulting equations, one with respect to x, the other with respect to y, so that the pressure terms can be easily eliminated. The result of the elimination should be a partial differential equation of ψ; write it down, together with the conditions (3.1a,b,c,d,e, f) expressed in terms of ψ. Conditions (3.1e,f) are quoted in Problem 3.1.1.

3.4.5 Show that the ansatz $\psi = \psi\langle y; t\rangle$ works for the system governing ψ in Problem 3.4.4. Show that the result is identical with that of Section 3.1. (What about p?)

3.4.6 Take the result of Problem 3.4.4 again. Restrict it to the steady case. Now, apply it to the flow treated in the present section. What boundary conditions in terms of ψ must be used in place of those associated with boundary conditions (3.1c,d)? What ansatz for ψ would you use? Show that your ansatz works and that the result is equivalent to the one presented in the text.

4.1.1 At what distance from the plate is the velocity of the fluid in Rayleigh's problem only one percent of U? Calling this the 99 percent thickness of the (boundary) layer (where the flow velocity reaches, within 99 percent, that at infinity) near the plate, and denoting it by $\delta_{99\%}$, show that

$$\delta_{99\%} \propto \sqrt{\nu t}$$

(That is, the layer increases its "thickness" as t increases according to a parabolic law; it is thin for liquids of small viscosity.)

4.1.2 Define the shear layer thickness δ_s as

$$\delta_s = \mu U/(\text{shear stress at } y = 0)$$

for the Rayleigh problem. Show that

$$\delta_s = \sqrt{\pi}\sqrt{\nu t}$$

Define also the mean thickness δ_m as

$$\delta_m = \int_0^\infty \frac{u\,dy}{U}$$

Show that

$$\delta_m = \frac{2}{\sqrt{\pi}}\sqrt{\nu t}$$

Indicate both δ_s and δ_m in the $u \sim y$ diagram at a fixed time.

4.1.3 Study Section 2.5-(iv) of Carslaw and Jaeger's *Conduction of Heat in Solids*, 2nd ed., Oxford University Press, London, 1960; and interpret it in terms of the Rayleigh problem with plate speed At^n.

4.1.4 With reference to Eqs. (4.20a,b), which of the following cases represents physically reasonable behavior at $y=\infty$: $f\langle\infty\rangle=0$, $f\langle\infty\rangle=A$, $f\langle\infty\rangle=B\neq A$?

4.1.5 Show that exp $(c\eta^2)$ is a solution of Eq. (4.19) if $n=-\frac{1}{2}$. What value must c assume? What condition at $\eta=\infty$ can be satisfied? What is the physical problem you have just solved?

4.1.6 In Problem 4.1.5, determine lim $u\langle y, t\rangle$ as $t\to 0$ for both $y=0$ and $y\neq 0$. Determine also

$$\lim_{t\to 0}\int_{-\infty}^{\infty} u\,dy$$

4.1.7 To the *really* ambitious (and *well*-prepared) student, we might mention that there are other methods of searching for similarity solutions—for example, that using the concept of invariant solutions under a group of transformations (initiated by A. Michal and A. J. A. Morgan), and that using dimensional analysis (initiated by G. Birkhoff and L. I. Sedov). As a starter, read Chapters 4 and 5 of A. G. Hansen's *Similarity Analysis of Boundary Value Problems in Engineering*, Prentice-Hall, 1964, and apply the above-mentioned methods to the Rayleigh problem.

4.1.8 If you are already familiar with the Laplace transform, obtain the solution, Eq. (4.25), by the transform method.

4.2.1 Verify the statement in the text about Eqs. (4.47a,b) immediately following them.

4.2.2 Take the corresponding system governing the stream function (see Problem 3.4.4). Show that the ansatz $\psi=xf\langle y\rangle$ works and is identical with the ansatz used in the text.

4.2.3 If you have already learned numerical analysis, outline how Eq. (4.35) is to be solved under the conditions stated in the text. Is this an initial-value problem? How do you handle $y=\infty$? What would the computer do if you fed in a value $b<0$?

5.1.1 Verify Eqs. (5.1) and (5.2a,b).

5.3.1 If you are familiar with the Fourier series, solve Eq. (3.8), with the terms in the brackets omitted, in a Fourier sine series under conditions (3.10c,d) and (3.11A). (If you have learned the method of finite integral transforms, use the finite Fourier sine transform instead.)

5.4.1 Determine the large-time solution of the flow between two fixed plates, $y=\pm Y/2$, under a pressure gradient that varies harmonically with time. Find the mean velocity $\bar{u}$ across the channel. Now, define $\delta=\sqrt{2\nu/\omega}$ where ω is the circular frequency of the pressure gradient. What is u for $Y/\delta\ll 1$ and $\gg 1$? What are the physical interpretations of your results?

5.4.2 If you are familiar with the operational method, solve for u_{makeup} by Laplace transform. (If you know Fourier sine transform, use that instead. What is the advantage in doing so?)

5.4.3 Determine the long-time solution of the flow between two plates, $y = 0$ and Y, where one plate is fixed and the other is oscillating harmonically.

5.5.1 For the straight-faced bearing, determine

(a) p_{max}
(b) location of p_{max}
(c) total load (per unit depth)
(d) maximum load
(e) ratio h_2/h_1 for which maximum load is sustained,
(f) shear force (drag) on the bearing guide
(g) drag when load is maximum
(h) drag/load when load is maximum.

6.1.1 Calculate the shear stress $\mu\, du/dy$ (in dimensional form) from Eq. (6.15). Then calculate the same quantity from Eq. (6.4). Compare the results as $Re \to \infty$.

6.1.2 In the discussion of the Friedrichs problem, the restriction $0 < C < 1$ is actually immaterial. Sketch u' vs. y' as Re increases for $C = 1$, >1, $=0$, and <0. What happens to the corresponding boundary layers?

6.1.3 Discuss the inner and outer limits and their matching for the case treated in Problem 3.4.2.

6.1.4 Take the damped vibration of a mass-spring system with mass m, spring constant k, and damping coefficient c. Discuss the inner and outer limits as $m \to 0$ with natural time and a properly stretched time fixed. (L. Prandtl, *Anschauliche und nützliche Mathematik*, 1931/32; see H. Schlichting, *Boundary-Layer Theory*, McGraw-Hill, 1968, and J. D. Cole, *Perturbation Methods in Applied Mathematics*, Blaisdell, 1968.)

6.2.1 Verify Eqs. (6.23a,b).

6.2.2 Show that Eq. (6.37) satisfies Eq. (6.36).

6.2.3 Find $\mathbf{q}'^o$ and q'^o from Eq. (6.37).

7.1.1 Show that there is no solution to Eq. (7.17) under condition (7.4A), in the back-flow case.

7.1.2 Suppose that a distribution of suction is available at the wall such that

$$y = 0, x \geqslant 0: \;\; v = v_0\langle x\rangle$$

What functional form must v_0 assume in order that similarity solutions exist? What are the corresponding ordinary differential equation and boundary conditions?

7.1.3 In the search for similarity solutions, what happens to the special case where $2\gamma - \lambda = 0$?

7.2.1 The shear stress τ_w on the wall in the Blasius flow

$$\mu \frac{du}{dy}\bigg|_{y=0}$$

can be nondimensionalized: $\tau'_w = \tau_w / \frac{1}{2}\rho\Upsilon^2$. Show that

$$\tau'_w = 0.66\sqrt{\Upsilon x/\nu}$$

for the Blasius flow.

7.2.2 Integrate τ'_w of Problem 7.2.1 over $x = 0$ to $x = L$ and show that the dimensionless drag $D' = 1.33\sqrt{\nu L/\Upsilon}$ (where the drag is on one side of the plate only). Show that the drag coefficient, based on one side of the plate of length L, $C_D = 1.33/\sqrt{Re_L}$, where $Re_L = \Upsilon L/\nu$ is the Re based on L.

7.2.3 Find out, from your computing center, how the Blasius solution is to be obtained on a modern computer.

7.3.1 Show that, for a Falkner-Skan flow,

$$\delta_1 \propto x^{(1-m)/2}$$

$$\tau_w \propto x^{(3m-1)/2}$$

(Note that $\delta_1 = $ constant when $m = 1$, $\tau_w = $ constant when $m = \frac{1}{3}$.)

7.3.2 Introducing $\psi''\langle x', y''\rangle$ such that Eq. (6.22) is satisfied, show that Eq. (6.28) can be transformed from (x', y'') to (ξ', ψ''):

$$\frac{\partial Z'}{\partial \xi'} = \sqrt{u_w'^{o2} - Z'}\,\frac{\partial^2 Z'}{\partial \psi''^2}$$

($\xi' = x'$ and $Z' = u_w'^{o2} - u'^2$). Show that this new equation must be solved with the conditions

$$\psi'' = 0{:}\;\; Z' = u_w'^{o}\langle \xi' \rangle$$

$$\psi'' = \infty{:}\;\; Z' = 0$$

Interpret this transformed problem in terms of heat conduction with temperature-dependent conductivity. Does the conduction problem need an initial temperature distribution to be completely determined? What is the equivalent requirement in the original Falkner-Skan flow problem? Is it not true that the influence of this initial temperature distribution decreases fast as time goes on? What is the counterpart interpretation of this fast decrease in the F-S flow problem? (R. von Mises, *ZAMM*, 7, 425, 1927; hence, the name *Mises transformation*.)

7.4.1 Verify the conditions listed right after Eq. (7.28).

7.4.2 Verify the conditions listed just before the example of Blasius flow.

7.4.3 Show that the Pohlhausen quartic form yields

$$\frac{dF}{d\eta} = (2 - 6\eta^2 + 4\eta^3) + \frac{\Lambda}{6}[(1 - 3\eta)^2 - 4\eta^3]$$

and

$$\frac{d^2F}{d\eta^2} = (1-\eta)\left[12\left(\frac{\Lambda}{6}-1\right)\eta - \Lambda\right]$$

Now, show that $(dF/d\eta)|_{\eta=0} < 0$ if $\Lambda < -12$; and that $d^2F/d\eta^2 = 0$ will occur at $\eta < 1$ if $\Lambda > 12$. Both situations must be excluded. Why?

7.4.4 Multiply the boundary-layer equation by $2u'$ and the continuity equation by $u_w'^{o2} - u'^2$. Subtract the first from the second. Integrate with respect to y'' from 0 to ∞, and obtain Leibenzon's integral relation:

$$\int_0^\infty \left(\frac{\partial u'}{\partial y''}\right)^2 dy'' = \frac{1}{2}\frac{d}{dx'}(u_w'^{o3}\delta_3'')$$

where δ_3'' is the dimensionless (stretched) energy thickness,

$$\delta_3'' = \int_0^\infty \left(\frac{u'}{u_w'^o} - \frac{u'^3}{u_w'^{o3}}\right) dy''$$

7.4.5 Take the Rayleigh problem:

$$\begin{cases} \dfrac{\partial u'}{\partial t'} = \dfrac{1}{Re}\dfrac{\partial^2 u'}{\partial y'^2} = \dfrac{\partial^2 u'}{\partial y''^2} \\ u'\langle 0\rangle = 1 \\ u'\langle \infty\rangle = 0 \end{cases}$$

Integrate the equation with respect to y'' from 0 to ∞. In the resulting integral relation, try the family of trial curves

$$u' = 1 - \text{erf}\,[y''/\delta''\langle t'\rangle]$$

Show that this choice yields the exact solution.

8.1.1 Derive Eqs. (8.12d,e,f).

8.1.2 Substantiate Eqs. (8.14a,b,c,d).

8.1.3 Derive Eqs. (8.15a,b,c).

8.1.4 Derive Eq. (8.16) by considering directly the mass conservation across a fixed element bounded by $\theta = \theta$, $\theta = \theta + d\theta$, $R = R$, and $R = R + dR$.

8.2.1 Derive Eqs. (8.35) and (8.36).

8.2.2 Calculate the torque (per unit length of the cylinder) exerted on (**a**) the outer cylinder when the inner one is at rest and (**b**) the inner cylinder when the outer one is missing and the fluid is tranquil at infinity.

8.2.3 Show that the ansatz (8.28) is also compatible with $q_R = c/R$ where $c \neq 0$. What is then the physical situation? Solve for this new (generalized) Couette flow.

8.2.4 Find the vorticity distribution for the Couette flow, Eq. (8.35). Under what condition would Ω be zero everywhere?

8.2.5 Show that Eq. (8.35) reduces to the simple shear velocity profile if ω_1 (or ω_2) $= 0$ and $(R_2 - R_1)/R_1 \ll 1$. (See Problem 3.2.2.)

8.2.6 Formulate and solve (if you can) the problem of unsteady start of the Couette flow from rest.

8.4.1 Show that the ansatz (8.38) is compatible not only with $q_\theta = 0$ but also with $q_\theta = c/R$, where c is a constant. What is the corresponding physical situation when $c \neq 0$? What is the modified form of Eq. (8.41)?

8.4.2 Derive Eqs. (8.56) and (8.58) and boundary conditions (8.57a,b) and (8.59a,b).

8.4.3 Derive Eqs. (8.66) and (8.67).

8.5.1 Show that δ_1 for the boundary layer in a convergent channel is proportional to R. Is this boundary-layer flow a member of the Falkner-Skan flows?

8.6.1 Derive Eq. (8.88).

8.6.2 Derive Eq. (8.92).

8.6.3 Show that Eq. (8.93) includes Eq. (6.16) as a special case.

8.6.4 Obtain Eq. (8.96) by matching q_θ.

8.6.5 Derive Eq. (8.97).

9.1.1 In Eqs. (9.13a,b,c), choose z in the vertically upward direction so that $\mathbf{g} = (0, 0, -g)$, where g is the constant gravitational acceleration. Introduce $\widehat{p} = p + \rho g z$ and rewrite Eqs. (9.13a,b,c) in terms of $\widehat{p}$. Together with the equation of continuity, the system yields u, v, w, and $\widehat{p}$ if proper conditions are given for them. After $\widehat{p}\langle x, y, z; t\rangle$ is obtained, $p = \widehat{p} - \rho g z$ can be easily calculated. Now show that this "absorption of the gravity term" is really useful when the boundaries of the problem are known (not necessarily fixed), but is of no value when a part of the boundaries has to be determined as a portion of the solution (for example, the sea surface around a ship).

9.2.1 Take the steady pseudoaxisymmetric flow (including the axisymmetric flow as a special case) in the presence of a flat plate $z = 0$. (Neglect gravity.) Develop the inner and outer flow problems in $z > 0$ for large Re. Note that, in comparison with Section 6.2, z plays the role of y here; and R, x.

9.2.2 Write Eqs. (9.27) and (9.28a,b,c) in the spherical coordinates.

9.2.3 Consider the flow in between two rotating coaxial and coapex cones similarly to the Couette flow. Set up the governing equations and boundary conditions. Can you solve the problem? (The solution is rather simple.) This solution is expected in practice when the two apex angles of the cones are close.

9.2.4 Solve Problem 9.2.3 with *proper* injection and suction at the walls.

9.3.1 Establish Eqs. (9.35a,b) by considering the force balance on a fluid element in the shape of a column with a ring-sectorial cross section, from z to $z + dz$, R to $R + dR$, and θ to $\theta + d\theta$.

9.3.2 Similarly to Section 3.1, establish the governing equation and the initial condition for the unsteady Poiseuille flow of a Newtonian fluid. Suggest a way of solving the equation. (F. Szymanski, *J. Math. Pures Appliqueés*, Series 9, **11**, 67, 1932.)

9.3.3 A very long circular tube has a thin layer of liquid on its inner surface. Water is introduced to flush the liquid out. Assuming that the liquid layer flows while maintaining a uniform thickness, set up the problem and solve it.

9.3.4 Solve for the pumped flow through an annulus between two concentric, fixed, circular cylinders.

9.3.5 Solve Problem 9.3.4 when the outer pipe moves along the axis with speed U.

9.3.6 Solve Problem 9.3.5 when the two pipes, in addition, rotate with angular speeds ω_1 and ω_2.

9.3.7 Find the stresses on an area element of the wall in the Poiseuille flow. Is there a shear stress on an area element at the wall, with a normal in the axial direction?

9.3.8 Solve Problem 9.3.5 with suction and injection through the two pipe walls. (K. N. Mehta, *AIAA J.*, **1**, 217, 1963).

9.4.1 Following Section 4.2, set up the problem of axisymmetric stagnation-point flow, show that the same ansatz works here, and discuss the situation to the same extent as in Section 4.2. (F. Homann, *ZAMM*, **16**, 153, 1936.)

9.4.2 Find out how Cochran solved Eqs. (9.56a,b,c).

9.4.3 The ansatz used for the axisymmetric stagnation-point flow happens to be the same as that for the rotating-disk problem. Show that, therefore, the same ansatz works for the (stagnation-point) flow toward a rotating disk. (D. M. Hannah, *Rep. Memor. Aero. Res. Coun.*, London, No. 2772, 1952.)

9.4.5 Referring to Problem 9.2.1, would the outer flow for a rotating disk (at large Re) in a quiescent fluid be just the surrounding fluid at rest? Compare the corresponding Prandtl problem with the rotating-disk problem as presented in the text. Are the two identical? (*Hint*: Remember the residual flow at infinity toward the disk? Compare Section 7.3, Case 1.)

9.4.6 Do a similar comparison between Problem 9.4.1 and the corresponding inner/outer-limit analysis at large Re.

9.4.7 Analyze the large-Re flow over $z = 0$ induced by the suction at a tiny hole at $(0, 0)$. Show that the corresponding outer flow yields $u_w{}^o \propto 1/R^2$. Show that a similarity solution exists for the boundary-layer flow with $q_R = F\langle\eta\rangle$ and $\eta \propto z/R^{3/2}$. (W. Mangler, *Ber. aerodyn. Ver-Anst. Göttingen*, 45/A/17, 1945.)

9.4.8 Show that the viscous torque on the disk from $R = 0$ to $R = R_0$ is $0.313\pi\rho R_0{}^4\sqrt{\nu\omega_0{}^3}$.

9.4.9 Show that the total volume (rate) of liquid thrown out by the disk across $R = R_0$ is $0.886\pi R_0^3\omega_0/\sqrt{Re}$, where $Re = R_0^2\omega_0/\nu$.

9.4.10 Does the ansatz for the rotating-disk problem work for the following cases?

(a) The fluid at infinity rotates with a constant angular speed ω_∞.
(b) The rotating disk is porous and fluid is sucked out at a uniform speed.
(c) The disk is of finite diameter.
(d) Two infinite disks at $z = 0$ and Z, rotate with speeds ω_0 and ω_1.

9.6.1 Show that

$$\int_0^\infty w''^2 R'\, dR'$$

is a constant.

9.6.2 Determine α for Eq. (9.79) in terms of a given parameter characterizing the strength of the jet.

9.7.1 Find the ordinary differential equation generated by the ansatz (9.84), and then solve it to get Eq. (9.86).

9.7.2 Show that the Stokes solution approaches the uniform oncoming flow at infinity.

9.7.3 Verify Eq. (9.89).

9.7.4 Referring to Problem 9.1.1, if the stream is vertically upward, p in this section should be replaced by $\widehat{p}$, and the *real* p should be $\widehat{p} - \rho gz$. In integrating the vertical components of the pressure and viscous forces on the surface of the sphere, the contributions of $\widehat{p}$ and the viscous stresses yield exactly the drag, Eq. (9.89). Now, show that the contribution of $(-\rho gz)$ is equal to the weight of fluid displaced by the sphere —that is, the buoyancy force that the sphere *would* experience if the fluid *were* at rest. (This conclusion is actually true for any body shape. Can you show this?)

10.2.1 Regarding the ensemble mean as just the arithmetic mean over N entries, prove Eqs. (10.3a,c).

10.2.2 Derive Eqs. (10.7a,b,c).

10.3.1 Derive Eq. (10.25).

10.3.2 Derive Eq. (10.27).

10.4.1 Derive Eq. (10.45).

11.1.1 **(a)** What is the meaning of $(\mathbf{A}\cdot\mathbf{B})\mathbf{C}$? Is it equal to $\mathbf{A}(\mathbf{B}\cdot\mathbf{C})$? **(b)** What is $\sqrt{\mathbf{A}\cdot\mathbf{A}}$ equal to? If $\hat{\mathbf{a}}$ is a unit vector in the direction a, what is the geometric meaning of $\hat{\mathbf{a}}\cdot\mathbf{A}$? Similarly, what is the meaning of $\hat{\mathbf{a}}\cdot\hat{\mathbf{b}}$? **(c)** If $\mathbf{A}\cdot\mathbf{B} = 0$, where $\mathbf{B}$ is an arbitrary, nonzero vector, what is $\mathbf{A}$? Why is the word "arbitrary" important for this conclusion?

11.1.2 **(a)** $\hat{\mathbf{x}}$, $\hat{\mathbf{y}}$, and $\hat{\mathbf{z}}$ are unit vectors in three mutually perpendicular directions. Establish the table of dot products among them. **(b)** Find the angle between the vectors $\mathbf{A} = \hat{\mathbf{x}} - \hat{\mathbf{y}}$ and $\mathbf{B} = 2\hat{\mathbf{y}} + \hat{\mathbf{z}}$.

11.1.3 (a) Show that $|\mathbf{A} \times \mathbf{B}|$ is twice the plane area enclosed by **A**, **B**, and the line connecting the two arrowheads. (b) Demonstrate that $\mathbf{A} \times (\mathbf{B} \times \mathbf{C}) \neq (\mathbf{A} \times \mathbf{B}) \times \mathbf{C}$ by taking $\mathbf{A} = \mathbf{B}$. (c) Establish a table of cross products for a right-handed orthogonal system of unit vectors $\hat{\mathbf{x}}$, $\hat{\mathbf{y}}$, and $\hat{\mathbf{z}}$.

11.1.4 (a) Show that a cyclic change $\mathbf{A} \to \mathbf{B} \to \mathbf{C} \to \mathbf{A}$ leaves $\mathbf{A} \cdot (\mathbf{B} \times \mathbf{C})$ unchanged; that is,

$$\mathbf{A} \cdot (\mathbf{B} \times \mathbf{C}) = \mathbf{B} \cdot (\mathbf{C} \times \mathbf{A}) = \mathbf{C} \cdot (\mathbf{A} \times \mathbf{B})$$

(b) Show that interchanging $\times$ with $\cdot$ leaves $\mathbf{A} \cdot (\mathbf{B} \times \mathbf{C})$ unchanged; that is, $(\mathbf{A} \times \mathbf{B}) \cdot \mathbf{C} = \mathbf{A} \cdot (\mathbf{B} \times \mathbf{C})$. (It is therefore acceptable just to write [**ABC**]. This is called the box product.) (c) Show that [**ABC**] is the volume of the parallelepiped with **A**, **B**, and **C** as the sides. (d) Show that $[\mathbf{ABB}] = 0$.

11.1.5 Use Problem 11.1.4b to prove that

$$\mathbf{A} \times (\mathbf{B} + \mathbf{C}) = \mathbf{A} \times \mathbf{B} + \mathbf{A} \times \mathbf{C}$$

(*Hint*: Set $\mathbf{X} = \mathbf{A} \times (\mathbf{B} + \mathbf{C}) - \mathbf{A} \times \mathbf{B} - \mathbf{A} \times \mathbf{C}$. Then, dot-multiply this by an *arbitrary* vector **Y**. What is the result? Show that this result leads to $\mathbf{X} = 0$.)

11.1.6 (a) Show that

$$\mathbf{A} \times (\mathbf{B} \times \mathbf{C}) = -\mathbf{A} \times (\mathbf{C} \times \mathbf{B}) = (\mathbf{C} \times \mathbf{B}) \times \mathbf{A}$$

(b) Show that

$$(\mathbf{C} \times \mathbf{A}) \cdot (\mathbf{A} \times \mathbf{B}) \times (\mathbf{B} \times \mathbf{C}) = [\mathbf{ABC}]^2$$

11.1.7 Prove Eq. (11.1). (*Hint*: First show that $\mathbf{A} \times (\mathbf{B} \times \mathbf{C})$ is equal to $P\mathbf{B} + Q\mathbf{C}$. Next, dot-multiply both sides by **A** to show that $P = \lambda(\mathbf{A} \cdot \mathbf{C})$ and $Q = -\lambda(\mathbf{A} \cdot \mathbf{B})$. Finally, since λ is a universal constant, find λ by taking the special case $\mathbf{A} = \hat{\mathbf{x}}$, $\mathbf{B} = \hat{\mathbf{x}}$, and $\mathbf{C} = \hat{\mathbf{y}}$.)

11.1.8 Show that

$$\mathbf{A} \times (\mathbf{B} \times \mathbf{C}) + \mathbf{B} \times (\mathbf{C} \times \mathbf{A}) + \mathbf{C} \times (\mathbf{A} \times \mathbf{B}) = 0$$

and that

$$(\mathbf{A} \times \mathbf{B}) \cdot (\mathbf{C} \times \mathbf{D}) = (\mathbf{B} \cdot \mathbf{D})(\mathbf{C} \cdot \mathbf{A}) - (\mathbf{B} \cdot \mathbf{C})(\mathbf{D} \cdot \mathbf{A})$$

11.2.1 Show that $\oint_S \hat{\mathbf{n}} \cdot \mathbf{q}\, dS$ is the rate of total volume outflow across S (if **q** stands for the flow velocity).

11.3.1 Show that $\mathbf{AB} = \mathbf{DE}$ if and only if $\mathbf{A} = k\mathbf{D}$ and $\mathbf{B} = \mathbf{E}/k$. (*Hint*: For the "only if" part, dot-multiply both sides by **C** in the "before" and "after" positions.)

11.3.2 Show that the dyadic multiplication is distributive; that is,

$$(\mathbf{A} + \mathbf{B})\mathbf{C} = \mathbf{AC} + \mathbf{BC}$$

and

$$\mathbf{C}(\mathbf{A} + \mathbf{B}) = \mathbf{CA} + \mathbf{CB}$$

11.3.3 Show that a dyadic $\tilde{\tau}$ can always be written in the form $\tilde{\tau} = \tilde{\mathbf{S}} + \tilde{\mathbf{A}}$, where $\tilde{\mathbf{S}}$ is a symmetric and $\tilde{\mathbf{A}}$ is an antisymmetric dyadic. (*Hint*: Use $\tilde{\tau}_c$ to construct $\tilde{\mathbf{S}}$ and $\tilde{\mathbf{A}}$.)

11.3.4 (a) Find the dot and double dot products of the two dyadics

$$\tilde{\tau} = \mathbf{A}_1\mathbf{B}_1 + \mathbf{A}_2\mathbf{B}_2 + \cdots + \mathbf{A}_m\mathbf{B}_m$$

$$\tilde{\sigma} = \mathbf{C}_1\mathbf{D}_1 + \mathbf{C}_2\mathbf{D}_2 + \cdots + \mathbf{C}_\ell\mathbf{D}_\ell$$

(b) Show that

$$\tilde{\tau} : \tilde{\sigma} = \tilde{\sigma} : \tilde{\tau} = \tilde{\sigma}_c : \tilde{\tau}_c = \tilde{\tau}_c : \tilde{\sigma}_c$$

(c) Show that

$$\tilde{\mathbf{S}} : \tilde{\mathbf{A}} = 0$$

where $\tilde{\mathbf{S}}$ is a symmetric and $\tilde{\mathbf{A}}$ is an antisymmetric dyadic.

11.3.5 Given

$$\mathbf{A} = 2\hat{\mathbf{x}} + \hat{\mathbf{y}} - 2\hat{\mathbf{z}}$$

$$\mathbf{B} = -2\hat{\mathbf{x}} + 3\hat{\mathbf{z}}$$

$$\tilde{\tau} = 5\hat{\mathbf{x}}\hat{\mathbf{x}} - \hat{\mathbf{x}}\hat{\mathbf{z}} + 3\mathbf{yz} + 2\mathbf{zx}$$

$$\tilde{\sigma} = 2\hat{\mathbf{x}}\hat{\mathbf{x}} + 3\hat{\mathbf{x}}\hat{\mathbf{z}} - \hat{\mathbf{z}}\hat{\mathbf{y}}$$

Find $\mathbf{A} \cdot \tilde{\tau}$, $\tilde{\tau} \cdot \mathbf{A}$, $\mathbf{A} \cdot \tilde{\tau} \cdot \mathbf{B}$, $\mathbf{B} \cdot \tilde{\tau} \cdot \mathbf{A}$, $\tilde{\sigma} \cdot \tilde{\tau}$, and $\tilde{\sigma} : \tilde{\tau}$.

11.3.6 Show that $(\mathbf{A} \cdot \tilde{\tau}) \cdot \mathbf{B} = \mathbf{A} \cdot (\tilde{\tau} \cdot \mathbf{B})$, and hence that $\mathbf{A} \cdot \tilde{\tau} \cdot \mathbf{B}$ is unambiguous.

11.4.1 Show that

$$\nabla \ast (\mathscr{F} \ast \mathscr{G}) = \nabla_{\mathscr{F}} \ast (\mathscr{F} \ast \mathscr{G}) + \nabla_{\mathscr{G}} \ast (\mathscr{F} \ast \mathscr{G})$$

(*Hint*: See L. M. Milne-Thomson, *Theoretical Hydrodynamics*, 5th ed., Macmillan, 1968.)

11.4.2 Show that

(a) curl grad $\phi = 0$
(b) curl $(f\mathbf{F}) = f\,\text{curl}\,\mathbf{F} - \mathbf{F} \times \text{grad}\,f$
(c) $(\mathbf{A} \cdot \nabla)(\mathbf{B} \cdot \mathbf{C}) = \mathbf{B} \cdot [(\mathbf{A} \cdot \nabla)\mathbf{C}] + \mathbf{C} \cdot [(\mathbf{A} \cdot \nabla)\mathbf{B}]$
(d) $(\mathbf{A} \cdot \nabla)(\mathbf{B} \times \mathbf{C}) = [(\mathbf{A} \cdot \nabla)\mathbf{B}] \times \mathbf{C} + \mathbf{B} \times [(\mathbf{A} \cdot \nabla)\mathbf{C}]$
(e) $2(\mathbf{A} \cdot \nabla)\mathbf{B} = \nabla(\mathbf{A} \cdot \mathbf{B}) - \nabla \times (\mathbf{A} \times \mathbf{B}) - \mathbf{B} \times (\nabla \times \mathbf{A})$
$\qquad - \mathbf{A} \times (\nabla \times \mathbf{B}) - \mathbf{B}(\nabla \cdot \mathbf{A}) + \mathbf{A}(\nabla \cdot \mathbf{B})$

(*Hint*: Solve $\mathbf{A} \cdot \nabla\mathbf{B}$ from Eqs. (11.22) and (11.23), and add up. *Note*: This can be looked upon as another interpretation of $(\mathbf{A} \cdot \nabla)\mathbf{B}$.)

(f) $(\mathbf{A} \cdot \nabla)\mathbf{A} = \frac{1}{2}\nabla A^2 - \mathbf{A} \times \text{curl}\,\mathbf{A}$

(*Note*: Therefore $(\mathbf{A} \cdot \nabla)\mathbf{A}$ can be regarded as simply an abbreviation for the right-hand side.)

(g) $\nabla(\phi\psi) = \phi\nabla\psi + \psi\nabla\phi$
(h) $\nabla^2(\phi\psi) = \psi\nabla^2\phi + 2(\nabla\phi) \cdot (\nabla\psi) + \phi\nabla^2\psi$

11.4.3 Establish the following formulas useful in fluid dynamics. ($\mathbf{\Omega}$ stands particularly for curl $\mathbf{q}$ in (c) and (d).)

(a) $(\mathbf{A} \cdot \nabla) q^2 = 2\mathbf{q} \cdot (\mathbf{A} \cdot \nabla)\mathbf{q}$

(*Hint*: Use Problem 11.4.2c.)

(b) $\mathbf{q} \cdot (\nabla \times \mathbf{\Omega}) = \Omega^2 - \nabla \cdot (\mathbf{q} \times \mathbf{\Omega})$

(*Hint*: Use Eq. (11.24).)

(c) $2\nabla \cdot (\mathbf{q} \times \mathbf{\Omega}) = \nabla^2 q^2 - 2\nabla \cdot (\mathbf{q} \cdot \nabla)\mathbf{q}$

(*Hint*: Use Problem 11.4.2f.)

(d) $\nabla \times \mathbf{\Omega} = \nabla(\nabla \cdot \mathbf{q}) - \nabla^2 \mathbf{q}$
(e) $\nabla \cdot [\mu(\nabla\phi)] = (\nabla\mu) \cdot (\nabla\phi) + \mu\nabla^2\phi$
(f) $\nabla \cdot (\mathbf{q}\,\text{div}\,\mathbf{q}) = (\text{div}\,\mathbf{q})^2 + (\mathbf{q} \cdot \nabla)(\text{div}\,\mathbf{q})$
(g) $\nabla \cdot (\phi\mathbf{q}) = \phi\,\text{div}\,\mathbf{q} + \text{grad}\,\phi \cdot \mathbf{q}$
(h) $\nabla \times (\phi\mathbf{\Omega}) = -\mathbf{\Omega} \times \nabla\phi + \phi\nabla \times \mathbf{\Omega}$
(i) $\nabla(\mu\,\text{div}\,\mathbf{q}) = (\text{div}\,\mathbf{q})(\text{grad}\,\mu) + \mu\,\text{grad}\,(\text{div}\,\mathbf{q})$
(j) $\nabla \cdot [\mu(\text{div}\,\mathbf{q})\mathbf{q}] = \mu(\text{div}\,\mathbf{q})^2 + \mu\mathbf{q} \cdot \text{grad}\,(\text{div}\,\mathbf{q}) + (\text{div}\,\mathbf{q})\mathbf{q} \cdot \nabla\mu$

11.4.4 Prove Eq. (11.32).

11.4.5 Prove Eq. (11.33). (*Hint*: Apply Eq. (11.32) to the vector $\mathbf{q} \times \mathbf{K}$, where $\mathbf{K}$ is an arbitrary constant vector. In the result, arrange to have $\mathbf{K}$ appear in the first position followed by a dot, by cyclic change of members and interchange of dot with cross and the box products.)

11.4.6 Deduce Eq. (11.31) from Eq. (11.32). (*Hint*: Set $\mathbf{q} = \phi\mathbf{K}$.)

11.5.1 Show that

(a) $$\int_V \nabla^2\phi\, dV = \oint_S \left(\frac{\partial\phi}{\partial n}\right) dS$$

(b) $$\int_V \nabla^2\mathbf{q}\, dV = \oint_S \hat{\mathbf{n}} \cdot \nabla\mathbf{q}\, dS$$

11.5.2 Prove the following two Green's theorems:

(a) $$\int_V \nabla\psi \cdot \nabla\phi\, dV = -\int_V \phi\nabla^2\psi\, dV + \oint_S \phi\frac{\partial\psi}{\partial n}\, dS$$

(b) $$\int_V (\psi\nabla^2\phi - \phi\nabla^2\psi)\, dV = \oint_S (\psi\nabla\phi - \phi\nabla\psi) \cdot \hat{\mathbf{n}}\, dS$$

(*Hint*: Apply the Gauss theorem to $\mathbf{q} = \phi\,\text{grad}\,\psi$.)

11.5.3 Show that

$$\int_V \nabla \ast \mathcal{F}\, dV = \oint_S \hat{\mathbf{n}} \ast \mathcal{F}\, dS - \int_\Sigma \hat{\mathbf{n}}_{12} \ast (\mathcal{F}_2 - \mathcal{F}_1)\, dS$$

where Σ is a surface of discontinuity across which $\mathcal{F}$ jumps from $\mathcal{F}_1$ on the 1-side to $\mathcal{F}_2$ on the 2-side, $\hat{\mathbf{n}}_{12}$ is the unit normal vector of Σ, positive when pointing from the 1-side to the 2-side. (Σ is actually a portion of the surface of discontinuity enclosed by the volume V.)

(*Hint*: Cut V into two parts, V_1 bounded by the 1-side of Σ and V_2 bounded by the 2-side of Σ, and apply Eq. (11.35) to V_1 and V_2.)

11.5.4 Show that

$$\oint_C g(\nabla f)\cdot d\mathbf{l} = -\oint_C f(\nabla g)\cdot d\mathbf{l}$$

11.5.5 In the xy-plane, show that

$$\oint_C (f\,dx + g\,dy) = \int_A \left(\frac{\partial g}{\partial x} - \frac{\partial f}{\partial y}\right) dx\,dy$$

where A is the area enclosed by C. (*Hint*: Apply Stokes theorem.)

11.5.6 Calculate

$$\oint_S \left(\frac{\hat{\mathbf{n}}\cdot\hat{\mathbf{r}}}{r^3}\right) dS$$

for *any* S surrounding $\mathbf{r} = 0$.

11.6.1 Write out $(dl)^2$ and h_i for the spherical coordinates.

11.6.2 Write out div $\mathbf{q}$ for cylindrical, spherical, and rectangular Cartesian coordinates.

11.6.3 Using Eq. (11.32), find $(\text{curl}\ \mathbf{q})_1$.

11.6.4 Repeat Problem 11.6.2 for curl $\mathbf{q}$.

11.6.5 Repeat Problem 11.6.2 for $\nabla^2\phi$ and $\nabla\phi$.

11.6.6 Show that $\nabla\mathbf{r} = \hat{\mathbf{x}}\hat{\mathbf{x}} + \hat{\mathbf{y}}\hat{\mathbf{y}} + \hat{\mathbf{z}}\hat{\mathbf{z}}$ in rectangular Cartesian coordinates.

11.6.7 (a) Prove that

$$\nabla\cdot(\mathbf{q}\mathbf{q} - \tfrac{1}{2}\tilde{\mathbf{I}}q^2) = \mathbf{q}(\nabla\cdot\mathbf{q}) - \mathbf{q}\times\boldsymbol{\Omega}$$

Hint: $\nabla\cdot\tilde{\mathbf{I}} = 0$. (Why?)

(b) Show that $\tilde{\mathbf{I}}:\tilde{\tau} = \tilde{\tau}_s$.

11.6.8 Carry out enough manipulations to get two representative terms of $\nabla\mathbf{q}$.

11.6.9 Prove that, in general orthogonal coordinates,

$$\hat{\zeta}_1\cdot\nabla\hat{\zeta}_1 = -\frac{\hat{\zeta}_2}{h_1h_2}\frac{\partial h_1}{\partial\zeta_2} - \frac{\hat{\zeta}_3}{h_1h_3}\frac{\partial h_1}{\partial\zeta_3}$$

$$\hat{\zeta}_1\cdot\nabla\hat{\zeta}_2 = \frac{\hat{\zeta}_1}{h_1h_2}\frac{\partial h_1}{\partial\zeta_2}$$

$$\hat{\zeta}_1\cdot\nabla\hat{\zeta}_3 = \frac{\hat{\zeta}_1}{h_1h_3}\frac{\partial h_1}{\partial\zeta_3}$$

and deduce the expression for $\hat{\zeta}_1\cdot\nabla\mathbf{q}$.

12.2.1 The velocity components of a plane flow are

$$u = A(x+y) + Ct$$

$$v = B(x+y) + Et$$

in the Eulerian language. Find the displacement of a fluid particle in the Lagrangian description.

12.2.2 Show that the streamlines and pathlines coincide for the flow $u = x/(1+t)$, $v = y/(1+t)$, $w = z/(1+t)$.

12.2.3 Find the pathlines and streamlines for the flow

$$\mathbf{q} = K\langle t\rangle x\hat{\mathbf{x}} - K\langle t\rangle y\hat{\mathbf{y}}$$

where $K\langle t\rangle$ is a given function of t; and then specialize to $K\langle t\rangle = kt^3$.

12.3.1 Repeat Problem 11.6.2 for $\mathbf{D}\mathbf{q}/Dt$.

12.4.1 Verify that the average angular speed about the z-axis of a fluid particle is

$$\frac{1}{2}\left(\frac{\partial v}{\partial x} - \frac{\partial u}{\partial y}\right)$$

12.4.2 Derive the components of $\tilde{\boldsymbol{\epsilon}}$ with respect to the cylindrical and spherical coordinates.

Answer:

$$\epsilon_{RR} = \frac{\partial q_R}{\partial R}$$

$$\epsilon_{\theta\theta} = \frac{1}{R}\frac{\partial q_\theta}{\partial \theta} + \frac{q_R}{R}$$

$$\epsilon_{zz} = \frac{\partial q_z}{\partial z}$$

$$\epsilon_{\theta z} = \epsilon_{z\theta} = \frac{1}{2}\left(\frac{1}{R}\frac{\partial q_z}{\partial \theta} + \frac{\partial q_\theta}{\partial z}\right)$$

$$\epsilon_{zR} = \epsilon_{Rz} = \frac{1}{2}\left(\frac{\partial q_R}{\partial z} + \frac{\partial q_z}{\partial R}\right)$$

$$\epsilon_{R\theta} = \epsilon_{\theta R} = \frac{1}{2}\left(\frac{\partial q_\theta}{\partial R} - \frac{q_\theta}{R} + \frac{1}{R}\frac{\partial q_R}{\partial \theta}\right)$$

$$\epsilon_{rr} = \frac{\partial q_r}{\partial r}$$

$$\epsilon_{\varphi\varphi} = \frac{1}{r}\frac{\partial q_\varphi}{\partial \varphi} + \frac{q_r}{r}$$

$$\epsilon_{\theta\theta} = \frac{1}{r}\left(\frac{1}{\sin\varphi}\frac{\partial q_\theta}{\partial \theta} + q_r + q_\varphi \cot\varphi\right)$$

$$\epsilon_{\theta\varphi} = \epsilon_{\varphi\theta} = \frac{1}{2}\left(\frac{1}{r}\frac{\partial q_\theta}{\partial \varphi} - \frac{q_\theta}{r}\cot\varphi + \frac{1}{r\sin\varphi}\frac{\partial q_\varphi}{\partial \theta}\right)$$

$$\epsilon_{\theta r} = \epsilon_{r\theta} = \frac{1}{2}\left(\frac{1}{r\sin\varphi}\frac{\partial q_r}{\partial \theta} + \frac{\partial q_\theta}{\partial r} - \frac{q_\theta}{r}\right)$$

$$\epsilon_{r\varphi} = \epsilon_{\varphi r} = \frac{1}{2}\left(\frac{\partial q_\varphi}{\partial r} - \frac{q_\varphi}{r} + \frac{1}{r}\frac{\partial q_r}{\partial \varphi}\right)$$

12.4.3 (**a**) Show that the $\mathbf{\Omega}$-field is divergence-free. (**b**) *Definition:* Strength of a vortex tube $= \int_S \mathbf{\Omega} \cdot \hat{\mathbf{n}}\, dS$ where S is any cross section of the tube. Show that the strength is constant along the tube. (This accounts for the word "any" in the definition.) *Hint:* Apply Gauss theorem to a portion of the tube; and note that $\mathbf{\Omega}$ is tangent to the wall of the tube. It is also instructive for the reader to prove this in the following manner: (a) From Stokes theorem, the strength equals Γ about a curve on and around the tube. (b) Take two such curves and bridge them with one curve on the tube. (c) Apply Stokes theorem again, this time to the big circuit made up of the two bridged curves.

12.4.4 Does Eq. (12.13) apply to the "free"-vortex flow? Why?

12.4.5 Show that, if (η_1, η_2, η_3) in the $\boldsymbol{\eta}$-space are three coordinates parallel to the principal axes at P, the quadric (12.12A) becomes

$$\epsilon_{11}\eta_1^2 + \epsilon_{22}\eta_2^2 + \epsilon_{33}\eta_3^2 = \text{constant}$$

where $(\epsilon_{11}, \epsilon_{22}, \epsilon_{33})$ are the three principal rates of strain. (Note that the shape of the quadric, Eq. (12.12), has nothing to do with the deformed shape of a fluid element. When it is a hyperbola, it does not mean that a fluid element is deformed into a hyperbola.) Show also that, for an incompressible fluid, the quadric cannot be an ellipsoid.

12.4.6 Consider $\tilde{\boldsymbol{\epsilon}}$ for the simple irrotational vortex flow described in Section 1.3, subsection (1), page 27, at an arbitrary point (away from the center). Show that the corresponding quadric is a hyperbolic cylinder. (*Hint*: Use cylindrical coordinates.) Locate also the principal axes of $\tilde{\boldsymbol{\epsilon}}$.

12.5.1 In a flow field $\mathbf{q} = \hat{\mathbf{x}} g\langle x \rangle$, take a material volume in the form of a cylinder parallel to the x-axis, with cross-sectional area A, confined in between $x = a\langle t \rangle$ and $x = b\langle t \rangle$ at time t. Apply Eq. (12.15A) to

$$\frac{D}{Dt}\int_V f\langle x, t\rangle\, dV$$

or

$$\frac{D}{Dt}\int_{b\langle t\rangle}^{a\langle t\rangle} f\langle x, t\rangle\, dx$$

The result is known as the Leibnitz rule in (advanced) calculus.

12.5.2 If $\mathscr{C}$ is an oriented material curve, show that

$$\frac{D}{Dt}\int_{\mathscr{C}\langle t\rangle} \mathbf{f}\langle \mathbf{r}, t\rangle \cdot d\mathbf{l} = \int_{\mathscr{C}\langle t\rangle} \left[\frac{\partial \mathbf{f}}{\partial t} - \mathbf{q}\times \text{curl}\,\mathbf{f} + \text{grad}\,(\mathbf{f}\cdot\mathbf{q})\right]\cdot d\mathbf{l}$$

($\mathscr{C}$ is not necessarily closed; $d\mathbf{l}$ is, of course, the line element of $\mathscr{C}$ in the oriented direction of $\mathscr{C}$.) *Hint*: As $t \to t + \delta t$, $\mathscr{C}\langle t \rangle$ moves to $\mathscr{C}\langle t + \partial t\rangle$, sweeping out a surface δS. Similarly to the proof in this section, split the integral on the left-hand side into two parts in the definition (a limiting process) of D/Dt. One part is easily seen to yield the first term on the right-hand side. The other part must then correspond to the rest of the terms on the right-hand side. Establish this correspondence by applying Stokes theorem to δS and its bounding curves. The proper orientation here can be handled this way: Denote the end points of $\mathscr{C}\langle t\rangle$ by A and B such that $A \to B$ is in the (given) oriented direction of $\mathscr{C}\langle t\rangle$, in which the line integration is to proceed. Let A' and B' be the corresponding end points of $\mathscr{C}\langle t+\delta t\rangle$. We can then take the unit normal vector $\hat{\mathbf{n}}$ to δS in such a way as to be in right-handed orientation with the circuit $A' \to B' \to B \to A \to A'$. Noting that δS in turn is made up of area elements

$$\hat{\mathbf{n}}\, dS = (\mathbf{q}\, \delta t) \times dl$$

the second term on the right-hand side is obtained. The third term on the right is best written as $\mathbf{f} \cdot \mathbf{q}$ evaluated at B minus that evaluated at A. It is then seen to come from

$$\int_B^{B'} \mathbf{f}\langle \mathbf{r}, t\rangle \cdot dl, \qquad \int_A^{A'} \mathbf{f}\langle \mathbf{r}, t\rangle \cdot dl$$

where the first, for example, behaves like

$$(\mathbf{f}|_{\text{at}\,B} \cdot \mathbf{q}|_{\text{at}\,B})\, \delta t$$

as $\delta t \to 0$. (The formula is of prime importance in the outer flow beyond a boundary layer. For a shorter proof, see Section 79 of C. Truesdell and R. Toupin's "The classical field theory," *Encyclopedia of Physics*, Vol. III/1, Springer-Verlag, Berlin, 1960. The seemingly simple proof quoted routinely in so many texts is, unfortunately, irreparably faulty.)

12.6.1 A balloon with radius $R\langle t\rangle$ is equipped with a small exit of area A_e. Air with density ρ_e is leaving the balloon with speed U_e. Find the time rate of change of average density of air in the balloon. Do this in two ways: (**a**) using Eq. (12.16), (**b**) using Eq. (12.17).

12.6.2 Apply Eq. (12.18) to a liquid tank *being filled* with an overhead pipe. Take S as the inside surface of the tank plus the cross-sectional area of the pipe at the junction of the pipe and tank. Now, introduce the fact that $\partial\rho/\partial t = 0$ since the liquid is incompressible. What absurd conclusion would you have from Eq. (12.18)? What is wrong?

12.7.1 Derive Eqs. (12.25)–(12.35).

12.7.2 Show that 2π times the difference of Stokes' stream function between two points in the representative plane is the net volume flow across the surface of revolution generated by any curve connecting the two points in the representative plane. Show also that, here,

$$q_l = \frac{1}{R}\frac{\partial \psi}{\partial l_1} \qquad \measuredangle\,(l, l_1) = +90^\circ$$

(Note that $R = r \sin \varphi$.)

12.7.3 Determine the stream functions for the following flows:

(**a**) Uniform flow: $w = C$. (Obtain both d'Alembert's and Stokes'.)

(b) Point-source flow: $q_r = A/r^2$

(c) Plane flow: $\begin{cases} u = Ax \\ v = -By \end{cases}$

(d) Plane flow: $\begin{cases} u = x/(1+t) \\ v = y \end{cases}$

13.1.1 Show that $\nabla \cdot (\rho \mathbf{qq}) = \mathbf{q}\nabla \cdot (\rho \mathbf{q}) + \rho \mathbf{q} \cdot \nabla \mathbf{q}$ by first distributing ∇ to ρ and $\mathbf{qq}$. (*Hint:* $\nabla \cdot (\mathbf{qq}) \neq 2\mathbf{q} \cdot \nabla \mathbf{q} \neq 2\mathbf{q}(\nabla \cdot \mathbf{q})$.)

13.1.2 Show that Eq. (13.2) holds for any flow quantity, scalar or vectorial.

13.1.3 For a fluid at rest,

$$\tilde{\sigma} = -\mathring{p}\tilde{\mathbf{I}}$$

where $\mathring{p}$ is the hydrostatic pressure. Show that the Archimedes principle (that is, a submerged body is buoyed up by a force equal and opposite to the weight of the displaced fluid, not necessarily incompressible, at rest) follows the integral balance of momentum.

13.3.1 The fundamental equation for liquid water is found to be

$$e = \left(a_1\rho^6 + \frac{a_2}{\rho}\right) + [b_1 \exp(b_2 s) + b_3]$$

where a_1, a_2, b_1, b_2, b_3 are constants. Obtain the equation of state (that is, the relation among Π, T, and ρ), c_v, and c_p. Incidentally, when Π is measured in atmospheres, $a_1 = 3001/6\rho_0{}^7$ and $a_2 = 3000$, where ρ_0 is the density of liquid water at 32°F and 1 atmosphere. (R. H. Cole, *Underwater Explosion*, Princeton University Press, 1948.)

13.4.1 Show that Eq. (13.13) can also be written as

$$\int_V \frac{\partial}{\partial t}[\rho(\tfrac{1}{2}q^2 + e)]\,dV + \oint_S \rho(\tfrac{1}{2}q^2 + h)\mathbf{q}\cdot\hat{\mathbf{n}}\,dS$$
$$= \oint_S \hat{\mathbf{n}} \cdot \tilde{\tau} \cdot \mathbf{q}\,dS + \int_V \rho\mathbf{g} \cdot \mathbf{q}\,dV$$
$$- \oint_S \mathbf{H} \cdot \hat{\mathbf{n}}\,dS + \int_V Q\,dV$$

13.4.2 Show that both Eqs. (13.23) and (13.25) lead to

$$\rho\frac{Ds}{Dt} = \frac{1}{T}(\Phi - \operatorname{div}\mathbf{H} + Q)$$

13.4.3 Show that Eq. (13.14) can also be written in the form

$$\rho\frac{D}{Dt}(\tfrac{1}{2}q^2 + h) - \frac{D\Pi}{Dt} = \Phi - \operatorname{div}\mathbf{H} + Q$$
$$+ \mathbf{q} \cdot \nabla \cdot \tilde{\sigma} + \rho\mathbf{g} \cdot \mathbf{q}$$

13.5.1 Show that

$$\nabla \cdot (\nabla\mathbf{q} + \mathbf{q}\nabla) = \nabla^2\mathbf{q} + \nabla(\nabla \cdot \mathbf{q})$$

13.5.2 Show that, for an ideal gas, Eq. (13.32A) has the alternative form

$$\rho c_p \frac{DT}{Dt} = \frac{D\Pi}{Dt} + \Phi + \text{div}\,(K \text{ grad } T) + Q$$

13.5.3 Show that, for an orthogonal coordinate system,

$$\frac{\Phi}{\mu} = 2 \sum_{\ell,m=1}^{3} \epsilon_{\ell m}^{2}$$

13.5.4 Regain the governing equation for various coordinate systems in Chapters 8 and 9, using the general equations established here.

13.5.5 Show that, for an incompressible Newtonian fluid, the normal stress on a fixed wall is just p. (L. M. Milne-Thomson, *Theoretical Hydrodynamics*, 5th ed., Section 21.06, Macmillan, 1968.) It is, therefore, preferable in principle to measure pressure in a flow of an incompressible Newtonian fluid with a probe made of a tiny solid surface backed up by a rather stiff spring (to make it practically unyielding in the normal direction), instead of the usual one with an opening. (See Section 2.3.)

13.5.6 Show that the conclusion in Problem 13.5.5 is also true for compressible fluids, if only the flow is steady. (P. A. Lagerstrom, "Laminar flow theory", Section B.4, *Theory of Laminar Flow*, edited by F. K. Moore, Princeton University Press, 1964.)

13.6.1 Regain the $\tilde{\boldsymbol{\sigma}} \sim \tilde{\boldsymbol{\epsilon}}$ relations quoted before in Sections 2.4, 8.1, 8.2, and 9.3.

13.7.1 Derive Eqs. (13.66), (13.67), and inequality (13.68).

13.7.2 Take Eq. (13.55A) for an incompressible fluid with constant μ. Take the curl on both sides and show that

$$\frac{D\boldsymbol{\Omega}}{Dt} = (\boldsymbol{\Omega} \cdot \boldsymbol{\nabla})\mathbf{q} - \nu \boldsymbol{\nabla} \times (\boldsymbol{\nabla} \times \boldsymbol{\Omega})$$

For a plane flow the first term on the right-hand side vanishes. Why? For an axisymmetric flow, it does not. Why not?

13.7.3 Referring to Problem 13.7.2, show that $\boldsymbol{\Omega}$ for the Rayleigh problem, that is,

$$\mathbf{q} = U\left[1 - \text{erf}\left(\frac{y}{2\sqrt{\nu t}}\right)\right]\hat{\mathbf{x}}$$

satisfies the equation governing $\boldsymbol{\Omega}$. Show also that $\boldsymbol{\Omega}$ here, at $t = 0$, is zero if $y \neq 0$, but $\sim 1/\sqrt{t}$ if $y = 0$. (Therefore, the sudden motion of the plate sets up initially a vorticity concentrated at the plate; this concentrated vorticity is then diffused into the fluid by viscosity as time goes on. In other words, this shows a vortex sheet at $t = 0$ gradually losing its status as a surface of discontinuity.)

13.8.1 Check the governing equations for the flat Couette flow.

13.8.2 Show that, for the flat Couette flow with $c_p = \text{constant}$ and $\mu \propto T$,

$$y' = \frac{\left[1+\left(\frac{\gamma-1}{2}\right)M^2Pr\left(1-\frac{u'^2}{3}\right)+(T'_w-T'_r)\left(1-\frac{u'}{2}\right)\right]u'}{1+\left(\frac{\gamma-1}{3}\right)M^2Pr+\frac{1}{2}(T'_w-T'_r)}$$

and

$$f\,(\text{the friction coefficient}) = \frac{\tau_w}{\frac{1}{2}\rho_\infty U^2}$$

$$= \frac{2}{Re}\left[1+\left(\frac{\gamma-1}{3}\right)M^2Pr+\tfrac{1}{2}(T'_w-T'_r)\right]$$

(P. A. Lagerstrom, "Laminar flow theory," in *Theory of Laminar Flow*, edited by F. K. Moore, Princeton University Press, 1964.)

13.8.3 In the result of Problem 13.8.2, take $M^2 \to 0$. Find u' explicitly in terms of y' for both $T'_w = 1$ and $\neq 1$. Show that u' is real and positive, that $u' \leqslant 1$ if $T'_w \geqslant 1$, and that u' overshoots the moving plate (that is, assumes values larger than 1) if $T'_w < 1$. Find also the temperature distribution.

13.8.4 Take $c_p = \text{constant}$, $\mu \propto T$, and $T_w = T_\infty$. Introduce u' and T' as series of M^2, truncated at the M^4-terms. Obtain the governing systems of $O(M^0)$, $O(M^2)$, and $O(M^4)$.

13.8.5 Solve the above three systems to get u' and T' to errors of $O(M^6)$.

13.8.6 Check the results for the flat Couette flow of an incompressible fluid.

13.8.7 Obtain the hypersonic limit of the flat Couette flow for $c_p = \text{constant}$, $\mu \propto T$, and insulated lower plate.

13.8.8 There are a few more rather simple flows of compressible fluids solvable in an exact manner: flow past a porous flat plate, the Couette flow (between rotating cylinders), and circulating flow around a circular cylinder with suction. Study them. (C. R. Illingworth, *Proc. Cambridge Phil. Soc.*, **46**, 469, 1950.) Note that Poiseuille or plane Poiseuille flow is not one of them.

13.9.1 Check the governing equations for the Rayleigh problem.

13.9.2 Show that Eqs. (13.69)–(13.73) do not admit a solution with $v' \equiv 0$. (*Hint*: The mechanical part calls for $\rho' = 1$ and $p' = 1$, whereas the energetic part dictates otherwise.)

13.9.3 Analyze the Rayleigh problem at large times for an incompressible fluid. Find the outer flow and show that the inner problem is mathematically the same as the original problem, so no actual benefit can be gained. (One might say that the boundary-layer solution here is so good that it is valid throughout the entire field!) For this reason, Section 4.1 was never reexamined in the framework of boundary-layer theory.

13.9.4 Prove Eqs. (13.79a,b) and (13.80a,b).

13.9.5 Derive Eq. (13.85).

13.9.6 Write out the explicit forms of Eqs. (13.86) and (13.87), and show that

$$\lim_{y'' \to \infty} v'' = \frac{T'_w - 1 + \frac{1}{2}(\gamma - 1) M_\infty^2(\sqrt{2} - 1)}{\sqrt{\pi t'}}$$

13.9.7 Take the Rayleigh problem for an incompressible fluid with $\mu \propto T$, c_{in} = constant, and $Pr = \mu c_{in}/K$ = constant. Using $\partial/\partial x \equiv 0$ and neglecting energy dissipation, write down the four governing equations. Show that $v = 0$, $p = p_\infty$. Next, nondimensionalize the remaining two equations, and repeat Problem 13.9.3 for this more general case (governed by two equations).

13.9.8 In Problem 13.9.7, show that, for $Pr = 1$, $T' = A + Bu'$. Determine A and B. Without actually solving for u', find out whether an analogy or correlation is possible between this and the compressible case with $M_\infty \to 0$. (*Answer*: No, unless the plate is insulated.)

INDEX

Page numbers followed by *n* indicate footnotes. A two-decimal number in parentheses refers to a Problem.